Wireless Communication in Underground Mines

L.K. Bandyopadhyay · S.K. Chaulya · P.K. Mishra

Wireless Communication in Underground Mines

RFID-Based Sensor Networking

Springer

Dr. L.K. Bandyopadhyay
Scientist
Central Institute of Mining
and Fuel Research
Dhanbad, Jharkhand
India, 826015
laxmikb@yahoo.com

Dr. S.K. Chaulya
Scientist
Central Institute of Mining
and Fuel Research
Dhanbad, Jharkand
India, 826015
chaulyask@yahoo.co.in

Dr. P.K. Mishra
Scientist
Central Institute of Mining
and Fuel Research
Dhanbad, Jharkhand
India, 826015
mishrapkapp@yahoo.co.in

ISBN 978-0-387-98164-2 e-ISBN 978-0-387-98165-9
DOI 10.1007/978-0-387-98165-9
Springer New York Dordrecht Heidelberg London

Library of Congress Control Number: 2009932740

Printed on acid-free paper

Springer is part of Springer Science+Business Media (www.springer.com)

Preface

Wireless communication has emerged as an independent discipline in the past decades. Everything from cellular voice telephony to wireless data transmission using wireless sensor networks has profoundly impacted the safety, production, and productivity of industries and our lifestyle as well. After a decade of exponential growth, the wireless industry is one of the largest industries in the world. Therefore, it would be an injustice if the wireless communication is not explored for mining industry.

Underground mines, which are characterized by their tough working conditions and hazardous environments, require fool-proof mine-wide communication systems for smooth functioning of mine workings and ensuring better safety. Proper and reliable communication systems not only save the machine breakdown time but also help in immediate passing of messages from the vicinity of underground working area to the surface for day-to-day normal mining operations as well as for speedy rescue operations in case of disaster. Therefore, a reliable and effective communication system is an essential requisite for safe working, and maintaining requisite production and productivity of underground mines. Most of the existing systems generally available in underground mines are based on line (wired) communication principle, hence these are unable to withstand in the disaster conditions and difficult to deploy in inaccessible places. Therefore, wireless communication is an indispensable, reliable, and convenient system and essential in case of day-to-day normal duty or disaster situations. However, at present there is no relevant book available in the global market regarding wireless communication technology in underground mines and hazardous areas. Therefore, a comprehensive textbook on wireless communication for underground mines is a vital need for mining and electronics engineers that provides a deeper understanding of the important subject and designing the intrinsic safe systems. Considering these facts, this book is written, incorporating various communication devices designed and developed by the authors for wireless communication in different areas of underground mines and their experiences in the fields.

The wireless communication systems used on surface cannot be applied straightaway in underground mines due to high attenuation of radio waves in underground strata, besides presence of inflammable gases and hazardous environment. Nonsymmetric mine topology, uneven mine structure, complex geological structures, and extensive labyrinths put further hindrance on the way of communication. Wireless communication in underground mine is a very complex technique that involves multidisciplinary subjects. Therefore, the main focus of the book is on the recent advancement in wireless communication technologies for underground mines and hazardous areas. In this book, a comprehensive treatment has been given that provides a unified foundation of wireless communication systems for transmission of data, voice, and video. The novelty of this book is that it covers up a wide range of the recent practices of wireless communication systems as well as wireless sensor networking, radio frequency identification devices, system-specific embedded software, application software, and designing and developing techniques for intrinsic safe circuits.

The book elaborates the technical aspects for design and development of intrinsically safe trapped miner communication system, shaft communication system, line-of-sight communication system, mine-wide communication system, and web-based information system. The book also describes the latest RFID technology, miner information and safety system, system-specific embedded software, and application software. Further, the book provides the deployment and operation of systems in underground mines. Thus, the book incorporates all the technical details starting from design, fabrication, assembling, intrinsic safety analysis, testing, laboratory and field trials, deployment, and operations of different wireless communication systems. The book is divided into 12 chapters to incorporate different aspects of wireless communication techniques for underground mines and the content of each chapter is briefly enumerated subsequently.

Chapter 1 – *Mine Communication Technique*: The different underground mine communication techniques available worldwide are summarized.

Chapter 2 – *Evaluation of Suitable Frequency*: Experiments carried out in the laboratory and fields are incorporated for characterizing the precise frequencies of the systems for communication in different underground mining zones.

Chapter 3 – *Trapped Miner Communication*: In underground mines, sometimes due to fissured strata, the roof or side walls of a gallery collapse and miners get trapped inside the sealed area. Many miners may also get trapped beneath the big chunk of fallen roof. A communication link between the trapped miner and rescue team is essential to find out the actual location of trapped miner for speedy rescue operation. Therefore, in this chapter, system description, technical description, block diagram, detailed circuit diagram, printed circuit board layout, power supply safety protection, safety analysis, and laboratory and field trials of intrinsic safe trapped miner communication system developed by the authors are discussed.

Chapter 4 – *Shaft Communication*: Bell signaling system is being used today in most of the underground mines in India and other countries. But this system is having its own drawbacks. A need exists for improved hoist communication between the persons in the moving skip and the hoist operator. Therefore, design and development of induction-based communication system using hoist/guide rope, as a current carrier is described for the reliable and real-time communication in shaft. The system description, technical description, block diagram, detailed circuit diagram, printed circuit board layout, power supply safety protection, safety analysis, and laboratory and field trials of the developed system are described.

Chapter 5 – *Line-of-Sight Communication*: Line-of-sight communication using ultra-high-frequency transceivers is discussed, with due emphasis on system description, technical details, detailed circuit diagram, and safety aspects.

Chapter 6 – *Mine-Wide Communication*: Leaky feeder-based communication technique for mine-wide communication is enumerated in detail. Technical details, safety aspects, and installation and commissioning procedures are also discussed.

Chapter 7 – *Web-Based Information and Decision Support System for Mining Industry*: The development of web-based system with different modules is presented for improving production and productivity, shift and personnel management, reduction of production discrepancy, maintenance of equipment, management of inventory, environment monitoring, information of mine, disaster forecasting, mine safety management, statutory requirements, postdisaster management, improvement in working by online record keeping, wireless communication in underground mine, decision making, training, and various other aspects. The chapter summaries the scope of development of web-based system by highlighting the existing problems and solution under different modules.

Chapter 8 – *ZigBee Technology: A Unique Wireless Sensor Networking Solution*: The worldwide advancement of the radio frequency identification (RFID) technology for underground applications is discussed in detail. The feasibility and exploration of the technology in underground mines are also discussed. The ZigBee technology is elaborated in detail.

Chapter 9 – *Miner Information and Safety System for Mines*: In an underground mine, many miners generally enter into underground in a shift for exploitation of coal/mineral from different working faces. As per the current practice, the identification and tracking of a miner is very difficult in case of disaster. Therefore, the identification and coding of the miners is a vital need for underground mine management in case of disaster as well as during normal operating conditions. Therefore, miner information and safety system using RFID technology has been developed by the authors and is discussed in detail. The installation and commissioning procedures both for underground and opencast mines is also highlighted along with its performance in underground mines.

Chapter 10 – *Programming of RFID Devices*: This chapter discusses the compatible system-specific embedded software for underground applications by forming dynamic wireless sensor network, which solves the redundancy problems in a large network. The software is designed to make the RFID devices as coordinator, routers, and end device for different underground use such as tracking of miners and machines, monitoring miners' unsafe practice and providing warning, monitoring the presence of methane and carbon monoxide gases, vehicle-collision prevention, fatal accident prevention, and messaging. Complete program codes are also provided for different devices, such as coordinator, router, end device, gas-monitoring device, proximity warning device, and message device.

Chapter 11 – *Tracking and Monitoring Software*: The application software for mining applications is discussed in detail. The programming aspects are also discussed. The software is especially designed for different purposes in mines and has different modules, namely, tracking of miners and vehicles, route tracking in opencast mines, preventing fatal accidents and vehicle collisions, environmental monitoring, observing miners' unsafe practice, sending alert message, message communication and preparing computerized miners' duty hours record. Algorithms of different modules of the software are also incorporated in the chapter.

Chapter 12 – *Intrinsic Safety for Hazardous Area*: The concept of intrinsic safe circuit development is the focus of this chapter. The technical aspects of designing and developing the intrinsically safe systems are described in detail. This chapter also enumerates the procedures for designing intrinsic safe circuits for hazardous areas.

All these chapters elaborately describe the different types of wireless communication systems for underground mines. Each chapter is complete in its particular aspects. Detailed technical aspects of various communication devices developed by the authors have been discussed in the respective chapters including implementation, installation, and commission procedures. The book is written for modern telecommunication, electronics and instrumentation, software and system, computer science, electrical, and mining engineers, scientists, and researchers in the field of mining communication and automation. The engineers devoted in the subject would be highly benefited from the book especially for intrinsically safe design aspects of systems. The book is also beneficial for researchers of universities, institutes, and R&D organizations, and business professionals engaged in the field.

Dhanbad, India

Dr. L.K. Bandyopadhyay
Dr. S.K. Chaulya
Dr. P.K. Mishra

Acknowledgment

Authors are thankful to Dr. Amalendu Sinha, Director, Central Institute of Mining and Fuel Research (CIMFR), Dhanbad, India, and Prof. S. K. Bramhachari, Director General, Council of Scientific and Industrial Research (CSIR), New Delhi, India, for their constant guidance, necessary help, and support. Authors are also grateful to Ms. M. Jennifer, Assistant Editor, Springer, USA, for her kind help and guidance to prepare the manuscript of the book. Authors also thank two anonymous reviewers for evaluating the book proposal and providing valuable suggestions.

Authors are grateful to Mr. B. M. Baveja, Sr. Director; Dr. R. C. Chopra and Mr. S. Basu, Group Coordinators; and J. K. Tyagi, Joint Secretary & Financial Advisor, Department of Information Technology (DIT), Ministry of Communication and Information Technology, Government of India, New Delhi, for funding various wireless communication projects for underground mines through which authors have developed different wireless communication devices for underground mines. Authors also thank Ministry of Coal and Central Mine Planning and Design Institute (CMPDI), India, for sponsoring R&D projects on mine control, automation, and communication.

Authors are also grateful to Prof. H. M. Gupta and Prof. B. N. Jain, Indian Institute of Technology (IIT), Delhi, India; Dr. K. R. Kini, Director, Society for Applied Microwave Electronics Engineering & Research (SAMEER), IIT Bombay Campus, India; Prof. U. B. Desai, IIT, Bombay, India; Prof. R. V. Raja Kumar and Prof. A. Bhattacharjee, IIT, Kharagpur, India; Prof. K. Rajanna and Prof. Anurag Kumar, Indian Institute of Science (IISc), Bangalore, India; Prof. B. Ramamurthy, IIT, Madras, India; Prof. B. K. Shrivastva, Institute of Technology, Banaras Hindu University, Varanasi, India; Prof. M. Radhakrishnan and Prof. U. S. Tiwary, Indian Institute of Information Technology (IIIT), Allahabad, India; Mr. J. K. Tyagi, DIT, India; Dr. A. S. Rao, Scientist "G," Department of Science and Technology, Government of India; Mr. B. N. Sarkar, Scientist, Department of Science and Industrial Research; Director General of Directorate of Mines Safety, Ministry of Labour, Dhanbad, India; Secretary, Ministry of Mines, India; Director, Indian School of Mines University; Prof. B. B. Dhar, ex-Director, CIMFR, Dhanbad, India; Mr. H. Ramakrishna, Advisor (R&D), Alpha Design Technologies Private Limited, Bangalore, India; and other expert members for their valuable suggestions and constant guidance during development and implementation of wireless devices by the authors.

Authors are grateful to the Chairman and Director (Technical), Coal India Limited, Kolkata, India; Chairman-Cum-Managing Director of Mahanadi Coalfields Limited (MCL), Central Coalfields Limited (CCL), Eastern Coalfields Limited (ECL), Bharat Coking Coal Limited (BCCL), Western Coalfield Limited (WCL), South Eastern Coalfields Limited (SECL), and Singareni Collieries Company Limited (SCCL), India, for partially funding the project and rendering their necessary help during implementation of the wireless devices in the fields.

Authors sincerely acknowledge the service rendered by Dr. A. L. Das, Mr. D. Dan, and Mr. A. Mazumdar, Scientists, SAMEER, Kolkata, India; and Mr. B. Deb and Mr. G. Das, Scientists, Electronics Regional Test Laboratory (East), Kolkata, India, for testing the devices regarding its radio frequency characteristics and intrinsic safety aspects. Authors also thank Prof. S. Bandyopadhyay, Mr. S. K. Banerjee, Mrs. S. Roy, Mr. A. Chakraborty, Mr. S. Chudhuri, and Mr. A. Choudhuri of Perv-Com Consulting Private Limited, Kolkata, India, for technical support in development of ZigBee-compliant wireless technology for underground mines. Authors are thankful to Texas Instrument, USA, and Tricoteco (Jiangxi) Technology Development Co. Ltd., China, for providing ZigBee-compliant chip and its technical information. Authors are also thankful to Mine Radio, Australia; GMG Techno Trades (P) Ltd., India; Simoco, India; Creative Solutions, India; and Halley Electronics, India, for providing technical support for field testing and fabrication of wireless communication devices. Authors also appreciate the effort of Bureau of Indian Standard (BIS) and International Electrotechnical Commission (IEC) for formulating excellent standards for intrinsic safety aspects of electrical equipment.

Authors are very much thankful to Mr. Ashutosh Narayan, Mr. Akhilesh Choure, Mr. Bhogesh Kumar, Mr. Basanta Mahato, Mr. Sunil Gautam, Ms. Sindhu Kumari, Ms. Vijeta Jha, Ms. Vijaylaxmi Singh, Ms. Arundhati Payra, Mr. Ajit Kumar, Mr. Rahul Sarkar, Mr. Pallav Sarkar, Mr. Manish Sinha, Mr. Manish Kumar, and Mr. Sarad Tiwary who had constantly worked hard for development of different wireless communication devices during their tenure as Project Assistants of CIMFR, Dhanbad, India, in different ministry-sponsored projects. Special thanks are due to Mr. Akhilesh Choure, Mr. Bhogesh Kumar, Ms. Sindhu Kumari, Ms. Vijeta Jha, Mr. Basanta Mahato, and Mr. Sunil Gautam, who had also helped in preparing the manuscript of the book.

Authors are grateful to Dr. S. Kumar, ex-Scientist, CIMFR, Dhanbad, India, for his technical help in designing and implementing the communication devices in underground mines. Authors thank Dr. P. Pal Roy, Mr. A. K. Ghosh, Dr. B. Kumar, Dr. B. K. Tewary, Dr. M. Prasad, and Mr. A. Roy, Scientists, CIMFR, Dhanbad, India, for their constant help and support for development and implementation of wireless communication devices in underground mines. Authors also thank Dr. A. K. Singh, Mr. R. K. Vishwakarma, Mr. B. Ahirwal, Mr. M. K. Dutta, and Mr. A. Kumar of Flame Proof Department of CIMFR, Dhanbad, India, for technical support in intrinsic safety testing of the devices. Finally, authors thank all the scientists and staffs of CIMFR, Dhanbad, India, who have directly or indirectly helped in designing, fabricating, and implementing the wireless communication devices in the fields.

Contents

1 Mine Communication Technique 1

1.1 Introduction 1

1.2 Wired Communication Services 2

1.2.1 Types of Wired Communication 3

1.2.2 Carrier Current System 7

1.2.3 Combined System 9

1.2.4 Fiber Optics Communication 9

1.2.5 Limitations of Wired Communication System . . . 9

1.3 Semi-wireless Communication System 10

1.3.1 Leaky Feeder-Based System 10

1.3.2 Location and Monitoring for Personal Safety 11

1.3.3 Digital Enhanced Cordless Telecommunications . . 12

1.3.4 Ethernet 13

1.4 Through-the-Earth Communication System 14

1.4.1 PED Communication System 15

1.4.2 Cover with Transmitter 18

1.4.3 Tram Guard Miner Track 18

1.4.4 Delta Electromagnetic Gradiometer Beacon Tracking System 18

1.4.5 Subterranean Wireless Electric Communication System 19

1.4.6 Emergency Broadcast Network 19

1.4.7 Very Low frequency and Low-Frequency Propagation 19

1.4.8 Research on TTE 22

1.5 Wireless Communication Services 23

1.5.1 Pocket Pagers 25

1.5.2 Walkie-Talkie System 26

1.5.3 Bluetooth 27

1.5.4 Wireless Fidelity 27

1.5.5 World Interoperability for Microwave Access . . . 29

1.5.6 Radio Frequency Identification Technique 36

1.5.7 Ultra-wide-Band Communication 41
1.5.8 Watcher-ATS 44
1.5.9 Tracker Tagging System 45
1.6 Rescue Systems for Disaster Management 45
1.6.1 Communication in Subterranean and Urban Environments 46
1.6.2 Tooth Microphone 49
1.6.3 Super-Low-Frequency Beacon 49
1.7 Conclusions 50
References 55

2 Evaluation of Suitable Frequency 65
2.1 Introduction 65
2.2 Wave Propagation Through Medium 66
2.3 Laboratory Experiment 67
2.3.1 Experimental Procedure 67
2.3.2 Analysis of Propagation Signals 69
2.4 Results and Discussion 70
2.5 Conclusions 70
References 71

3 Trapped Miner Communication 73
3.1 Introduction 73
3.2 System Description 73
3.3 Transmitter Unit 74
3.4 Receiver Unit 77
3.5 Technical Description 80
3.5.1 Power Supply and Safety Protection 80
3.5.2 Technical Specification 81
3.5.3 Infallible Components 81
3.5.4 Input and Output Details 82
3.6 Safety Analysis 82
3.6.1 Transmitter Unit 82
3.6.2 Receiver Unit 84
3.7 Component Details 85
3.7.1 Ratings of Safety Components 85
3.7.2 List of Components for Transmitter Unit 85
3.7.3 List of Components for Receiver Unit 87
3.8 Capabilities of the System 87
3.9 Field Trial 88
3.10 Conclusions 89
References 89

4 Shaft Communication 91
4.1 Introduction 91
4.2 Theory 91

4.2.1 Inductance 91
4.2.2 Current Clamp 93
4.2.3 Electromagnetic Coupling 93
4.2.4 Consequences 94
4.3 Principle 95
4.4 System Description 95
4.5 Technical Details 99
4.5.1 Transmitting Unit 99
4.5.2 Receiving Unit 99
4.5.3 Specifications 100
4.6 Field Installation Procedure 100
4.7 Laboratory and Field Trials 101
4.7.1 Laboratory Trial 101
4.7.2 Field Trial 102
4.8 Capabilities of the System 103
4.9 Conclusions 104
References 104

5 Line-of-Sight Communication 105
5.1 Introduction 105
5.2 Propagation of UHF Radio Waves in Coal Mine Tunnels . . . 105
5.2.1 Waveguide Modes 106
5.2.2 Loss due to Surface Roughness 110
5.2.3 Loss due to Tilt of Tunnel Walls 112
5.3 Expected Communication Range in a Mine and Propagation Loss 113
5.4 System Description 114
5.5 Circuit Diagram 115
5.6 Specification 116
5.7 Field Trial 117
5.8 Capabilities 118
5.9 Conclusions 119
References 119

6 Mine-Wide Communication 121
6.1 Introduction 121
6.2 The System 122
6.2.1 Basic Principle 122
6.2.2 System Description 123
6.3 Specifications of Leaky Feeder System 124
6.4 Working Technique 126
6.5 Advantages 127
6.6 Types of Cables 127
6.6.1 Long-Wire Antennas 127
6.6.2 Twin-Wire Feeders 127
6.6.3 Delogne System 128

6.6.4 Slotted Shield Cables 128
6.6.5 Loose-Braided Cables 128
6.7 Drawbacks 129
6.8 Field Trial 129
6.9 Development of Passive Amplifier 131
6.9.1 Description 131
6.9.2 Circuit Diagrams 132
6.9.3 Specifications 134
6.9.4 Field Installation 135
6.9.5 Testing of Antenna 135
6.9.6 Capabilities 137
6.10 Conclusions 137
References 137

7 Web-Based Information and Decision Support System for Mining Industry 141
7.1 Introduction 141
7.2 Required Areas of IT Application in Coal Mines 141
7.2.1 Improving Production and Productivity 141
7.2.2 Shift and Personnel Management 142
7.2.3 Reduction of Production Discrepancy 142
7.2.4 Maintenance of Equipment 142
7.2.5 Management of Inventory 143
7.2.6 Environment Monitoring 143
7.2.7 History of Mine 143
7.2.8 Disaster Forecasting and Mine Safety Management 144
7.2.9 Statutory Requirements 145
7.2.10 Postdisaster Management 145
7.2.11 Improvement in Working by Online Record Keeping 145
7.2.12 Wireless Communication in Underground Mine . . 146
7.2.13 Decision-Making 146
7.2.14 Training 146
7.3 System Description 146
7.3.1 Software and Database 147
7.3.2 Middleware 148
7.3.3 Hardware 149
7.4 Conclusions 149
References 151

8 ZigBee Technology: A Unique Wireless Sensor Networking Solution 153
8.1 Introduction 153
8.2 ZigBee Technology 155
8.2.1 ZigBee Stack 155
8.2.2 The Network Layer 155
8.2.3 The Data Link Layer 157

8.2.4 The General MAC Frame Structure 158
8.2.5 MAC Protocol Overview in IEEE 802.15.4 160
8.2.6 The Physical Layer 162
8.2.7 Sensitivity and Range 164
8.2.8 Security 164
8.2.9 Limitations 165
8.3 Specialty of ZigBee Technology 165
8.4 Comparison of ZigBee with Other Wireless Technologies . . 166
8.5 Features of ZigBee 169
8.6 Application Areas 170
8.7 Manufacturers of ZigBee Devices 170
8.8 Conclusions 170
References 173

9 Wireless Information and Safety System for Mines 175
9.1 Introduction 175
9.2 Function 175
9.3 The System 178
9.3.1 Coordinator 178
9.3.2 Router/End device 192
9.3.3 Carbon-Monoxide-Monitoring Device 206
9.3.4 Methane-Monitoring Device 214
9.3.5 Message Device 225
9.4 Principle and Operating Procedure 237
9.4.1 Tracking and Monitoring System 237
9.4.2 Reduction in Fatal Accident 245
9.4.3 Collision Prevention System 245
9.4.4 Efficiency and Productivity Monitoring System . . 246
9.4.5 Monitoring Miners' Unsafe Practice and Warning System 246
9.4.6 Message Communication System 247
9.5 System Installation Procedure 247
9.6 Capabilities of the System 250
9.7 Performance Analysis of RFID Devices in Underground Mines 251
9.7.1 Experiments for Evaluating Packet Delivery Ratio . 251
9.7.2 Experiments for Evaluating Beacon Rate 258
9.8 Conclusions 260
References 261

10 Programming of RFID Devices 263
10.1 Introduction 263
10.2 Function of RFID Devices 263
10.3 CC2430 Chip 264
10.4 CC2430 Modules 266
10.5 Programming of RFID Devices 268

10.5.1 Debugging Using USB Interface 268
10.5.2 Debugging with CC2430 Debug Connector 269
10.6 Network Creation 269
10.7 Network Topologies 270
10.7.1 Star Topology 270
10.7.2 Peer-to-Peer Topology 270
10.7.3 Cluster-tree Topology 271
10.8 Network Layers 271
10.8.1 Application (Layer 7) 272
10.8.2 Presentation (Layer 6) 272
10.8.3 Session (Layer 5) 272
10.8.4 Transport (Layer 4) 272
10.8.5 Network (Layer 3) 272
10.8.6 Data Link (Layer 2) 273
10.8.7 Physical (Layer 1) 273
10.9 Wireless Application Protocol 273
10.10 Medium Access Control 275
10.10.1 MAC Application 275
10.10.2 MAC Operation 275
10.10.3 MAC Application Design 276
10.10.4 Programming with TI-MAC Library 277
10.11 Programming and Functionality of RFID Devices 279
10.11.1 Wireless Network Program 279
10.11.2 Function of Different Devices 279
10.11.3 Programming Procedure 281
10.12 Conclusions 281
References 281

11 Tracking and Monitoring Software 283
11.1 Introduction 283
11.2 Application Software 283
11.3 Operating Mode 284
11.4 Flowchart and Algorithm 285
11.5 Installation and Operation Procedures 285
11.5.1 Installation Procedures 285
11.5.2 Operation 291
11.6 Conclusions 298

12 Intrinsic Safety for Hazardous Area 299
12.1 Introduction 299
12.2 Concept of Intrinsic Safety 300
12.3 Classifications 302
12.3.1 Hazardous Area 302
12.3.2 Temperature of Ex-Equipment 304
12.4 Standards Related to Intrinsic Safety 305
12.4.1 ATEX Directive 305

12.5 Important Definitions 309
12.6 Grouping and Classification of IS and Associated Apparatus . 311
12.6.1 Categories of Electrical Apparatus 311
12.6.2 Simple Apparatus 312
12.7 Designing of Intrinsically Safe Apparatus 312
12.7.1 Construction Features 312
12.7.2 Description of the Various Assemblies 313
12.7.3 Designing of Simple Circuits 324
12.7.4 Analysis by the Parameters 332
12.8 Parameters Considered for Intrinsic Safety 347
12.8.1 Ratings of Components 347
12.8.2 Connectors 347
12.8.3 Fuses 347
12.8.4 Cells and Batteries 348
12.8.5 Semiconductors 349
12.8.6 Failure 350
12.9 Infallible Components 351
12.9.1 Transformers 351
12.9.2 Current-Limiting Resistors 353
12.9.3 Blocking Capacitors 353
12.10 Use of Shunt Safety Assemblies 353
12.10.1 Safety Shunts 354
12.10.2 Shunt Voltage Limiters 354
12.10.3 Wirings and Connection 354
12.10.4 Galvanically Separating Components 355
12.10.5 Safety Barriers 356
12.11 Electromagnetic and Ultrasonic Energy Radiating Equipment 359
12.11.1 Radio Frequency Sources 360
12.12 Advice to the First-Time Designer 360
12.12.1 Basic Design Concepts 361
12.12.2 Safety Analysis 362
12.12.3 Factors Affecting Intrinsic Safety 365
12.12.4 Maintenance and Handling of the Ex-Equipment . . 365
12.13 Guidelines for Designing Intrinsic Safety Circuit 367
12.14 Basic Protection Concepts of Intrinsic Safety 370
12.15 Conclusions 372
References 374

Appendix A Program for RFID Devices 379
1 Programming for Coordinator 379
1.1 Programming with CC2430 RFID Chip (48 Pins) . 379
1.2 Programming with CC2430 Chip (22 Pins) 401
2 Programming for Router 402
2.1 Programming with CC2430 Chip (48 Pins) 402
2.2 Programming with CC2430 Chip (22 Pins) 414

3 Programming for End Device 415
3.1 Programming with CC2430 Chip (48 Pins) 415
3.2 Programming with CC2430 Chip (22 Pins) 426
4 Programming for Proximity Warning Device 426
4.1 Programming with CC2430 Chip (48 Pins) 426
4.2 Programming with CC2430 Chip (22 Pins) 428
5 Programming for Router with Sensor Device 429
5.1 Programming with CC2430 Chip (48 Pins) 429
5.2 Programming with CC2430 Chip (22 Pins) 439
6 Programming for End Device with Data Logger 439
6.1 Programming with CC2430 Chip (48 Pins) 439
6.2 Programming with CC2430 Chip (22 Pins) 441
7 Programming for Messaging Device 442
7.1 Use of Keil uVersion3 Software 442
7.2 Use of UniPro Software 442
7.3 Program Code 442

Appendix B Algorithm of Different Modules of TMS Software 465

Index 471

Abbreviations

ADC	Analog to digital converter
AES	Advanced encryption standard
AF	Audio frequency
AM	Amplitude modulation
AO	Area office
ATEX	Atmosphere explosive
BAM	Bundesanstalt für Materialforschung und Prüfung
BCCL	Bharat Coking Coal Limited
BE	Back-off exponent
BIS	Bureau of Indian Standard
BLE	Battery life extension
BO	Beacon order
BPF	Band pass filter
BU	Branch unit
CAP	Contention access period
CCK	Complimentary code keying
CDMA	Code division multiple access
CENELEC	European Committee for Electrotechnical Standardization
CESI	Centro Elettrotecnico Sperimentale Italiano
CFP	Contention-free period
CIMFR	Central Institute of Mining and Fuel Research
CLR	Current-limiting resister
CPU	Central processing unit
CRT	Cathode ray tube
CSA	Canadian Standard Association
CSMA-CA	Carrier sense multiple access with collision avoidance
CSMA/CD	Carrier sense multiple access with collision detection
CW	Contention window
DAS	Data Acquisition Software
dB	decibel
DECT	Digital enhanced cordless telecommunications Ethernet
DeltaEM	Delta Electromagnetic
DFT	Discrete Fourier Transform

DGMS	Directorate General of Mines Safety
DLL	Data link layer
DSP	Digital signal processing
DSSS	Direct sequence spread spectrum
DSRC	Dedicated short-range communication
EBN	Emergency broadcast network
EEPROM	Erasable programming read only memory
EM	Electromagnetic
EMF	Electromotive force
EMI	Electromagnetic interference
ERTL	Electronic Regional Testing Laboratory
FCC	Federal Communication Commission
FCS	Frame check sequence
FDM	Frequency division multiplexing
FFD	Full-function device
FFT	Fast Fourier Transform
FLP	Flameproof
FM	Frequency modulated
FSA	Forschungsgesellschaft für angewandte Systemsicherheit
GPR	Ground penetrating radar
GSM	Global system for mobile communications
GTS	Guaranteed time slots
HO	Head office
IC	Integrated circuit
IEC	International Electrotechnical Commission
IEEE	Institute of Electrical and Electronic Engineers
INERIS	Institute National de l'Environment Industriel et des Risques
IP	Ingress Protection
IS	Intrinsic safety
ISM	Industrial, scientific, and medical
ISO	International Organization for Standardization und Arbeitsmedizin mbH
IT	Information Technology
ITU-T	International Telecommunication Union – Telecommunication
LAMPS	Location and monitoring for personal safety system
LAN	Local area network
LCIE	Laboratoire Central des Industries Électriques
LEL	Lower explosive limit
LF	Low frequency
LLC	Logical link control
LMDS	Local multipoint distribution service
LNA	Low-noise amplifier
LO	Local oscillator
LOM	Laboratorio Oficial José María de Madariaga
LQI	Link quality indication

LR-WPAN	Low rate wireless personal area network
LT	Line terminator
MMDS	Metropolitan multipoint distribution service
MAC	Medium access control
MAN	Metropolitan area network
MCIT	Ministry of Communication and Information Technology
MCL	Mahanadi Coalfields Limited
MCPS	MAC common part sublayer
MFR	MAC footer
MHR	MAC header
MIC	Minimum igniting current
MIE	Minimum ignition energy
MLME	MAC layer management entity
MPDU	MAC protocol data unit
MSHA	Mine Safety and Health Administration
MSDU	MAC service data unit
NB	Number of back off
NEC	National Electrical Code
OFDM	Orthogonal frequency-division multiplexing
OFDMA	Orthogonal frequency-division multiple accesses
OMS	Output per man shift
OSI	Open systems interconnection
PC	Power coupler
PCB	Printed circuit board
PED	Personal emergency device
PHY	Physical
PMP	Point-to-multipoint
PPDU	PHY protocol data unit
PSDU	PHY service data unit
PSK	Phase shift keying
PSS	Physical signaling sublayer
PTB	Physikalisch-Technische Bundesanstalt Braunschweig
PTH	Point to hole
P2P	Peer-to-peer
QAM	Quadrature amplitude modulation
QOS	Quality of service
RAS	Remote access server
RF	Radio frequency
RFD	Reduced functionality device
RFID	Radio frequency identification
Rx	Receiver unit
SA	Standard Association
SAP	Service access points
SABS	South African Bureau of Standard
SEV	Schweizerischer Elektrotechnischer Verein

SNR	Signal-to-noise ratio
SO	Superframe order
SoC	System on chip
SP	Sveriges Provnings-Och Forskingsinstitut
SSCS	Service-specific convergence sublayer
SWECS	Subterranean wireless electric communication system
TCP/IP	Transmission Control Protocol/Internet Protocol
TEM	Transverse electromagnetic
TI-MAC	Texas instruments – medium access layer
TMS	Tracking and Monitoring Software
TTW	Through-the-wire
TTA	Through-the-air
Tx	Transmitter unit
UEL	Upper explosive limit
UHF	Ultra-high frequency
UL	Underwriters' Laboratories Inc.
ULF	Ultra-low frequency
UPS	Uninterrupted power supply
USBM	U.S. Bureau of Mines
UWB	Ultra wide band
VB	Visual Basic
VHF	Very high frequency
VLF	Very low frequency
VoIP	Voice internet protocol
WAP	Wireless application protocol
Wi-Fi	Wireless fidelity
WiMAX	World interoperability for microwave access
WLAN	Wireless local area networks
WNP	Wireless network program
WPAN	Wireless personal area network
WSN	Wireless sensor networks

Symbols

A	Ampere
A_p	Power attenuation
A_v	Voltage attenuation
bps	bytes per second
C	Capacitance
dB	Decibel
D	Diode
GHz	Giga Hertz
F	Farad
H	Henry
Hz	Hertz
I	Current
Ii	Maximum input current
Io	Maximum output current
J	Joule
kbps	Kilo byte per second
kHz	Kilo Hertz
L	Inductance
m	Meter
Mbps	Mega byte per second
MHz	Mega Hertz
P	Power
Pi	Maximum input power
Po	Maximum output power
P_d	Destination power
P_s	Signal power
R	Resistance
Um	Maximum r.m.s a.c. or d.c. voltage
Ui	Maximum input voltage
Uo	Maximum output voltage
V	Volt
V_d	Destination voltage
V_s	Signal voltage

W	Watt
μs	Microsecond
ϵ	Dielectric constant
ω	Angular frequency
σ	Conductivity
μ_0	Permeability
Ω	Ohm
λ	Wavelength
ϕ	Grazing angle
θ	Tilt

About the Authors

Dr. L. K. Bandyopadhyay has completed M.Sc. (Physics) with specialization in Electronics with First Class from National Institute of Technology (NIT), Rourkela, India, and joined as Research Fellow at Central Institute of Mining and Fuel Research (CIMFR), Dhanbad, India, in 1976 and worked for about 3 years on the development of miners' safety devices. During the period, he has developed intrinsically safe methane, air velocity, and machine status sensors.

Subsequently, in 1979, he joined Oil and Natural Gas Commission, Dehradun, as Geophysicist for development of digital oil exploration devices. In 1980, he proceeded to Poland with a Government scholarship for Ph.D. and he had developed digital control system for longwall mining machines, which had already been incorporated in the modern longwall mining machine worldwide. After returning India, he worked as lecturer initially at NIT, Rourkela, and carried R&D studies on propagation of electromagnetic waves. Subsequently, in 1985 he joined Indian School of Mines, Dhanbad, for teaching under and post-graduate engineering, and R&D on longwall instrumentation. In May 1986, he has again joined CIMFR, Dhanbad, as Scientist "C" and till now continuing in the same institute, gradually promoted to the position of Scientist "G" in May 2006. He has been working as the head of Instrumentation and Miner's Safety Division.

At CIMFR, he has worked in the various diversified field of Physical, Earth & Environmental Sciences. He has worked in several National Mission Interlaboratory projects, sponsored by different ministries. He has conducted several studies on electrical properties of coal/rock and propagation characteristics of radio frequency waves in mines. He has acquired significant knowledge and expertise on the development of mining automation, monitoring, control, and communication systems.

As a leader he has designed, developed, patented, and implemented number of indigenous technologies in underground mines and transferred these know-how to different manufacturers for commercialization. He has taken a lead role in getting sanctioned a number of R&D and consultancy projects from different ministries and industries and managed to successfully implement the projects in the field. He has published more than 40 research papers in the peered reviewed and cited national and international journals, and received several awards. He has been nominated as member of different committees at national and international level for policy formulation and product standardization. He has developed good linkage with different academic and research institutes nationally and internationally for technical collaborations. He has organized several training programmes and workshops. He has attended many national and international (UK, Poland, German, Russia, etc.) conferences and workshops to present the institute's R&D findings and to have interaction to build technical cooperation with other organizations. He has been constantly trying to develop and introduce latest technologies in the mining and fuel industry.

Dr. S. K. Chaulya is a senior scientist at Central Institute of Mining and Fuel Research (CIMFR), Dhanbad, India. He has B.Tech. (Mining Engineering) degree from National Institute of Technology, Karnataka. He also holds M.Tech. degree (Mine Planning) and Ph.D. (Mining Engineering) from Institute of Technology, Banaras Hindu University, Varanasi, India. He has worked on various national mission R&D projects sponsored by the various central and state government departments during last 15 years of service at CIMFR under aegis of Council of Scientific and Industrial Research, India. In addition to these, he has worked on around 30 research projects sponsored by various public and private sectors. His field of research is monitoring, assessment, control and management of mining environment and wireless communication in underground mines. He has been associated with development of mining monitoring, control, automation, and wireless communication technologies for mining industries. He has patented 11 technologies on mine monitoring and communication, and some of the technologies have already been transferred to different manufacturing companies for commercialization in mining industry. Based on the research carried out during the period, he has published 23 research articles in international journals, 26 papers in national journals, 19 papers in international conferences and 27 papers in national conferences, and 6 chapters in edited books. He has also awarded with (i) Whitaker Golden Jubilee Award, 1997; (ii) Research Paper Publication Awards, 2000–2007; and (iii) Award for filing Patent, 2003, 2005, 2006, 2007, and 2008. He is a Member of Editorial Board of two international journals. He is a reviewer for various national and international journals. Presently he is working in the various wireless communication projects sponsored by the Ministry of Communication and Information Technology, Government of India.

Dr. P. K. Mishra is a senior scientist at Central Institute of Mining and Fuel Research (CIMFR), Dhanbad, India. He has completed his M.Sc. (Physics) from Magadh University, Bodh Gaya and Ph.D. (Applied Physics) from Institute of Technology, Banaras Hindu University, Varanasi, India, in 2000. He participated in space science technology programme for Centre for Space Science Technology and Education – Asia and Pacific, funded and affiliated by United Nations. Subsequently, he worked as Research Associate at Indian Institute of Tropical Meteorology, Pune, India, and actively involved in a Joint Indo-UK collaborative research programme on climate change impacts in India. He joined CIMFR under aegis of Council of Scientific and Industrial Research, India, as Scientist C in 2004 and actively involved in various national R&D projects sponsored by the various ministries. He has conducted several studies on electrical properties of coal/rock and propagation characteristics of radio frequency waves in mines. His field of interest is wireless sensor networking, wireless communication, laser technology in underground mines, climate modeling, free electron lasers, and quantum electronics. He has been associated with development of mining monitoring, control, automation, and wireless networking, and communication technologies for mining industries. He has patented eight technologies and some of the technologies have already been transferred to different manufacturing companies for commercialization in mining industry. He has published 10 research articles in national and international peered review cited journals, and 13 papers in national and international conferences. He has also been awarded for Best Research Paper Publication Award for Quality Research from 2005 to 2007 and Award for filing Patent in 2007 and 2008. He is a referee of Physics of Plasmas, USA. He has been constantly trying to develop and introduce latest technologies in the mining and fuel industry.

Chapter 1
Mine Communication Technique

1.1 Introduction

Mining, being a dynamic and intricate operation, needs a multifaceted continuous stream of information from surface to underground and vice versa. This two-way interaction is mandatory for risk assessment, coordinated team work, and effective safety management for underground miners. Communication becomes more important in underground mines as mining, by its inherent nature, is a hazardous activity. An effective and real-time communication can avoid or mitigate potentially dangerous situations (Young, 2002). Although the technology involved in removing material from below the earth's surface has a long history, communication systems in underground workings are rather new to the industry. Coal mines often rely on several independent communication systems to support different areas of their activity. The search of effective and reliable techniques and technologies to overcome the operational problems of underground mine communications started in 20th century because of repeated underground accidents (Brown, 1984; Chao and Chung, 1994; Chung, 1994; Deshpande et al., 1998; Yin and Chung, 1998; Miller et al., 1999; Haykin, 2001). Communication equipment did begin appearing in underground mines in the early 1900s. Telephone sets based on European electric standard for underground mines have been used in 1913. These early phones were essentially the same as those used above ground, except that they were enclosed in cast-iron boxes for protection against moisture, acid fumes, and gases. In the 1950s, a telephone set was introduced for use in explosive atmospheres which was designed around the philosophy of explosion containment. Both of these telephone sets required pipe or conduit, which was not at all practical for a coal mine beyond a limited range. From the early 1970s, work began on the design of modern communication systems for use in underground mines (Murphy and Parkinson, 1978; IC 8955, 1984; Drabich et al., 2002; Nichoga et al., 2005; Robinson et al., 2006; Li and Chen, 2006; Kennedy and Foster, 2006; Nie et al., 2008). Mine Safety and Health Administration (MSHA), USA, has studied over 60 proposals of communication and tracking systems from various vendors up to 2005 (Chirdon, 2006). Underground coal mining employs a diverse mix of communication devices (Kohler and Jeffery, 1992). The conditions which the system ought to fulfill are that it (i) must

L.K. Bandyopadhyay et al., *Wireless Communication in Underground Mines*,
DOI 10.1007/978-0-387-98165-9_1, © Springer Science+Business Media, LLC 2010

be intrinsically safe and explosion proof; (ii) should adhere to the ingress protection (IP) standards; (iii) must be rugged in structure; (iv) must be size flexible; (v) must have totality in design including cables, power supply unit, base stations, etc.; (vi) must be value-added priced; and (vii) must be robust, inexpensive, easy to expand, and enable fast and secure connections.

Underground mining operations can be tied up with communication like other industrial enterprises. Adequate communication within a mine and between surface and underground working areas is a vital part of the proper operation of any underground facility (Large et al., 1973; Forrest, 1975; Schiffbauer and Brune, 2006; http://origin.cdc.gov/niosh/mining/pubs/pdfs). This communication capability not only is an important factor in the concept of safety, production, and productivity but also is an aid to the day-to-day operations and helps in extracting and moving products to the surface. From a coal miner survivability standpoint, it only makes sense to enable all of the voice communication devices to interact with all other voice systems. This would be especially true during disaster situations (Ristenbatt et al., 1988; Dilmaghani and Rao, 2008). A properly designed system should be capable of providing this ability. If a rapid, accurate, and continuous flow of data from the subsurface is automatically presented to mine management, then decisions can be taken immediately and more accurately (Cocheril et al., 2008). It is observed by productive trends that a large amount of time is wasted due to unpredictable daily events. This can be minimized by clubbing the mine operations with communication. If management is aware of breakdowns within minutes, attention could be more quickly focused on solving the problems, which ultimately increases the long-term production. Safety can certainly be enhanced by accurate fast communication means. Quicker medical assistance, faster evaluation of the situation in underground, and accurate location of problems will be direct benefits of communication.

The options for communication signaling in underground mine include through-the-wire (TTW), through-the-earth (TTE), and through-the-air (TTA).

- TTW — Through-the-wire communication in an underground mine can travel over twisted pair, coaxial, trolley, leaky feeders, and fiber optic cables. Each of these cables has unique properties and limitations. Therefore, cables should suit the characteristics of the signals being conveyed.
- TTE — Through-the-earth system requires a portable or person-worn TTE. It requires no cabling between surfaces to underground and thus has some advantages over TTW systems (Asimov, 1979; Ghosh et al., 2008).
- TTA — It includes wireless communication systems.

Table 1.1 summarizes the frequency spectrum designations with their wavelength ranges which can be used for communication within different underground mining zones and between surface and underground working areas.

1.2 Wired Communication Services

Wired phone systems are those that depend on a wire connection among phones with voice-carrying signal wire. In this type of system, each phone is connected

Table 1.1 Frequency spectrum designations

Acronym	Description	Wavelength	Frequency range
VF	Voice frequency	10^6–10^5 m	3 Hz–3 kHz
VLF	Very low frequency	10^5–10^4 m	3 kHz–30 kHz
LF	Low frequency	10^4–10^3 m	30 kHz–300 kHz
MF	Medium frequency	10^3–10^2 m	300 kHz–3 MHz
HF	High frequency	10^2–10 m	3 MHz–30 MHz
VHF	Very high frequency	10–1 m	30 MHz–300 MHz
UHF	Ultra high frequency	1–0.1 m	300 MHz–3 GHz
SHF	Super high frequency	10–1 cm	3 GHz–30 GHz
EHF	Extremely high frequency	1–0.1 cm	30 GHz–300 GHz

by its own individual pair of wires to a central switch or telephone exchange. To establish a call between two phones in the system, the lines between the two phones must be connected (switched together) within the telephone exchange.

1.2.1 Types of Wired Communication

The following types of phones are used in underground mines: (i) magneto type, (ii) sound powered, (iii) bell signaling system, (iv) paging, (v) dial, and (vi) dial and page.

1.2.1.1 Magneto-type Phones

Magneto or crank ringer phones are probably the earliest underground communication instruments and can well be called the gadgets of 20th century. Magneto telephone consists of a battery, hand generator (magneto), bells, hook switch, line coupler, transmitter, and receiver (Fig. 1.1). The battery supplies direct current (DC) to transmitter and hand generator impresses alternate current (AC) on the line where it is sensed by bells. The magneto generates about 100 V at 15–20 Hz and is responsible for ringing other phone. Magneto phones are placed in a ladder fashion with a code of short and long rings for identification of called station (Murphy and Parkinson, 1978). The disadvantage of magneto phones is that it has inadequate signal strength for a large number of phones. They are not compatible with pager phones, and since they do not amplify the transmitted voice, they often receive a feeble and strident signal.

1.2.1.2 Sound-Powered Phones

As the name itself suggests, a sound-powered set provides voice communication which is powered by the speaker's voice without use of external energy, like batteries or power supply units. These phones have highly efficient transmitters and receivers for converting voice into electrical signals. The sound-powered handset is small in size. Two of these phones are usually connected by a single line to constitute

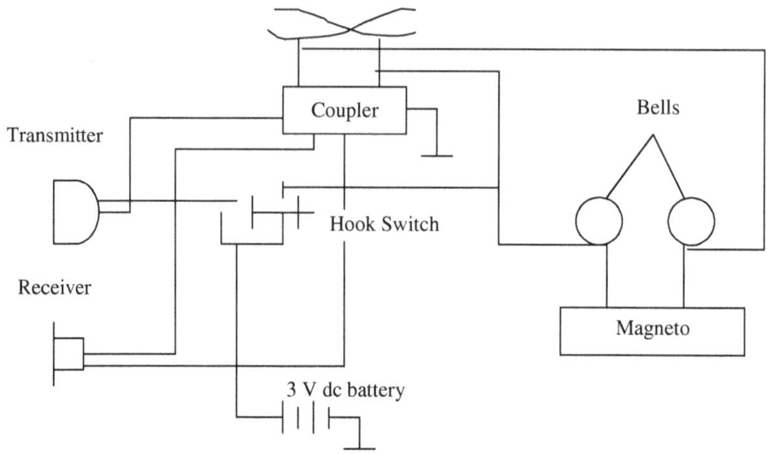

Fig. 1.1 Schematic diagram of magneto-type phone

an intercom circuit. Such a circuit reproduces speech with reasonable good quality for only short distances in a quiet condition. Sound-powered sets are used in underground mines as independent intercom systems. They are of two types:

(i) *Headset*: Here, the distance of the transmitter should be about 1–2.5 cm from the mouth. They should always be unplugged when not in use.
(ii) *Handset*: Here, a button is pushed down for talking and depressed to listen.

1.2.1.3 Bell Signaling

Bell signaling systems have gained widespread acceptance and are still in use on many hoists. Bell signaling was the only technique in the past to communicate between hoist operator and a man requesting a cage in underground. A single twisted-pair wire is run down the mine shaft and tapped off at those shaft stations where signaling is required (Fig. 1.2). To signal the hoist operator, a miner at any level pulls the pull bottle, causing the associated switch to close. This applies voltage to the buzzer at that level, and also to the hoist room buzzer and all other buzzers, through the switched line of the twisted pair (IC 8955, 1984).

Bell signaling system although a reliable system has some severe shortcomings. First, special message other than codes cannot be communicated to the hoist man. Second, the bell codes required becomes quite long in case of a mine having many levels. Long or complicated bell codes are obviously more difficult to remember and can become a source of confusion during disaster situation. Another disadvantage of bell signaling system is that communication with the cage cannot be established between two levels.

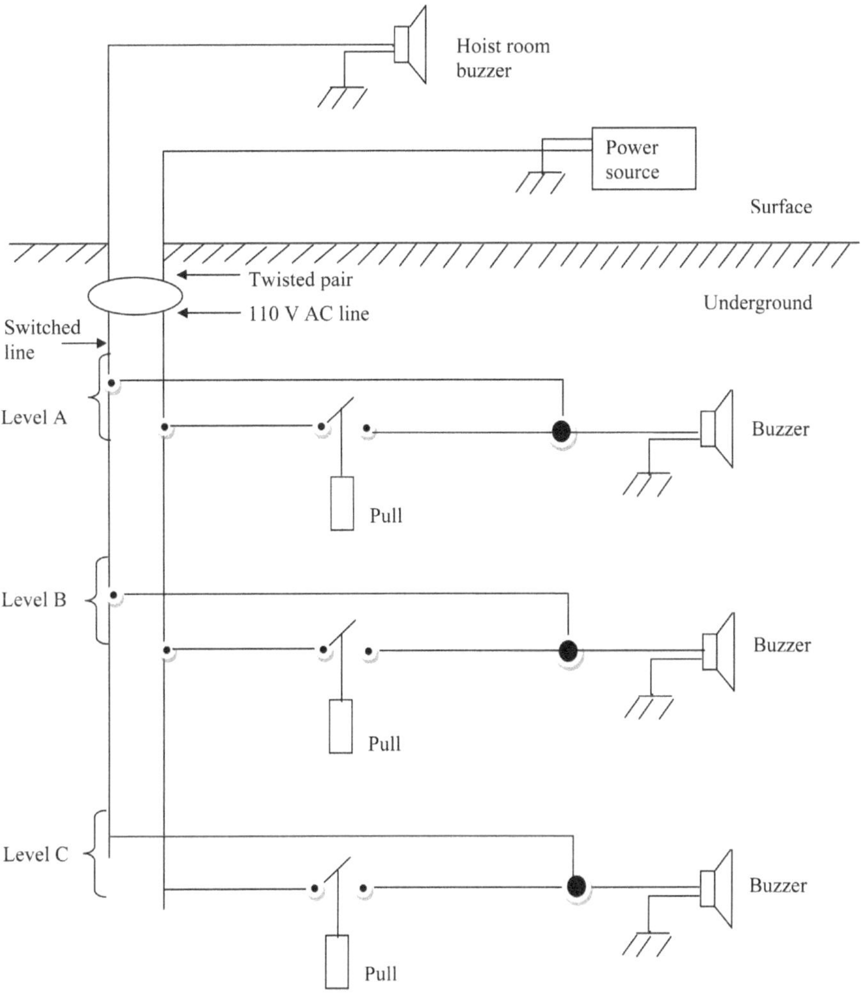

Fig. 1.2 Bell signaling system

1.2.1.4 Paging Phones

The most common type of communication system used in underground mines is paging telephone system (Fig. 1.3). A pager operates on a two-wire cable in party line fashion, i.e., only one conversation can be maintained at one time (Murphy and Parkinson, 1978). They are simple to install, reliable in operation, and easy to maintain. All paging phones have a self-contained battery, a paging speaker associated with signaling, and switching circuits. The handset is used for party line communication between stations. A DC voltage is impressed upon all phones when a page switch is actuated. The DC voltage actuates a solid-state switch to give signal at the paging amplifier. When the page switch is activated, the person initiating a call

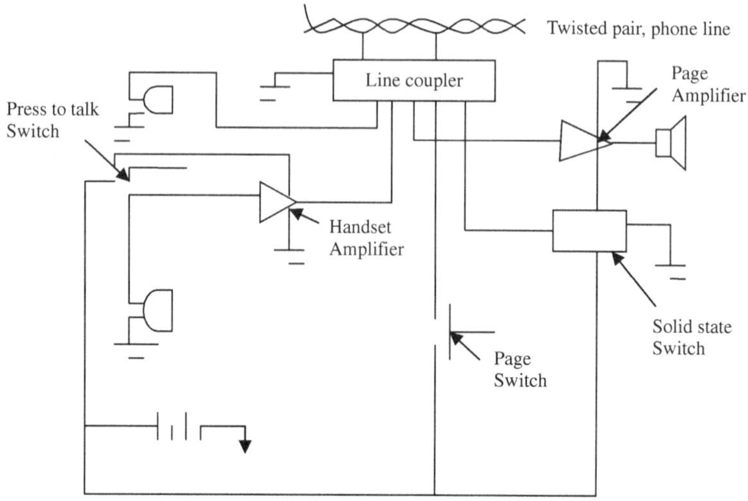

Fig. 1.3 Paging phone

presses the press-to-talk switch for calling the desired person or location. The press-to-talk switch applies signal to the handset transmitter amplifier. Now the amplified voice signal is given as an input signal to the activated paging amplifiers. In this way a call is broadcast throughout the phone system. Two-way party line conversations are possible between all phones by pressing the press-to-talk switch. A power supply of 12 V is provided through the solid-state switches for both signaling and powering the local amplifiers. The disadvantage of paging telephone system is that the telephone line must be used in a party line fashion. This prevents simultaneous conversations in the system.

1.2.1.5 Dial and Page Phones

The use of normal surface-type telephones in underground mines has two disadvantages: potential hazard in a methane environment from the 120 V at 20 Hz bell ringing voltage; and inability to locate a person who is not in his assigned work area. The ringing-type telephone has two advantages, namely (i) selective call feature and (ii) multiple private lines. Therefore, a unique phone system was developed which combines dial telephone with pager phone (Fig. 1.4). In its features, an interface is provided to isolate the potentially hazardous voltages from the mine line. A converter is used to convert the ring voltage into low-voltage DC required for operating the electronic switch that actuates the paging amplifier in the dial-selected pager phone. A tone signal notifies the caller when to begin the page. A handset switch accomplishes all functions normally accomplished by lifting the handset of a conventional phone from cradle. A common page line feature permits the paging of all phones, as it is required when searching for moving miner. The system uses a multiple-pair cable.

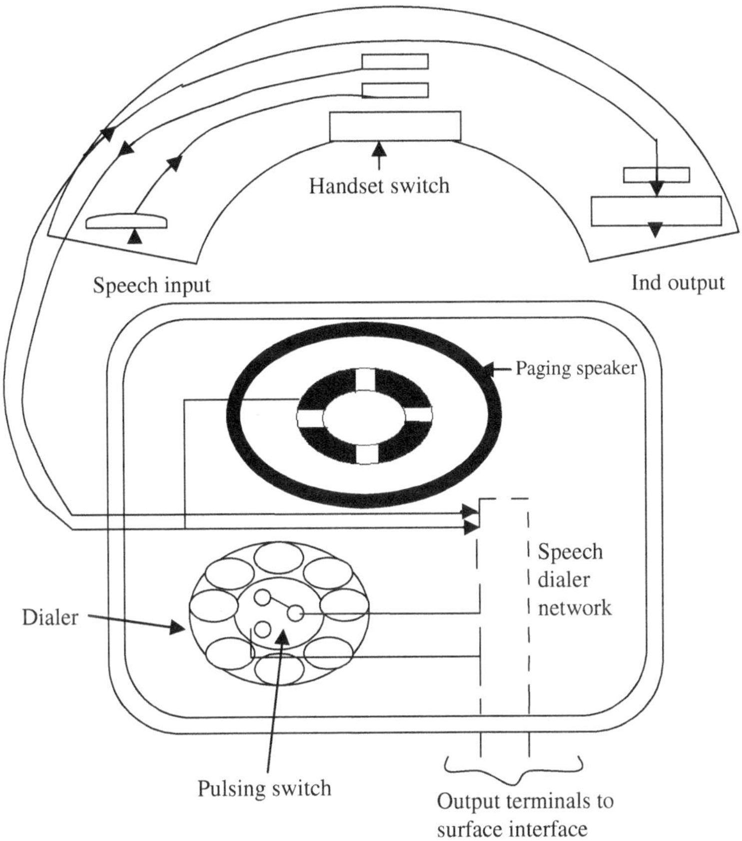

Fig. 1.4 Mine dial phone

1.2.2 Carrier Current System

When a radio frequency (RF) signal is fed to any underground wire or cable, it tends to distribute that signal throughout its length. Carrier current systems establish communication paths using existing mine wiring (Holloway et al., 2000; Wei et al., 2008). The wire used may be AC or DC power lines (Balanis, 1973; Anatory et al., 2008), neutral lines such as hoist rope, existing phone lines, or other wirings. Carrier current devices are basically frequency modulation (FM) radio transceivers that transmit and receive over existing mine wiring instead of using an antenna system. The low-frequency (LF) and medium-frequency (MF) RF signals ranges propagate best in carrier current systems (Chufo, 1978; Stolarczyk, 1991; Stolarczyk et al., 1991). The carrier current system, in many cases, is the most dependable communication system. Compared with the telephone circuits, it has a better insulation and an appreciably higher mechanical strength. There are two types of carrier current systems, namely (i) trolley carrier phone and (ii) hoist rope radio.

1.2.2.1 Trolley Carrier Phone

Trolley carrier phones (Chufo, 1977) are unique to the underground mining industry and are still used in many mines with trolley haulage. They operate at 60–140 kHz using frequency-modulated (FM) narrow band. In a trolley carrier current phone system, the trolley wire is transmission medium, and receiver and transmitter are connected to trolley wire through a coupler capacitor (IC 8955, 1984). Coupler capacitor acts as a short circuit at frequency of FM voice signals, but as an open circuit to the trolley wire DC power voltage. High voltage levels on trolley wire are thus blocked from entering receiver and transmitter sections of carrier phone, while FM voice signals pass freely through coupler capacitor. A simplified block diagram of a typical trolley carrier phone is shown in Fig. 1.5. It contains a power supply unit that converts high voltage of trolley wire to low voltage levels and provides power to carrier phone circuits. Power supply may also contain a battery backup in case of power failure. Such a system operates in push-to-talk, release-to-listen mode. Inductive coupling in these phones provides improved range and clarity.

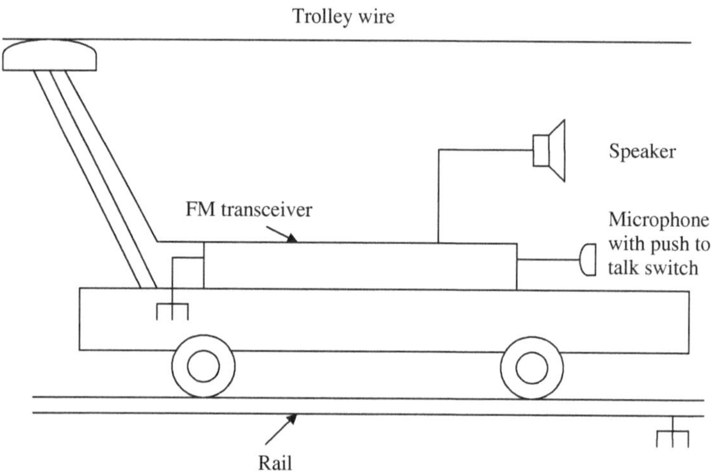

Fig. 1.5 Trolley carrier phone

1.2.2.2 Hoist Rope Radio

The hoist rope radio consists of two signal couplers and two transceivers. Each unit is of push-to-talk, release-to-listen design. During transmission, sending unit feeds its coupler with a FM carrier. Coupler induces a signal in hoist rope, which is then picked up by the coupler of the second unit. Both couplers are electrically identical, and each operates both as a transmitting and as a receiving element. Operation of hoist radio is the same as trolley carrier phone, except that hoist radio signal is inductively coupled to propagation medium (hoist rope). Transceivers of hoist room and cage are identical, except for battery required in cage. Hoist room power supply provides power for surface equipment. Surface equipment also may include a boom-type microphone and a foot-actuated push-to-talk switch to facilitate hands-free operation.

1.2.3 Combined System

There are distinct advantages in using a total system integrated from subsystems. A carrier system interconnected to wired telephones greatly extends range between fixed station and motormen on the locomotives. Nominal carrier frequency is either 88 or 100 kHz. Coupler interconnects audio from carrier, placing it on phone line, and audio is again applied to carrier. Another system combines features of pager and dial phones. An interconnector joins external surface dial phones with underground permissible phone system.

1.2.4 Fiber Optics Communication

A fiber optics cable system ensures data reliability. Fiber optic cables and connectors are unaffected by noise, lightning, interference from RF, electromotive force (EMF), electromagnetic interferences (EMI), and harmonics. The general parts of the fiber optic linked system include (i) central control rooms for mining; (ii) complex conveyor belt system controls; (iii) automation of remote system controls; (iv) conveyor belt scale system; (v) in-line coal-ash analysis; (vi) fire-detection system; (vii) conveyor belt video inspection system; and (viii) video camera system (Kohler and Jeffery, 1992; www.fibersystems.com/products). With these advancements in technology, fiber optic implementations help to improve processes, quality, and safety.

1.2.5 Limitations of Wired Communication System

Though wired communication works well during normal operations and are inexpensive and easy to use, still, sometimes they fail to satisfy the extreme conditions of mine environment. The very dependence of these systems on a wire connection makes them prone to the following obvious drawbacks:

- Wired communication is susceptible to failure during disasters as cable breakage interrupts communication.
- A wee bit of imperfection in design and development of wired communication systems can produce sparks or flames which are potential source of ignition for gassy atmosphere.
- Communication cannot be established from anywhere in an underground mine, but only from the point at which the phone set is available.
- Inadequacy of wired phone systems in terms of capacity and privacy and inability to communicate with moveable men is the limitation of wired phone.

1.3 Semi-wireless Communication System

1.3.1 Leaky Feeder-Based System

There are various methods available for underground communication for distribution of voice and data, but the most versatile and reliable wired cum wireless system is the so-called leaky feeder system. The leaky feeder cable is analogous to a surface antenna system. A good leaky feeder system will provide the basis for basic two-way voice and data applications (Bandyopadhyay et al., 2002).

Fig. 1.6 Leaky feeder cable

The leaky feeder (Fig. 1.6) basically contains a normal coaxial cable designed to pass RF signals from one end of the cable to the other with varying degrees of loss, depending on the quality (www.becker-mining.com). The purpose of the cable's outer copper shield is to keep as much signal as possible inside the cable and prevent external RF signals and/or interference from entering the cable. The leaky feeder is called so as its external copper shield has "holes" or "gaps," allowing the RF signals to leak out evenly along its length, so that they can be picked up by radios nearby (Fisahn et al., 2005). The holes in the leaky feeder cable also enable external RF signals to enter and run along the cable. In large mines, repeaters may also be used to amplify and retransmit incoming and outgoing signals to roving miners carrying portable radios. The spacing of these repeaters along the cable is governed primarily by the receiver sensitivity, the longitudinal attenuation rate of the cable, the coupling loss from the cable to the portable unit (usually very high frequency sets), and the transmitter power. Since the portable unit's transmitter power is generally lower than that available for fixed repeater or base stations, portable units set the coverage limits for two-way communication. The basic function of a leaky feeder system is to maintain a constant level of signal along a considerable length of leaky feeder cable, a solution which is both effective and reliable. The leaky feeder radio system is virtually noise free and is capable of both voice and data communication on all channels simultaneously. The system also has standard video band-pass capabilities and onboard diagnostics.

The system has a limited range and because of the high frequency it uses, transmissions cannot pass through solid rock, which limits the system to a line-of-sight application. It does, however, allow two-way mobile communication.

1.3.2 Location and Monitoring for Personal Safety

Location and monitoring for personal safety system (LAMPS) has three types of components, viz. tags (also known as personal transponders), readers (also known as network beacons), and control/monitoring facility component. Tag built into cap-lamp battery covers and routinely transmits a unique identification number. A network reader provides communication and staff location information through control/monitoring facility component.

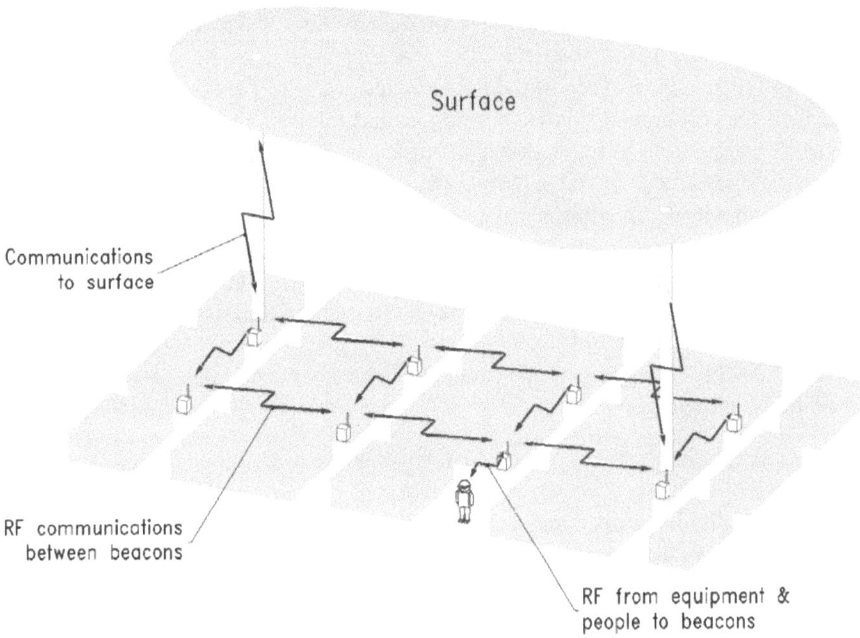

Fig. 1.7 Depiction of an underground LAMPS network

A LAMPS network is depicted in Fig. 1.7. Eight readers are shown, which provide communication along multiple, redundant paths within underground tunnels (www.australiancoal.csiro.au/pdfs/mallett.pdf). The figure also depicts a personnel transponder, reporting vital signs data to nearby readers. A separate control and monitoring subsystem at the surface monitors the communication traffic and displays the staff location information. In emergencies, the control and monitoring subsystem serves to communicate escape-route information to the individual transponders.

Although the development of LAMPS was motivated by underground coal mine safety, it equally has other personnel/equipment location-monitoring applications. The tags perform the same function as other active radio frequency identification (RFID) systems (Finkenzeller, 2003; Dobkin, 2007; Nikitin and Rao, 2008; http://www.icwcuca.ca/schedule.asp) in that they communicate unique identification numbers to a communication network. Additional readers can be added to the network at any place or time provided their serial numbers and positions are registered in the control and monitoring computer(s).

The LAMPS system offers independence from mine power as equipped with internal batteries, tolerates failures in sections of a mine by communicating over redundant pathways, supports self-rescue, aided rescue, and mine recovery. The LAMPS is a status- and location-monitoring system for personnel and resources within underground mines during normal and emergency conditions. Alarm conditions can be triggered whenever personnel data are not received or regions are unable to sustain communication. During mine emergencies, when there may be a loss of mine power, LAMPS supports emergency response as the LAMPS readers are equipped with internal batteries. The readers provide a communication pathway with high redundancy to report the location and status of personnel. An evacuation signal can be communicated to underground staff by flashing the cap lamps. Finally, LAMPS can be used to relay sensor data for gas- and air-quality monitoring from appropriate sensors deployed in hazard areas. This information can be accessed from any point in a mine. In addition, the technology can permit the display of the location of personnel and equipment on images of mine maps and layouts. The LAMPS tags can be used to a variety of sensor-monitoring applications. Equipment monitoring could include vibration, temperature, mine gas analysis data, and roof loading measurements.

The LAMPS communication system has been specifically designed for extreme conditions and can provide

- a robust communication system which will survive even if the main mine services are lost;
- links to individual miners inside the underground mine and provides information like person is alive or not and their location; and
- simple instructions to trapped miners.

1.3.3 Digital Enhanced Cordless Telecommunications

Digital enhanced cordless telecommunications (DECT) technology has established itself as a global standard technology in the field of communication system, not only on surface, but also in hazardous areas like underground mine (http://www.ericssoncs.com/advEric.htm). It is flexible and reliable, ideal for use in any environment with high volume of calls. Presently DECT is the most successful digital standard for cordless telecommunications worldwide. DECT standards are the most advanced, enabling the implementation of all kinds of different services (voice, fax, data, multimedia) and ensuring the greatest interoperability through the appropriate DECT profiles. Improved voice quality, secure communication, and an ample set of features have become synonymous for the outstanding range of DECT telephones. Products based on this system range from digital cordless telephones to fully featured systems including answering machines, handsets with menu-driven user interfaces, or cordless terminal adapters. DECT-based systems provide (http://www.call-systems.com/dect.php) (i) text and data messaging, (ii) full two-way speech communication, (iii) mobile telephony, (iv) free internal calls, (v) traditional functionality of desk phones, (vi) make and receive external calls, (vii) personal alarm and lone worker protection, (viii) incident location, (ix) monitoring remote equipment, (x) third-party interfacing, (xi) IP64-rated handsets, and (xii) various handset options.

1.3.4 Ethernet

Ethernet is fundamentally a communication standard (IEEE 802.3) used on LANs. An ethernet LAN typically uses coaxial cable, fiber optic cables, or special grades of twisted-pair wires (CAT5 or better). The most commonly installed ethernet systems are called 10BASE-T or 100Base-T and provide transmission speeds up to 100 Mbps. Devices are connected to the cable and compete for access using a carrier sense multiple access with collision detection (CSMA/CD) protocol. This is a system where each computer listens to the cable before sending anything through the network. If the network is clear, the computer will transmit. If some other node is already transmitting on the cable, the computer will wait and try again when the line is clear. Sometimes, two computers attempt to transmit at the same instant. When this happens a collision occurs. Each computer then backs off and waits a random amount of time before attempting to retransmit. With this access method, it is normal to have collisions. However, the delay caused by collisions and retransmitting is very small and does not normally affect the speed of transmission on the network. Ethernets can employ a version of Transmission Control Protocol/Internet Protocol (TCP/IP) (a collection of protocols that support computer network communication). Basically, it is a common set of rules that helps to define the complex process of transferring data (Casad, 2001) and in some cases is even used on leaky feeder transmission lines (Fisahn et al., 2005). Newer versions of ethernet can support 1,000 Mbps or more.

Ethernet LANs, in principle, can support a wide range of applications (Rowan et al., 2003). There are numerous application layer protocols for transporting sensor and equipment data over ethernet. These include Modbus/TCP, Ethernet/IP (or CIP), Profibus on ethernet, and Foundation fieldbus high-speed ethernet. Ethernet networks support IP telephones, also known as voice over IP (VoIP) telephones. VoIP telephones are in use within urban organizations throughout the world. Typically the VoIP telephones are line powered and connect directly to a RJ-45 wall socket, ethernet hub, or switch. An international VoIP standard exists and is fully compatible with public networks. The open architecture allows a mix of VoIP phones from different sources to be used, instead of relying on phones from one PBX manufacturer. VoIP telephony can be used over 802.11b-equipped laptop, palm, and industrial GP-104 computers. A mine trial of VoIP technology for emergency mine communication (Ralston and Hainsworth, 1997; Turoff et al., 2004) is a planned outcome of the parent project. Some mine operators have nominated video applications as one of the top three ethernet services priorities – after equipment/sensor monitoring and personal communications.

1.4 Through-the-Earth Communication System

TTE communication system in underground mines uses low-frequency electromagnetic waves that can deeply penetrate into the earth. These low-frequency waves typically penetrate several hundred meters into solid rock in a mine environment (Reagor et al., 1997). Digital signal processors (DSPs) are used to compress voice audio signals into narrow bandwidths that are compatible with these low-frequency carriers. The overall system uses low-cost components that are easily reconfigured for changing mine conditions. The low cost of the radios allows for large number of such stations to relay the data throughout the mine.

Underground radio is a TTE communication mechanism that offers high-level security. It uses very low frequency (VLF) electromagnetic radiation and digital audio compression technology to carry voice and text data (Gogoi and Raghuram, 1996, 1997a, b). The VLF signals also can transmit tracking and location data for radio users in case they are unable to respond. Underground radio provides two-way voice reception that can be used to alert individuals of underground conditions during blasts, fires, or collapses or to locate trapped miners. The technology provides convenient, portable underground communication and a data link to robotic machines. A TTE system likely will have the best chance of providing contact with miners since it offers the best resistance to damage from roof falls, fires, and explosions (Kononov and Higginson, 1994). There are several companies who offer TTE systems. Most are limited to communication from surface to underground. Only one system provides communication both surface to underground and underground to surface, but it is not a portable system. A brief description of each product is provided below:

- *Flex Alert* – Flex alert is a one-way, portable, and surface-to-underground communication system. It is an emergency evacuation system. It employs a low-frequency electromagnetic field to convey information to miners wearing special cap-lamp receivers. It is a one-way TTE transmission system. It consists of a 10–120 m wire loop antenna that is strategically placed above the mine. The receiver is contained inside a miner's cap lamp. A received evacuation signal causes the cap lamp to flash. It also illuminates a secondary light emitting diode (LED).
- *Personal Emergency Device (PED)* – PED communication system is a one-way TTE transmission system that enables communication of specific text messages to individuals underground without any dependence on cables or wiring underground. It does not provide underground-to-surface TTE communication, but communication to the surface is facilitated via a separate leaky feeder system.

- *TeleMag* – TeleMag is a wireless TTE two-way voice and data communication system. It operates between 3,000 and 8,000 Hz. It employs a single sideband-modulated carrier technique. It uses a DSP-based tracking cum filter for attenuating harmonic-induced noise, which improves the signal-to-noise ratio (SNR) thus improving the range of the system. It is a fixed station-to-station system. It is not portable. The underground and surface antennae consist of a wire loop. It has been tested to a depth of 100 m. Extended communication from the surface and underground fixed stations using wireless handsets is possible. Calculations indicate that 330 m of ground penetration is possible. The first demonstration of the system was in August 2000 (Conti, 2000) at the NIOSH Lake Lynn Laboratory mine. Other mine installations were not known at the time.

  Advantages:

  - Provides clear two-way voice communication TTE.
  - Interfaces to other Transtek communication systems in the mine and above the mine which enables extended voice communication through the earth.

  Disadvantages:

  - System is not MSHA permissible and is not portable.
  - System is restrained by not having a portable loop antenna. A fixed place loop antenna will be subject to destruction from a roof-fall, fire, or explosion.

Among all the TTE systems available till date, PED is the most successful and widely used.

1.4.1 PED Communication System

The PED communication system is based on ultra-low-frequency (ULF) transmission that propagates through rock strata, and that is why it is called TTE transmission system (http://www.minesite.com.au). The main difference between PED and other so-called TTE systems is that PED is proven and is operating in many mines continuously throughout the year. PED has been installed in over 150 mines since 1990 in Australia, USA, Canada, China, and Sweden. The use of ULF signals enables PED to transmit directly through rock strata, so the message can be sent to the miner wherever he is available. The mine-wide signal coverage of PED also means it is very useful for day-to-day communication system. Hence, PED can also be called as productivity enhancement device. Following are the capabilities of PED:

- *Paging* – PED can send a 32-character text message to an individual wherever they are in underground.
- *Emergency evacuation warning* – in an emergency, an evacuation instruction can be sent simultaneously to all personnel in only 15 seconds.
- *Overall communication* – PED complements existing phone and radio systems to maximize benefits to mine operator.

- *Safer blasting* – the blast PED system uses the proven PED transmission system to provide a safe and reliable remote blast initiation system.
- *Remote control* – ventilation fans, etc., can be remotely switched to reduce energy usage and manage pre- and postblast fan use.

1.4.1.1 Transmission System

The PED transmission system (Fig. 1.8) consists of (i) windows PC, (ii) proprietary modulator card, (iii) PEDCALL software package, (iv) PED ULF transmission head-end, and (v) loop antenna. The different components of PED system are as follows:

(i) *Modulator board* – The modulator unit is a standalone unit that connects to the PC through a standard RS-232 com port/modem.
(ii) *PEDCALL PC and Software (operator interface)* – The PEDCALL software provides the interface from the operator to the system in a simple and efficient manner.
(iii) *PED ULF transmission head-end* – The main role of the head-end is to boost the signal into a high-power output capable of driving up to 250 V at 5 A into a large loop antenna. The transmitter within the head-end utilizes an efficient

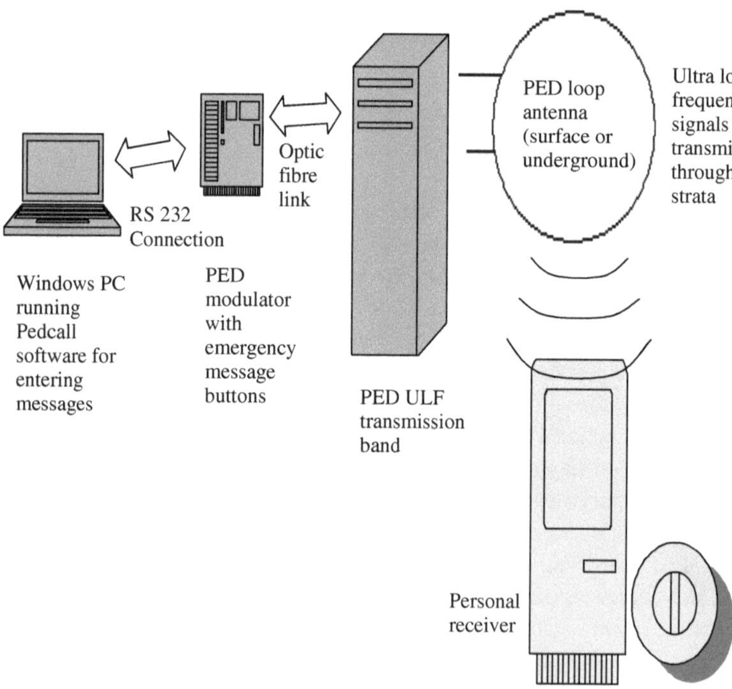

Fig. 1.8 PED transmission system

technique to achieve this role continuously; it is also protected from thermal overload and short-circuit conditions. The head-end also contains antenna safety unit and loop-matching transformer (Fig. 1.8) to optimize signal quality and safety.

(iv) *Loop antenna* – The layout will determine the range of signal transmission. Generally, the larger the loop (up to a maximum of 12 km), the better the coverage.

1.4.1.2 Receiver System

A number of receivers can receive the signals transmitted from ULF transmission system. These are (i) personal receiver, (ii) blast PED, (iii) control PED, and (iv) auto PED.

(i) *Personal receiver* – Personal receiver is integrated with a miner's cap lamp. The cap lamp is an ultra-lightweight lithium ion battery pack, known as integrated communications cap lamp (ICCL). On receipt of a message, the cap lamp flashes, a buzzer sounds, and the 32-character text message is illuminated on a liquid crystal display. Receiver versions are also available to retrofit to some existing cap-lamp batteries such as Koehler-Wheat, Oldham, Northern Lights, and MSA.
(ii) *Blast PED* – Blast PED is a receiver/exploder unit that allows for remote initiation or firing of blasts. Specially coded signals are sent via PED system that ensures blast PED receivers only operate when required. This coding, and several other levels of physical and software security, ensures the total safety of the system. Blast PED is approved for use in a number of countries including Australia, USA, and Canada and is the only "radio" remote blasting system in general use in underground mines.
(iii) *Control PED* – Control PED is a receiver that allows remote switching of equipment, such as fans, pumps, etc. The control PED receiver is typically interfaced to stop–start contacts in a device's control panel.
(iv) *Auto PED* – Auto PED is a vehicle-mounted receiver to ensure that people traveling in a vehicle receive messages. The large display on auto PED is clearly visible to all occupants.

The PED system has been proven to give significant productivity and safety benefits to a mining operation, large or small. The ability of PED to transmit through rock strata means it can truly deliver complete signal coverage to an underground mine. This is achieved without the need of installing antenna cable in every part of a mine. A relatively small antenna on the surface, or underground, provides complete signal coverage. This signal coverage is achieved at a fraction of the cost than any other type of radio system. Where other systems are vulnerable to rock fall, fire, and general wear and tear, PED greatly reduces these typical problems of unreliability and maintenance.

1.4.2 Cover with Transmitter

The cover with transmitter GLON-W is a component part of a cap lamp. It consists of a locating transmitter and system to switch-off light bulb when battery voltage drops below threshold value. The transmitter is a part of GLON-LOP system for locating people in case of disaster. The GLON-W transmitter provides precise location of a person supplied with this transmitter. Location goes on by means of LOP receiver that belongs to rescue brigade equipment. Integration of transmitter and circuit breaker of bulb prolongs run time of locating transmitter over 7 days after the bulb has been switched off. This time depends on battery condition.

The transmitter with circuit breaker and cap-lamp cover forms the whole system. The elements of these two subassemblies are placed on one printed circuit encapsulated in epoxide resin-type Epidian in side chamber of the cover. The cover with reflector is placed on the battery of lamp. The GLON-W includes two subassemblies: transmitter and circuit breaker. The transmitter generates nonmodulated electromagnetic field detectable at distances of up to 25 m. Power of this field depends on battery voltage. Thus, there is a need for maintenance of effective supply source. In case of breaking down and miners in breaking zone, battery is discharged by a bulb, and run time of the transmitter is prolonged to over 7 days due to the presence of automatically switched-off bulb.

The GLON-W electronic system keeps controlling the battery supply voltage and when battery voltage drops below threshold value, set at 2.25 V, the bulb automatically switches off, and due to self-regeneration the battery voltage goes up. The transmitter continues its operation consuming little current up to the battery discharge to voltage below 1 V.

1.4.3 Tram Guard Miner Track

Geo Steering (http://www.cdc.gov/niosh/topics/minerescue/minecomms.html) presently markets an MSHA-approved proximity warning system called tram guard for continuous mining machines. Information from the system is archived in the system and can be locally accessed with appropriate hardware. Geo Steering has been engineering a method to provide the data via a TTE connection with the surface. The data include the identification of all local miners, their distance from the system, and other useful data. This part of the system will be called tram guard miner track. The system is portable and includes a backup battery.

1.4.4 Delta Electromagnetic Gradiometer Beacon Tracking System

A system consisting of a beacon transmitter and a delta electromagnetic (DeltaEM) wave gradiometer (receiver) has been developed. The DeltaEM receiver is portable and is used on the surface of the mine to locate the beacon transmitter. The beacon transmitter generates a 2,000 kHz EM signal. The system is a prototype and is not yet MSHA approved. The DeltaEM receiver antenna consists of three ferrite-core antennas. There are 3 beacon antennas: 76 cm diameter loop, 15 cm ferrite-rod antenna, and large loop antenna.

1.4.5 Subterranean Wireless Electric Communication System

Subterranean wireless electric communication system (SWECS) is being developed under an SBIR contract to the US Army CERDEC program. It is expected to be a fully portable system with TTE capability. The relative location of the underground device can be determined. Connections to other underground communication systems are planned. The US Army originally commissioned the project for wartime use in such places as Afghanistan or to communicate with soldiers who may be in a collapsed building. But in response to the 14 miners who died in West Virginia coal mines in the recent past, government officials recently directed Kutta to adapt the device for commercial use in the mining industry.

SWECS consists of a PDA-type device with a screen and keypad, an 8-ounce radio, and a foot-long antenna. It has push-to-talk capability similar to a walkie-talkie and fits into a small backpack. The device has been tested in caverns and Arizona mines and can send voice communication through 265 m of solid rock and a digital photo through at least 133 m of rock.

1.4.6 Emergency Broadcast Network

Vital Alert has created an emergency broadcast network (EBN), called Canary 2, which is a two-way, TTE, voice and text messaging technology for use in urban, subterranean, and ocean environments. The technology was developed with government support under contract no. W-7405-ENG-36 awarded by the US Department of Energy. The technology was licensed to Vital Alert. The network's two-way voice system has the ability to penetrate the earth to depths of up to a thousand feet. Vital Alert claims that its EBN text messaging system can penetrate to 3,000 m. The equipment consists of mobile surface units and employs a ferrite rod as an antenna. Each surface mobile unit can communicate with several underground base units. Preliminary tests suggest the system can penetrate up to 133 m of overburden.

1.4.7 Very Low frequency and Low-Frequency Propagation

Existing warning system for underground mines, such as horn, siren, stench gases through the ventilation system, or messengers (other miners), can be slow and ineffective. This prompted the development of wireless, ultra-low-frequency electromagnetic signaling technology for warning and paging systems for the mining industry (Allen and Linefield, 1973; Zamel, 1990; Kumar et al., 2004).

As early as 1899, Nicola Tesla, France, suggested the use of extremely low frequency communication system using an earth medium (Wheeler, 1961). Pioneering research was conducted by the US Bureau of Mines (USBM) on the propagation of radio waves through the earth and detection of trapped miners (Westinghouse Georesearch Laboratory, 1973; Powell, 1976; Shope et al., 1982; Dobroski and Stolarczyk, 1982; Lagace et al., 1980 and 1982). Subsequently, USBM research showed that ultra-low-frequency electromagnetic signals from 630 Hz to 2 kHz could be transmitted through mine rock for distance as great as 1,645 m to an intrinsically safe receiver. The prototype wireless system of the previous research used off-the-shelf components and state-of-the-art technology to ensure high reliability and low cost. The technology enabled simultaneous and instantaneous warning of all personnel, regardless of their underground location or work activities, by flashing their cap lamp (Wadley, 1949).

The communication systems capable of detecting and locating trapped miners using waves of low frequency are principally of two types, namely, (i) seismic and (ii) electromagnetic. The seismic techniques have been developed to sense the vibration caused by a miner trapped in the part of a mine and the electromagnetic technique includes narrowband signaling, where a periodic pulse is transmitted. Both techniques are described below.

1.4.7.1 Seismic Method

In the seismic system, in case of disaster, the trapped miners strike a part of the mine with any heavy object they could find nearby. The resulting vibrations would then be detected on the surface by the use of seismic transducers (seismometer) which will be referred to as geophones. The geophones convert seismic signals to voltages that are then amplified, filtered, and recorded in a seismic van placed on the surface. The location of a trapped miner at the underground can be found by the signals obtained from various geophone arrays through suitable processing scheme (Ruths, 1977).

Principle: The seismic rescue system uses an array composed of several subarrays, rather than the same number of individual geophones to receive seismic signals; the reason is that a subarray will give a better SNR than a single geophone. The miner is instructed to do the following in the event he is trapped underground:

(i) When all possible escape is cut off, the miner is to erect barricade for protection from possible toxic gases and wait for a signal from the surface before beginning to signal the seismic system.
(ii) As soon as the system is in a state of readiness, the surface crew detonates three explosive charges that can be easily heard by the trapped miner.
(iii) After hearing these three shots, the miner is to pound 10 times on any hard part of the mine, preferably the roof or a roof bolt, with any heavy object that can be found nearby; a heavy timber is best.

(iv) Following this, the miner is to rest for 15 minutes and then repeat the pounding. While resting, if the miner hears five shots from the surface he knows the signaling has been detected and help is on the way.
(v) If the miner hears no shots, he repeats signaling after every 15 minutes.

Probability of detection: It is desirable to determine the probability that a surface array will detect an underground source. The configuration normally used seven subarrays placed on the surface to monitor a portion of the subsurface. A method has been developed to calculate the probability that m subarrays or more, with $1 \leq m \leq 7$, will detect a miner's signal. The detection of a signal by one subarray may be sufficient to identify the signal as coming from an underground miner. However, identification can be more certain if several subarrays can detect the signal. To locate, at least three subarray detections are required, and five or more are desirable for better accuracy (Ruths, 1977).

Seismic noise: Seismic noise can at times be a major problem when detecting small-level seismic signals, since the signal from a trapped miner can be of the order of a few microcentimeters per second, which is a normal background noise level. The predominant noise sources during a rescue operation are surface-generated noise from moving equipment, people walking, and power lines. Thus, information is needed on the types of noise sources, the expected amplitude ranges, and the amplitude variation with frequency and time. There are various signal noise processing techniques available to enhance the SNR.

Three common noise sources are typically encountered in the field: (i) natural seismic background noise, (ii) man-made seismic noise and (iii) man-made EMI coupled into the field equipment. For noise reduction, the seismic rescue system uses an array, composed of subarrays rather than seven individual geophones to receive seismic signals. The reason is that a subarray will give a better SNR than a single geophone. This improvement is achieved principally in three ways. First, noise that is uncorrelated between the geophones will be reduced in amplitude by the cancellation that occurs when zero mean random numbers are averaged. Second, noise that is propagating at a slow horizontal velocity will be reduced on the output of the subarray because, if the subarray is thought of as an antenna, the noise will be outside of the antenna's main beam. Finally, adverse effects that would result if a single badly planted geophone was used will be alleviated by the averaging of all the subarray geophone outputs. From theoretical consideration of SNR improvement by the 24 and 7 geophone subarrays, it is to be expected that the 24 geophone subarrays can offer a significant SNR gain over the 7 geophone subarrays (Ruths, 1977).

1.4.7.2 Electromagnetic Method

The electromagnetic (EM) detection and TTE communication require narrowband transmitter, which would be used underground by a trapped miner to send signal to the surface (Olsen and Farstad, 1973; Durkin, 1984, Kravitz et al., 1994; Kumar et al., 2004; Vong et al., 2006). In an underground mine, a radio wave can propagate a useful distance only if the environment has the necessary electrical and physical properties (Higginson, 1992; Beiter et al., 1998; Boutin et al., 2008). Attenuation depends primarily upon the physical properties of the entry such as cross-sectional area, wall roughness, entry tilts, and obstacles in the propagation path (Wait, 1989). Secondary effects such as the dielectric constants and the earth conductivity also influence attenuation. The low-frequency electromagnetic field can penetrate kilometers of soil and rock to reach the most remote shaft or tunnel which makes it ideal for underground application. A large number of fundamental investigations have been carried out regarding the propagation of EM waves through rock in the frequency band ranging from 600 Hz to 60 MHz in South Africa and USA (Austin, 1978; Durkin and Greenfield, 1981; Durkin, 1982; Ayuso et al., 2006). Reliable undersurface communication at 200 m depth was proved in South Africa's coal mines (Kononov and Higginson, 1994; Kononov and Smit, 1997; Kononov, 1998a, b). One of the latest technology developments uses the principle of modulation of EM wave by the heart beat and/or the movement of human chest while breathing. It employs a microwave transmitter with directional antenna and a Doppler receiver to detect a reflected cardiac/breathing signal from a living person. One such system was developed by Selectronics (Germany) in 1995 and manufactured under the name SIRUS. The system is capable of identifying a living person at a distance of 90 m in the air and about 20 m through walls and debris.

Principle: This type of system consists of a low-frequency transmitter that can be strategically placed at selected location to create an electromagnetic signal, which can completely envelop most mines without the use of repeater system. The system consists of two parts: a transmitter and a surface system for detection and communication. The transmitter can be carried on miner's belt and is powered by a cap-lamp battery. The transmitter serves as a radio beacon to help rescue personnel and locate trapped miner under or behind roof falls or barricades or in other inaccessible places. A compact loop antenna which can be either carried by the miner or strategically located in the mine is used to generate EM field that is sensed by a mobile receiver on the surface. On the surface, sensitive receivers detect the signal and locate the source. Once detection and location are made, a large surface transmitter is deployed above the trapped miner. The transmitter is powerful enough to send voice message by radio, directly down through the earth (Conti and Yewen, 1997).

In an emergency, the miner would connect his tone transmitter to the deployed loop antenna. The transmitter supplies a current to the antenna with low-frequency range. The known interval of the intermittent signal allows the receiver on the surface to distinguish between the signal and the EM noise or interference.

1.4.8 Research on TTE

Research on TTE has been conducted by different universities and government agencies around the world and resulted in a few commercially available products. Some of the significant developments are elaborated below:

(i) *CSIR Miningtek, South Africa:* Miningtek developed a trapped-miner-locating device. A prototype was successfully tested underground where the device provided detection and location of a trapped miner at a distance of more than 30 m through rock. The system consists of a uniquely coded belt-wearable miner's tag and portable search unit. The tag is built into a metal buckle and includes a LED and buzzer.

(ii) *Institute for Advanced Physics, University of Innsbruck, Austria:* Research at the University of Innsbruck resulted in the development of a system which was composed of a beacon contained in a miner's cap lamp and a handheld location receiver which could search for the trapped miner's beacon. Field tests at the Schwaz/Tirol mine demonstrated a detection accuracy of 50 cm (Nessler and Norbert, 2000).

(iii) *U.S. Bureau of Mines, United States:* From the mid-1970s to the early 1980s, the U.S. Bureau of Mines conducted extensive electronic communications research over a broad spectrum of frequencies and system types. Frequencies investigated ranged from extremely low frequencies (ELF) to a few GHz. Most significant was their TTE research at frequencies between 600 and 3,000 Hz. The research resulted in development of several communication devices (Lagace et al., 1980). The above-the-mine part of the system consisted of a transmitter and long wire-loop antenna and a handheld locator receiver with a 38 cm loop antenna. The miner-carried part of the system was a compact belt-worn device with a voice receiver and a wire-loop antenna that would be unfolded during an emergency to provide a beacon signal to the surface. A statistical analysis of the data concluded that, at 250 m depth, there was a 68% probability of signal detection.

1.5 Wireless Communication Services

Through-the-air (TTA), i.e., wireless network refers to any type of network that is wireless (Stolarczyk, 1984; Pahlavan and Levesque, 1995; Rappaport, 1996; Akyildiz et al., 2002; Pahlavan and Krishnamurthy, 2006; http://www.ari.uta.edu). It refers to a communication network which is interconnected between nodes that are implemented without the use of wires (Zhang, 2008). The inadequacies of wired communication gave rise to wireless technology. The advanced wireless network architecture was, in the long run, proved to be an efficient means of communication in complex confined media such as the one in mine galleries. An obvious advantage of any wireless communication system is that the system does not require transmission lines or cables. These systems are immune to communication outages caused by line breaks due to roof falls or damage from machinery. There is an ever-increasing demand for wireless communication as well as accurate and real-time data information in underground mines. Wireless communication technique concept in underground mine describes a new and reliable wireless communication and data-transfer solutions (Sydanheimo et al., 2000; Vasquez et al., 2004; Srinivasan et al., 2005; Akyildiz and Stuntebeck, 2006). This new concept will provide wireless and real-time connection with full coverage of a mine up to the surface. But before discussing the wireless communication systems, the subject of EMIs and signal attenuation in underground mines must be understood.

EMI: It is the disruption of operation of an electronic device when it is in the vicinity of an EM field in the RF spectrum that is caused by another electronic device. During normal mining operation, the machinery used in underground randomly creates a wide range of many types of intense EMIs, which is a major limiting factor in the range of a radio communication system. The internal circuits of personal computers generate EM fields in the RF range. Also, cathode ray tube (CRT) displays generate EM energy over a wide band of frequencies. These emissions can interfere with the performance of sensitive wireless receivers nearby. Moderate or high-powered wireless transmitters can produce EM fields strong enough to upset the operation of electronic equipment nearby. Problems with EMI can be minimized by ensuring that all electronic equipment is operated with a good electrical ground system. In addition, cords and cables connecting the peripherals in an electronic or computer system should, if possible, be shielded to keep unwanted RF energy from entering or leaving. Specialized components such as line filters, capacitors, and inductors can be installed in power cords and interconnecting cables to reduce the EMI susceptibility of some systems. Electromagnetic noise amplitude decreases with increase in frequency (Ndoh, 2004).

Signal attenuation: Attenuation is a process of reduction of signal strength during transmission. As a radio signal travels down the haulage or tunnel, its strength decreases (Ndoh et al., 2003). It occurs with any type of signal, whether digital or analog (Proakis, 1995). Attenuation is a natural consequence of signal transmission over long distances. The extent of attenuation is usually expressed in units called decibels (dBs). If P_s is the signal power at the transmitting end (source) of a communication circuit and P_d is the signal power at the receiving end (destination), then $P_s > P_d$. The power attenuation A_p in decibels is given by the formula: $A_p = 10 \log_{10}(P_s/P_d)$. Attenuation can also be expressed in terms of voltage. If A_v is the voltage attenuation in decibels, V_s the source signal voltage, and V_d the destination signal voltage, then $A_v = 20 \log_{10} (V_s/V_d)$. Transmission loss may be combined directly with transmitter power and antenna gains. Attenuation increases nearly linearly with increasing distance. Transmission loss decreases significantly at a given distance as the frequency is increased. Signal strength around a corner is also considerable. Because of the high attenuation of signal at corner, propagation around multiple corners is even more severely attenuated. Although it is an advantage to operate at a higher frequency in a straight haulage or tunnel for more coverage, the higher frequencies suffer the greatest loss in turning a corner. Therefore, the choice of frequency is often dictated by type of coverage desired. The presence of stoppings for direction of airflow, passage blocked by machinery, or blockage caused by a roof fall seriously limits wireless communication range. When it is necessary to transmit signals over long distances via cable, one or more repeaters can be inserted along the length of cable. The repeaters boost signal strength to overcome attenuation. This greatly increases the maximum attainable range of communication. Attenuation is caused by absorption, scattering, and bending losses.

There are different types of wireless communication system, such as (i) pocket pagers, (ii) walkie-talkie, (iii) Bluetooth, (iv) wireless fidelity (Wi-Fi), (v) world interoperability for microwave access (WiMAX), (vi) RFID, (vii) watcher-ATS, (viii) tracker tagging; (ix) ultra wide band (UWB) communication, etc. All these communication systems are described subsequently.

1.5.1 Pocket Pagers

The pocket pager is a receiver only and cannot be used to talk back to dispatcher. Therefore, the system should be used only for giving instructions. The pocket pager is activated by tones and receives a message, so that the dispatcher can selectively radio-page any individual. In an emergency, a special tone can activate all pagers at once (IC 8955, 1984). The system is shown in Fig. 1.9. Only the dispatcher can initiate a page, because it is having a carrier phone equipped with an encoder. However, other encoders could be used with other carrier phones, if necessary. A system is also available in which encoder is remotely accessed by a dial telephone line. Thus, any dial telephone associated with mine switchboard (PBX) could be used to initiate a page without ever being near encoder. The advantage of such a system is that many people can page into a mine from several surface locations. Existing pager receivers are equipped with a small internal timer that automatically turns the device off after a preselected time, usually 15 seconds. A continuous "on" mode is usually not desirable because it wastes battery power. A radio paging system can be operated on a special channel (frequency), or the regular channel. But the only difference is that if both are included on the same regular channel, all the carrier phones will hear the paging traffic, but the pagers will hear only what is sent to them directly.

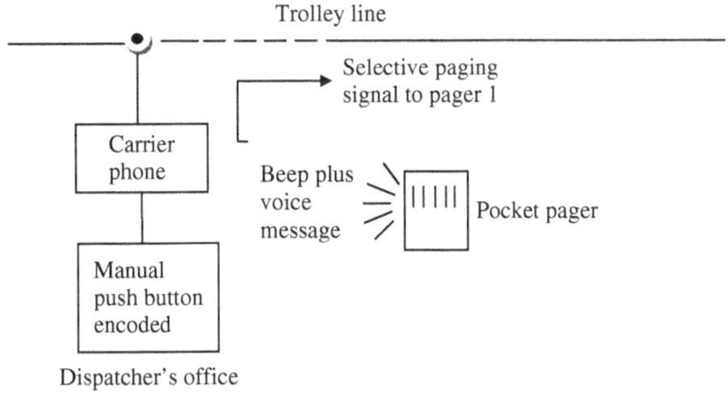

Fig. 1.9 Block diagram of radio paging system

1.5.2 Walkie-Talkie System

A walkie-talkie is a handheld, portable, two-way radio transceiver. It includes a half-duplex channel (only one radio transmits at a time, though any number can listen) and a push-to-talk switch that starts transmission. This system uses ultra-high-frequency (UHF) portable radio equipment. Typical walkie-talkies resemble a telephone handset, possibly slightly larger but still a single unit, with an antenna sticking out of the top (Fig. 1.10). Where a phone's earpiece is only loud enough to be heard by the user, a walkie-talkie's built-in speaker can be heard by the user and those in his immediate vicinity. Handheld transceivers may be used to communicate between each other or to vehicle-mounted or base stations. Because of the curvature of the loop tunnel, propagation of radio waves at UHF is severely restricted (Reed and Russell, 1966; Zhang and Mao, 2006). In fact, direct radio communication between the two individuals may not be possible in some cases. However, this deficiency can be overcome with a dual-frequency radio repeater connected to

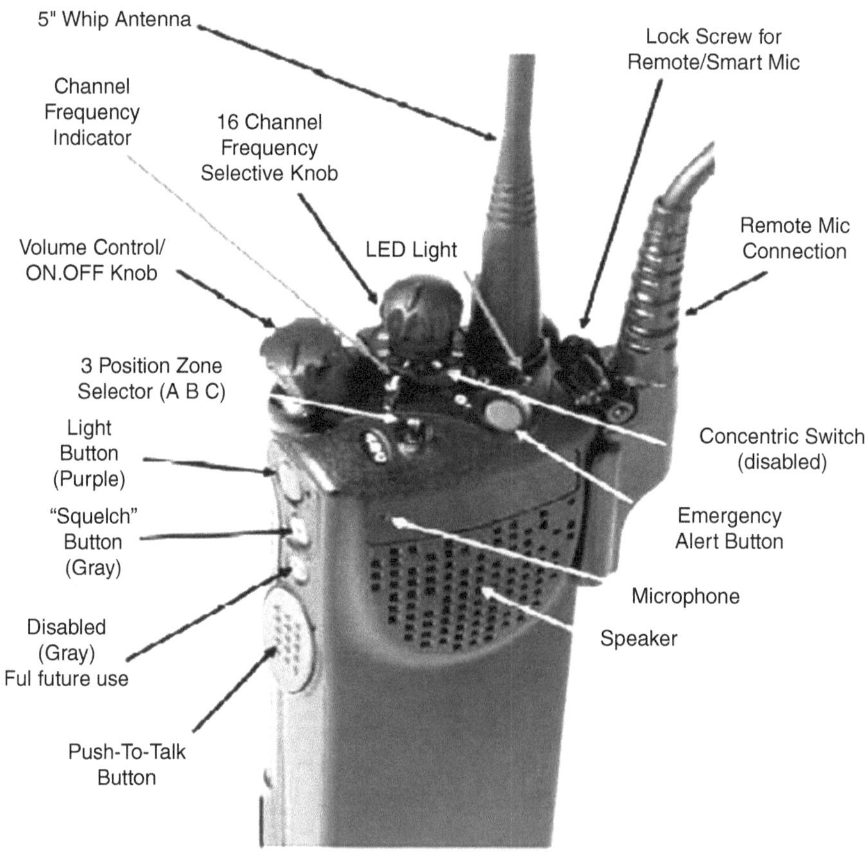

Fig. 1.10 Walkie-talkie system (http://en.wikipedia.org/wiki/Walkie-talkie)

a radiating cable (leaky feeder cable). The cable carries the radio signals, and the repeater effectively boosts them to a higher power level. The coaxial cable extends along the loading track and down the main haulage way far enough to assure communication coverage to the motorman. Cables several hundred feet shorter are connected at the end. Commercially available portable radio transceiver and repeater can be used to implement this system. Another drawback of push-to-talk system is that it only works when people have phones using the same carrier's network.

1.5.3 Bluetooth

Bluetooth was initiated in 1998 and standardized by the Institute of Electrical and Electronic Engineers (IEEE) as wireless personal area network (WPAN) specification IEEE 802.15. Bluetooth is a short-range RF technology aimed at facilitating communication of electronic devices between each other and with the Internet allowing for data synchronization that is transparent to the user. Supported devices include PCs, laptops, printers, joysticks, keyboards, mice, cell phones, and consumer products. Mobile devices are also supported (Sydanheimo et al., 2000; Leeper, 2001). Discovery protocols allow new devices to be hooked up easily to the network. Bluetooth uses the unlicensed 2.4 GHz band and can transmit data up to 1 Mbps, can penetrate solid nonmetal barriers, and has a nominal range of 10 m that can be extended to 100 m. A master station can service up to seven slave links simultaneously.

1.5.4 Wireless Fidelity

Wireless fidelity (Wi-Fi) is a wireless technology that uses radio frequency to transmit data through the air. It is a relatively new concept, which opens the door to a new level of wireless communication. Wi-Fi is a generic term that refers to a type of ethernet specified under the IEEE 802.11a, b, and g standards for LANs operating in 5 and 2.4 GHz unlicensed frequency bands (Aniss et al., 2004). 802.11a transmits at 5 GHz and can move up to 54 Mb of data per second. It also uses orthogonal frequency division multiplexing (OFDM), a more efficient coding technique that splits the radio signals into several subsignals before they reach a receiver. This greatly reduces interference. 802.11b is the slowest and least expensive standard. For a while, its cost made it popular, but now it is becoming less common as faster standards become less expensive. 802.11b transmits in 2.4 GHz frequency band of the radio spectrum. It can handle up to 11 Mb of data per second, and it uses complimentary code keying (CCK). 802.11 g is a modified version of 802.11b and transmits at 2.4 GHz like 802.11b, but it is faster than 802.11b. It can handle up to 54 Mb of data per second. 802.11 g is faster because it uses the same OFDM coding as 802.11a.

Wi-Fi carries two-way radio signals between a fixed base station and one or more laptop computers or some other device, such as a personal digital assistant. The Wi-Fi alliance, a nonprofit research and education group, claims that transmissions have a range of some 116 m outside and 27 m indoors at full speed and longer ranges at reduced speeds. Wi-Fi works much like a cordless phone with its base station, but is typically connected to the Internet rather than the phone system. Wi-Fi signals usually use 2.4 GHz band (Boutin et al., 2005) on which many cordless phones operate. Wi-Fi requires strategically placed wireless repeaters. Interestingly, these systems are digital, which opens up a new realm of possibilities, including simultaneous delivery of voice (VoIP), data, and video over the link. There has also been a merging of technologies which combine leaky feeder, ethernet, and Wi-Fi. A few cell phone vendors now market a phone that combines standard cell phone communication protocols such as code division multiple access (CDMA) or global system for mobile communication (GSM) with Wi-Fi. With the appropriate software installed in a PC at an opencast mine office (Fig. 1.11), and a Wi-Fi network installed in a mine, a miner can walk into a mine and continue to use his cell phone (http://www.dreamhistory.org). A significant development in underground longwall coal mining automation has been achieved with successful implementation of WLAN technology for communication on a longwall shearer (Hargrave et al., 2007). Wi-Fi can be selected to meet the bandwidth requirements of the underground data network, and several configurations can be installed on operating longwalls to evaluate their performance (www.sm.luth.se/csee/courses/smd). Although WLAN technology in a longwall operation is feasible, it is clear that new research and development is required in order to establish optimal, full-face

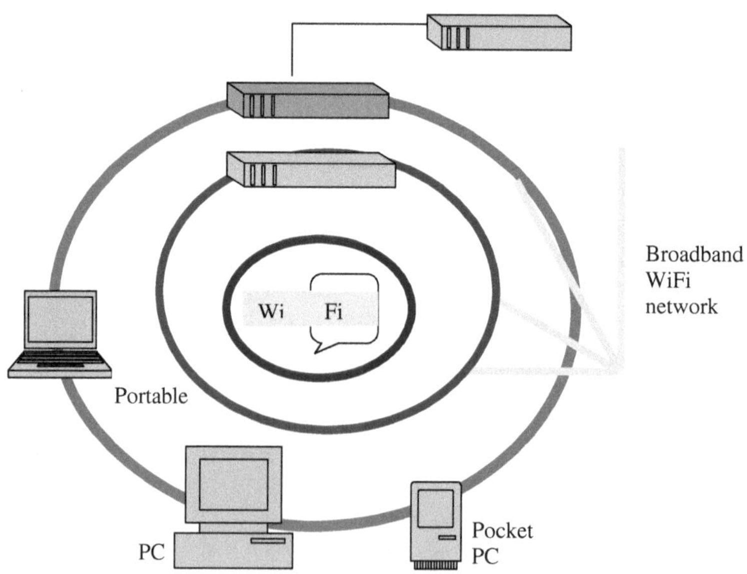

Fig. 1.11 Wi-Fi technology

coverage (Barber et al., 1996; Einicke et al., 1997, 2002). By undertaking an accurate characterization of target environment, it is possible to achieve great improvements in WLAN performance over a nominal Wi-Fi installation. Wi-Fi is fast, convenient, and ubiquitous. It is also more cost-effective for industry to set up a wireless network than a wired connection.

One of the few limitations of Wi-Fi is that it is not 100% secure. Wi-Fi connections can be encrypted and there are many different versions of encryption. Also, the systems can interfere with each other. The disadvantages like poor coverage, security concerns, high power consumption (especially for handsets), lack of traffic prioritization, and unlicensed/class license spectrum can sometimes make Wi-Fi deployment difficult.

1.5.5 World Interoperability for Microwave Access

World interoperability for microwave access (WiMAX) is the only wireless standard today that has the ability to deliver true broadband speeds (Hui, 1988). If Wi-Fi is similar to a cordless phone, WiMAX is analogous to a cell phone. It is a term coined to describe standard, interoperable implementations of IEEE 802.16 wireless networks, similar to the way the term Wi-Fi is used for interoperable implementations of the IEEE 802.11 wireless LAN standard. However, WiMAX is very different from Wi-Fi in the way it works. The WiMAX Forum is an industry-led, not-for-profit organization formed to certify and promote the compatibility and interoperability of broadband (Schwartz, 1996) wireless products based on WiMAX (http://www.wimaxforum.org/about/). Different variants of WiMAX standards are available for various purposes (Table 1.2).

1.5.5.1 The System

A WiMAX system consists of a WiMAX tower and a WiMAX receiver (Fig. 1.12). WiMAX tower is similar in concept to a cell phone tower (http://prabu.files.wordpress.com). A single WiMAX tower can provide coverage to a very large area as big as 8,000 km^2. A WiMAX receiver and antenna could be a personal computer memory card or they could be interfaced with a laptop in the same way as presently Wi-Fi is accessed.

1.5.5.2 Wireless Architectures

There are two scenarios (Fig. 1.13) for a wireless deployment: (i) point-to-point and (ii) point-to-multipoint (http://www.wimax.com/education/wimax/wireless_architectures).

Point-to-point (P2P): Point-to-point is used where there are two points of interests: one sender and one receiver (Xiaogang et al., 2008). This is also a scenario for backhaul or transport from data source to a subscriber or for a point of distribution using point-to-multipoint architecture. Backhaul radios comprise an industry of

Table 1.2 Overview of the different variants within the 802.16 standard

Aspects	802.16	802.16a / Rev d	802.16e
Completion time	December 2001	802.16a January 2003; 802.16 Rev d June 2004	Established in mid-2005
Spectrum	10–66 GHz	2–11 GHz	2–6 GHz
Application	Backhaul	Wireless digital subscriber line (WDSL) and backhaul	Mobile Internet
Channel conditions	Line of sight only	Nonline of sight	Nonline of sight
Bit rate	32–134 Mbps at 28 MHz channelization	Up to 75 Mbps at 20 MHz channelization	Up to 15 Mbps at 5 MHz channelization
Modulation	Quadrature phase shift keying (QPSK), 16 quadrature amplitude modulation (QAM), and 64 QAM	OFDM 256 subcarriers, QPSK, 16 QAM, and 64 QAM	Scalable orthogonal frequency division multiple accesses (OFDMA)
Mobility	Fixed	Fixed	Pedestrian mobility – regional roaming
Channel bandwidth	20, 25, and 28 MHz	Scalable channel bandwidths between 15 and 20 MHz	Same as 802.16a with UL subchannels to conserve power
Typical cell radius	1.6–4.5 km	6.5–10 km, maximum range 77 km based on tower height, antenna gain, and power transmit	1.6–4.5 km

their own within the wireless industry. As the architecture calls for a highly focused beam between two points, range and throughput of point-to-point radios will be higher than that of point-to-multipoint products.

Point-to-multipoint (PMP): As seen in Fig. 1.13, point-to-multipoint is synonymous with distribution. One base station can service hundreds of dissimilar subscribers in terms of bandwidth and services offered.

1.5.5.3 Line-of-Sight or Non-line-of-Sight

Earlier wireless technologies such as local multipoint distribution service (LMDS) and metropolitan multipoint distribution service (MMDS) were unsuccessful in the mass market as they could not deliver services in non-line-of-sight scenarios (Fig. 1.14). This confined the number of subscribers they could reach, and given the high cost of base stations and CPE, those business plans failed (http://www.wimax.com/education/wimax/wireless_architectures). WiMAX functions best in line-of-sight situations and, unlike those earlier technologies, offers

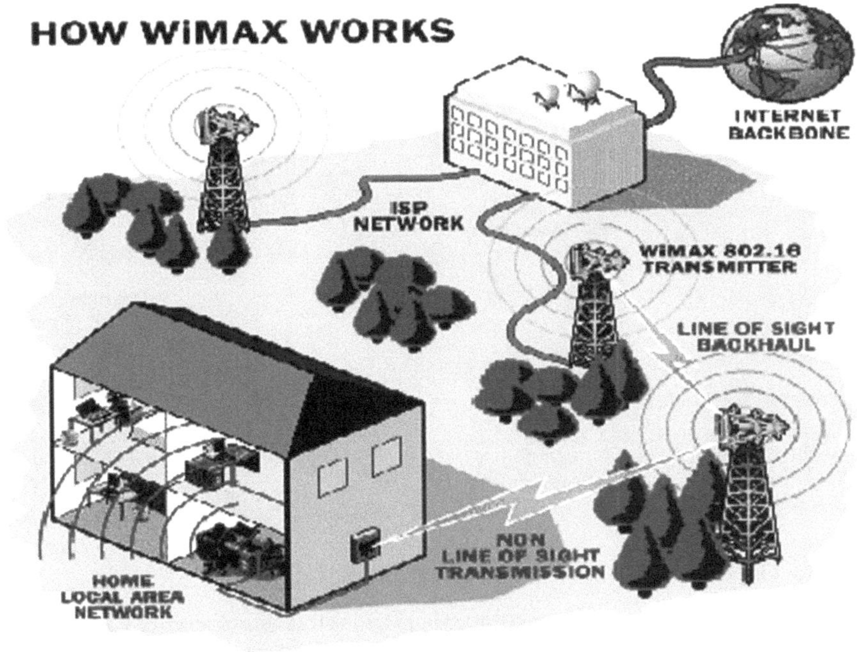

Fig. 1.12 Working procedures of WiMAX

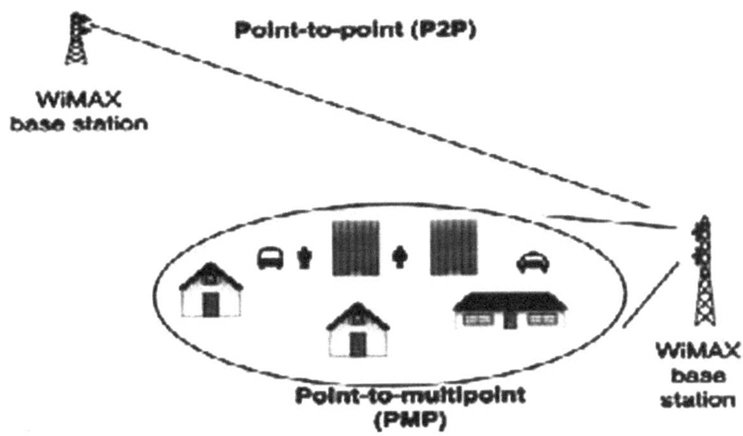

Fig. 1.13 Point-to point and point-to-multipoint configurations

acceptable range and throughput to subscribers who are not in line-of-sight to the base station. Buildings between base station and subscriber diminish the range and throughput, but in an urban environment, the signal will still be strong enough to deliver adequate service. Given WiMAX's ability to deliver services non-line-of-sight, the WiMAX service provider can reach many customers in high-rise office

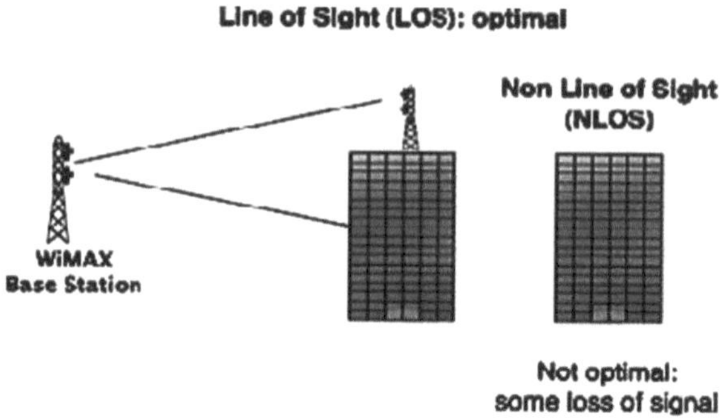

Fig. 1.14 The difference between line-of-sight and non-line-of-sight

buildings to achieve a low cost per subscriber because so many subscribers can be reached from one base station (Fig. 1.14).

1.5.5.4 Orthogonal Frequency Division Multiplexing

Orthogonal frequency division multiplexing (OFDM) is a multicarrier transmission technique that has been recently recognized as an excellent method for high-speed bidirectional wireless data communication. Its history dates back to the 1960s, but it has recently become popular because economical integrated circuits that can perform the necessary high-speed digital operations have now become available. OFDM effectively squeezes multiple modulated carriers tightly together, reducing the required bandwidth but keeping the modulated signals orthogonal so they do not interfere with each other. Today, the technology is used in such systems as asymmetric digital subscriber line (ADSL) as well as wireless systems such as IEEE 802.11a/g (Wi-Fi) and IEEE 802.16 (WiMAX). It is also used for wireless digital audio and video broadcasting.

It is based on frequency division multiplexing (FDM), which is a technology that uses multiple frequencies to simultaneously transmit multiple signals in parallel. Each signal has its own frequency range (subcarrier), which is then modulated by data. Each subcarrier is separated by a guard band to ensure that they do not overlap (Fig. 1.15). These subcarriers are then demodulated at the receiver by using filters to separate the bands (http://www.intel.com/netcomms/technologies/wimax).

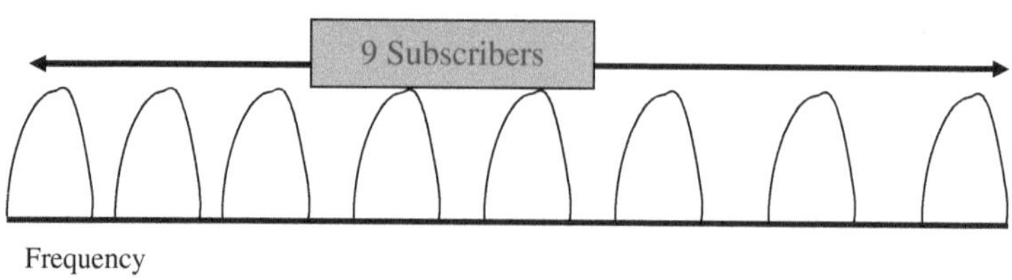

Fig. 1.15 FDM with nine subcarriers using filters

OFDM is similar to FDM but much more spectrally efficient by spacing the subchannels much closer together (until they are actually overlapping). This is done by finding frequencies that are orthogonal, which means that they are perpendicular in a mathematical sense, allowing the spectrum of each subchannel to overlap another without interfering with it. The effect of this is seen as the required bandwidth is greatly reduced by removing guard bands and allowing signals to overlap (Fig. 1.16).

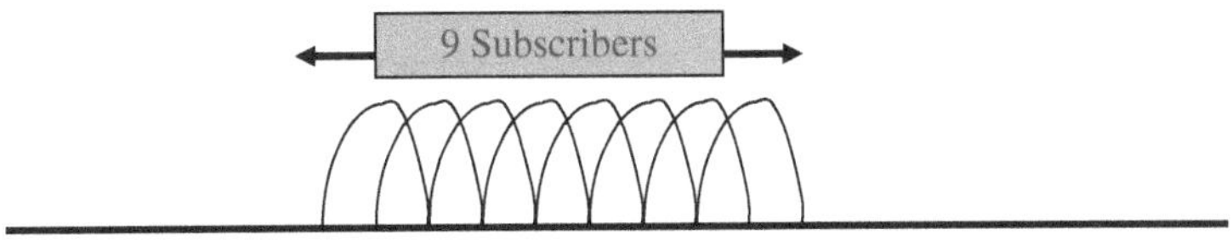

Fig. 1.16 OFDM with nine subcarriers

In order to demodulate the signal, a discrete Fourier transform (DFT) is needed. Fast Fourier transform (FFT) chips are commercially available, making this a relatively easy operation.

In Fig. 1.17 we can see 256 OFDM with 192 data subcarriers, 8 pilot subcarriers, and 56 nulls. In its most basic form, each data subcarrier could be on or off to indicate a 1 or 0 bit of information. However, either phase shift keying (PSK) or quadrature amplitude modulation (QAM) is typically employed to increase the data throughput. Therefore, in this case, a data stream would be split into n (192) parallel data streams, each at $1/n$ (1/192) of the original rate. Each stream is then mapped to the individual data subcarrier and modulated using either PSK or QAM. Pilot subcarriers provide a reference to minimize frequency and phase shifts during the transmission, while null carriers allow for guard bands and the DC carrier (center frequency).

Orthogonal frequency division multiple accesses (OFDMA) allow some subcarriers to be assigned to different users. For example, subcarriers 1, 3, and 7 can be assigned to user 1 and subcarriers 2, 5, and 9 to user 2. These groups of

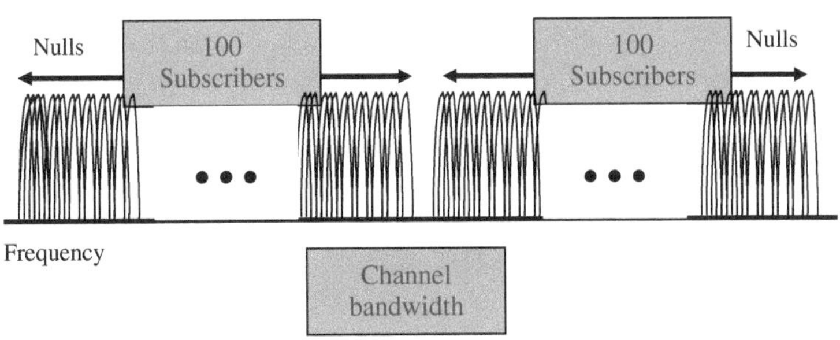

Fig. 1.17 OFDM with 256 subcarriers

subcarriers are known as subchannels. Scalable FDMA allows smaller FFT sizes to improve performance (efficiency) for lower-bandwidth channels. This applies to IEEE 802.16-2004, which can now reduce the FFT size from 2,048 to 128 to handle channel bandwidths ranging from 1.25 to 20 MHz. This allows subcarrier spacing to remain constant independent of bandwidth, which reduces complexity while also allowing larger FFT for increased performance with wide channels. Another advantage of OFDM is its resilience to multipath, which is the effect of multiple reflected signals hitting the receiver. This results in interference and frequency-selective fading which OFDM is able to overcome by utilizing its parallel, slower bandwidth nature. This makes OFDM ideal to handle the harsh conditions of the mobile wireless environment. OFDM's high spectral efficiency and resistance to multipath make it an extremely suitable technology to meet the demands of wireless data traffic. This has not only made it ideal for such new technologies like WiMAX and Wi-Fi but also currently one of the prime technologies being considered for use in future as fourth generation (4G) networks. The advantages of OFDM (http://urgentcomm.com/mag/radio_moving_front_pack/) are (i) immunity to delay spread and multipath, (ii) resistance to frequency-selective fading, (iii) simple equalization, and (iv) efficient bandwidth. The disadvantages of OFDM are (i) synchronization, (ii) need FFT units at transmitter and receiver, (iii) sensitive to carrier frequency offset, and (iv) high peak to average power ratio.

1.5.5.5 WiMAX Regulatory

The most important regulatory aspect of WiMAX is the availability and subsequent use of spectrum (a frequency band on which the service provider will broadcast). Unlicensed spectrum allows any one to broadcast on that frequency within certain power limits. Licensed spectrum protects the broadcaster by giving them exclusive right to broadcast on that frequency (spectrum). Other broadcasters who interfere with that frequency are subject to governmental sanction. Licensed spectrum can be obtained directly through a national government or via a sublease from someone who has obtained spectrum from the national government (http://www.wimax.com/education/wimax/regulatory). Spectrum is often allocated by location and frequency (Table 1.3).

The media access control (MAC) in Wi-Fi works on a random interrupt basis, i.e., all subscriber stations sending data through wireless access point (AP) compete for the AP attention which causes distant stations interrupted by the closer ones, reducing their throughput. This in turn makes quality of service (QOS)–dependent services difficult to maintain for more than a few simultaneous users. WiMAX, on the other hand, uses a scheduling algorithm in which the subscriber station, though initially competes for a base station, is allocated an access slot which means it remains assigned to the subscriber station and other subscribers cannot use it. This scheduling algorithm gives WiMAX some obvious advantages over Wi-Fi as given below:

Table 1.3 Overview of WiMAX-related spectrum

Frequency	Uses
2–11 GHz	As specified by IEEE 802.16-2004 as the operating range for point to multipoint operations
2.4–2.483 GHz	ISM and Federal Communication Commission (FCC) Part 15, largely unlicensed, used for Wi-Fi; to be avoided by WiMAX operators on concerns of interference from Wi-Fi
2.5 GHz	BRS/EBS projected as being a popular licensed WiMAX spectrum choice
3.5 GHz	Unlicensed in much of the world outside US; some vendors have completed this testing and product is shipping
3.65 GHz	FCC issued an announcement in 2004 promoting opening spectrum here for quasi-unlicensed use, it is yet to be finalized; many products made for 3.5 GHz may work well in 3.65 GHz US application
4.9 GHz	Used for "Public Safety" in US, intended for use by first responders (police, fire, ambulance, and other emergency services)
5.72–5.85 GHz	US unlicensed

- Wi-Fi is generally a shorter-range system, typically 100 m, though its range can be extended to over a kilometer using directional antennas. WiMAX is a long-range system, covering up to 50 km.
- Wi-Fi supports only around 10 users, while WiMAX supports around 1,000 users.
- The frequency channel bandwidth in 802.11 is 20 MHz wide, whereas in 802.16a, channel bandwidth can be chosen by the operator and can be 1.5–20 MHz wide.
- The maximum data rate in 802.16a is 100 Mbps as compared to 54 Mbps in 802.11.
- Maximum spectral efficiency, i.e., bps/Hz is 5.0 bps/Hz in 802.16a and 2.7 bps/Hz in 802.11.
- Wi-Fi is designed to handle indoor multipath delay spread of 0.8 μs, while WiMAX is designed to tolerate greater multipath delay spread (signal reflections) up to 10.0 μs.
- While no mesh topology support is there in Wi-Fi, WiMAX standards support mesh topologies and advanced antenna techniques.
- The WiMAX standard itself incorporates much better and more flexible security support than the Wi-Fi deployments. Wi-Fi operates on unlicensed 2.4 GHz frequencies, making it more vulnerable to scanning and packet interception, while WiMax network operates at a licensed 2.5 GHz frequency. Licensed frequencies and OFDM make a more secured connection.

In spite of being a technology worth the wait, WiMAX is underutilized at present as it is not widespread yet. In order to make WiMAX successful, new products must be researched and developed that incorporate WiMAX, which will require the enthusiasm and R&D investment of major companies. A comparison of WiMAX technology with other technologies is given in Table 1.4 (http://www.neatware.com/wscn/technology.html).

Table 1.4 Wireless technology comparison

Technology	ZigBee (802.15.4)	Wi-Fi (802.11a/b/g)	Bluetooth	Ultra wide band (UWB)
Data rate	250 kbps	11 and 54 Mbps	1 Mbps	100–500 Mbps
Range	10–100 m	50–100 m	10 m	<10 m
Topology	Ad hoc, peer-to-peer, star, mesh	Point-to-hub	Ad hoc	Peer-to-peer
Frequency	868 MHz (Europe), 900–928 MHz (North America), 2.4 GHz (Worldwide)	2.4 or 5 GHz	2.4 GHz	3.1–10.6 GHz
Complexity	Low	High	Medium	Low
Power consumption	Ultra low	High	Medium	Low
Security	128 AES plus application layer security	64/128 bit encryption		
Applications	Industrial control and monitoring, sensor networks, building automation, home control and automation, toys, games, etc.	Wireless LAN connectivity, broadband Internet access	Wireless connectivity between devices such as phones, laptops, headsets, etc.	Streaming video, home entertainment applications

1.5.6 Radio Frequency Identification Technique

Radio frequency identification (RFID) is a data collection technology that uses electronic tags to store identification data and a wireless transmitter or reader to capture it. It is an automatic wireless data collection technology with long history roots (Brown, 1984; Landt, 2005) and uses electromagnetic or electrostatic coupling in the RF portion of the electromagnetic spectrum to uniquely identify an object.

1.5.6.1 Principle

The vast majority of RFID systems operate according to the principle of inductive coupling (Johnson and Jasik, 1984). Radio waves in the classic sense are used in RFID systems that operate beyond 30 MHz. Every moving charge is associated with a magnetic field. The intensity of the magnetic field can be demonstrated experimentally by the forces acting on a magnetic needle (compass) or a second electric current (Moutairou et al., 2006). The magnitude of the magnetic field is described by the magnetic field strength (H) regardless of the material properties of the space. In general form we can say that the contour integral of magnetic field strength along a closed curve is equal to the sum of the current strengths of the currents (I) within it:

$$\sum I = \oint H \cdot ds \tag{1.1}$$

We can use this formula to calculate the field strength H for different types of conductor. In a straight conductor, the field strength H along a circular flux line at a distance r is constant (Finkenzeller, 2004) and is given by

$$H = \frac{1}{2\pi r} \tag{1.2}$$

1.5.6.2 RFID Tags

An RFID tag is an object that can be applied to or incorporated into a product, animal, or person for the purpose of identification and tracking using radio waves. Most RFID tags contain at least two parts: First is an integrated circuit for storing and processing information, modulating and demodulating a RF signal, and other specialized functions. The second is an antenna for receiving and transmitting the signal. To communicate, tags respond to query-generating signals that must not create interference with the readers, as arriving signals can be very weak and must be differentiated. Besides backscattering, load modulation techniques can be used to manipulate the reader's field. Typically, backscatter is used in the far field, whereas load modulation applies in the near field, within a few wavelengths from the reader (http://en.wikipedia.org/wiki/RFID). RFID tags come in four general varieties: passive, active, semi-passive, and beacon types.

Passive tags require no internal power source, thus being pure passive devices (Curty, 2007; http://www.rfid-asia.info). The minute electrical current induced in the antenna by the incoming RF signal provides just enough power for the complementary metal oxide semiconductor (CMOS) integrated circuit (IC) in the tag to power up and transmit a response. Most passive tags transmit signal by backscattering the carrier wave from the reader. This means that the antenna has to be designed both to collect power from the incoming signal and also to transmit the outbound backscatter signal. The response of a passive RFID tag is not necessarily just an identification number. The tag chip can contain nonvolatile data, possibly writable electronically erasable programming read only memory (EEPROM) for storing data. Passive tags have practical read distances ranging from about 11 cm with near-field up to approximately 10 m with far-field and can reach up to 183 m (http://www.rfidradio.com) when combined with a phased array. Basically, the reading and writing depend on the chosen RF and the antenna design/size. Due to their simplicity in design, they are also suitable for manufacture with a printing process for the antennas. The advantages of lack of an onboard power supply are (i) device can be quite small, (ii) commercially available, and (iii) can be embedded in a sticker, or under the skin in the case of low-frequency RFID tags.

First functional passive RFID systems with a range of several meters appeared in early 1970s (Koelle et al., 1975). Since then, RFID has significantly advanced and experienced a tremendous growth (Dobkin and Wandinger, 2005; Zhao and Gan, 2006; Boglione, 2007). Multiple articles and several books dedicated to RF and other aspects of passive UHF RFID systems have been published in the recent years (Bolomey and Gardiol, 2001; Zhang et al., 2001; Finkenzeller, 2004, Wang et al., 2007; Wenfeng et al., 2007; Wu and Li, 2008; Wang et al., 2008). In 2006, Hitachi, Ltd., developed a passive device called the μ-Chip measuring 0.15 × 0.15 mm without antenna and thinner than a sheet of paper (7.5 μm) (http://www.hitachi.com; http://www.eetimes.com). In February 2007, Hitachi unveiled an even smaller RFID device measuring 0.05 × 0.05 mm and thin enough to be embedded in a sheet of paper (http://news.bbc.co.uk).

Unlike passive RFID tags, active RFID tags have their own internal power source, which is used to power the integrated circuits and to broadcast the response signal to the reader. Communication from active tags to readers is typically much more reliable than those from passive tags due to the ability for active tags to conduct a "session" with a reader. Active tags, due to their onboard power supply, also may transmit at higher power levels than passive tags, allowing them to be more robust in "RF-challenged" environments with humidity and spray or with RF-dampening targets (including humans and cattle, which contain mostly water) and reflective targets from metal (shipping containers, vehicles). In turn, active tags are generally bigger (due to battery size) and more expensive to manufacture due to onboard power source. Many active tags today have operational ranges of hundreds of meters and a battery life from several months to 10 years. Active tags may include larger memories than passive tags and may include the ability to store additional information received from the reader. Special active RFID tags may include specialized sensors. For example, a temperature sensor can be used to record the temperature profile during the transportation and storage of perishable goods. Other sensor types used include humidity, shock/vibration, light, nuclear radiation, pressure, and concentrations of gases such as ethylene. The United States Department of Defence has successfully used active tags to reduce search and loss in logistics and to improve supply chain visibility (http://www.gartner.com).

Semi-passive tags are similar to active tags in that they have their own power source, but the battery only powers the microchip and does not power the broadcasting of a signal. The response is usually powered by means of backscattering the RF energy from the reader, where energy is reflected back to the reader as with passive tags (Fig. 1.18). An additional application for the battery is to power data storage. Semi-passive tags have three main advantages: (i) greater sensitivity than passive tags, (ii) longer battery-powered life cycle than active tags, and (iii) active function performance under their own power, even when no reader is present for powering the circuitry (www.convex.stc.arizona.edu). If energy from the reader is collected and stored to emit a response in the future, the tag is operating active.

1.5.6.3 Tag Attachment

There are three different kinds of RFID tags based on their attachment with identified objects, i.e., attachable, implantable, and insertion tags (http://www.rfid-asia.info/2006/12/rfid-tag-attachments.htm). In addition to these conventional

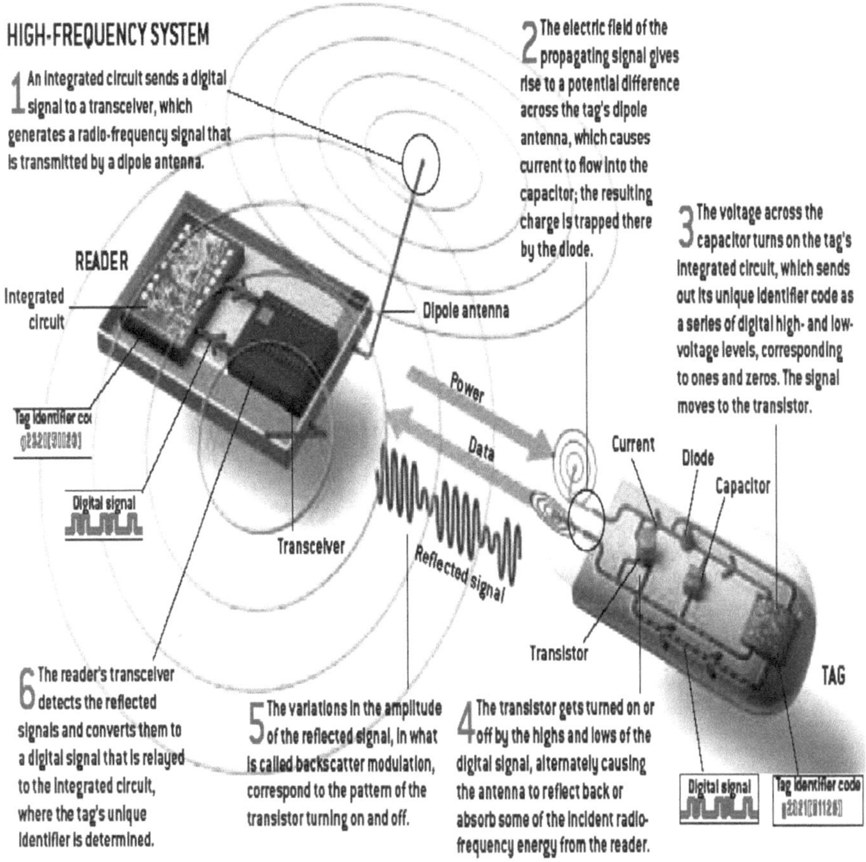

Fig. 1.18 Communication between tag and reader

RFID tags, Eastman Kodak Company has filed two patent applications for monitoring ingestion of medicine based on a digestible RFID tag (http://www.rfid-asia.info/2007/02/digestible-rfid-tag-alternative-for.htm).

The RFID system could profit some of its features by introducing the ZigBee technology into the existing RFID architectures, such as having extended effective range, improving network flexibility, and having compatibility with other ZigBee-enabled environment systems. ZigBee technology (IEEE 802.15.4 wireless protocol) is an important wireless sensor networking solution for short- and medium-range communication (http://Zigbee.org). It uses 2.4 GHz unlicensed industrial, scientific, and medical (ISM) band frequency for data transmission.

1.5.6.4 Application in Mining

In mining industry, RFID can provide improved response to downtime, identification of workers entering or leaving a mine, control of personnel traffic into hazardous areas and warning to them, identification of vehicles entering or leaving production units or passing specific locations in the mine, tracking of supplies and materials, reducing the fatal accidents due to collision, monitoring of underground gases, and maintenance scheduling. The general objectives can be summed up as follows:

- Tracking and monitoring miners and equipment using ZigBee-enabled active RFID devices;
- Monitoring equipment locations and their operation to improve productivity and reduce fatal collision accident;
- Locating and tracking the miners in case of disaster for speedy rescue operation;
- Monitoring miner's unsafe practices and warning;
- Online monitoring of gases and other parameters; and
- Message communication.

1.5.6.5 Global Scenario of RFID Devices in Mining Industry

Tracking and monitoring of miners and mining equipment are basic needs in hazardous environment like underground mines. Several pilot projects have been taken up in Australia, South Africa, and America in this context. Few mines use satellite support for tracking tailors and other heavy vehicles and equipment. But, price of such satellite links is exorbitant. Apart from that, satellite tracking is only possible in case of opencast mines because signal from satellite cannot penetrate and reach underground mines. In order to track underground miners, RFID technology is being explored worldwide as a cost-effective and viable alternative. Active tags are physically attached to the mobile equipment, vehicles, or personnel to be tracked and readers are placed at strategic locations covering the entire area to be tracked in order to monitor the presence of a tagged item in the vicinity of a reader. Each reader serves a zone depending on its read range. At any point of time, the location of a tag can be identified by the corresponding reader's zone under which the tag is currently present.

South African mines are using RFID tags for rescue, gas detection, and first-aid equipment. Local large mines need automated processes to generate reliable and real-time management information for about 6,000 employees a shift, all using a combination of equipment that also needs to be tracked. They have implemented locally developed dual-frequency (DF) RFID technology from Pretoria-based iPico Holdings (http://www.ipico.co.za/). iPico's DF RFID technology uses low-frequency radiation together with the high-speed data-carrying ability. This allows for the reading of multiple tags even in a crowd of miners, as well as achieving improved accuracy through fast reading times and longer tag-reading distances. MSHA has been working on accident prevention through the use of new technologies. Many fatal and nonfatal accidents have occurred at underground mines to continuous mining machine operators and helpers who are in close proximity to machine pinch points. After analysis of these accidents, MSHA believes that proximity detection and protection systems could have prevented a large number of these accidents. A proximity protection system provides automatic proximity detection and machine shutdown to protect personnel from being run over, crushed, or pinned when they are positioned in a hazardous area in close proximity to the machine. The system is also capable of giving the operator an audiovisual warning when entering a protection zone before the machine shuts down. The system consists of a personally worn electronic unit, machine electronics, and antenna. Additional personnel may be protected if they are wearing a portable protection unit. MSHA has undertaken a special project to identify and investigate existing technologies providing proximity protection. RFID technology was identified as the best technology to pursue for underground application.

1.5.7 Ultra-wide-Band Communication

Wireless sensor networks (WSN) have attracted a great attention in the last few years (Stuntebeck et al., 2006; Li et al., 2007; Vuran and Akyildiz, 2008; Zheng and Hu, 2008). WSN offer several advantages over the traditional sensor networks, such as elimination of costly wires, security, and larger area coverage (Cramer et al., 2002; Lewis, 2004; Chehri et al., 2007). In the last few years, there has been great interest in commercial applications based on ultra wide band (UWB) (Baldi et al., 2002; Cuomo et al., 2002). It has potentially low complexity, low cost, and an excellent time domain resolution, which facilitates location and tracking applications. Therefore, UWB provides a good combination of high performance with low complexity for WSN applications (Choi and Stark, 2002; Arslan and Benedetto, 2005; Molisch, 2005; Arslan et al., 2006).

Traditionally, various hard wired and wireless mine communication systems are available in the market. These systems differ in the use of their physical communication infrastructure as well as in the communication protocols used. Data communication is performed using phone lines, mobile radio systems, leaky feeder cables, and fiber optic backbones together with wireless communication based on mine radio and mobile phone system technology. Different philosophies exist regarding the use of analog or digital communication and big differences are visible in the available bandwidth. In mining, infrastructure cost is essential. Thus, a communication system should be integrated as a multipurpose system capable of transferring all types of information as data, voice, and video on an identical infrastructure (Hongxian et al., 2006).

An obvious advantage of wireless transmission is a significant reduction and simplification in wiring and harness. It has been estimated that typical wiring cost in industrial installations is US $ 130–650 per meter and adopting wireless technology would eliminate 20–80% of this cost. Another advantage of wireless terminal is their mobility (these terminals can be placed in transporting vehicles, rotating equipment, etc.). In fact, the motivation for using wireless communication in industrial mining automation is generally twofold: economy and safety.

Wireless communication technologies in mining will have a significant impact on mine operations in the coming decades, giving mine managers and staff much greater understanding and control over mining processes in underground to monitor and optimize mining operations. Increase in wireless communication capabilities helps in establishing the technology base necessary to support remote and autonomous mining operations. In a mine gallery, there is a requirement for many types of communication. Out of them voice communication among mine workers is very critical. Video surveillance through infrequent snapshots in mine gallery is another application of interest and is used for data analyses. On top of applications for improved public safety through the use of vehicular radar systems for collision avoidance, remote control applications are also of interest to the mine operators so that machinery can operate in extreme conditions.

Wireless sensor monitoring is another application that is very crucial for the safety of mine workers (www.icwcuca.ca/documents/SerhanYarkanok_000.pdf). In the proposed applications for IEEE 802.15.4, several companies mention sensor networks, which stand to derive huge benefits from the low power and location aware properties of UWB. Since UWB has excellent spatial resolution it can be advantageously applied in the field of localization and tracking (Chehri and Fortier, 2006a, b; Chehri et al., 2008; Nedil et al., 2008). There are a number of applications that would take advantage from precise positioning and navigation such as automatic storage and tracking of various targets (Ghavami, 2004; Chehri et al., 2006a, b, c; Salih et al., 2008). All these types of communications can use UWB technology. A modern underground UWB communication system integrating both communication and automation is in high demand.

An example of simplified diagram representing global UWB wireless communication in underground mines is shown in Fig. 1.19. Three major components are shown: an application layer system, an integration layer, and a communication layer.

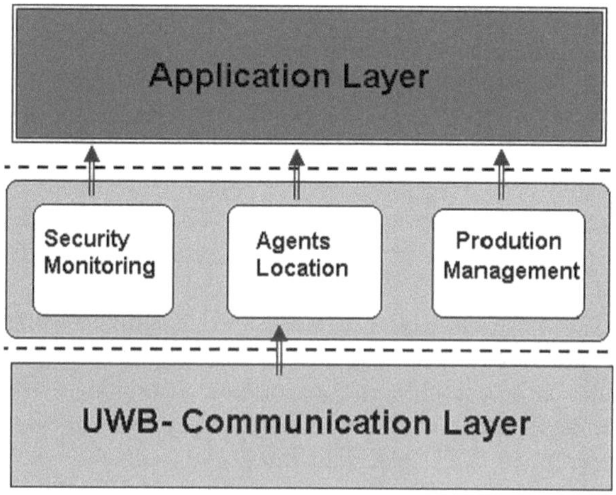

Fig. 1.19 Simplified diagram of the monitoring application for mining industry

The application layer includes, for example, the graphical user interface that is displayed on the end user's computer. The display will provide an overall picture of the current risk profile of the mine and show a specific sensor data for security monitoring, vehicle location, and process data (production management, working place etc.) when requested. All the data displayed can use UWB communication (PHY layer).

Compared to narrowband systems UWB has several advantages (Cramer et al., 2002; Stoica et al., 2006). Because of the combination of wide bandwidth and low power, UWB signals have a low probability of detection and intercept (Qiu, 2002; Zhao and Haimovich, 2002; Nekoogar, 2005; Yarkan and Arslan, 2007; Yarkan et al., 2008; http://www.icwcuca.ca). Additionally, the wide bandwidth gives UWB excellent immunity to interference from narrowband systems and from multipath effects (Qiu and Lu, 1999; Cassioli et al., 2002; Ramirez, 2002; Qiu, 2004; Fares et al., 2004, 2006a, b). Another significant advantage of UWB is its high data rate. Also, the carrierless nature of UWB gives it potential for simple circuit implementation without intermediate oscillators and mixers. UWB devices may have a nearly all digital implementation in CMOS without minimal analog RF electronic. This simple architecture can translate to low power dissipation and low cost, which opens a variety of possible mobile applications. In general, UWB technology has many benefits due to its ultra-wide-band nature, which include the following:

- Coexistence with current narrowband and wideband radio services,
- Large channel capacity: it can support real-time high-definition video streaming,
- Ability to work with low SNR: offers high performance in noisy environments,
- High performance in multipath channels: delivers higher signal strengths in adverse conditions,
- Simple transceiver architecture: which can enable ultra-low power, smaller form factor, and better mean time between failures, all at a reduced cost.

UWB technology for wireless networks is not all about advantages. In fact, there are many challenges involved in using nanosecond-duration pulses for communication. Some of the main difficulties of UWB communication are summarized in following Table 1.5.

Table 1.5 Some challenges and problems associated with UWB systems

Challenge	Problem
Pulse-shape distortion	Low performance using classical matched filter receivers
Channel estimation	Difficulty in predicting the template signals
High-frequency synchronization	Very fast ADCs required
Low transmission power	Information can travel only short distances

1.5.8 Watcher-ATS

Watcher-ATS system can correctly read the tags of a whole bus-load of workers passing at 40 kmph (www.sensorsmag.com/sensors). The equipment and Watcher software have to be carefully tuned for the application. Watcher, a software application at the core of the Watcher-ATS system, is installed on a server, and PC clients are defined that connect to the server and configure the RFID equipment in the software. More systems can be integrated with the system, namely video surveillance and fire alarms. The tag sends a one of a kind of ID to the RFID interrogator, and Watcher receives the information, updates tag position in real time, and "notes" the time of the previous reading. All data, reports, and statistics are handled in the Watcher database. The open communication interface allows units to exchange data via TCP/IP, RS-232, RS-422, UHF, and VHF. Users can define and adjust zones while the system runs, and zones can overlap, while field strength measurement still locates the tag. The function of the system is depicted in Fig. 1.20. The interrogator uses its antennas to communicate with the tags over 868/915 MHz RF. The carrier also communicates with the tags and uses a wireless local area network (WLAN) to communicate with the interrogator and watcher server. Either TCP/IP or RS-232/-485 links the interrogator and the watcher server.

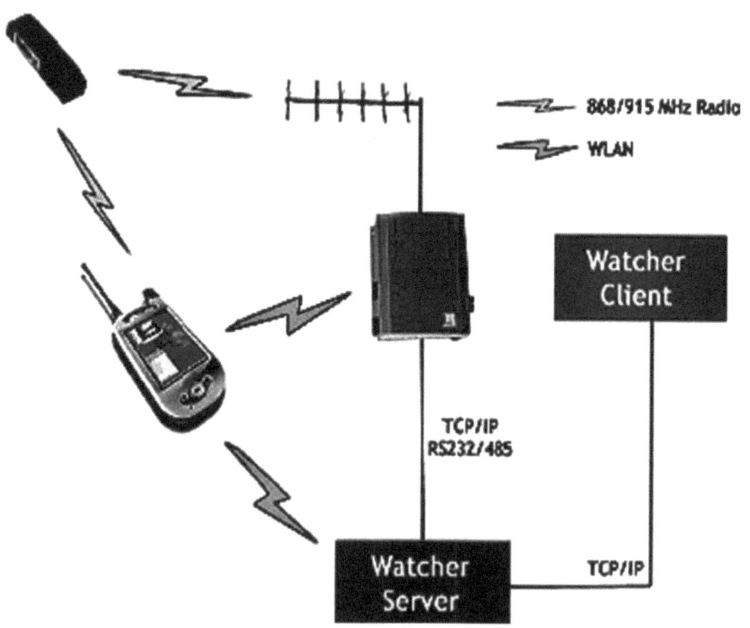

Fig. 1.20 Watcher-ATS

1.5.9 Tracker Tagging System

Tracker is an underground tracking system developed specifically for use in underground mines and tunneling applications. The system tracks active tags (Fig. 1.21) carried by personnel or attached to vehicles and other equipment (http://www.minesite.com.au). The core components of the tracker mine tagging system are (i) RFID tags that transmit an unique ID and readers/beacons that record those IDs, (ii) readers/beacons that communicate data to a PC at a central office or control/communication room. Readers can be either Wi-Fi access points or dedicated UHF readers depending on the version of tracker, and (iii) tracker software that keeps a record of all tag logins and known locations in real time. The software provides users with powerful sorting, filtering, and searching tools to allow for the realization of data, comprehensive logging, as well as extensive report-generating tools.

Fig. 1.21 Tracker tagging system

The tracker system was developed to meet the stringent requirements needed in the underground industry. It is rugged, simple to use, highly reliable, and compliant with all underground safety standards.

1.6 Rescue Systems for Disaster Management

Survival of an underground miner during disaster, fire, or other emergency can be measured in terms of minutes. An emergency warning that arrives late can result in a tragedy. Therefore, proper and reliable rescue systems are required for speedy and effective disaster management in underground mines. Various rescue systems have already been developed as mentioned earlier. Following are few more rescue systems which can be used for disaster management in underground mines.

1.6.1 Communication in Subterranean and Urban Environments

1.6.1.1 Overview

Communication in subterranean and urban environments (CSUE) function as an ideal voice and data communication system for underground mine rescue personnel. This system is built upon a hardware platform of small (450 g), self-contained communication relays that can be placed on the mine floor during emergency maneuvers, and handheld voice-and-data communication terminals (essentially two-way radios) carried by each rescuer. Wireless protocols enable these relays and handheld terminals to form an easily deployed, robust, self-forming communication network supporting simultaneous reliable and secure two-way voice and data communication at rates up to 11 Mbps (http://www.msha.gov/regs/ comments/06-722/AB44-COMM-83.pdf). Deployable and adaptive mobile ad hoc networks (DAMAN) protocol enables formation of self-organizing, self-routing, and self-maintaining communication networks supporting continuous data communication between many highly mobile users, ideal for underground rescue operations (Iwata et al., 1999; Chakrabarti and Mishra, 2001; Cuomo et al., 2002; Moutairou et al., 2006; Guo et al., 2007; Jafarian and Jaseemuddin, 2008; Wu et al., 2008). DAMAN is the communication technology that is the backbone of the CSUE system and it is a multihop mobile ad hoc routing protocol designed from first principles to operate in situations where mobility is the rule rather than exception.

1.6.1.2 System Description

The CSUE system is based upon both IEEE 802.11 and DAMAN network protocols (http://www.msha.gov/regs/comments/06-722/AB44-COMM-83.pdf). It consists principally of intelligent terminals carried by each rescuer, and deployable relays that are placed at key mine corridor intersections. The terminals provide a voice-and-data interface for the rescuers. The relays, with extended line-of-sight range in all directions, form the network grid. The inherent DAMAN protocol in both terminals and relays provides superior network performance in dynamic high-load environments where other protocols break down. Figure 1.22 shows a block diagram of a simple DAMAN CSUE network as it would be used for underground mine rescue.

1.6.1.3 The CSUE Relay

The CSUE relay is built around a commercial single board computer with an Intel XScale PXA 255 400 MHz CPU. The unit uses a 30 mW, IEEE 802.11b radio modified to use an externally mounted 3 dB omnidirectional rubberduck antenna. The relay also contains a relay node interface (RNI) board to interface the computer board to its environment. The RNI board controls startup, shutdown, battery voltage monitoring, and battery charging functionality. The relay also contains an integral battery pack and an onboard battery charger, supplying 7 hours of continuous

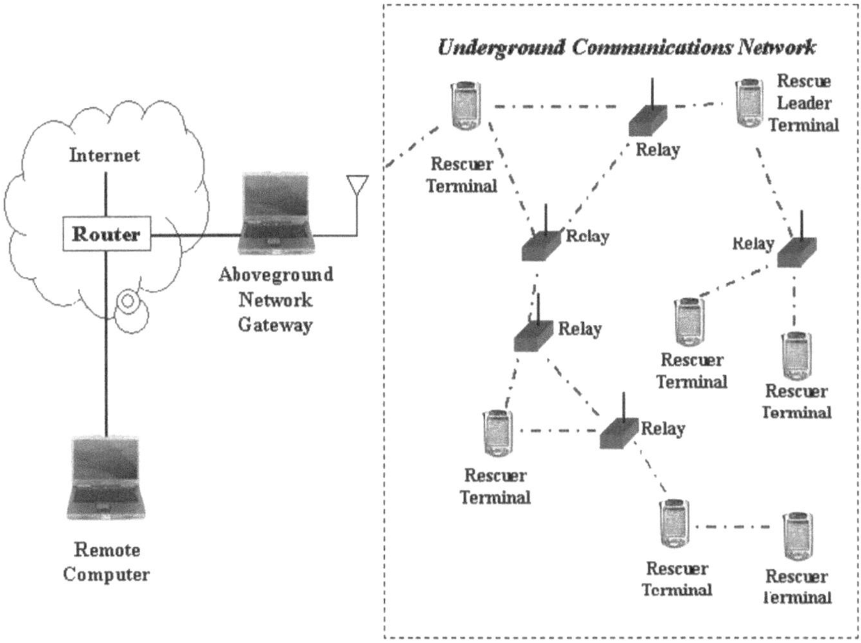

Fig. 1.22 A simple DAMAN CSUE network for underground mine rescue

operation in its present form. LED indicators and momentary switches on the top of unit provide a simple user interface. The switches provide startup and shutdown control, while LED functions are software programmable. The relay enclosure is ruggedized and environmentally sealed for safe and reliable operation. The relay antenna is an omnidirectional folding swivel antenna; a recess in the enclosure provides storage space for the antenna that snaps securely in place.

1.6.1.4 The CSUE Terminal

The CSUE terminal is a voice-and-data terminal designed to function as the rescuer's squad radio, and is based on an iPAQ h5500 series handheld computer. It is equipped with a COTS IEEE 802.11b WLAN card, modified to use an external antenna to create a stronger, more uniform RF radiation pattern. The antenna is semi-permanent and can be removed if necessary. The iPAQ-based CSUE terminal provides:

- Active support by the open source software community. Linux operating system and application development is well supported for this platform.
- Advanced performance: The h5500 series iPAQ is based upon the 400 MHz Intel Xscale, newer, faster, more power-efficient processor architecture than previous iPAQ handhelds.

- External headset jack: The iPAQ h5500 contains an onboard microphone and headphone jack for easy interfacing to a wide variety of COTS audio transducers.

The CSUE terminal can be operated in either voice-operated (VOX) or push-to-talk (PTT) mode, and can be equipped with either a throat microphone (useful for gasmask-equipped rescuers) or a conventional boom-mounted microphone.

1.6.1.5 System Operation

Typical system operation involves the rescue team placing one relay at the mine entrance and placing other relays as needed thereafter to establish a minimal number of line-of-sight point-to-point links between the rescuers and aboveground support. When a rescuer is leaving the network coverage area, he receives an audio indication from his terminal that instructs him to deploy another relay at his current location. In this manner, a network is established providing maximum effective coverage using the fewest number of relays. The grid nature of the room-and-pillar underground coal mine lends itself quite readily to the deployment of CSUE system, with long line-of-site typical in one or both dimensions. Rescuers maintain constant voice contact with other rescuers and with the aboveground support personnel as they work their way through the mine. As DAMAN CSUE system is designed to support all types of digital data, it can also function as a general-purpose data network. For example, network-enabled image sensors and emergency medical equipment can be operated from the surface to deep within the mine.

1.6.1.6 Expansion of Emergency CSUE Communication Network

The CSUE system can readily be expanded to enable its communication relays to interface to a wide variety of sensors and secondary communication devices. A modified CSUE system could form the backbone of an underground mine sensor and communication network for operational as well as emergency use. This system would have many advantages over currently approved underground mine communication systems:

- The CSUE network could be configured as a redundant mesh providing communication from surface to underground even in the event of relay failure or loss of relay connectivity (e.g., due to roof collapse).
- Battery backed-up communication relays would keep the underground network functional for hours or even days (depending upon specific requirements) following removal of power.
- In emergency circumstances, fully isolate a section of the network from the surface, a single relay could be lowered through a small borehole to a location with line-of-sight access to any relay in the isolated subnetwork, thus allowing reestablishment of communication from the isolated subnetwork to the surface.

- Failed and nonresponding relays detected following a fire, collapse, explosion, etc., could rapidly provide information as to the location of emergency.
- The CSUE network could be programmed to provide location information to the surface for terminal-carrying miners. Data would indicate approximate distance from a miner to the nearest relay in near real time. Thus, present miner location could be read, and "last-known location" information would be available in the event of a network-isolating underground event.

1.6.2 Tooth Microphone

A dental bone conduction voice communication system (the tooth microphone) can be used for improving communication – a rescuer is wearing a breathing apparatus and talking through a diaphragm in the mask. It provides high voice intelligibility independent of the ambient noise level. The system consists of a small intra-oral transducer (IOT) that clips to the user's upper left molars, and a body- or head-mounted transceiver that receives a wireless signal from the IOT and outputs it to a traditional two-way radio. The tooth microphone system provides outstanding speech intelligibility in demanding environments ill-suited to traditional air-path microphones, such as from within a gasmask, self-contained self-rescuer (SCSR), or SCUBA facemask. The most obvious application of the tooth microphone is as a means to improve the rescuer's ability to communicate while wearing breathing apparatus. An IOT interfacing wirelessly with a transceiver-equipped breathing mask would implement an unobtrusive voice communication system for ready use with a voice-activated two-way radio.

A secondary application would integrate the tooth microphone with the SCSR worn by the miner during emergencies and/or in the presence of smoke or hazardous gases. The IOT and a transceiver-equipped SCSR could be worn by the miners, driving their voice-activated two-way radios. This would allow for safe, intelligible, and unobtrusive communication during emergency situations.

1.6.3 Super-Low-Frequency Beacon

In many recent mine disasters, miners were trapped underground but healthy, and waiting for emergency workers to find and rescue them. A simple tool that could aid in locating such trapped miners is a super-low-frequency (SLF) beacon (Durkin, 1984). The SLF beacon would be used to indicate the underground location of the miners in the event of a communication system failure. Low-cost battery-operated beacons with hand-cranked auxiliary power could be stored in underground caches and rescue chambers for use during an isolation emergency. These beacons would consist of a small (desktop telephone size) main unit containing a battery, a hand crank and generator, and beacon-generating electronics interfaced to a large detachable loop antenna. The loop antenna would be a robust yet flexible jacketed cable 16–33 m in length with physical characteristics similar to those of an outdoor extension cord. The cable itself would be made of multiple insulated small-gauge wires jacketed in a bundle, forming a multiturn loop antenna. This loop would be permanently connected to the main beacon unit to ensure safe, simple, and reliable operation. To use the SLF beacon, the miner would first spread out the antenna to achieve the largest loop area possible. A simple on–off switch on the main unit would then be toggled "on" to activate the beacon, and an LED would begin blinking at a low-rate/low-duty cycle to indicate an actively pulsing beacon. Extreme caution would be taken in the development of the system to render it fully explosion-proof.

The SLF beacon would produce harmonic rich pulses of super-low-frequency energy with a fundamental frequency between 20 and 200 Hz at a rate of approximately 1 pulse per second. Rescuers would employ portable SLF receivers at the surface to monitor signal strength and thus determine the surface point closest to the beacon. The SLF receivers would feature a significantly smaller ($\sim$1 m) diameter multiturn loop antenna interfaced to an analog receiver front end, an analog-to-digital converter, and a DSP-based SLF demodulation/detector. Any number of rescuers equipped with such SLF receivers could thus cover the surface of the mine either on foot or from within ground vehicles to rapidly determine the location of the underground beacon.

1.7 Conclusions

From the above discussions, it may be concluded that there are various communication and tracking technologies for mining applications (Mark and Zhuang, 2003; Nerguizian et al., 2005; Nutter, 2007). Each technology has their own communication and/or tracking capabilities as well as advantages and disadvantages as listed in Table 1.6 (http://www.wvminesafety.org). Therefore, for selecting the site-specific communication technology for underground mines, proper consideration should be given on techno-economical feasibility of different communication technologies. Further, mining communication technology is developing fast and various researches are continuing throughout the world for development of RFID-based communication and tracking system for underground mines. With the advent of new technology, time is not far away when underground mines will have mobile communication as common as today's surface application.

Table 1.6 Comparison of existing mine communication and tracking technology

Technology	Description	Communication capability	Tracking capability	Advantages	Disadvantages
Ethernet (TCP/IP)	Ethernet communication system for tracking, paging, voice, video, and data transmission. Combination of wireless and wired networks usually used. Some systems can use a leaky feeder system for voice and data transmission.	Voice, data, and video	Yes Accuracy limited to zones defined by wireless access point placement.	Open architecture, mine monitoring from any remote location via Internet. Two-way voice, data, and video on one system.	System uses combination of wired and wireless ethernet. Damage to lines or equipment may disrupt service.
Leaky feeder	Leaky feeder cable and amplifiers are strung through all areas of mine. Voice, data, and video can be transmitted by RF coupling to leaky feeder cable.	Voice, data, and video	Yes If optional RFID readers are coupled to leaky feeder. Accuracy limited to zones.	Two-way voice, data, and video.	Damage to leaky feeder cable or amplifiers may disrupt service. No inherent tracking capability. Radios must be in line-of-sight of leaky feeder cable.

Table 1.6 (continued)

Technology	Description	Communication capability	Tracking capability	Advantages	Disadvantages
Through-the-earth (TTE)	Loop antennas on surface of mine transmit low-frequency signal to receivers integrated into cap lamps. Text message or flashing cap lamp alerts miner to emergency.	Most systems are one-way alarming and text messaging to underground.	Yes	Wireless systems with transmitting loops on surface are not disrupted by explosions, fires, or rock falls.	Large or deep mines may require transmitting loop antennas underground which may be damaged by rock falls or explosions. No tracking. No voice communication for more popular systems.
Medium frequency (MF)	Radio at 280–520 kHz using signal propagation on existing pipes, wiring, etc. Requires repeaters for full mine coverage.	Voice and data	No	Less-dedicated wiring needed than leaky feeder.	Repeaters necessary to cover all areas of mine. Damage to repeaters, conductors, or metal structures that supply transmission paths could disrupt service.

Table 1.6 (continued)

Technology	Description	Communication capability	Tracking capability	Advantages	Disadvantages
Radio frequency identification (RFID)	Active (powered) RF tags are worn by workers or installed on mobile equipment. Tags are interrogated by tag readers placed in zones throughout the mine.	Limited to data exchange and messaging.	Yes Accuracy limited to zones as defined by tag reader placement.	Real-time tracking of miners and equipment. Other safety applications possible, such as controlled access, proximity warning, sensors, etc.	Systems require separate communication infrastructure. Damage to communication lines or readers could disrupt service.
Distributed antenna system	RF antennas are tapped into a coaxial cable backbone where coverage is needed.	Voice, data, and video	No	Reconfigurable as mine layout changes. Antennas are placed only where needed which reduce cost. Relatively simple to install.	System requires coaxial cable backbone to all areas where communication is needed. Damage to cable or equipment may disrupt service.
Trolley phone	High-voltage trolley line used as signal path.	Voice	No		
Pager phone	Party line-wired phone system.	Voice	No		
Phone	Traditional wired phone system.	Voice	No	Inexpensive and easy to maintain and use.	Damage to phone line will disrupt service. Little or no mobility at limited number of phone locations.

Table 1.6 (continued)

Technology	Description	Communication capability	Tracking capability	Advantages	Disadvantages
Seismic monitoring	Portable system using seismic monitoring sensors on the surface that can detect sounds generated by trapped miners.	One-way communication from miners to surface, limited to simple codes generated by tapping.	Yes If trapped miners signal as they move.	Trapped miner locating system that does not depend on integrity of mine communication. Portable in size.	Emergency system only. Limited to mines 500 m deep or less. Depends on trapped miners signaling the surface by tapping on mine structures.
TTE – Miner locator system	Miner carries GLON transmitter which is embedded in cap lamp. Transmitter uses 8 channels (4100–5850 kHz). GLOP receiver used by rescuers to locate miner.	No communication capability. Transmitter is only used for miner location purposes.	Receiver GLOP enables location of transmitters (i) by measuring the distance from two measurement points or (ii) by measuring distance and determining direction of the transmitters from one measurement point	Lamp shuts down allowing 7 days of power to the locator transmitter, if voltage drops to certain level locatable through 25–50 m of roof fall or solid rock. In case of line-of-sight, it would enable location by rescue teams at greater distances. Also, LED cap lamp (including GLON transmitter) is lighter than standard cap lamp.	No communication capability. Transmitter only used for miner location purposes. Maximum range is 50 m.

References

Akyildiz IF and Stuntebeck EP (2006) Wireless underground sensor networks: research challenges. *Ad Hoc Networks*, **4**(6): 669–686.

Akyildiz IF, Su W, Sankarasubramaniam Y and Cayirci E (2002) Wireless sensor networks: a survey. *Computer Networks*, **38**(4): 393–422.

Allen JW and Linefield RF (1973) Trapped miners location and communication system. *Westinghouse Electric Corporation*, May 1973, Vol. 1–4.

Anatory J, Theethayi N, Thottappillil R, Kissaka MM and Mvungi NH (2008) The influence of load impedance, line length, and branches on underground cable power-line communications systems. *IEEE Transactions on Power Delivery*, **23**(1): 180–187.

Aniss H, Tardif PM, Ouedraogo R and Fortier P (2004) Communications network for underground mines based on the IEEE 802.11 and DOCSIS standards. *Proceedings of IEEE 60th Vehicular Technology Conference*, Los Angeles, USA, Vol. 5, 26–29 September 2004, pp. 3605–3609.

Arslan A, Chen AN and Benedetto MG (2006) *Ultra-wideband wireless communication*. Wiley Interscience, Hoboken, New Jersey.

Arslan H and Benedetto MGD (2005) *Introduction to UWB*. Book Chapter, Ultra Wideband Wireless Communications (ed. Arslan H), John Wiley & Sons, USA.

Asimov I (1979) Communicating through the earth. *IEEE Communications Magazine*, **17**(5): 9–19.

Austin BA (1978) Underground radio communication techniques and system in South African mines. *Proceedings of the Workshop on Electromagnetic Guided Waves in Mine Environment*, Boulder, CO, Institute for Telecommunication Science, 28–30 March 1978, pp. 87–102.

Ayuso N, Cuchi JA, Lera F and Villarroel JL (2006) Through-the-earth magnetic field propagation: modelling and experimental validation. *Proceedings of IEEE International Symposium Antennas and Propagation Society*, Albuquerque, New Mexico, 9–14 July 2006, pp. 680–683.

Bandyopadhyay LK, Chaulya SK and Kumar S (2002) A proposed wireless communication system for underground mines. *Proceedings of the International Conference on Mineral Industry, Issues on Economics, Environment and Technology*. Mining, Geological and Metallurgical Institute of India, Kolkata, India, 2002, pp. 317–324.

Balanis C (1973) Line sources above lossy media for low-frequency underground communication. *IEEE Transactions on Antennas and Propagation*, **21**(1): 127–128.

Baldi P, Nardis LD and Benedetto MD (2002) Modeling and optimization of UWB communication networks through a flexible cost function. *IEEE Journal on Selected Areas in Communications*, **20**(9): 1733–1744.

Barber TJ, Carvey P and Chandrakasan A (1996) Designing for wireless LAN communications. *IEEE Circuits and Devices Magazine*, **12**(*4*): *29-33.*

Beiter S, Poquette R, Filipo BS and Goetz W (1998) Precision hybrid navigation system for varied marine applications. *Proceedings of IEEE Symposium on Position Location and Navigation*, California, USA, 20–23 April 1998, pp. 316–323.

Boglione L (2007) RFID technology—are you ready for it? *IEEE Microwave Magazine*, **8**(6): 30–32.

Bolomey JC and Gardiol FE (2001) *Engineering Applications of the Modulated Scatterer Technique*. Artech House, London, UK.

Boutin M, Affes S, Despins C and Denidni T (2005) Statistical modelling of a radio propagation channel in an underground mine at 2.4 and 5.8 GHz. *Proceedings of IEEE Vehicular Technology*, Stockholm, Sweden, Vol. 1, 30 May–1 June 2005, pp. 78–81.

Boutin M, Benzakour A, Despins CL and Affes S (2008) Radio wave characterization and modeling in underground mine tunnels. *IEEE Transactions on Antennas and Propagation*, **56**(2): 540–549.

Brown WC (1984) The history of power transmission by radio waves. *IEEE Transactions on Microwave Theory and Techniques*, **32**(9): 1230–1242.

Casad and Joseph (2001) *Sams Teach Yourself TCP/IP in 24 Hours*, 2nd Edition. Sams Publishing Co., Indianapolis, IN 46290.

Cassioli D, Win MZ and Molisch AF (2002) The ultra-wide bandwidth indoor channel: from statistical model to simulations. *IEEE Journal of Selected Areas Communication*, **20**(6): 1247–1257.

Chakrabarti S and Mishra A (2001) QoS issues in ad hoc wireless networks. *IEEE Communications Magazine*, **39**(2): 142–148.

Chao RY and Chung KS (1994) A low profile antenna array for underground mine communication. *Proceedings of International Conference on Conceptual Structures*, Singapore, Vol. 2, 14–18 November 1994, pp. 705–709.

Chehri A and Fortier P (2006a) Frequency domain analysis of UWB channel propagation in underground mines. *Proceedings of IEEE 64th Vehicular Technology Conference*, Montreal, Canada, 25–28 September 2006, pp. 1–5.

Chehri A and Fortier P (2006b) Measurements and modeling of line-of-sight UWB channel in underground mines. *Proceedings of IEEE Global Telecommunications Conference*, San Francisco, USA, 27 November–1 December 2006, pp. 1–5.

Chehri A, Fortier P and Tardif PM (2006a) Deployment of ad-hoc sensor networks in underground mines. *Proceedings of Conference on Wireless and Optical Communication, and Wireless Sensor Network*, Alberta, Canada, 3–4 July 2006, pp. 13–19.

Chehri A, Fortier P and Tardif PM (2006b) Geolocation for UWB Networks in underground mines. *Proceedings of IEEE Wireless and Microwave Technology Conference,* Florida, USA, 4–5 December 2006, pp. 1–4.

Chehri A, Fortier P and Tardif PM (2007) Security monitoring using wireless sensor networks. *Proceedings of 5th Annual Conference on Communication Networks and Services Research*, Fredericton, Canada, 14–17 May 2007, pp. 13–17.

Chehri A, Fortier P, Aniss H and Tardif PM (2006c) UWB spatial fading and small scale characterization in underground mines. *Proceedings of 23rd Biennial Symposium on Communications,* Kingston, ON, pp. 213–218.

Chehri A, Fortier P and Tardif PM (2008) An investigation of UWB-based wireless networks in industrial automation. *International Journal of Computer Science and Network Security*, **8**(2): 179–188.

Chirdon DC (2006) *Mine Safety and Health Administration*, Triadelphia, WV: Personal Communication, February 2006 (www.cdc.gov/niosh/topics/minerescue/minecomms.html).

Choi JD and Stark WE (2002) Performance of ultra-wideband communications with suboptimal receivers in multipath channels. *IEEE Journal on Selected Areas in Communications*, **20**(9): 1754–1766.

Chufo RL (1977) Vehicular communications 2500 feet underground. *Proceedings of 27th IEEE Conference on Vehicular Technology*, Florida, USA, Vol. 27, 16–18 March 1977, pp. 161–165.

Chufo RL (1978) Medium frequency mine communications. *Proceedings of 28th IEEE Conference on Vehicular Technology Conference,* Colorado, USA, Vol. 28, 22–24 March 1978, pp. 261–266.

Chung KS (1994) Underground microcellular communications network. *Proceedings of International Conference on Conceptual Structures*, Singapore, Vol. 1, 14–18 November 1994, pp. 343–346.

Cocheril Y, Langlais C, Berbineau M and Moniak G (2008) Advantages of simple MIMO schemes for robust or high data rate transmission systems in underground tunnels. *Proceedings of IEEE 68th Vehicular Technology Conference*, Calgary, Canada, 21–24 September 2008, pp. 1–5.

Conti R (2000) *White Paper on Technologies for Communications and Locating Trapped Miners.* (www.wvminesafety.org/PDFs/MSTTF%20Report%20Final.pdf).

Conti RS and Yewen RG (1997) Evaluation of a signaling and warning system for underground mines. *Report of Investigations* 9641, US Department of Health and Human Services, National Institute of Occupational Safety and Health, Pittsburgh, May 1997, p. 17.

Cramer RJ, Scholtz RA and Win MZ (2002) Evaluation of an ultra-wideband propagation channel. *IEEE Transactions of Antennas Propagations*, **50**(5): 561–570.

Cuomo F, Martello C, Baiocchi A and Capriotti F (2002) Radio resource sharing for ad hoc networking with UWB. *IEEE Journal on Selected Areas in Communications*, **20**(9): 1722–1732.

Curty JP (2007) *Design and Optimization of Passive UHF RFID systems*. Springer, New York.

Deshpande ND, Deshpande DA and Rangole PK (1998) *Communication electronics*. Tata McGrew-Hill Publishing Ltd., New Delhi, India.

Dilmaghani RB and Rao RR (2008) Hybrid wireless mesh network with application to emergency scenarios. *Journal of Software*, **3**(2): 52–60.

Dobkin DM (2007) *The RF in RFID: Passive UHF RFID*. Elsevier, Burlington.

Dobkin D and Wandinger T (2005) A radio-oriented introduction to RFID. *High Frequency Electronics*, **8**: 46–55.

Dobroski H Jr and Stolarczyk LG (1982) Medium frequency radio communication system for mine rescue. *Proceedings of Post Disaster Survival and Rescue Research*, Bureau of Mines Technology Transfer Seminar, Pittsburgh, PA, US Department of the Interior, Bureau of Mines, 1C 8907, 16 November 1982, pp. 39–48.

Drabich O, Drabich P, Kychma A, Koval R and Javorskyj I (2002) Methods of indication and operative detection of underground communication elements. *Proceedings of the International Conference on Modern Problems of Radio Engineering, Telecommunications and Computer Science*, Lviv-Slavska, Ukraine, 18–23 February 2002, pp. 112–113.

Durkin J (1984) Electromagnetic detection of trapped miners. *IEEE Communication Magazine*, **22**(2): 37–46.

Durkin J and Greenfield RJ (1981) Evaluation of the seismic for locating trapped miners. *Report Investigations* 8567, US Bureau of Mines, 1981. 55 p.

Durkin J (1982) Performance evaluation of electromagnetic techniques for location of trapped miners. *Report of Investigations* 87II, US Bureau of Mines, 1982.

Einicke G, Duff E, Reid D, Ralston J, Cunningham J, Hainsworth D and Corke P (2002) The application of wireless LANs in mine automation. *Proceedings of Evans Conference on Creating the Virtual Enterprise*, Sydney, Australia, Vol. 2, pp. 1–10.

Einicke G, Dekker D and Gladwin M (1997) A robust WLAN for survivable emergency communications. *Proceedings of IEEE Region 10 Annual Conference on Speech and Image Technologies for Computing and Telecommunications*, Brisbane, Australia, Vol. 1, 2–4 December 1997, pp. 101–104.

Fares SA, Denidni TA and Affes S (2004) Sequential blind beamforming algorithm using combined CMA/LMS for wireless underground communications. *Proceedings of IEEE 60th Vehicular Technology Conference*, Los Angeles, USA, Vol. 5, 26–29 September 2004, pp. 3600–3604.

Fares SA, Denidni TA, Affes S and Despins C (2006a) CMA/fractional-delay sequential beamforming for wireless multipath communications. *Proceedings of IEEE 63rd Vehicular Technology Conference*, Melbourne, Australia, 7–10 May 2006, Vol. 6, pp. 2793–2797.

Fares SA, Denidni TA, Affes S and Despins C (2006b) Efficient sequential blind beamforming for wireless underground communications. *Proceedings of IEEE 64th Vehicular Technology Conference*, Montreal, Canada, 25–28 September 2006, pp. 1–4.

Finkenzeller K (2003) RFID-handbook, *Fundamentals and Applications in Contactless Smart Cards and Identification*, 2nd Edition. Wiley and Sons, New York.

Fisahn S, Camp M and Garbe H (2005) Transient emission behavior of leaky section cables for wireless applications. *Proceedings of International Symposium on Electromagnetic Compatibility*, Chicago, USA, Vol. 2, 8–12 August 2005, pp. 477–482.

Finkenzeller K (2004) *RFID Handbook: Radio-Frequency Identification Fundamentals and Applications*. John Wiley & Sons, USA.

Forrest RT (1975) A practical approach to radio propagation measurements – as used in the design of mobile radio communications systems. *IEEE Transactions on Vehicular Technology*, **24**(4): 46–53.

Ghavami M (2004) *Ultra Wideband Signals and Systems in Communication Engineering*. John Wiley, USA.

Ghosh D, HongSik M and Sarkar TK (2008) Design of through-the-earth mine communication system using helical antennas. *Proceedings of IEEE International Symposium of Antennas and Propagation Society*, California, USA, 5–11 July 2008, pp. 1–4.

Gogoi AK and Raghuram R (1996) Analysis of VLF loop antennas on the earth surface for underground mine communication. *Proceedings of IEEE International Symposium of Antennas and Propagation Society*, Maryland, USA, Vol. 2, 21–26 July 1996, pp. 962–965.

Gogoi AK and Raghuram R (1997a) Closed form expression for input impedance of VLF small antenna located on earth surface. *Proceedings of IEEE International Symposium of Antennas and Propagation Society*, Montreal, USA, Vol. 3, 13–18 July 1997, pp. 1612–1615.

Gogoi AK and Raghuram R (1997b) Variation of field strength in underground mine area on the size of VLF loop antennas laid on the earth surface. *Proceedings of IEEE International Symposium of Antennas and Propagation Society*, Montreal, USA, Vol. 3, 13–18 July 1997, pp. 1792–1795.

Guo Y, Chen C and Zhou S (2007) A topology visualization tool for large scale communications networks, *Electronics Letters*, **43**(10): 597–598.

Hargrave CO, Ralston JC and Hainsworth DW (2007) Optimizing wireless LAN for longwall coal mine automation. *IEEE Transactions on Industry Applications*, **43**(1): 111–1117.

Haykin S (2001) *Communication Systems*. 4th Edition, John Wiley & Sons, New York.

Higginson M (1992) *Radio Propagation Experiments Durban Deep Goldmine*. Draft report on investigation, COMRO, September 1992, 4 p.

Holloway CL, Hill DA, Dalke RA and Hufford GA (2000) Radio wave propagation characteristics in lossy circular waveguides such as tunnels, mine shafts, and boreholes. *IEEE Transactions on Antennas and Propagation*, **48**(9): 1354–1366.

Hongxian T, Wei Y, Haibo Z and Gongxun Y (2006) Full direction search (FDS) based fast motion estimation algorithm for video encoding in coalmining underground. *Proceedings of International Conference on Communication Technology*, Guilin, China, 27–30 November 2006, pp. 1–5.

Hui JY (1988) Resource allocation for broadband networks. *IEEE Journal on Selected Areas in Communications*, **6**(9): 1598–1608.

Information Circular (IC 8955) (1984) *Underground mine communications, control and monitoring*. Bureau of Mines Information Circular, United States Department of Interior, USA.

Iwata A, Chiang C, Pei G, Gerla M and Chen T (1999) Scalable routing strategies for ad hoc wireless networks. *IEEE Journal on Selected Areas in Communications*, **17**(8): 1369–1379.

Jafarian M and Jaseemuddin M (2008) Routing of emergency data in a wireless sensor network for mines. *Proceedings of IEEE International Conference on Communications*, Beijing, China, 19–23 May 2008, pp. 2813–2818.

Johnson RC and Jasik H (1984) *Antenna Engineering Handbook*, 2nd Edition. McGraw-Hill, London.

Kennedy GA and Foster PJ (2006) High resilience networks and microwave propagation in underground mines. *Proceedings of the 9th European Conference on Wireless Technology*, Vienna, Austrin, 10–12 September 2006, pp. 193–196.

Koelle AR, Depp SW, and Freyman RW (1975) Short-range radiotelemetry for electronic identification, using modulated RF backscatter. *Proceedings of the IEEE*, **63**(8): 1260–1261.

Kohler and Jeffery (1992) Monitoring, control, and communications. Chapter 12.6 in Hartmann HL (ed.), *Mining Engineering Handbook*, 2nd Edition, Society for Mining, Metallurgy and Exploration, Denver.

Kononov VA and Smit JJ (1997) global mining warning and monitoring system. *Proceedings of the 27th International Conference of Safety in Mines Research Institute*, New Delhi, India, 20–22 February 1997, pp. 947–953.

Kononov VA (1998a) Develop a trapped miner location system and an adequate rescue strategy and associated technologies. *Final Project Report No. GEN 502*, CSIR Division of Mining Technology, South Africa, 36 p.

Kononov VA (1998b) Investigation into and report on early warning and communication systems for irrespirable atmospheres in underground operations. *SIMRAC Final Project Report GEN413*, Pretoria: Department of Minerals and Energy, 60 p.

Kononov VA and Higginson MR (1994) Trapped miner locator (marketing feasibility). *CSIR Miningtek*, Final report Y5I62, Johannesburg, South Africa.

Kravitz JH, Kovac JG and Duerr WH (1994) Advances in mine emergency communication. *Proceedings of the International Emergency Management and Engineering Conference*, Hollywood Beach, Florida, pp. 23–26.

Kumar A, Chaulya SK, Kumar S and Bandyopadhyay LK (2004) Trapped miners detection, location and communication system. *Minetech*, **24**(6): 3–13.

Large D, Ball L and Farstad A (1973) Radio transmission to and from underground coal mines—theory and measurement. *IEEE Transactions on Communications*, **21**(3): 194–202.

Leeper D (2001) A long-term view of short-range wireless. *IEEE Computer*, **34**(6): 39–44.

Lagace RL, Dobbie, JM and William S (1980) *Detection of trapped miner electromagnetic signals above coal mines*. US Bureau of Mines, Contract Report No. J0188037.

Lagace RL, Dobbie JM, Raisbeck G, Spencer RH, Degatano MA, Lambiase FJ and Moussa NA (1982) System study of mine rescue through electromagnetic means. Cambridge, MA: Arthur D. Little, Inc. US Bureau of Mines, Contract No. J0113043.

Landt J (2005) The history of RFID. *IEEE Potentials*, **24**(4): 8–11.

Lewis F (2004) Wireless sensor networks—smart environments: technologies, protocols, and applications. John Wiley, New York.

Li C and Chen C (2006) Topology analysis of Gnutella by large scale mining. *Proceedings of International Conference on Communication Technology*, Guilin, China, 27–30 November 2006, pp. 1–4.

Li L, Vuran MC and Akyildiz IF (2007) Characteristics of underground channel for wireless underground sensor networks. *Proceedings of Med-Hoc-Net*, Corfu, Greece, 13–15 June 2007, pp. 67–71.

Mark JW and Zhuang W (2003) *Wireless communications and networking*. Prentice-Hall, Upper Saddle River, NJ.

Miller MHB, Hancke GP and Horvath L (1999) Underground communications infrastructure design. *Proceedings of IEEE Africon*, Africa, Vol. 1, 28 September–1 October 1999, pp. 365–369.

Molisch AF (2005) Ultra wideband propagation channels-theory, measurement, and modeling. *IEEE Transactions on Vehicular Technology*, **54**(5): 1528–1545.

Moutairou M, Aniss H and Delisle GY (2006) Wireless mesh access point routing for efficient communication in underground mine. *Proceedings of IEEE International Symposium of Antennas and Propagation Society*, Washington, USA, 9–14 July 2006, pp. 577–580.

Murphy JN and Parkinson HE (1978) Underground mine communications. *Proceedings of the IEEE*, **66**(1): 26–50.

Ndoh M (2004) *Modélisation de la propagation des ondes électromagnétiques dans un environnement minier*. Thèse de Ph. D., Département de génie électrique et génie informatique, Université de Laval, Québec, Canada.

Ndoh M, Delisle GY and Le R (2003) A novel approach to propagation prediction in a confined rough surfaces. *The International Journal of Numerical Modelling, Electronical Network, Devices and Fields*, **16**: 535–555.

Nedil M, Denidni TA, Djaiz A and Habib AM (2008) A new ultra-wideband beamforming for wireless communications in underground mines. *Progress in Electromagnetics Research*, **4**: 1–21.

Nerguizian C, Despins CL, Affes S and Djadel M (2005) Radio-channel characterization of an underground mine at 2.4 GHz. *IEEE Transactions on Wireless Communications*, **4**(5): 2441–2453.

Nekoogar F (2005) *Ultra-Wideband Communications: Fundamentals and Applications*. Prentice Hall, London.

Nessler and Norbert H (2000) Electromagnetic location system for trapped miners. *Subsurface Sensing Technologies and Applications*, **1**(2): 229–246.

Nichoga VO, Dykmarova P, Drabych PP and Dub PB (2005) Diagnostics of technical state of stretched engineering underground communications by electromagnetic methods. *Proceedings of 7th International Conference on Telecommunications in Modern Satellite, Cable and Broadcasting Services*, Antwerp, Belgium, Vol. 1, 28–30 September 2005, pp. 279-286.

Nie D, Liu W and Wei S (2008) A study on analyzing the channel characteristics of low-voltage powerline under coal mine. *Proceedings of Second International Intelligent Information Technology Application*, Shanghai, China, Vol. 2, 20–22 December 2008, pp. 483–487.

Nikitin PV and Rao KVS (2008) Antennas and propagation in UHF RFID systems. *Proceedings of IEEE International Conference on RFID*, Las Vegas, USA, 16–17 April 2008, pp. 277–288.

Nutter RS (2007) Underground coal mine communications and tracking status SAGO plus one year. *Proceedings of IEEE Industry Applications Conference*, Louisiana, USA, 23–27 September 2007, pp. 2086–2089.

Olsen RG and Farstad AJ (1973) Electromagnetic direction finding experiments for location of trapped miners. *IEEE Transactions on Geoscience Electronics*, **11**(4): 178–185.

Pahlavan K and Krishnamurthy P (2006) *Principles of Wireless Networks*. Prentice-Hall of India Private Limited, New Delhi.

Pahlavan K and Levesque AH (1995) *Wireless Information Networks*. John Wiley & Sons, New York.

Powell JA (1976) An electromagnetic system for detecting and locating trapped miners. *Report of Investigations 8159*, US Bureau of Mines, 1976.

Proakis JG (1995) *Digital Communications*. McGraw-Hill, New York, 1995.

Qiu RC (2002) A study of the ultra-wideband wireless propagation channel and optimum UWB receiver design. *IEEE Journal on Selected Areas in Communications*, **20**(9): 1628–1637.

Qiu RC (2004) A generalized time domain multipath channel and its applications in UWB wireless optimal receiver design: wave-based system analysis. *IEEE Transactions on Wireless Communications*, **3**: 2312–2324.

Qiu RC and Lu IT (1999) Multipath resolving with frequency dependence for wide-band wireless channel modeling. *IEEE Transactions on Vehicular Technology*, **48**(1): 273–285.

Ralston JC and Hainsworth DW (1997) The Numbat: a remotely controlled mine emergency response vehicle. *Proceedings of the International Conference on Field and Service Robotics*, Canberra, Australia, pp. 48–55.

Ramirez MF (2002) Signal design for ultra-wide-band communications in dense multipath. *IEEE Transactions on Vehicular Technology*, **51**(6): 1517–1521.

Rappaport TS (1996) *Wireless Communications: Principles and Practices*. Prentice-Hall, London.

Reagor D, Yan F, Mombourquette C, Quanxi J and Stolarczyk LA (1997) High-temperature superconducting receiver for low-frequency radio waves. *IEEE Transactions on Applied Superconductivity*, **7**(4): 3845–3849.

Reed HR and Russell CM (1966) *Ultra High Frequency Propagation*. Chapman and Hall, London.

Ristenbatt MP, Holland-Mortiz E and Metzger K (1988) A new post-disaster mine communication system. *IEEE Transactions on Industry Applications*, **24**(2): 204–211.

Robinson MP, Clegg J and Marvin AC (2006) Radio frequency electromagnetic fields in large conducting enclosures: effects of apertures and human bodies on propagation and field-statistics. *IEEE Transactions on Electromagnetic Compatibility*, **48**(2): 304–310.

Rowan GJ, Glynn P, Einicke G, Beitz A, Smartt P, Balusu R and Guo G (2003) Mine communication and information systems for real-time risk management. *Proceedings of 30th International Conference of Safety in Mines Research Institutes*, South African Institute of Mining and Metallurgy, South Africa.

Ruths MA (1977) The reference correction method for improving accuracy in the seismic location of trapped coat miners. M.S Thesis, Pennsylvania State University, College of Earth and Mineral Sciences, University park, PA, November 1977, 141 p.

Salih AY, Despins C and Affes S (2008) Ultra-wideband fast acquisition system for positioning in an underground mining environment. *Proceedings of IEEE International Conference on Communications*, Beijing, China, 19–23 May 2008, pp. 473–479.

Schiffbauer WH and Brune JF (2006) Underground coal mine communications for emergencies and everyday operation. *Proceedings of Symposium on the Capabilities and Availability of Wireless Communication and Tracking Systems for Underground Coal Mines in Abingdon*, VA, February 13, 2006 (http://www.nrcce.wou.edu/energyforum/docs/Brune.pdf)

Schwartz M (1996) *Broadband Integrated Networks*. Prentice-Hall, Upper Saddle River, New Jersey.

Srinivasan K, Ndoh M and Kaluri K (2005) Advanced wireless networks for underground mine communications. *Proceedings of First International Workshop on Wireless Communications in Underground and Confined Areas*, Val-d'Or, Québec, Canada, 6–7 June 2005, pp. 51–54.

Stoica L, Rabbachin A and Oppermann I (2006) Impulse radio based non-coherent UWB transceiver architectures – an example. *Proceedings of IEEE International Conference on Ultra Wideband Systems and Technologies*, Waltham, USA, pp. 483–488.

Stolarczyk LG (1991) Emergency and operational low and medium frequency band radio communications system for underground mines. *IEEE Transaction on Industrial Application*, **27**: 780–790.

Stolarczyk LG, Stolar I and Raton NM (1991) Emergency and operational low and medium frequency band radio communications system for underground mines. *IEEE Transactions on Industry Applications*, **27**(4): 780–790.

Stolarczyk GL (1984) *A Medium Frequency Wireless Communication System for Underground Mines*. NIOSHTIC-2 No. 10004817, Bureau of Mines, U.S. Department of Interior.

Stuntebeck EP, Pompili D and Melodia T (2006) Wireless underground sensor networks using commodity terrestrial motes. *Proceedings of 2nd IEEE Workshop on Wireless Mesh Networks*, Virginia, USA, 25–28 September 2006, pp. 112–114.

Sydanheimo L, Keskilammi M and Kivikoski M (2000) Reliable mobile computing to underground mine. *Proceedings of IEEE International Conference on Communications*, Louisiana, USA, Vol. 2, 18–22 June 2000, pp. 882–888.

Shope S, Durkin J and Greenfield R (1982) "Finding and communication with trapped miners". Information Circular 8907, *Proceedings of Post Disaster Survival and Rescue Research*, Bureau of Mines Technology Transfer Seminar, Pittsburg, 16 November 1982, pp. 49–78.

Turoff M, Chumer M, Walle VB and Yao X (2004) The design of a dynamic emergency response management information system. *Journal of Information Technology Theory and Application*, **5**(4): 1–36.

Vasquez J, Rodriguez V and Reagor D (2004) Underground wireless communications using high-temperature superconducting receivers. *IEEE Transactions on Applied Superconductivity*, **14**(1): 46–53.

Vong PK, Lai, HC and Rodger D (2006) Modeling electromagnetic field propagation in eddy-current regions of low conductivity. *IEEE Transactions on Magnetics*, **42**(4): 1267–1270.

Vuran MC and Akyildiz IF (2008) Cross-layer packet size optimization for wireless terrestrial, underwater, and underground *sensor networks. Proceedings of IEEE 27th Conference on Computer Communications*, Arizona, USA, 13–18 April 2008, pp. 226–230.

Wadley TL (1949) Radio communication through rock on the Witwatersrand mines. *CSIR report* CTR-4, August 1949.

Wait JR (1989) Electromagnetic and electrochemical response of geological conductors. *Proceedings of IEEE International Symposium of Antennas and Propagation Society*, California, USA, Vol. 2, 26–30 June 1989, pp. 1156–1159.

Wang Y, Zhao X, Wu Y and Xu P (2008) The research of RFID middleware's data management model. *Proceedings of IEEE International Conference on Automation and Logistics*, Qingdao, USA, 1–3 September 2008, pp. 2565–2568.

Wang Z, Sun X, Zhang C and Li Y (2007) Issues in integrated circuit design for UHF RFID. *Proceedings of IEEE International Workshop on Radio-Frequency Integration Technology*, Singapore, 9–11 December 2007, pp. 322–328.

Wei S, Cao J, Chen Y, Cheng F, Nie D and Li H (2008) A canceling noise research for underground mine powerline carrier communication based on adaptive theory. *Proceedings of Workshop on Power Electronics and Intelligent Transportation System*, Guangzhou, China, 2–3 August 2008, pp. 345–348.

Wenfeng L, Jie G and Peng B (2007) Mine multimedia emergency communication system. *Proceedings of International Conference on Wireless Communications, Networking and Mobile Computing*, Dalian, China, 21–25 September 2007, pp. 2865–2868.

Westinghouse Georesearch Laboratory (1973) EM location system prototype and communication station modification. *Boulder Co., U.S. Bureau of Mines* Contract No. H0232049. 1973.

Wheeler HA (1961) Radio-wave propagation in the earth's crust. *Journal of Research of the National Bureau of Standards*, **65**(2):189–191.

Wu D, Li R and Bao L (2008) A holistic routing protocol design in underground wireless sensor networks. *Proceedings of 4th International Conference on Mobile Ad-hoc and Sensor Networks*, Wuhan, China, 10–12 December 2008, pp. 187–194.

Wu J and Li X (2008) A dynamic mining algorithm of association rules for alarm correlation in communication networks. *Proceedings of 3rd International Conference on Communication Systems Software and Middleware and Workshops*, Bangalore, India, 6–10 January 2008, pp. 799–802.

Xiaogang P, Zhen J, Zongwei L, Wong EC and Tan CJ (2008) A P2P collaborative RFID data cleaning model. *Proceedings of 3rd International Conference on Grid and Pervasive Computing Workshop*, Kunming, China, 25–28 May 2008, pp. 304–309.

Yarkan S and Arslan H (2007) Statistical wireless channel propagation characteristics in underground mines at 900 MHz. *Proceedings IEEE Military Communications Conference*, Orlando, Florida, 29–31 October 2007, pp. 235–242.

Yarkan S, Guzelgoz S and Arslan H (2008) Wireless channel propagation characteristics in underground mines: a statistical analysis and a radio controlled robot experiment. *Proceedings of IEEE International Conference on Wireless Communications in Underground and Confined Areas*, Val-d'Or, Canada, 25–27 August 2008, pp. 272–279.

Yin ZJ and Chung KS (1998) Multiple access protocol for an underground wireless communication network. *Proceedings of 48th IEEE Conference on Vehicular Technology*, Ottawa, Canada, Vol. 2, 18–21 May 1998, pp. 1528–1532.

Young A (2002) Project designed to reduce risk to miners. *Engineering Dimensions*, **23**(2): 20–21.

Zamel G (1990) A breakthrough in underground communication for enhanced safety and productivity. *Proceedings of Mine safe International, An International Conference on Occupational Health and Safety in the Mineral Industry, Chamber of Mines and Energy of Western Australia*, Inc., Perth, Australia, 1990, pp. 763–766.

Zhang C and Mao Y (2006) Effects of cross section of mine tunnel on the propagation characteristics of UHF radio wave. *Proceedings of 7th International Symposium on Antennas, Propagation & EM Theory*, Guilin, Canada, 26–29 October 2006, pp. 1–5.

Zhang Y (2008) Design and implementation of a wireless sensor network node. *Proceedings of IEEE 4th International Conference on Wireless Communications, Networking and Mobile Computing*, Dalian, China, October 12–14, 2008, pp. 42282–422301.

Zhang YP, Zheng GX and Sheng JH (2001) Radio propagation at 900 MHz in underground coal mines. *IEEE Transaction Antennas Propagation*, **49**: 757–762.

Zhao L and Haimovich AM (2002) Performance of ultra-wideband communications in the presence of interference. *IEEE Journal on Selected Areas in Communications*, **20**(9): 1684–1691.

Zhao YZ and Gan OP (2006) Distributed design of RFID network for large-scale RFID deployment. *Proceedings of IEEE International Conference on Industrial Informatics*, Singapore, 16–18 August 2006, pp. 44–49.

Zheng Z and Hu S (2008) Research challenges involving cross-layered communication protocol design for underground WSNS. *Proceedings of 2nd International Conference on Anti-counterfeiting, Security and Identification*, Guiyang, China, 20–23 August 2008, pp. 120–123.

Websites

http://en.wikipedia.org/wiki/RFID (RFID).

http://en.wikipedia.org/wiki/Walkie-talkie

http://news.bbc.co.uk

http://news.bbc.co.uk (World's tiniest RFID tag unveiled, BBC News, 23 February 2007).

http://origin.cdc.gov/niosh/mining/pubs/pdfs

http://www.arri.uta.edu/acs/networks/WirelessSensorNetChap04.pdf, (Wireless sensor networks by Lewis FL).

http://www.australiancoal.csiro.au/pdfs/mallett.pdf (Mine communication and information for real time risk analysis).

http://www.becker-mining.com

http://www.call-systems.com/dect.php

http://www.cdc.gov/niosh/topics/minerescue/minecomms.html (Focus on coal mining: brief overview of mine communication systems).

http://www.convex.stc.arizona.edu/RCR/RCR_UW_RFID_syncd_files/Borriello-051202 RFID.ppt

http://www.dreamhistory.org/sitebuildercontent/sitebuilderfiles/4_wi-ficomputing.ppt

http://www.eetimes.com (Hara Y "Hitachi advances paper-thin RFID chip", EE Times, 6 February 2006).

http://www.ericssoncs.com/advEric.htm

http://www.fibersystems.com/products/pdf_configurator/Mining/Coal%20Agg%20Prep.pdf (Improved safety, communications, and productivity using fiber optics in coal prep and aggregate processing).

http://www.gartner.com, (Paolo M "A world of smart objects: the role of auto identification technologies", 2001).

http://www.hitachi.com (News release: World's smallest and thinnest 0.15 × 0.15 mm, 7.5 μm thick RFID IC chip, Hitachi Ltd.).

http://www.icwcuca.ca/documents/SerhanYarkanok_000.pdf (Wireless channel propagation characteristics in underground mines: a statistical analysis and a radio controlled robot experiment, by Yarkan S, Güzelgöz S and Arslan H).

http://www.icwcuca.ca/schedule.asp (2nd International conference on wireless communications in underground and confined areas, Val-d'Or, Québec, Canada, August 25–27, 2008).

http://www.intel.com/netcomms/technologies/wimax/303787.pdf

http://www.ipico.co.za/

http://www.minesite.com.au/coal_mines_ped_system

http://www.minesite.com.au/coal_mines_tracker_tagging

http://www.msha.gov/regs/comments/06-722/AB44-COMM-83.pdf (Underground mine rescue equipment and technology: a communications system plan)

http://www.neatware.com/wscn/technology.html

http://www.prabu.files.wordpress.com/2007/08/4sashi_wimaxf.pdf

http://www.rfid-asia.info/2006/12/rfid-tag-attachments.htm (Tedjasaputra A "RFID tag Attachments", RFID Asia).

http://www.rfid-asia.info/2007/02/digestible-rfid-tag-alternative-for.htm (Tedjasaputra A "Digestible RFID tag: an alternative for your internal body monitoring", RFID Asia, 15 February 2007).
http://www.rfidradio.com (Mojix redefines the range for passive RFID, RFID radio).
http://www.sensorsmag.com/sensors
http://www.sm.luth.se/csee/courses/smd/147/pages05/projects/LKAB_project.pdf (Measure- ment based study and evaluation of IEEE 802.11 a/b/g wireless communication network in underground mines, by LeBlanc J and Phanse K)
http://www.urgentcomm.com/mag/radio_moving_front_pack/
http://www.wimax.com/education/wimax/regulatory
http://www.wimax.com/education/wimax/wireless_architectures
http://www.wimaxforum.org/about/
http://www.wvminesafety.org/comtraclibrary.htm
http://zigbee.org

Chapter 2
Evaluation of Suitable Frequency

2.1 Introduction

The radio wave propagation through coal and rock strata suffers from dispersion, absorption, and scattering of electromagnetic waves due to its natural properties and space limitations (Bandyopadhyay et al., 2007; Wait, 1971; Wait and Hill, 1972). The heterogeneous and complex structure of coal and rock strata further complicates the process of radio propagation. Radio frequency (RF) waves get attenuated significantly when traversing through coal strata due to absorption. The attenuation of signal mainly depends upon the dielectric constant and conductivity of coal strata. The dielectric constant of different types of coal available in Indian underground mines is given in Table 2.1. The dielectric constant for coal with 15% moisture content is 4. The conductivity of coal varies from 10^{-8} to 0.02 mho/m depending upon the physicochemical properties of the coal (Singh, 2006).

To establish appropriate radio communication system in underground coal mines, studies of radio propagation inside mine's gallery and through coal and rock strata are of paramount importance (Bandyopadhyay et al., 2005a, b; IC 8955: 1984). Table 2.2 describes the evaluation of frequencies for different communication purposes in India and a comparison has been presented with the international practices. Apart from mine-wide communication, study of radio propagation through coal strata is itself important to establish communication through coal barrier. Although propagation characteristics of radio waves through tunnel have been studied by few researchers in some developed countries (Mariage et al., 1994; Zhang and Hwang, 1998), rigorous studies have not been carried out so far in developing countries such as India for radio wave propagation through coal strata.

Accident due to roof fall and collapse of side gallery is a regular occurrence in coal mines. The radio propagation through coal strata is an important technique to establish communication with the miners, trapped under coal debris. Therefore, the detecting system to locate the trapped miner is a useful device for rescue and relief operation. To meet the intrinsic safety criteria for hazardous zone (Indian Standard, IS 5780: 2002), the power restriction of transceiver to be used in underground mine is 2 W, which further limits the communication range. Therefore, it is important to find out the suitable frequency, which attenuates minimum

L.K. Bandyopadhyay et al., *Wireless Communication in Underground Mines*,
DOI 10.1007/978-0-387-98165-9_2, © Springer Science+Business Media, LLC 2010

Table 2.1 Dielectric constant of different types of coal available in Indian underground mines

Type of coal	Dielectric constant
Anthracite coal	3.2
Bituminous coal	2.8
Coal dust	2.5
Coal with 15% moisture content	4.0

Table 2.2 Comparison of frequencies used in India and other countries

Communication purpose	Frequency used in India	International practiced frequency
Trapped miner communication	457 kHz	457 kHz and 6.0 MHz
Shaft communication	32 kHz	30–60 kHz
Line-of-sight communication	410–500 MHz	400–1000 MHz
Mine-wide communication	146–170 MHz	136–174 MHz

when passing through strata. This will ultimately help in designing appropriate trapped miner locator and other wireless communication devices for underground mines.

An experimental study has been carried out to analyze the radio wave propagation characteristics and to find out the suitable frequency for getting maximum signal strength while passing through coal. The theories of electromagnetic propagation and laboratory experimental procedure along with the results are discussed in subsequent sections.

2.2 Wave Propagation Through Medium

In underground mine, the low-frequency refractive index is predominately real, and is also greater than unity. Suppose some fraction (f_0) of the electrons are free in the sense of having initial frequency $\omega_0 = 0$. In this situation, the low-frequency dielectric constant takes the form (Elliott, 1966; Bagguley, 1973)

$$\varepsilon(\omega) = n^2(\omega) = n_0^2 + iNe^2 f_0 \Big/ \varepsilon_0 m\omega \left(\Gamma^0 - i\omega\right)^2 \tag{2.1}$$

where, n_0 is the contribution to the refractive index from all the other resonance, N the number density of electrons, $\Gamma^0 = \lim_{\omega 0 \to 0} \omega_0 g_0$, where g_0 is the dimensionless damping constant. But for a medium, the contribution to the refractive index from the free electrons is singular at $\omega = 0$. Thus, using the Maxwell's field equation, the dielectric constant is given by

$$\varepsilon(\omega) \equiv n^2(\omega) = n_0^2 + i\frac{\sigma}{\varepsilon_0 \omega} \tag{2.2}$$

A comparison of this term with Eq. (2.1) yields the following expression for conductivity:

$$\sigma = \frac{f_0 N e^2}{m\,(\tau_0 - i\omega)} \tag{2.3}$$

Thus, at low frequencies conductors possess predominately real part of conductivity. However, at higher frequencies the conductivity becomes complex. At these frequencies, there is little meaningful distinction in coal barriers, since the conductivity contribution to $\varepsilon(\omega)$ appear as resonant amplitude just like the other contributions. The conventional way to represent the complex refractive index of a conducting medium (in the low-frequency limit) is to write it in terms of a real normal dielectric constant, ($\varepsilon = n_0^2$) and a real conductivity (σ). Thus, from Eq. (2.2), the following equation is derived:

$$n^2(\omega) = \varepsilon + i\frac{\sigma}{\varepsilon_0 \omega} \tag{2.4}$$

It indicates that the field energy is almost entirely magnetic in nature. It is clear that an electromagnetic wave propagating through a good coal block has markedly different properties to a wave propagating through a conventional dielectric. For a wave propagating in the x-direction, the amplitudes of the electric and magnetic fields attenuate according to the expression $\exp\left(-{}^{x}/_{d}\right)$ where

$$d = \sqrt{\frac{2}{\mu_0 \sigma \omega}} \tag{2.5}$$

and is called the skin depth. Equations (2.4) and (2.5) and associated parameters govern the RF wave propagation through rock and coal strata. These parameters also vary from place to place depending upon the geographical region, geological formation, properties of strata, and local conditions. Therefore, these parameters must be evaluated properly in a particular region for designing effective wireless communication system to be applied in underground mines.

2.3 Laboratory Experiment

2.3.1 Experimental Procedure

The laboratory setup to find the field strength of radio waves passing through a coal block is shown in Fig. 2.1. Different RF signals of fixed amplitude were generated by standard RF generator and fed to the matched directional transmitting loop antenna. The high "Q" matched directional antennae of different RF waves were designed in the laboratory. The strength of input RF signals fed to the antennae at different frequencies is depicted in Fig. 2.2. The transmitting RF signals were

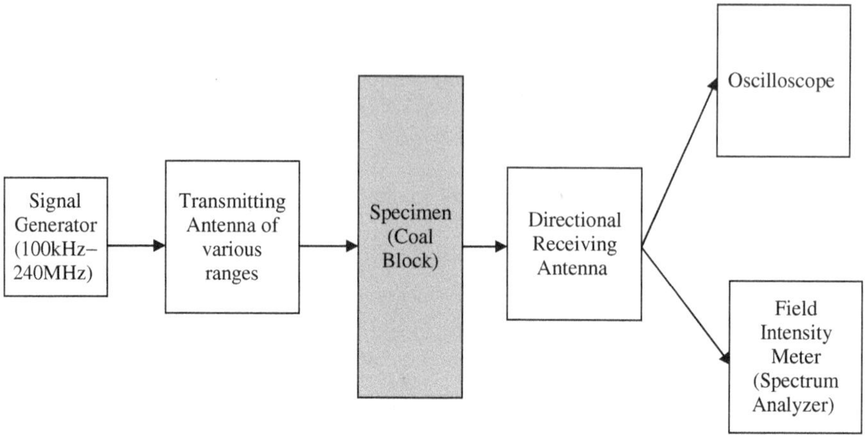

Fig. 2.1 Experimental setup of RF signal propagation

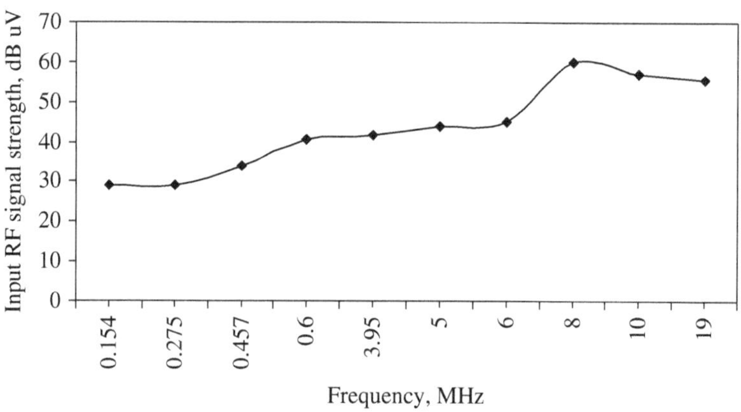

Fig. 2.2 Strength of input RF signals at different frequencies

passed through the smoother side of a coal block (having dimension 1 × 1 × 0.7 m) placed about 15 cm away from the transmitting antenna and the field attenuated waves were received by a compatible receiving antenna placed at the same distance as transmitting antenna, i.e. at around 15 cm from another side of the coal block. It is evident from Fig. 2.2 that the signal strength is high at around 457 kHz. The amplitude of attenuated RF signals and field strength of the attenuated waves were measured by Digital Oscilloscope (Gould Electronics, Model: 1425 M) and Spectrum Analyzer (Anritsu, Model: MS2661C), respectively. The values of attenuation of RF signal with respect to the variation in transmitting frequencies were recorded. The same experiment was repeated for another coal block having the dimension 1 × 1 × 0.5 m.

2.3.2 Analysis of Propagation Signals

From the experimental data, a graph was prepared showing the value of output RF signal strength with respect to the frequency variation for the two different coal blocks (Fig. 2.3). The graph represents a polynomial function. Thus, the following methodology was adopted to evaluate the best suitable frequency, which generates maximum signal strength while passing through the coal blocks.

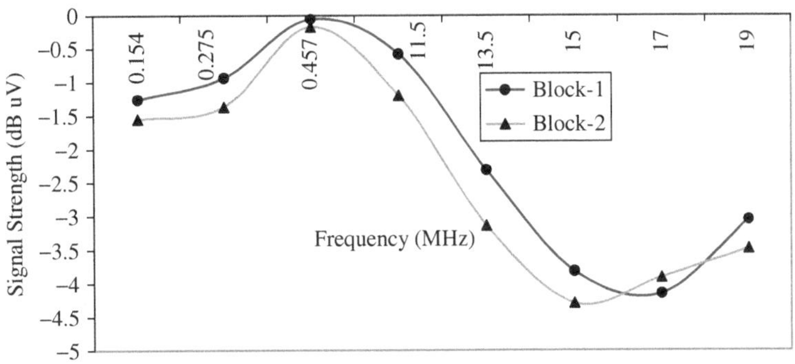

Fig. 2.3 Plot of output RF signal strength versus transmitting frequency

Let $f(x) = ax^3 + bx^2 + cx + d$, be a polynomial function, which rightly fits into the experimental data of change in signal strength (dB μV) with respect to variation of RFs (MHz). The negative sign of $f(x)$ indicates the loss in signal strength and the positive sign indicates the amplification in the signal transmitted through the coal blocks. The suitable frequency for getting optimized signal may be mathematically estimated using the maxima and minima method as explained below.

The solutions of the equation $f'(x) = 3ax^2 + 2bx + c = 0$ are

$$x_1 = \frac{-b + \sqrt{b^2 - 3ac}}{3a} \tag{2.6}$$

and

$$x_2 = \frac{-b - \sqrt{b^2 - 3ac}}{3a} \tag{2.7}$$

Now, $f''(x_1) = 2\sqrt{b^2 - 3ac} > 0$, i.e., f(x) is minimum at $x = x_1$ and $f''(x_2) = -2\sqrt{b^2 - 3ac} < 0$, i.e., f(x) is maximum at $x = x_2$.

Therefore, the recommended frequency for both transmitting and receiving the maximum signal passing through the coal barrier using transceiver is

$$x = \left(-b - \sqrt{b^2 - 3ac}\right) \Big/ 3a \quad \text{MHz} \tag{2.8}$$

Based on the statistical analysis of data using Statistica Software (SPSS 15.0), it was found that the graph as shown in Fig. 2.3 was best fitted at $x = 457$ kHz for both the coal blocks. Thus, it may be concluded that the best suitable frequency for getting maximum signal strength through coal blocks is around 457 kHz. It is also evident from Fig. 2.3 that the signal strength is maximum at around 457 kHz.

2.4 Results and Discussion

The experiment was carried out in the frequency range 100 kHz–20 MHz. Fascinatingly, it was found that there was appreciable high signal strength at 457 kHz. The frequency level, at which the absorption is critically low, the signal strength decreases with increase in frequency due to respective low gain of antennae and high attenuation of RF waves (Fig. 2.4). Based on the experimental results, it was found that 457 kHz frequency is the best suitable frequency for generating maximum signal strength while transmitting through coal strata. Therefore, the detecting system for underground mine worker should be designed at 457 kHz frequency.

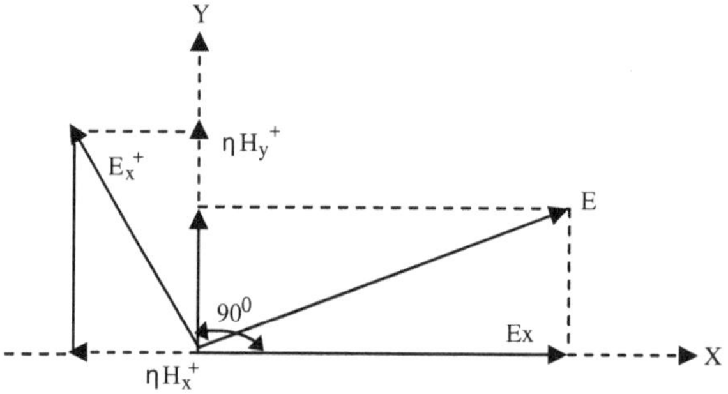

Fig. 2.4 Radio wave traveling in a given direction

2.5 Conclusions

Propagation of electromagnetic wave through strata is a complex phenomenon. The same can be analyzed using modeling technique and experimental studies. Based on the experimental results, it was found that 457 kHz frequency is the best suitable frequency for generating maximum signal strength while transmitting through coal strata. Therefore, the wireless communication system, which will be developed for communication through coal strata, should be designed at 457 kHz frequency. The polarization in coal molecules at this particular frequency may be one of the reasons for amplified signal strength. Therefore, a deliberative study on polarization mechanism of coal molecules with varying frequency is warranted to establish the relationship between the signal strength and polarization in coal molecules.

References

Bagguley DMS (1973) *Electromagnetism and Linear Circuits*. Van Nostrand Reinhold Company, London.

Bandyopadhyay LK, Mishra PK, Kumar S and Narayan A (2005a) Radio frequency communication system in underground mines. *Proceedings of International Seminar on 28th General Assembly of International Union of Radio Science*, Vigyan Bhawan, New Delhi, India, 23–29 October 2005.

Bandyopadhyay LK, Kumar S, Mishra, PK, Narayan A and Sinha M K (2005b) Studies on wireless communication systems for underground coal mines. *Proceedings of International Seminar on Coal Science and Technology – Emerging Global Dimensions* (eds. Singh AK and Sen K) Allied Publishers, New Delhi, India, pp. 56–64.

Bandyopadhyay LK, Mishra PK, Kumar S, Selvendran D and Chaulya SK (2007) Studies on radio frequency propagation characteristics for underground coalmine communications. *Indian Journal of Radio and Space Physics*, **36**: 418–422.

Elliott RS (1966) *Electromagnetics*. McGraw-Hill International Book Company, New Delhi, India.

IC 8955 (1984) *Underground Mine Communications, Control and Monitoring*. Bureau of Mines Information Circular, USA.

IS 5780 (2002) *Electrical Apparatus for Explosive Gas Atmospheres – Intrinsic Safety Specification*. Bureau of Indian Standard, New Delhi, India.

Mariage Ph, Lienard M and Degauque P (1994) Theoretical and experimental approach of the propagation of light frequency waves in road tunnels. *IEEE Transaction Antennas and Propagation*, **42**: 75–81.

Singh KKK (2006) Using resistivity to delineate surface areas at risk of subsidence over Kamptee Colliery, Maharastra, India. *The Leading Edge*, **25**: 1066–1072.

Wait JR (1971) Influence of earth curvature on the subsurface electromagnetic fields of a line source. *Electronics Letters*, **7**(23): 697–699.

Wait JR and Hill DA (1972) Transient magnetic fields produced by a step-function-excited loop buried in the earth. *Electronics Letters*, **8**(11): 294–295.

Zhang YP and Hwang Y (1998) Theory of radio wave propagation in railway tunnels. *IEEE Transaction Vehicle Technology*, **47**: 1027–1032.

Chapter 3
Trapped Miner Communication

3.1 Introduction

Accident due to roof fall and collapse of side gallery is a regular occurrence in underground coal mines which causes death of huge number of miners. To save valuable miners' life, a suitable system is required for detecting the precise location of a trapped miner and helping the rescue team or mine management in displacing the debris from the right place and at the right time without wasting the valuable time after disaster (Bandyopadhyay et al., 2002). Radio propagation through coal strata is important to establish communications with trapped miner beneath coal block/chunk (Durkin, 1984). Therefore, the gadget for detecting of underground mine worker (transceivers) which locates him beneath coal chunk is an important device for speedy rescue and relief operation. The power restriction for transceiver is up to 2 W so that the intrinsic safety (IS) criteria for underground gassy coal mines is met without using flameproof (FLP) enclosure. This power restriction further limits the communication range in underground mines. Therefore, transceivers should be designed with suitable frequency, which attenuates less when passing through coal block (Balanis, 1973; Reagor, 1997). Based on the findings of the experimental results, as described in Chapter 2, a frequency of 457 kHz is selected for designing the transceiver to locate the trapped miner. The developed system is named as "Detecting System for Underground Mine Worker."

3.2 System Description

The detecting system for underground mine worker is meant for detection of those workers who may get trapped under coal debris in case of roof collapse in an underground mine. The system works on the principle of wireless transmission of radio frequency (RF) energy (Large et al., 1973; Hill and Wait, 1974; Kennedy and Foster, 2006). There are two different units, which work in conjunction with each other, namely (i) transmitter unit attached with mine workers and (ii) receiver unit carried by rescue team.

L.K. Bandyopadhyay et al., *Wireless Communication in Underground Mines*,
DOI 10.1007/978-0-387-98165-9_3, © Springer Science+Business Media, LLC 2010

The transmitter unit is to be incorporated with cap-lamp batteries and will be carried by each miner. The transmitter associated with the cap-lamp battery or independent battery emits a low-frequency (LF) radio wave of 457 kHz through a bidirectional coil antenna, which is modulated by a fixed audio tone of 2 kHz frequency. This LF radio wave is capable of transmitting through rock strata and coal debris.

The receiver unit, which will be carried by the rescue team members, is designed to catch the said frequency and amplify it several times before input to the final alarm circuit. As soon as the receiver unit gets a signal, the rescue team has to move/turn the receiving antenna in the direction of the maximum signal strength. In this way the rescue member will try to move toward the direction where the loudness keeps increasing and then remove the debris in the particular place only to save time during rescue operation. A functional diagram of the system is shown in Fig. 3.1.

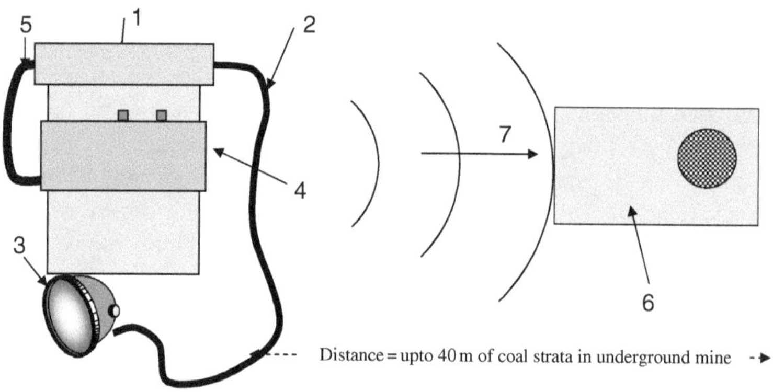

LEGEND:
(1) Cap-lamp battery (4V), (2) Connecting cable for powering the Cap-lamp, (3) Cap-Lamp, (4) Transmitter Unit attached with battery assembly, (5) Connecting cable for Transmitter (for powering the unit and switching of cap-lamp), (6) Receiver unit with in built battery for power, and (7) Direction of RF propagation.

Fig. 3.1 Functional diagram of detecting system for underground mine worker

3.3 Transmitter Unit

The transmitter unit consists of the following subsystems as depicted in Fig. 3.2: (i) RF section, (ii) mixer stage, (iii) modulated signal amplifier, (iv) final-stage amplifier, (v) coil antenna, (vi) cap-lamp switching section, and (vii) DC–DC converter.

The detailed circuit diagram of the transmitter unit is shown in Fig. 3.3. The circuit is fabricated on a double side point to hole (PTH) printed circuit board (PCB).

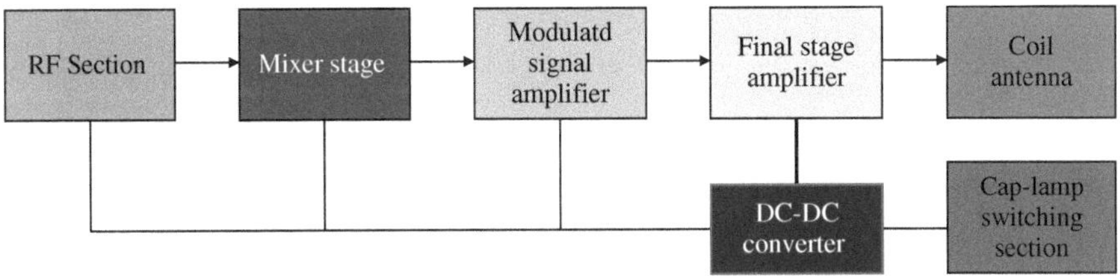

Fig. 3.2 Block diagram of transmitter unit

Top layer of PCB layout is shown in Fig. 3.4 and bottom layer in Fig. 3.5. The component layout is illustrated in Fig. 3.6. The PCB layout along with component mounting is depicted in Fig. 3.7.

The RF section consists of the crystal oscillator, buffer amplifier, mixer, and voltage amplifier. The crystal oscillator is a Colpitt crystal oscillator which provides output of 457 kHz carrier frequency. This RF wave is fed to the input of buffer amplifier and provides a low-impedance output. The output from this section is coupled to the next section which cuts the negative peaks and amplifies positive peaks only. In this way, this section provides almost digital-type output RF of 4 Vpp. This signal is further conditioned with the help of a Schmitt trigger that gives a pure digital signal of 457 kHz frequency.

The mixer stage 1 consists of an Astable multivibrator producing a tone of 2 kHz frequency and a NAND gate. The output of the RF section is mixed with the output of the Astable multivibrator with the help of NAND gate. The output from NAND gate is a pure digital signal of 457 kHz getting digitally modulated from the tone of 2 kHz. This signal is further switched ON and OFF with the help of another Astable multivibrator at a frequency of approximately 1 Hz. In this way, output from the mixer stage becomes a digitally modulated interrupted continuous wave.

The current output from the mixer stage is first amplified using NAND gates and finally fed to a complementary symmetry power amplifier. The output from this amplifier is tapped from collectors of the amplifying transistors and the DC is decoupled with the help of DC-decoupling capacitor. The final output from DC-decoupling capacitor is connected with the matched antenna coil, which radiates the tone-modulated RF power. The antenna coil is a ferrite core antenna tuned to transmit a single frequency of 457 kHz.

The power is fed to the transmitter unit (Tx) via a DC–DC converter, which transforms the 4 V DC input from the cap-lamp to a regulated 12 V DC power. This 12 V is further step-downed through LM7805 regulator to 5 V and is fed to all the sections except the final power amplifier section. The final power amplifier circuit is powered directly from the 12 V line. The housing of the complete Tx unit is done inside a polycarbonate thermoplastic box covered with leather jacket, which does not affect performance of the unit.

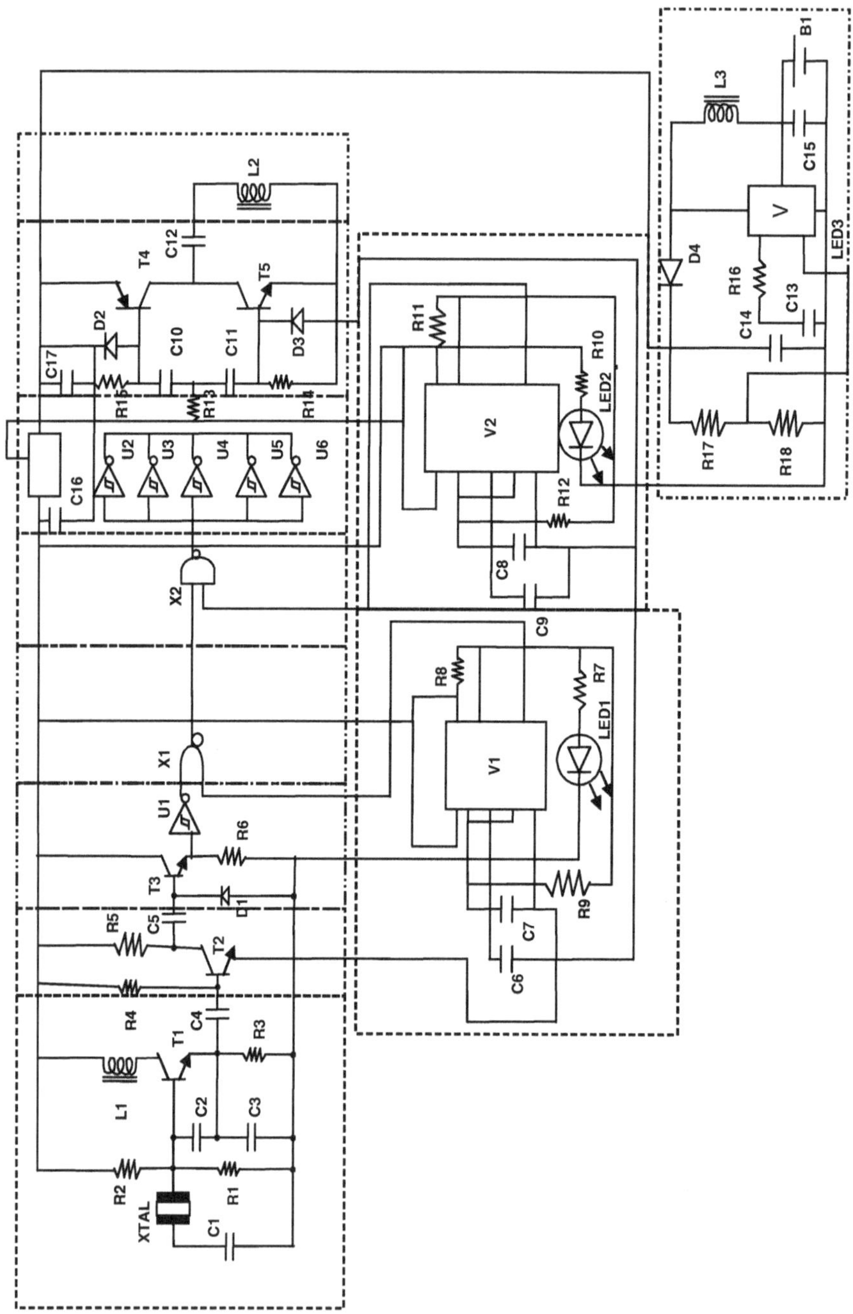

Fig. 3.3 Circuit diagram of transmitter unit

Fig. 3.4 Top layer of PCB for transmitter unit

Fig. 3.5 Bottom layer of PCB for transmitter unit

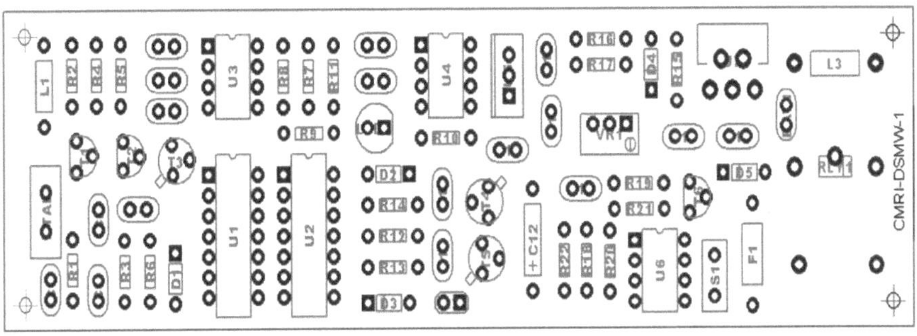

Fig. 3.6 Component layout of PCB for transmitter unit

3.4 Receiver Unit

The receiver unit consists of the following subsystems as depicted in Fig. 3.8: (i) tuned receiver antenna coil, (ii) AM receiver and detector IC, (iii) amplifier and alarm section, and (iv) power supply.

The detailed circuit diagram of the receiver unit is shown in Fig. 3.9. The circuit is fabricated as a single-side PCB and the PCB layout is shown in Fig. 3.10. The component layout is illustrated in Fig. 3.11. The PCB layout with component mounting is depicted in Fig. 3.12.

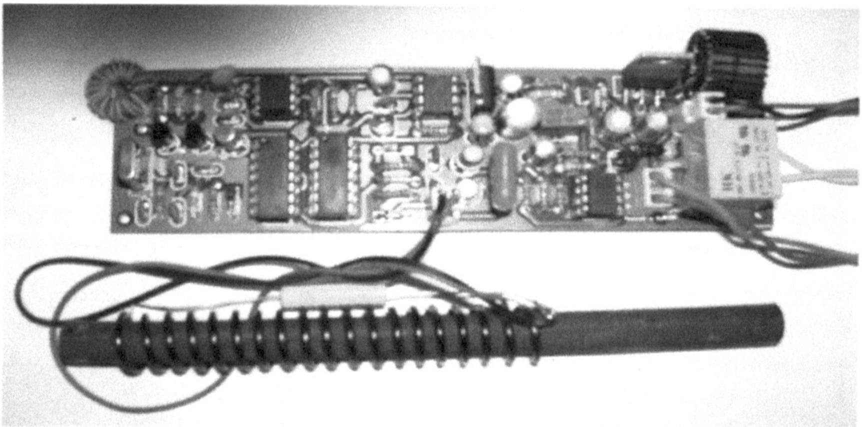

Fig. 3.7 Photograph of mounted PCB for transmitter unit

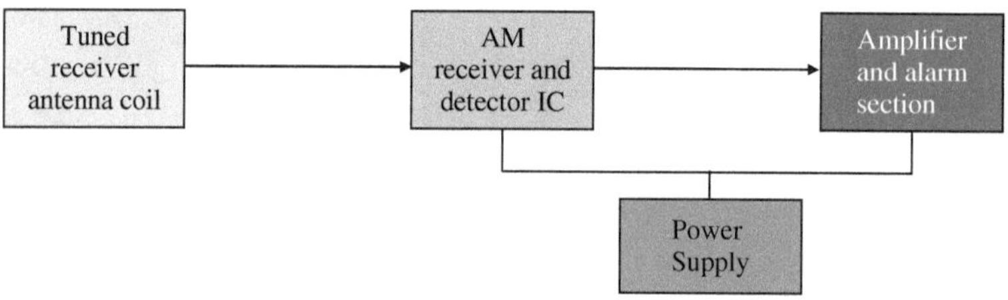

Fig. 3.8 Block diagram of receiver unit

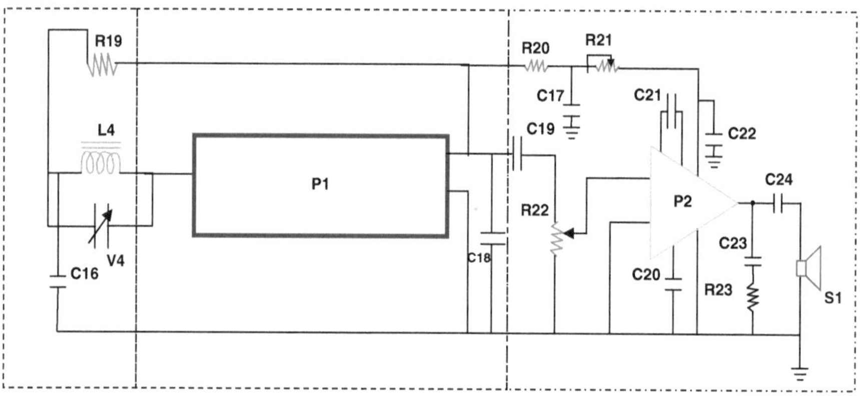

Fig. 3.9 Circuit diagram of receiver unit

The transmitted signal is received at the receiver end via a tuned ferrite core antenna. The antenna is a directional type. Which ensures maximum reception when placed perpendicular to the direction of spreading electromagnetic (EM) waves. The voltage induction across antenna coil turns to minimum when it is

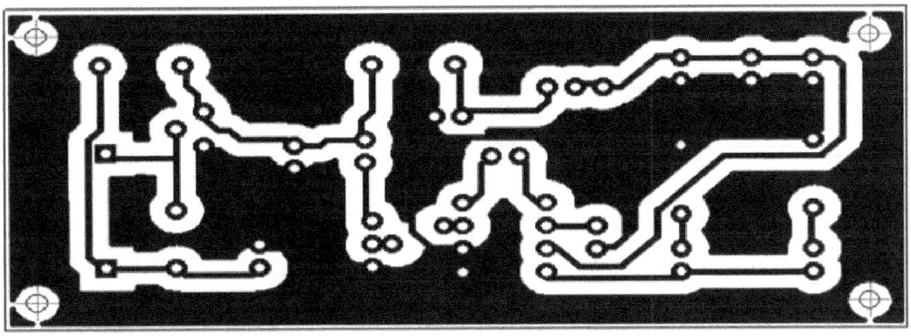

Fig. 3.10 PCB layout for receiver unit

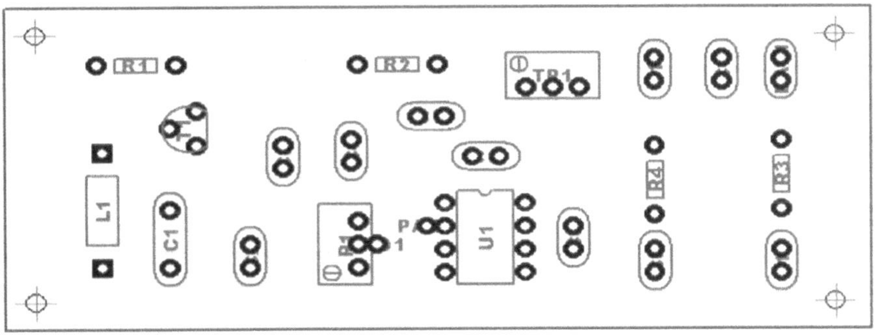

Fig. 3.11 Component layout of receiver unit

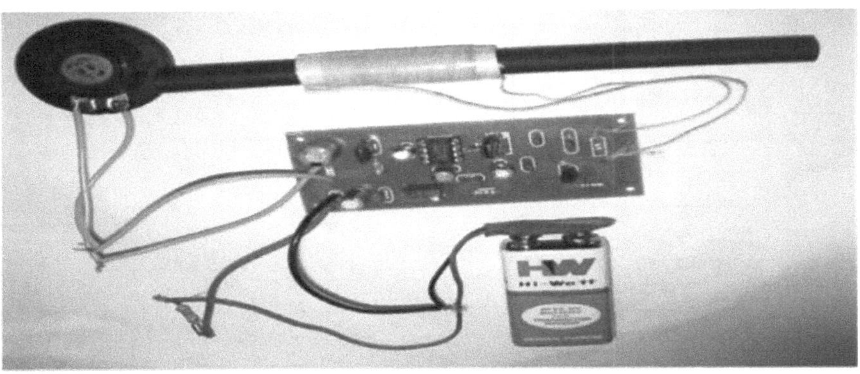

Fig. 3.12 Photograph of receiver unit

parallel to direction of spreading EM waves. The induced voltage is applied to the high-input impedance AM receiver IC that has inbuilt automatic amplification regulation. The signal gets amplified and demodulated in the inbuilt detector and an audio tone signal is obtained at the output of the IC. This signal is further amplified with the help of the audio amplifier and then fed to the alarm-generating speaker.

3.5 Technical Description

3.5.1 Power Supply and Safety Protection

(i) *Transmitter unit*
Power supply requirement for transmitter unit is 4 V/400 mA maximum. The power is fed via a 4 V cap-lamp battery with inbuilt fuse of 1 A (Fig. 3.13). At the input end of the transmitter unit a current-limiting resistance (CLR) R24 of 0.5 ω/5 W, a fast fuse F1 of 500 mA rating, and a diode D5 have been used to provide safety to the transmitter circuit. In case of any internal short circuit or incorrect polarity, the fuse is blown off and the circuit gets disconnected from the battery. Further, the CLR R24 prevents any excess current from the battery. This power first enters the DC–DC converter unit, which boosts the DC voltage from 4 to 12 V. The current requirement by rest of the transmitter circuit is around 200 mA. Since the unit is powered by a battery which is a fixed voltage source and cannot increase in any circumstances, the circuit remains protected from over-current.

(ii) *Receiver unit*

Power supply requirement for receiver unit is 9 V/100 mA max. The power to the receiver unit is fed via a current-limited power supply from a small and portable 6F22 9 V battery (Fig. 3.14). There is a CLR of 10 ω/1 W connected in series to the battery that limits over-current in case of any short circuit within the circuitry.

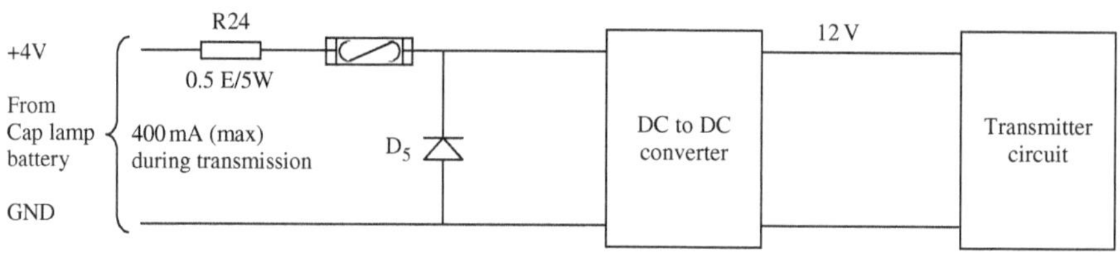

Fig. 3.13 Power coupling arrangement for transmitter unit with cap-lamp battery

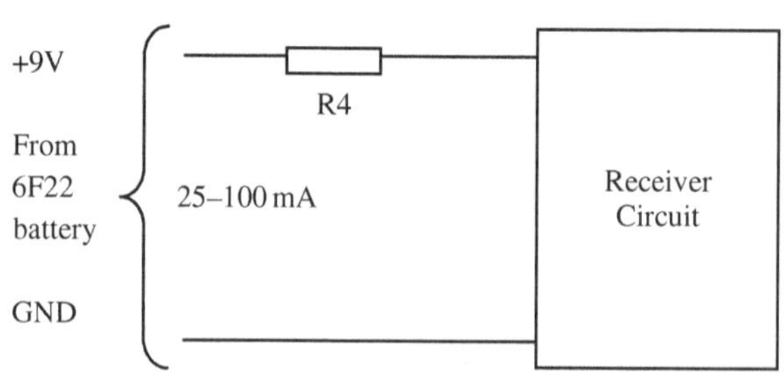

Fig. 3.14 Power supply arrangement for receiver unit

3.5.2 *Technical Specification*

(i) *Transmitter unit*

- Center frequency: 457 kHz
- Maximum field strength: 80 dBuV @ 1 meter
- Operation type: Pulsed RF
- Pulse duration: 300 msec ON/1700 msec OFF
- Modulating frequency: 2 kHz
- Average power consumption: 1.1 W
- Peak power consumption: 1.6 W
- Minimum power consumption: 1 W

(ii) *Receiver unit*

- Tuned frequency: 457 kHz
- Sensitivity: –70 dBm
- Output: 8 Ω/0.25 W
- Maximum power consumption: 1 W
- Minimum power consumption: 0.27 W

3.5.3 *Infallible Components*

(i) *Transmitter unit*

- Fuse F1 500 mA
- Diode D7 1N4007
- Diode D4 1N4148
- Diode D5 1N4148
- Diode D6 1N4148
- Resistor R24 0.5 Ω/5 W

(ii) *Receiver unit*

1. Resistance – R2 – 1 kΩ/¼ W
2. Resistance – R4 – 10 Ω/1 W

3.5.4 *Input and Output Details*

(i) *Input data*

Parameter	Transmitter unit	Receiver unit
U_m	4.2 V	9.2 V
U_i	4 V	9 V
I_i	400 mA	110 mA
P_i	1.6 W	1W
C_i	343 μF	340 μF
L_i	6 mH	300 μH

(ii) *Output data*

Parameter	Transmitter unit	Receiver unit
U_o	2.7 V_{RMS}	350 mV_{RMS}
I_o	1.7 mA_{RMS}	50 mA_{RMS}
P_o	22 dBm in 50 Ω termination of spectrum analyzer	20 mW
C_o	4.7 nF	<4 pF
L_o	27 μH	<1 μH

3.6 Safety Analysis

The safety analysis has been carried out for the components in Tx and Rx units with respect to IS criteria for gas-group I environment (IEC60079-11, 2007).

3.6.1 *Transmitter Unit*

The power source of the Tx unit is a lead acid battery used with the cap lamp and is fitted with a 1 A fuse between the two cells as shown in Fig. 3.13. This voltage does not rise beyond 4.2 V, even when fully charged. Therefore, V_{max} becomes 4.2 V.

The current requirement for the Tx unit is 4 V/400 mA. The permitted short circuit current for gas-group IIA with a safety factor of ×1.5 at 12.1 V is 3.33 A (Table 12.11 of Chapter 12). For group I, the current of 400 mA is a very low value at 4 V and therefore the safety is ensured.

If anywhere in the Tx circuit, the supply gets shorted due to any reason, then the maximum inrush current (I_{in1}) through the battery will be (Ref. Fig. 3.3):

$$\begin{aligned} I_{in1} &= V_{max}/R24 \\ &= 4.2\,V/0.5\Omega = 8.4\,A \end{aligned}$$

However, the current versus voltage graph (Fig. 12.7 of Chapter 12) starts from a minimum voltage of 10 V and even at that voltage the current depicted is very high for gas-group I. If the graph is extrapolated, then the current at 10 V will be even more than 10 A. So, the $I_{\text{in1}} = 8.4$ A at 4.2 V is still lower than 10 A at 10 V. So, this current should be safe for gas-group I.

If this current is allowed to flow continuously, it will drain the battery and hence to prevent any such unnecessary discharge of battery, a fast blow fuse of 500 mA is used in series with the CLR R24 as the preventive measure.

The diode D7 is used for reverse polarity protection. If the battery terminals are connected incorrectly, the diode conducts heavily due to forward bias and the short circuit current in this case becomes

$$I_{\text{in2}} = (4.2 - 0.7)/0.5 = 7\,\text{A}$$

This current is as safe as discussed above for gas-group I.

The maximum current I_{max} that the fuse F1, 500 mA, is assumed to conduct as per the IEC60079-11, 2007 standard, before blowing off:

$$\begin{aligned} I_{\text{max}} &= 1.7 \times I_{\text{n}} \\ &= 1.7 \times 0.5 \\ &= 850\,\text{mA} \end{aligned}$$

This current is much lower than the short circuit currents I_{in1} and I_{in2}, the short circuit currents flow momentarily and the fuse blows off immediately protecting the rest of the circuit.

The power rating of the CLR R24 (as per the IEC60079-11, 2007) should be

$$\begin{aligned} P &= 1.5 \times (1.7 \times I_{\text{n}})^2 \times \text{R24} \\ &= 1.5 \times (1.7 \times 0.5)^2 \times 0.5 \\ &= 0.54\,\text{W} \end{aligned}$$

The used R24 has a rating of 5 W, which is approximately 10 times the above; therefore the power rating of the CLR is safe to use.

The diodes D4, D5, and D6 are high-conductance fast diodes (1N4148) each capable of delivering up to 200 mA of average rectified forward current. Internally, the chip U7 provides boosted voltage of approximately 12 V, and output current 130 mA is consumed by rest of the circuit. This current is approximately 2/3 times less than the rated 200 mA current for the diodes D4, D5, and D6. Therefore, the rating of a single diode with safety factor $\times 1.5$ ($=130$ mA $\times$ 1.5 $=$ 195 mA) is sufficient for delivering the full current required to the circuitry. Three diodes in parallel provide more safety and capable of sustaining multiple faults.

The capacitor C15 is used in parallel to the incoming supply of 4 V for filtering out high-frequency noise generated in the DC–DC converter section. Figure 12.6 of Chapter 12 contains information about minimum igniting voltage across a capacitor for gas-group I. Since the value of capacitance is more than 3,000 μF at 10 V and the capacitor C15 is only 100 μF, therefore its value is very low and can be treated as safe.

The inductor L3 being used in the DC–DC converter section has $L = 5$ mH, $R = 0.1\ \Omega$, and oscillates at $f = 100$ kHz.

The maximum voltage spike across L3 as seen on oscilloscope is $V_{\text{max}} = 60$ V and normal voltage is approximately 13 Vpp. Hence, the maximum current being switched through inductor

$$
\begin{aligned}
&= V_{\text{max}}/\text{Z} \\
&= V_{\text{max}}/\surd(R^2 + \text{X}_\text{L}^2) \\
&= V_{\text{max}}/\surd(R^2 + (2\pi f L)^2) \\
&= 60/\surd((0.01 + (6.28 \times\ 100\text{e} + 3 \times 0.005\,)^2) \\
&= 19.11\text{mA}
\end{aligned}
$$

With safety factor of $\times 1.5$, this current would increase to 28.67 mA, say 30 mA, and hence the maximum energy contained during maximum current spike in L3

$$
\begin{aligned}
E &= 1/2 \times L \times (I)^2 \\
&= 1/2 \times 5 \times 10\text{e} - 3 \times (0.03)^2 \\
&= 2.25\mu\text{J} \ll 525\mu\text{J}
\end{aligned}
$$

Therefore, the maximum energy contained in L3 at any point is far less than the safety limit of 525 μJ for group I and hence is safe.

3.6.2 Receiver Unit

The power source of the Rx unit is a small 6F22 9 V battery and is fitted with a CLR as shown in Fig. 3.14. The maximum voltage of a new battery is 9.2 V. Therefore,

$$V_{\text{max}} = 9.2\text{V}$$

The CLR R4 is 10 Ω/5 W and therefore, the short circuit current is

$$
\begin{aligned}
I_{\text{SC}} &= V_{\text{max}}/\text{R4} \\
&= 9.2/10 = 0.92\ \text{A(say 1 A)}
\end{aligned}
$$

The maximum current requirement for the Rx unit is 9 V/ 110 mA and the short circuit current is 1 A. The permitted short circuit current for gas-group IIA with a safety factor of × 1.5 at 12.1 V is 3.33 A (Table 12.7 of Chapter 12). For gas-group I, the short circuit current of 1 A is a very low value at 9 V and therefore may be regarded as safe.

The power rating of the CLR R4 (as per the IEC60079-11, 2007) should be:

$$
\begin{aligned}
&= 1.5 \times (0.11)^2 \times 10 \\
&= 0.18\ \text{W}
\end{aligned}
$$

The used R4 has a rating of 5 W, which is approx. 25 times the above, therefore, the power rating of the CLR is safe to use.

3.7 Component Details

3.7.1 Ratings of Safety Components

Component designation	Component location	Value	Maximum rating used by circuit (W2)	Commercial rating (W1)	W1/W2
R24	Tx enclosure	0.5 Ω/5 W/ ±10%	0.54 W	5 W	≈10
R4	Rx enclosure	10 Ω/5 W/ ±10%	0.18 W	5 W	≈25
F1	Tx PCB	500 mA	400 mA	500 mA	≈1.25
D4, D5, D6	Tx PCB	1N4148	135 mA	200 mA	≈1.5

3.7.2 List of Components for Transmitter Unit

Category	Component code	Value	Voltage/wattage
Battery	BAT1	4 V Cap-lamp battery	
Capacitors	C1	1 kpF	16 V
	C2	470 pF	16 V
	C3	470 pF	16 V
	C4	220 pF	16 V
	C5	1 kpF	¼ W
	C6	0.01 μF	16 V
	C7	0.1 μF	16 V
	C8	10 μF	25 V
	C9	0.01 μF	16 V
	C10	0.01 μF	16 V
	C11	0.01 μF	16 V
	C13	1 μF	25 V
	C14	220 μF	25 V
	C15	100 μF	25 V
	C16	10 μF	25 V
	C17	10 μF	25 V
	C18	1 μF	25 V
Diodes	D1	1N4148	
	D2	1N4148	
	D3	1N4148	
	D4	1N4148	
	D5	1N4007	
Fuse	F1	500 mA	Fast fuse

Category	Component code	Value	Voltage/wattage
Inductors	L1	180 μH	
	L2	Ferrite core antenna (Core diameter = 10 mm, 15 turns of 18 SWG, widely spaced)	
	L3	6 mH	
Light emitting diodes	LED1	Red color, 5 mm	
Resistors	R1	22 k	¼ W
	R2	22 k	¼ W
	R3	30 k	¼ W
	R4	470 k	¼ W
	R5	1 k	¼ W
	R7	1 k	¼ W
	R8	220 E	¼ W
	R8	3.3 k	¼ W
	R9	330 E	¼ W
	R10	15 k	¼ W
	R12	180 E	¼ W
	R12	100 k	¼ W
	R13	10 k	¼ W
	R14	10 k	¼ W
	R15	1 k	¼ W
	R16	1 k	¼ W
	R17	1.2 k	¼ W
	R18	3.3 k	¼ W
	R19	10 k	¼ W
	R20	220 k	¼ W
	R21	470 E	¼ W
	R22	10 k	¼ W
Relay	RLY1	400 E	12 V
Switch	S1	Push to On switch	
Transistors	T1	BC548	
	T2	BC548	
	T3	BC109	
	T4	2N2907	
	T5	2N2222	
	T6	BC548	
Integrated circuits	U1	74LS14	
	U2	74LS00	
	U3	NE555	
	U4	NE555	
	U5	LM7805	
	U6	LM393	
	U7	LT1170	
Potentiometer	VR1	1 M	
Crystal	XTAL1	457 kHz	

3.7.3 List of Components for Receiver Unit

RefDes	Type	Part	Number of pins	Pattern	Quantity
BAT1	CAP	Part 1	2	CAP100	1
C1	CAP250	Part 1	2	CAP250	1
C2	CAP	Part 1	2	CAP100	1
C3	CAP	Part 1	2	CAP100	1
C4	CAP	Part 1	2	CAP100	1
C5	CAP	Part 1	2	CAP100	1
C6	CAP	Part 1	2	CAP100	1
C7	CAP	Part 1	2	CAP100	1
C8	CAP	Part 1	2	CAP100	1
C9	CAP	Part 1	2	CAP100	1
C10	CAP	Part 1	2	CAP100	1
C11	CAP	Part 1	2	CAP100	1
L1	INDI	Part 1	2	IND500	1
P1	POT	Part 1	3	POT3T	1
R1	RES	Part 1	2	RES300	1
R2	RES	Part 1	2	RES300	1
R3	RES	Part 1	2	RES300	1
R4	RES	Part 1	2	RES300	1
TP1	POT	Part 1	3	POT3T	1
U1	LM386N-1	Part 1	8	DIP-8	1
U2	ZN414	Part 1	3	BCY-3/D5.2	1

3.8 Capabilities of the System

The main capabilities of the detecting system are as follows:

(i) The system detects the precise location of an underground mine worker trapped in case of roof fall/collapse of gallery side.
(ii) The transmitter unit of the system is incorporated with the miner's cap-lamp battery and powered by the same battery.
(iii) The transmitter unit is capable of working in case the cap lamp automatically switches off due to low battery, which enables the rescue team to detect the trapped miner for longer duration.
(iv) The system is intrinsically safe and can be used in underground mines.
(v) The system is very light and low cost and can be provided to each underground mine worker.
(vi) The system is capable in helping the rescue team or mine management to identify the coal chunk/coal debris to be displaced at right time to save the valuable human lives after knowing the precise location of the worker trapped underneath the coal chunk/coal debris.

3.9 Field Trial

The laboratory-scale detecting system was designed and fabricated and tested in the laboratory. Subsequently, the prototype system was developed (Fig. 3.15). The prototype system was tested in the laboratory with the simulated conditions. Then the technical documents were prepared containing detailed circuit diagram, PCB layouts, safety analysis, specifications, and technical descriptions. Then the developed prototype system was forwarded to an authorized testing institute, Electronics Regional Testing Laboratory (ERTL), Kolkata, India, along with a bare PTH-PCB, an assembled PCB, and technical document for its IS criteria fulfillment test. After receiving the ERTL IS testing certificate, application for field trial permission was forwarded to the Directorate General of Mines Safety (DGMS), Ministry of Labour, Government of India, Dhanbad, along with the ERTL test document. Simultaneously, an application was also submitted to the Ministry of Communication and Information Technology (MCIT), Government of India, New Delhi, for getting permission to use the required frequency band at the mines of Mahanadi Coalfields Limited (MCL) and Bharat Coking Coal Limited (BCCL), India. On receipt of field trial permission from DGMS and license from MCIT, the developed system was tested in the underground mines.

A prototype of the detecting system was tested at Nandira Colliery, Talcher Area of MCL, Sambalpur, Orrisa, India. The transmitter unit was kept beneath a chunk of fallen coal block of about 5 m dimensions. The signal could be detected by receiver at 40 m distance away from the coal chunk. Further, the transmitter unit was put inside a heap of coal dust having 10 m radius. The signal could be detected by receiver around 35 m away from the coal dust heap. Further, the cap lamp and transceiver were operated together for 9 hours, and then the cap-lamp light was

Fig. 3.15 View of receiver and transmitter of trapped miner detecting system

switched off and subsequently the transmitter was operated for more than 60 hours using the same cap-lamp battery.

Another experiment was performed with the prototype system in Bagdiggi Mine of BCCL in the 12th pit of the mine. The transmitter and receiver units were kept at two levels having parting thickness of around 25 m and the signal was heard by the receiver up to an angular distance of around 35 m from the transmitter. The transmitter was kept below the broken coal in a working face and the signal could be detected by the receiver up to a distance of 40 m from the transmitter.

The detecting system essentially enabled to precisely locate the position of the transmitter placed under the debris, coal block, broken coal, coal dust, and at different levels. The system enabled to transmit and receive RF signal through rock strata or coal debris or coal chunk in the underground mines. The system enabled to operate for longer time with the same cap-lamp battery switched off due to low battery power. The prediction of precise location by the system also helped the mine management to enhance safety and reduce the fear of the underground mine workers. The provision of use of the present system carried by the underground mine worker in any place of underground mine enabled mine management to locate the worker in any conditions.

3.10 Conclusions

The detecting system of underground mine worker enables in detecting the location of trapped mine worker based on the radio wave transmission through rock strata or coal debris. Use of the system would help in precisely locating the position of the mine worker trapped in case of roof fall/collapse of gallery side. This would help the rescue team or mine management to identify the coal chunk/coal debris to be displaced at the right time and right place. This would also help in saving valuable life of men working in underground mines.

References

Balanis C (1973) Line sources above lossy media for low-frequency underground communication. *IEEE Transactions on Antennas and Propagation*, **21**(1): 127–128.

Bandyopadhyay LK, Chaulya SK and Kumar S (2002) A proposed wireless communication system for underground mines. *Proceedings of the International Conference on Mineral Industry: Issues on Economics, Environment and Technology*, Mining, Geological and Metallurgical Institute of India, Kolkata, India, pp. 317–324.

Durkin J (1984) Electromagnetic detection of trapped miners. *IEEE Communication Magazine*, **22**: 37–46.

Hill D and Wait J (1974) Diffusion of electromagnetic pulses into the earth from a line source. *IEEE Transactions on Antennas and Propagation*, **22**(1): 145–146.

Kennedy GA and Foster PJ (2006) High resilience networks and microwave propagation in underground mines. *Proceedings of 9th European Conference on Wireless Technology*, Manchester, UK, 10-12 September 2006, pp. 193–196.

Large D, Ball L and Farstad A (1973) Radio transmission to and from underground coal mines – theory and measurement. *IEEE Transactions on Communications*, **21**(3): 194–202.

Reagor D, Fan Y, Mombourquette C, Jia Q and Stolarczyk L (1997) A high-temperature superconducting receiver for low-frequency radio waves. *IEEE Transaction Application Superconductor*, **7**: 3845–3849.

Chapter 4
Shaft Communication

4.1 Introduction

Bell signaling system is commonly being used in underground mines. But this system has its own drawbacks. Therefore, a need exists for improved hoist communications between the skip and the hoist operator. An induction-based communication system using hoist/guide rope as a current carrier can be used for reliable and real-time communication in the shaft. The system transmits and receives energy over a transmission line through hoist/guide rope. Low-frequency transceivers are to be used as transmitting and receiving media. Ferrite current coupler can be used as antenna for the communication. An induction-based hoist communication system has been developed for establishing communication among the mine personnel present at pit top, in moving cage, and at pit bottom. The system consists of a low-frequency electromagnetic wave transceivers unit powered by the external battery and attached with a current coupler clamped with the guide rope, which induces the current throughout the rope (Bandyopadhyay et al., 2002).

4.2 Theory

4.2.1 Inductance

Theory of inductance can be explained by representing two loops of wire at rest as shown in Fig. 4.1.

If we run a steady current I_1 around loop 1, it will produce a magnetic field B_1. Some of the field lines pass through loop 2; let ϕ_2 be the flux of B_1 through loop 2. Unless the shape of loop 1 is particularly simple, calculating B_1 may be difficult, but a glance at the Biot–Savart law (Griffiths, 1995; Bueno and Assis, 1997)

$$B_1 = \frac{\mu_0}{4\pi} I_1 \oint \frac{dl_1 \times \hat{r}}{r^2} \tag{4.1}$$

L.K. Bandyopadhyay et al., *Wireless Communication in Underground Mines*,
DOI 10.1007/978-0-387-98165-9_4, © Springer Science+Business Media, LLC 2010

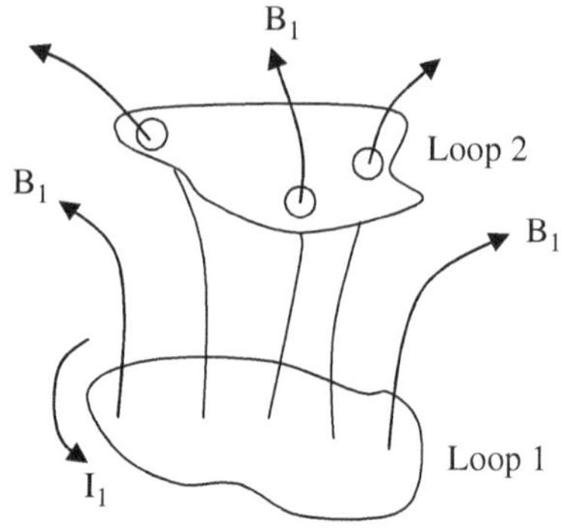

Fig. 4.1 Inductance for two loops of wire at rest

reveals one significant fact about this field: it is proportional to the current I_1, where μ_0 is the permeability of free space and r the radius of Amperian loop. The flux through loop 2 also follows the same principle, since

$$\phi_2 = \int B_1 . da_2 \tag{4.2}$$

Thus

$$\phi_2 = M_{21} I_1 \tag{4.3}$$

where M_{21} is the constant of proportionality; it is known as the mutual inductance of the two loops. Although the mutual inductance is difficult to calculate in most practical cases, there is a simple formula for calculating it by expressing the flux in terms of the vector potential:

$$\phi_2 = \int B_1 \cdot da_2 = \int (\nabla \times A_1) \cdot da_2 = \oint A_1 \cdot dl_2 \tag{4.4}$$

Now,

$$A_1 = \frac{\mu_0 I_1}{4\pi} \oint \frac{dl_1}{r} \tag{4.5}$$

and hence

$$\phi_2 = \frac{\mu_0 I_1}{4\pi} \oint \left(\oint \frac{dl_1}{r} \right) \cdot dl_2 \tag{4.6}$$

Therefore, comparing (4.3)

$$M_{21} = \frac{\mu_0}{4\pi} \int \int \frac{dl_1 \cdot dl_2}{r} \tag{4.7}$$

This is known as Neumann formula (Griffiths, 1995); it involves a double integral, one integration around loop 1 and the other around loop 2. It illuminates two important things about mutual inductance:

i. M_{21} is a purely geometrical quantity, which depends on sizes, shapes, and relative positions of the two loops.
ii. The integral in Eq. (4.7) is unchanged if we switch the roles of loops 1 and 2; evidently

$$M_{21} = M_{12} \tag{4.8}$$

i.e., whatever the shapes and positions of the loops, the flux through loop 2 when we run a current I around loop 1 is exactly the same as the flux through loop 1 when we send the same current I around loop 2.

4.2.2 Current Clamp

A current clamp or current probe is an electrical device having two jaws that open to allow clamping around an electrical conductor. This allows the electrical current in the conductor to be measured, without making physical contact with it, or to disconnect it for insertion through the probe. Some types of current clamp are used to induce current in the conductor (http://en.wikipedia.org/wiki/current_clamp).

4.2.3 Electromagnetic Coupling

Electromagnetic fields can result in currents that are coupled into other wires (Butler et al., 1978; www.wvminesafety.org; www.camese.org). But the electromagnetic field must be changing for this coupling to happen. Changing electromagnetic fields are caused by changing currents. Thus, DC currents cannot couple signals into other wires. Only AC currents can do that. Further, coupling can be done in two ways, namely electric and magnetic couplings.

4.2.3.1 Electric Coupling

Similar charges repel each other. Therefore, if a charge density exists at a point along a wire, it will tend to repel like charges away from that point in adjacent wires. Those elements of charge that are repelled away are moving. By definition this is a current. So, a changing electric field in one wire (as charge density changes) causes a changing current in another adjacent wire. This effect is often referred to as electric coupling, charge coupling, or capacitive coupling.

4.2.3.2 Magnetic Coupling

If a current flowing along a wire causes a magnetic field around the wire, then a changing current in a wire causes a changing magnetic field around the wire. Michael Faraday (Faraday's Law) showed that a changing magnetic field causes an electric field that is perpendicular to the magnetic field (Griffiths, 1995). This electric field can cause a current to flow in an adjacent wire. This is the fundamental principle behind a transformer; a changing current in the primary winding causes a changing magnetic field and thus a current to flow in the secondary winding. This effect is often referred to as magnetic coupling or inductive coupling.

4.2.4 Consequences

There is nothing inherently good or bad about electromagnetic coupling. Whether the consequences of this coupling are beneficial or not depends entirely on the circumstances. There are many positive effects. Electromagnetic coupling is the primary principle behind a transformer. The magnetic aspect of electromagnetic coupling provides the principle behind a motor or a relay. Faraday's Law is the fundamental principle behind an electrical generator. Transmitters and receivers (radio, television, CB radio, automotive wireless door locks, etc.) work by electromagnetic radiation coupling into a receiving antenna on a specific frequency.

But electromagnetic coupling is also at the heart of almost all the negative signal integrity issues that board designers face. Therefore, it is important to understand how and where electromagnetic coupling manifests itself in our circuits. For example, electromagnetic coupling results in a coupled current in another wire or trace. This is a bad thing if we are coupling at an undesired frequency into adjacent traces on our boards. We call it crosstalk. We commonly call the effect of this type of coupling electromagnetic interference (EMI). But in the specific case of differential signals, the coupling into an adjacent trace is beneficial (www.mentor.com/pcb/tech_papers.cfm). Due to this reason, we route differential signals close to each other to maximize this coupling effect (Douglas, 2003).

When a signal trace is routed over a plane, the return signal "wants" to be on the plane directly under the trace. Although this is an example of a single-ended trace, in a sense it is a differential pair (involving differential coupling) because the return signal (equal to and opposite from the driven signal) automatically travels as close as possible to the signal itself. The reason is that this path is the lowest impedance path. The core phenomenon here is the electromagnetic coupling between the trace and the plane.

A concept that is difficult for many people to understand is that if a signal on one wire can electromagnetically couple into another wire, then it can also electromagnetically couple into itself. This creates an induced (coupled) signal on the wire that opposes the driving signal that creates it in the first place. It may take a brief period of time for the driving signal to overcome this coupled signal. We describe this effect as inductance. The core cause of inductance in a wire or trace is the magnetic field created by the driving signal that induces a current in the same wire or trace in the opposite direction.

All wires and traces have some inductance, so this is not necessarily a bad thing. But the effect of inductance (a noise voltage) is typically a function of rise time (*di/dt*), i.e., how fast the signal changes in a given increment of time. This is the cause behind ground bounce on our circuit boards when we use fast rise-time devices, a problem we try to correct with bypass capacitors.

4.3 Principle

Basic principle of hoist communication scheme is electromagnetic coupling (Leviatan, 1988; Manning, 1998; Nowakowski, 1998; Kováčová and Kováč, 2007; Xie et al., 2007; Kováč and Kováčová, 2008). A sensitive current clamp couples voice-modulated RF signal to and from the guide rope. Because the hoist rope is terminated with electrical short circuit, the signal will form standing waves along the hoist rope. Null occurs at every quarter wavelength apart. Therefore, frequency is chosen, 32 kHz for present scheme, so that length of the rope is less than quarter wavelength ($\lambda = 9375$ m) at signal frequency.

Condenser microphone converts audio signal into an electrical signal. This audio frequency (AF) electrical signal is preamplified and fed to the base of transistor modulator. A Colpitt oscillator generates 32 kHz weak signal, which is also fed to the base of the modulator. Modulator section also provides needed gain of 40 dB. Because the current coupler offers high reactive impedance to the amplifier, a buffer amplifier is used to match the modulator output to the sensor. Current coupler will encircle guide rope to form a transformer where guide rope acts as single-turn secondary winding of the transformer. Therefore, 32 kHz modulated signal will be induced in the guide rope. To ensure flow of current through the rope, both ends of the rope must be electrically continuous. Coupling clamp in receiving side will interact with the magnetic field created by the current induced in the rope due to the current in clamp at transmitting side. Here guide rope will act as single-turn primary and coil wound on the ferrite core will act as secondary of the transformer. Due to the current in the rope, a proportionate voltage will develop at output terminal of the coupler. This signal voltage is the input to the preamplifier of the receiver. Because the signal is an AF-modulated electrical signal, AF is detected and goes through a power amplification stage prior to feeding it to the speaker.

4.4 System Description

The system consists of two modules, a current coupler magnetically coupled with the guide rope and a transceiver of 32 kHz that transmits and receives the RF signal of 32 kHz. Transceiver connected with current coupler is placed at pit top, in moving cage, and at pit bottom.

Figure 4.2 depicts the block diagram of transmitting unit of the transceivers. The transmitting unit consists of a Colpitt oscillator of 32 kHz, which is subsequently connected one after another to modulator, tuned amplifier coupled with transformer, class A voltage amplifier, buffer amplifier, and current coupler. The modulator is connected with the AF amplifier and microphone.

Figure 4.3 shows the block diagram of receiving unit of the transceivers. The receiving unit consists of current coupler, which is connected sequentially through a RF amplifier, a RF-tuned amplifier, RF amplifier, diode detector, AF voltage amplifier, AF power amplifier, and finally speaker.

Figure 4.4 illustrates the circuit diagram of transmitting unit of the transceivers. A Colpitt oscillator generates 32 kHz signal, which is fed to the base of the modulator. An AF electrical signal is also preamplified and fed to the base of the said modulator. Modulator section modulates both of the signals and provides the needed gain of 40 dB. Since the current coupler offers high reactive impedance to the amplifier, a buffer amplifier is used to match the modulator output to the coupler. Current coupler is encircled around the guide rope to form a transformer where guide rope acts as single-turn secondary winding of the transformer and 32 kHz modulated signal is induced in the guide rope.

Figure 4.5 illustrates the circuit diagram of receiving unit of the transceivers. Current coupler in receiving side interacts with the magnetic field created by the current induced in the said guide rope due to current coupler at transmitting side.

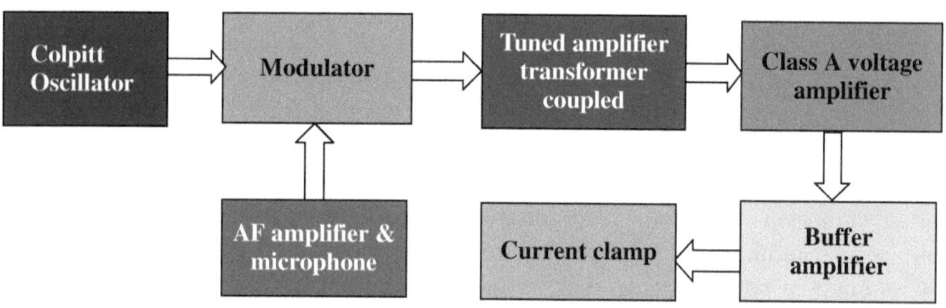

Fig. 4.2 Block diagram of the transmitting unit

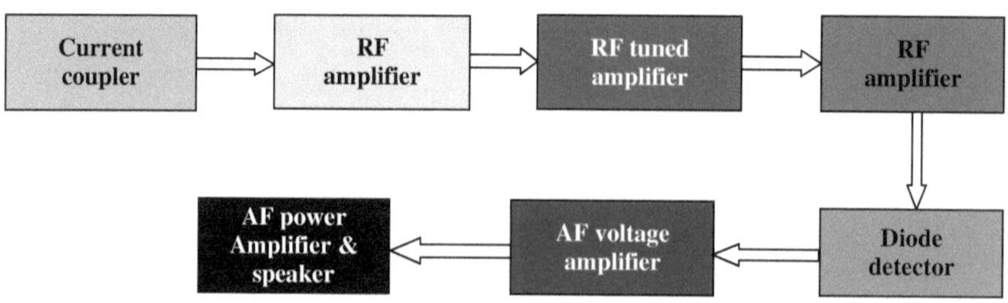

Fig. 4.3 Block diagram of receiving unit

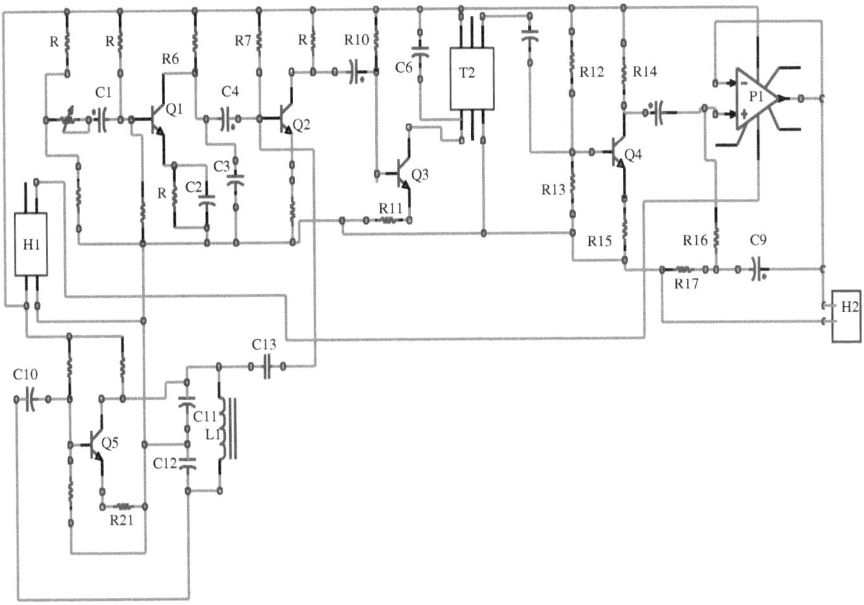

Fig. 4.4 Circuit diagram of transmitting unit

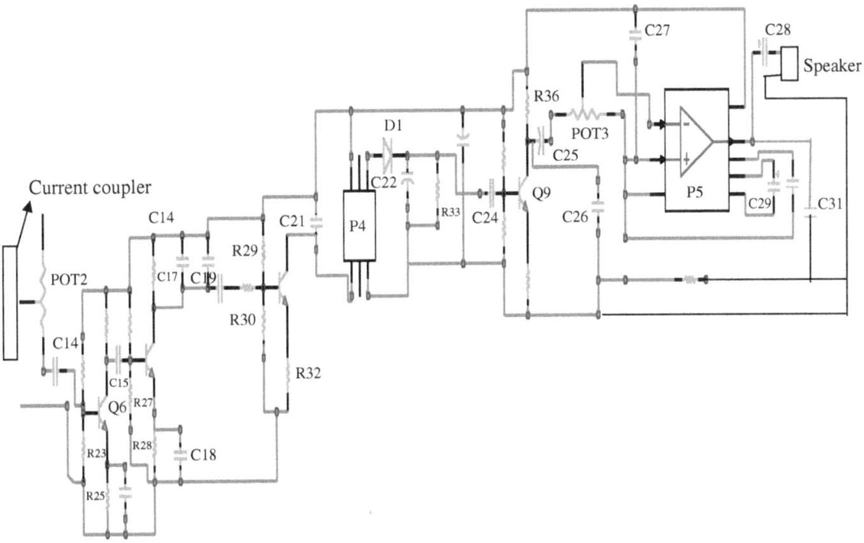

Fig. 4.5 Circuit diagram of receiving unit

Here the guide rope acts as single-turn primary coil and the coil wound on current coupler acts as secondary coil of the transformer. Due to the current in the rope, a proportionate voltage is developed at the output terminal of the current coupler. This signal voltage is the input to the preamplifier of the receiver. Because the signal is an AF-modulated electrical signal, AF is detected and goes through a power amplification stage prior to feeding it to the speaker.

Figure 4.6 shows a schematic diagram of a shaft and installation details. The current coupler should be installed around the guide rope. The two guide ropes are connected with each other using conducting wires in the shaft top and bottom, to form a loop as a whole.

The specification of the different units of the system are given below, however, equivalent components can also be used wherever required.

Colpitt oscillator as shown in Fig. 4.2: resistors: R1 – 10 kΩ, R2 – 150 kΩ, R3 – 100 kΩ, R4 – 8.2 kΩ, R18 – 100 kΩ, R19 – 1 kΩ, R20 – 10 kΩ, R21 – 220 Ω; capacitors: C1 – 2.2 μF/63 V, C10 – 1 μF/63 V, C11 – 0.047 μF, C12 – 0.22 μF; primary transformer: T2 – 900 μH, 24 T, 30 SWG, POT1 – 150 kΩ; transistor: Q5 – BC547B, Header H1– 2 × 2H.

Modulator as shown in Fig. 4.2: resistors: R5 – 220 Ω, R6 – 3.3 kΩ, R7 – 100 kΩ, R8 – 100 kΩ, R9 – 5.1 kΩ; capacitors: C2 – 10 μF/63 V, C3 – 0.33 μF, C4 – 0.01 μF, C13 – 0.001 μF; transistors: Q1 – BC547B, Q2 – BC547B.

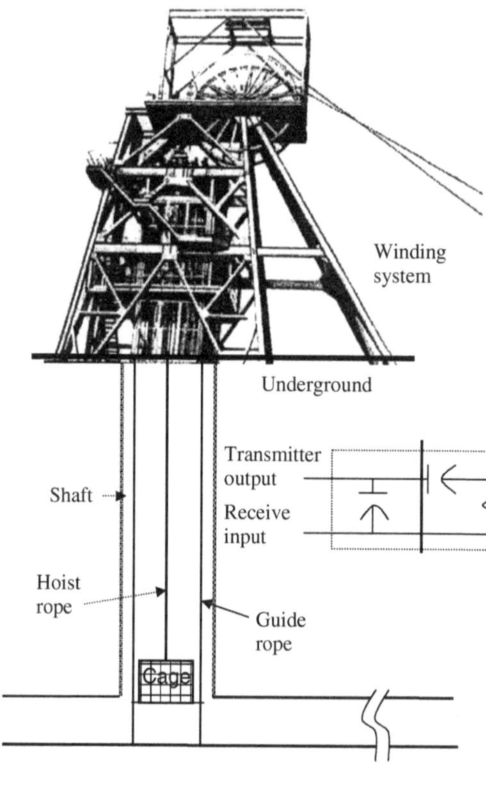

Fig. 4.6 Schematic diagram of shaft communication system

Tuned amplifier as shown in Fig. 4.2: resistor: R10 – 330 kΩ, R11 – 100 kΩ; capacitors: C5 – 0.1 μF, C6 – 0.047 μF; transistor: Q3 – BC547B; transformer: T1 – primary coil 17 T, 30SWG, 560 μH; secondary coil 5 T, 30 SWG, 60 μH.

Class "A" voltage amplifier, buffer amplifier, and AF amplifier and microphone as shown in Fig. 4.2: resistors: R12 – 100 kΩ, R13 – 8.2 kΩ, R14 – 5.1 kΩ, R15 – 220 Ω, R16 – 33 kΩ, R17 – 33 kΩ; capacitors: C7 – 100 pF, C8 – 2.2 μF/63 V, C9 – 10 μF/63 V; transistors: Q4 – BC547B, IC P1 – T108L, Header H2.

Current coupler as shown in Figs. 4.2 and 4.3: two U-shaped ferrite cores, housed in a hard 0.5-mm thick fiber enclosure.

RF amplifier as shown in Fig. 4.3: resistors: R22 – 100 kΩ, R23 – 8.2 kΩ, R24 – 5.1 kΩ, R25 – 220 Ω, POT2 – 470 kΩ; capacitors: C14 – 1 μF/63 V, C15 – 22 μF, C16 – 22 μF; transistor Q6 – BC549C.

RF-tuned amplifier as shown in Fig. 4.3: resistors: R26 – 100 kΩ, R27 – 8.2 kΩ, R28 – 220 Ω; capacitors: C17 – 0.022 μF, C18 – 0.22 μF, C19 – 0.0047 μF, C20 – 0.033 μF; inductor: L1 – 1 MH; transistor: Q7 – BC547B.

RF amplifier as shown in Fig. 4.3: resistors: R29 – 100 kΩ, R30 – 10 kΩ, R31 – 8.2 kΩ, R32 – 150 Ω; capacitor: C21 – 0.047 μF; transistor: Q8 – CL100.

Diode detector as shown in Fig. 4.3: Diode D1 – BAT86; resistors: R33 – 100 kΩ; capacitors: C22 – 0.1 μF, C23 – 100 μF/63 V, C24 – 0.22 μF; transformer: T3 – primary coil 21T, 30 SWG, 628 μH; secondary coil 12T, 30 SWG, 194 μH.

AF voltage amplifier, AF power amplifier, and speaker as shown in Fig. 4.3: resistors: R34 – 8.2 kΩ, R35 – 100 kΩ, R36 – 5.1 kΩ, R37 – 220 Ω, R38 – 10 Ω, POT3 – 100 kΩ; capacitors: C25 – 1 μF, C26 – 0.01 μF, C27 – 0.01 μF, C28 – 220 μF/63 V, C29 – 10 μF/63 V, C30 – 10 μF, C31 – 47 nF; IC: P2 – LM386; transistor: Q9 – BC547, speaker.

4.5 Technical Details

4.5.1 Transmitting Unit

The transmitting unit of the transceivers consists of various subsystems for generating and transmitting signal of 32 kHz (Figs. 4.2 and 4.4). The oscillator is a Colpitt oscillator, which provides an output of 32 kHz. The RF signal is fed to modulator, which modulates the carrier signal of 32 kHz coming from the oscillator and the audio signal coming from the audio amplifier and gives the needed gain of 40 dB. The output of modulator is coupled with tuned amplifier, class A voltage amplifier, and a buffer amplifier, which amplify the signals. Current coupler couples voice-modulated RF signal to guide rope.

4.5.2 Receiving Unit

The receiving unit of the transceivers consists of different subsystems for receiving RF signal coming from the transmitting unit (Figs. 4.3 and 4.5). The current coupler in receiving side interacts with the magnetic field created by the current induced in the guide rope due to the current in current coupler at transmitting side. Due to the current in the guide rope, a proportionate voltage develops at output terminal of the coupler. This signal voltage is input to the RF amplifier, RF-tuned amplifier, and RF amplifier. The diode detector detects the AF-modulated signal. The detected signal passes through the AF voltage amplifier and then AF power amplifier for power amplification prior to feeding to the speaker.

4.5.3 Specifications

(i) Transmitting unit

- 32 kHz carrier frequency
- $\pm$ 2 kHz bandwidth
- 100 μV sensitivity
- 1.2 W (maximum) power consumption
- +12 V power supply

(ii) Receiving unit

- 32 kHz carrier frequency
- 6.5 V peak-to-peak O/P approximately
- +12 V and –12 V power supply
- 200 mW power consumption approximately

4.6 Field Installation Procedure

Transceivers are housed in hard and tough enclosure to sustain vibration and other mechanical disturbances of the surroundings after placement of the system at pit top, in cage, and at pit bottom. The current couplers are also housed in a hard and tough fiber case to sustain vibration of guide rope and other mechanical disturbances of the surroundings. The transceivers are powered by the external battery, which is intrinsically safe. The transceivers are connected with the current coupler through a cable (Fig. 4.5). The current couplers are clamped with the guide rope of the cage at pit top, with the moving cage, and at pit bottom. Once the transceivers are placed at pit top, in the cage, and at pit bottom and are properly connected with the current couplers encircled around guide rope to form a transformer where guide rope acts as single-turn secondary winding of the transformer, then RF-modulated signal is induced in the guide rope. To ensure flow of current through the guide rope, both the ends of the guide rope are electrically shorted. The signal coming from the transmitting unit is sensed by the receiving unit. Current couplers on the receiving side interact with the magnetic field created by the current induced on the rope due to the current in current coupler at transmitting side. Here guide rope acts as single-turn primary and coil wound on the ferrite core acts as secondary of the transformer. Due to the current in the guide rope, a proportionate voltage is developed at output terminal of the current coupler. This signal voltage is the input to the preamplifier of the receiver. Because the signal is an AF-modulated electrical signal, AF is detected and passed through a power amplification stage prior to feeding it to the speaker.

4.7 Laboratory and Field Trials

4.7.1 Laboratory Trial

Laboratory testing of the prototype system was carried out at Central Institute of Mining and Fuel Research, Dhanbad, India. A 500-m cable was laid on the ground and shorted (Fig. 4.7). Then transceivers attached with the current couplers were placed at different locations of the cable. The transceivers were powered by external power supply of ±12 V battery. Subsequently, communication was established

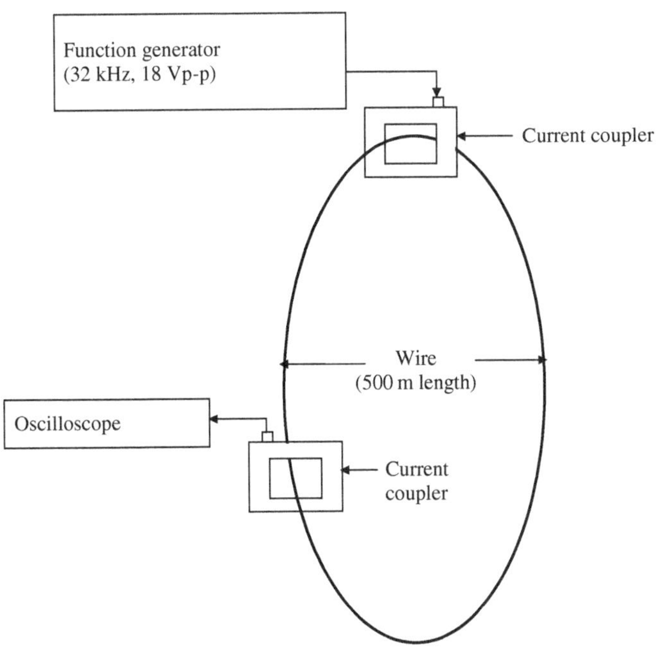

Fig. 4.7 Laboratory test setup for current coupler evaluation

among the transceivers. The frequency of the system was measured by Spectrum Analyzer and it was found that throughout the cable it was 32 kHz.

4.7.2 Field Trial

The induction-based hoist communication system (Fig. 4.8) was first experimented at Bagdiggi underground mine of Bharat Coking Coal Limited, India (in 9th pit of the mine). The depth of the shaft is around 165 m. The current couplers were fixed in the guide rope of the cage at pit top, inside the cage, and at pit bottom. The transceivers were connected with the current coupler and are powered by the external battery. The audio signal could be detected in the moving cage as well as at pit bottom and pit top. The voice communication was clear and loud.

The system was again experimented at Bagdiggi underground mine (in 12th pit of the mine). The depth of the shaft is around 225 m (Fig. 4.9). The current couplers were fixed in the guide rope of the cage at pit top, inside the cage, and at pit bottom. The transceivers were connected with the current coupler and are powered by the external battery. The audio signal could be detected in the moving cage as well as at pit bottom and pit top. The communication was established from pit top, pit bottom, and different levels of the shaft.

Fig. 4.8 View of shaft communication system

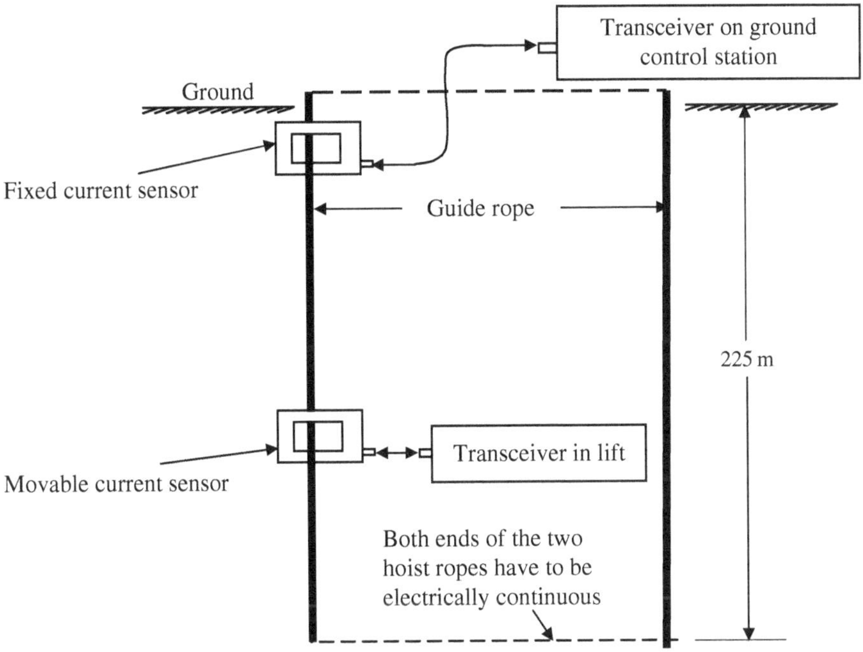

Fig. 4.9 Field trial test setup for establishing shaft communication

4.8 Capabilities of the System

The main capabilities of the induction-based shaft communication system are

(i) The system is capable of establishing communication among the mine personnel present at pit top, in moving cage, and at pit bottom.
(ii) The system establishes communication throughout shaft of an underground mine.
(iii) The system has provision to provide the messages from the moving cage to pit top and pit bottom in case of maintenance work of a shaft.
(iv) The system is capable of providing messages during the moving cage to pit top and pit bottom in case of any installation work taking place in a shaft.
(v) The system is able to provide messages during the moving cage to pit top and pit bottom in case of stopping the cage at the particular level in case of two or more working seams available in a particular underground mine.
(vi) The system is capable of helping the mine management to establish communication from pit top to pit bottom or from anywhere from the shaft in case of emergency.
(vii) The system is of low cost.

4.9 Conclusions

The induction-based hoist communication system for underground mine consists of a low-frequency electromagnetic wave transceivers unit powered by external battery and attached with a current coupler clamped with the guide rope, which induces the current throughout the rope. Use of the system in an underground mine shaft would help in establishing the voice communication throughout the shaft. This would help the mine personnel during the maintenance of the shaft and other installation work in a shaft. The system would also help the mine management to establish communication from pit top to pit bottom or from anywhere from the shaft in case of emergency.

References

Bandyopadhyay LK, Chaulya SK and Kumar S (2002) A proposed wireless communication system for underground mines. *Proceedings of the International Conference on Mineral Industry: Issues on Economics, Environment and Technology*, Mining, Geological and Metallurgical Institute of India, Kolkata, India, pp. 317–324.

Bueno M and Assis AKT (1997) Equivalence between the formulas for inductance calculation. *Canadian Journal of Physics*, **75**: 357–362.

Butler CM, Rahmat-Samii Y and Mittra Y (1978) Electromagnetic penetration through apertures in conducting surfaces. *IEEE Transactions on Antennas and Propagation*, **26**: 82–93.

Douglas B (2003) *Signal Integrity Issues and Printed Circuit Board Design*. Prentice Hall, New Jersey, USA.

Griffith DJ (1995) *Introduction to Electrodynamics*. 2nd Edition, Prentice Hall, New Delhi, India.

Kováč D and Kováčová I (2008) Electromagnetic coupling of the electrical drive – EMC (Part II). *Electrical Power Quality and Utilization*, **14**(1): 89–93.

Kováčová I and Kováč D (2007) Electromagnetic coupling of the electrical drive – EMC (Part I). *Eletrical Power Quality and Utilization*, **13**(2): 81–87.

Leviatan Y (1988) Electromagnetic coupling between two half-space regions separated by two slot-perforated parallel conducting screens. *IEEE Transactions on Microwave Theory and Techniques*, **36**(1): 44–52.

Manning KV (1998) *Electromagnetic Induction*. McGraw Hill, London, UK.

Nowakowski M (1998) The electromagnetic coupling in Kemmer-Duffin-Petiau theory. *Physics Letters*, **244**(5): 329–337.

Xie P, Tan Z, Liu S and Zhang R (2007) Electromagnetic coupling effect emulation analysis of metal body with small hole. *Proceedings of 8th International Conference on Electronic Measurement and Instruments*, Vol. 4, 18 July–16 August, 2007, pp. 286–291.

Websites

http://en.wikipedia.org/wiki/Current_clamp

www.camese.org/20012002compendium.pdf

www.mentor.com/pcb/tech_papers.cf Douglas Brooks, "Differential Trace Design Rules; Truth vs. Fiction")

www.wvminesafety.org/PDFs/communications/Additional%20Documents/Communication_Systems_in_Underground_Mines.pdf

Chapter 5
Line-of-Sight Communication

5.1 Introduction

The rock and coal, surrounding a coal mine tunnel, act as relatively low-loss dielectric media in the frequencies of range 200–4,000 MHz and dielectric constant of 5–10. Under these conditions a reasonable hypothesis is that the transmission takes the form of wave propagation, since the wavelength of ultra-high-frequency (UHF) waves are smaller than the tunnel dimensions (Reed and Russell, 1966; Emslie et al., 1975). An electromagnetic wave traveling along a rectangular tunnel in a dielectric medium can propagate in any one of a number of allowed waveguide modes (Alfred et al., 1974). All of these modes are lossy modes because any part of the wave that impinges on tunnel wall is partly refracted into the surrounding dielectric and partly reflected back into the waveguide (Holloway et al., 2000). The reflected part propagates away from the waveguide and represents a power loss. The overall loss in strength of the signal, in a straight tunnel, is the sum of propagation loss and the insertion loss of the transmitting and receiving antennas (Zheng et al., 2008). It has been found that the total loss is minimal in the range of 400–1,000 MHz depending on the desired communication distance (Zhang et al., 2001; Bandyopadhyay et al., 2002). Hence, the UHF communication system is used in straight gallery for line-of-sight communication.

5.2 Propagation of UHF Radio Waves in Coal Mine Tunnels

Radio waves in underground mines suffer substantial perturbation due to relatively poor propagation characteristics of mine tunnels (Emslie et al., 1975; Zelley and Constantinou, 1999; Zheng et al., 2008). The underground tunnel has been modeled as an imperfect waveguide whose walls partially absorb and partially reflect incident electromagnetic radiation (Porrat and Cox, 2004; Talbi, 2001). Propagation, therefore, is a function of both the tunnel dimensions and the surrounding material comprising the tunnel floor, ribs, and roof (Kurner et al., 1991; Bertoni et al., 1994; Zhang and Mao, 2006; Rak and Pechac, 2007). Measurements of radio propagation characteristics along straight tunnels correlate to the characteristics of such a

L.K. Bandyopadhyay et al., *Wireless Communication in Underground Mines*,
DOI 10.1007/978-0-387-98165-9_5, © Springer Science+Business Media, LLC 2010

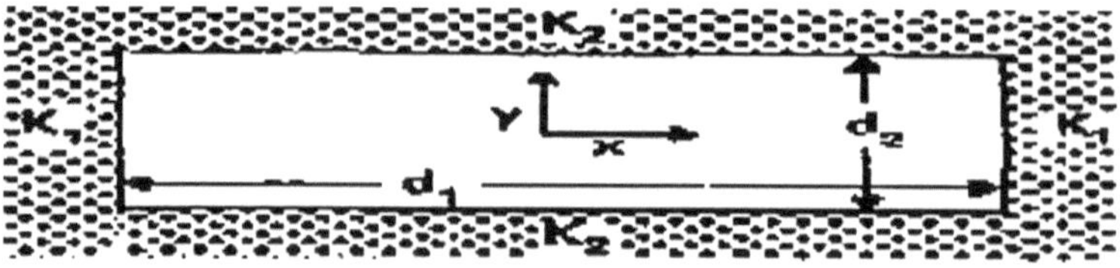

Fig. 5.1 Waveguide geometry

waveguide (Kurner et al., 1996; Wang et al., 2007). Figure 5.1 shows the waveguide geometry of an electromagnetic wave in a rectangular tunnel. The coordinate system is centered in the tunnel with x – horizontal, y – vertical, and z – along the tunnel.

5.2.1 Waveguide Modes

The propagation modes with the lowest attenuation rates in a rectangular tunnel in a dielectric medium are the two fundamental modes, which have the electric field (E) polarized predominantly in the horizontal (E_h) and vertical directions (E_v), respectively. Referring to these two modes as the E_h and E_v modes, the main field components of the E_h mode in a tunnel are

$$E_x = E_0 \cos k_1 x \cos k_2 y \exp(-ik_3 z) \tag{5.1}$$

$$H_y = \frac{(k_3)}{\omega\mu_0} E_0 \cos k_1 x \cos k_2 y \exp(-ik_3 z) \tag{5.2}$$

where E_0 is the initial electric field (at $z = 0$). In addition to these transverse field components there are small longitudinal components of E_z and H_z and a small transverse component H_x. For the frequencies of interest here propagation constants k_1 and k_2 are small compared with k_3, which means that the wave propagation is mostly in the z direction. From a geometrical optics point of view, the ray makes small grazing angles with the tunnel walls. In the dielectric surrounding of tunnel, the wave has the form of progressive waves in the transverse as well as the longitudinal directions. The propagation constant k_3 for the fundamental mode is an Eigen value determined by the boundary conditions of continuity of the tangential components of E and H at the walls of the tunnel. Owing to the simple form of the wave given by Eqs. (5.1) and (5.2) these conditions can be satisfied only approximately. However, a good approximation to k_3 is obtained. The imaginary part of k_3, which arises owing to the leaky nature of the mode, gives the attenuation rate of the wave.

The fundamental (1, 1) E_h mode is approximately a transverse electromagnetic (TEM) wave given by Eqs. (5.1) and (5.2). The complete field of this form in the tunnel ($-d_1/2 \le x \le d_1/2$, $-d_2/2 \le y \le d_2/2$), satisfying Maxwell's equations, is

$$E_x = E_0 \cos k_1 x \cos k_2 y \exp(-ik_3 z) \tag{5.3}$$

$$E_y = 0 \tag{5.4}$$

$$E_z = \frac{ik_1}{k_3} E_0 \sin k_1 x \cos k_2 y \exp(-ik_3 z) \tag{5.5}$$

$$H_x = \frac{k_1 k_2}{\omega \mu_0 k_3} E_0 \sin k_1 x \sin k_2 y \exp(-ik_3 z) \tag{5.6}$$

$$H_y \frac{\left(k_1^2 + k_3^2\right)}{\omega \mu_0 k_3} E_0 \cos k_1 x \cos k_2 y \exp(-ik_3 z) \tag{5.7}$$

$$H_z = \frac{ik_2}{\omega \mu_0} E_0 \cos k_1 x \sin k_2 y \exp(-ik_3 z) \tag{5.8}$$

where

$$k_1^2 + k_2^2 + k_3^2 = k_0^2 = \frac{4\pi^2}{\lambda^2} \tag{5.9}$$

Since the wavelengths of interest are small compared with the tunnel dimensions, the wave vector components k_1 and k_2 are small compared with k_3, which is close to $k_0 = 2\pi/\lambda$. Therefore, H_y reduces to the expression given in Eq. (5.2) and E_z, H_x, and H_z are very small. In the roof ($y \geq d_2/2$) of dielectric constant K_2, the field must represent an outgoing wave in the y direction and therefore has the form

$$E_x = B \cos k_1 x \exp\left(-ik_2' y\right) \exp(-ik_3 z) \tag{5.10}$$

$$E_y = 0 \tag{5.11}$$

$$E_z = \frac{ik_1}{k_3} B \sin k_1 x \exp(-ik_2' y) \exp(-ik_3 z) \tag{5.12}$$

$$H_x = \frac{ik_1 ik_2'}{\omega \mu_0 k_3} B \sin k_1 x \exp(-ik_2' y) \exp(-ik_3 z) \tag{5.13}$$

$$H_y = \frac{\left(k_3^2 + k_1^2\right)}{\omega \mu_0 k_3} B \cos k_1 x \exp\left(-ik_2' y\right) \exp(-ik_3 z) \tag{5.14}$$

$$H_z = -\frac{k_2}{\omega \mu_0} B \cos k_1 x \exp\left(-ik_2' y\right) \exp(-ik_3 z) \tag{5.15}$$

which satisfies Maxwell's equations. The wave number component k_2' in the dielectric is given by the relation:

$$k_1^2 + k_2'^2 + k_3^2 = k_2 k_0^2 \tag{5.16}$$

The boundary conditions at $y = d_2/2$ are that the tangential components of E and H are continuous. These conditions require that

$$E_0 \cos\left(\frac{k_2 d_2}{2}\right) = B \exp\left(\frac{-ik_2' d_2}{2}\right) \tag{5.17}$$

and

$$k_2 E_0 \sin\left(\frac{k_2 d_2}{2}\right) = ik'_2 B \exp\left(\frac{-ik'_2 d_2}{2}\right) \tag{5.18}$$

from which we obtain the condition

$$k_2 \tan\frac{k_2 d_2}{2} = ik'_2 \tag{5.19}$$

Since k_1 and k_2 are small compared with k_0 we find from Eqs. (5.9) and (5.16) that k'_2 is given approximately by

$$ik'_2 = k_0(k_2 - 1)^{1/2} \tag{5.20}$$

Therefore, from Eqs. (5.19) and (5.20) we obtain the following mode condition for k_2, for modes that are even functions of y:

$$k_2 \tan\frac{k_2 d_2}{2} = ik_0(k_2 - 1)^{1/2} \tag{5.21}$$

This equation could readily be solved numerically for the Eigen values of k_2, for given values of the other quantities. However, we prefer here to obtain an approximate closed form expression that provides a better insight into the problem.

Since $k_2 d_2 << 1$ we find for the lowest E_h mode:

$$k_2 \cong \frac{\pi}{d_2} + \frac{i\lambda}{d_2^2(k_2 - 1)^{1/2}} \tag{5.22}$$

This result shows that, except for a small imaginary part, k_2 has the same value as for a metal waveguide. The imaginary part arises from the power loss due to the outgoing refracted wave. The approximation involved in Eq. (5.22) is valid provided that λ is small compared with $\pi d_2(K_2 - 1)^{1/2}$. This criterion is very well satisfied, for example, for $\lambda = 0.33$ m, $d_2 = 2.33$ m, $K_2 = 10$.

In the side wall ($x \geq d_1/2$) of dielectric constant K_1, the field has the form

$$E_x = A \exp\left(-ik'_2 x\right) \cos k_2 y \exp\left(-ik_3 z\right) \tag{5.23}$$

$$E_y = 0 \tag{5.24}$$

$$E_z = -\frac{k'_1}{k_3} \exp\left(-ik'_1 x\right) \cos k_2 y \exp\left(-ik_3 z\right) \tag{5.25}$$

$$H_x = \frac{ik'_1 k_2}{\omega \mu_0 k_3} A \exp\left(-ik'_1 x\right) \cos k_2 y \exp\left(-ik_3 z\right) \tag{5.26}$$

$$H_y = \frac{\left(k_1^{'2} + k_3^{'2}\right)}{\omega\mu_0 k_3} A \exp\left(-ik_1' x\right) \cos k_2 y \exp\left(-ik_3 z\right) \tag{5.27}$$

$$H_z = \frac{ik_2}{\omega\mu_0} A \exp\left(-ik_1' x\right) \sin k_2 y \exp\left(-ik_3 z\right) \tag{5.28}$$

where

$$k_1^{'2} + k_2^2 + k_3^2 = K_1 k_0^2 \tag{5.29}$$

Continuity of the tangential E-field gives the condition

$$k_1 E_0 \sin\left(\frac{k_1 d_1}{2}\right) = ik_1' A \exp\left(\frac{-ik_1' d_1}{2}\right) \tag{5.30}$$

Continuity of H_y and H_z requires that

$$\left(k_3^2 + k_1^2\right) E_0 \cos\left(\frac{k_1 d_1}{2}\right) = \left(k_3^2 + k_1^{'2}\right) A \exp\left(\frac{-ik_1' d_1}{2}\right) \tag{5.31}$$

and

$$k_2 E_0 \cos\left(\frac{k_1 d_1}{2}\right) = k_2 A \exp\left(\frac{-ik_1' d_1}{2}\right) \tag{5.32}$$

Since Eqs. (5.31) and (5.32) are inconsistent we can only satisfy the H boundary condition approximately. We note that since $K_1 >> 1$, k_1' is of the same order as k_0, whereas $|k_2|$ is much smaller.

Therefore, we may ignore Eq. (5.32) and also, from Eqs. (5.9) and (5.29), we may write

$$k_1^2 + k_3^2 \approx k_0^2 \tag{5.33}$$

$$k_1^{'2} + k_3^2 \approx K_1 k_0^2 \tag{5.34}$$

Then from Eqs. (5.30), (5.31), (5.33), and (5.34), we obtain, approximately,

$$k_1 \tan\left(\frac{k_1 d_1}{2}\right) = \frac{ik_0 (K_1 - 1)^{1/2}}{K_1} \tag{5.35}$$

Again taking advantage of the smallness of k_1 and k_2 relative to k_0, we find for the lowest E_h mode that

$$k_1 \cong \frac{\pi}{d_1} + \frac{iK_1 \lambda}{d_1^2 (K_1 - 1)^{1/2}} \tag{5.36}$$

which shows that the mode shape in the x direction is also the same as for a metal waveguide, except for a small imaginary part. The approximation is valid if λ is small compared with $\pi d_1\{(K_1 - 1)/K_1\}^{1/2}$. The criterion is well satisfied for $\lambda = 0.33$ m, $d_1 = 4.67$ m, and $K_1 = 10$. On substituting for k_1 and k_2 from Eqs. (5.36) and (5.22) into Eq. (5.9) we find, on neglecting second-order terms, that the propagation constant in the z direction is

$$k_3 = k_0 - \frac{i\lambda^2}{2}\left(\frac{K_1}{d_1^3(K_1 - 1)^{1/2}} + \frac{1}{d_2^3(K_2 - 1)^{1/2}}\right) \tag{5.37}$$

The power loss in dB for the (1, 1) E_h mode for a distance z is therefore

$$\begin{aligned} L_{E_h} &= -8.686\mathrm{Im}(k_3) \\ &= 4.343\lambda^2 z\left(\frac{K_1}{d_1^3(K_1-1)^{1/2}} + \frac{1}{d_2^3(K_2-1)^{1/2}}\right) \end{aligned} \tag{5.38}$$

We obtain the loss for the (1, 1) E_v mode by interchanging the subscripts 1 and 2 in Eq. (5.38)

$$L_{E_v} = 4.343\lambda^2 z\left(\frac{1}{d_1^3(K_1 - 1)^{1/2}} + \frac{K_2}{d_2^3(K_2 - 1)^{1/2}}\right) \tag{5.39}$$

As a check on these formulae, we find that exactly the same results are obtained if one adds the losses for two infinite slot waveguides of slot widths d_1, d_2 and dielectric constants K_1, K_2, respectively.

One can readily generalize Eqs. (5.38) and (5.39) to the case of a higher mode (n_1, n_2) with approximately n_1 half-wave loops in the x direction and n_2 in the y direction. The results are

$$L_{E_h}(n_1,n_2) = 4.343\lambda^2 z\left(\frac{n_1^2 K_1}{d_1^3(K_1 - 1)^{1/2}} + \frac{n_2^2}{d_2^3(K_2 - 1)^{1/2}}\right) \tag{5.40}$$

$$L_{E_v}(n_1,n_2) = 4.343\lambda^2 z\left(\frac{n_1^2}{d_1^3(K_1 - 1)^{1/2}} + \frac{n_2^2 K_2}{d_2^3(K_2 - 1)^{1/2}}\right) \tag{5.41}$$

Table 5.1 shows the loss rates for a number of modes for $f = 1{,}000$ MHz, $\lambda = 0.32$ m, $d_1 = 4.67$ m, $d_2 = 2.33$ m, $K_1 = K_2 = 10$, $z = 33$ m.

The validity criteria for the approximate solutions of the mode conditions (5.21) and (5.22) are the same as for the (1, 1) modes, i.e., they do not depend on n_2 or n_1.

5.2.2 *Loss due to Surface Roughness*

When a parallel beam of radiation of intensity I_0 strikes a rough surface at normal incidence, the reflected radiation consists of a parallel beam of reduced intensity I

Table 5.1 Loss rate for various modes

n_1	n_2	L_{E_h} (dB/33 m)	L_{E_v} (dB/33 m)
1	1	0.9	4.1
1	2	2.1	16.3
2	1	2.4	4.3
2	2	3.6	16.4
1	3	4.2	36.5
3	1	5.0	5.0
2	3	5.7	36.7
3	2	6.2	16.7
3	3	8.2	36.9

together with a diffuse component. If the surface is a perfect reflector:

$$I = I_0 \exp\left[-2\left(\frac{2\pi}{\lambda}h\right)^2\right] \tag{5.42}$$

where λ is the wavelength and h the root-mean-square roughness. For incidence at a grazing angle ϕ one may assume that the effective roughness is now $h \sin\phi$, so the loss factor becomes

$$f = \exp\left[-2\left(\frac{2\pi}{\lambda}h\sin\phi\right)^2\right] \tag{5.43}$$

In the case of the dominant mode in a dielectric waveguide we can write for the roughness loss factors per reflection for the vertical and horizontal walls:

$$f_1 = \exp\left[-2\left(\frac{\pi h}{d_1}\right)^2\right] \tag{5.44}$$

$$f_2 = \exp\left[-2\left(\frac{\pi h}{d_2}\right)^2\right] \tag{5.45}$$

The loss factor for a distance z is, therefore, written as

$$f = \exp\left[-2N_1\left(\frac{\pi h}{d_1}\right)^2 - 2N_2\left(\frac{\pi h}{d_2}\right)^2\right] \approx \exp\left[-\pi^2 h^2\lambda\left(\frac{1}{d_1^4}+\frac{1}{d_2^4}\right)z\right] \tag{5.46}$$

Therefore, the loss in dB is

$$L_{\text{roughness}} = 4.343\pi^2 h^2\lambda\left(\frac{1}{d_1^4}+\frac{1}{d_2^4}\right)z \tag{5.47}$$

5.2.3 Loss due to Tilt of Tunnel Walls

Suppose that a ray of the E_h mode encounters a portion of a side wall that is tilted through a small angle θ about a vertical axis. Then the reflected beam is rotated through an angle 2θ. This means that the electric field is changed from

$$E_x = F(x, y) \exp(-ik_3 z) \tag{5.48}$$

to

$$E'_x = F(x, y) \exp[-ik_3(z \cos 2\theta + x \sin 2\theta)] \tag{5.49}$$

The power coupling factor g_1 of the disturbed field (5.49) back into the mode (5.48) is given by

$$g_1 = \frac{|\int\int E_x \bar{E}'_x dxdy|^2}{\int\int |E_x|^2 dxdy \int\int |E'_x|^2 dxdy} \tag{5.50}$$

where the integrations are over the cross section of the tunnel. The bar over E'_x indicates complex conjugate. Since θ is small, we can replace cos 2θ by 1 and sin 2θ by 2θ. Then Eq. (5.50) becomes

$$g_1 = \frac{|\int\int |F|^2 \exp(2ik_2 x\theta) dxdy|^2}{(\int\int |F|^2 dxdy)^2} \tag{5.51}$$

where F is a Gaussian function. Instead of using the actual function cos $k_1 x$ cos $k_2 y$ for F, we find it more convenient to use an equivalent Gaussian function:

$$F = F_0 \exp\left[-\left(\frac{x^2}{a^2} + \frac{y^2}{b^2}\right)\right] \tag{5.52}$$

and integrate over infinite limits. The result is

$$g_1 = \exp\left(-\frac{1}{2}k_3^2 a^2 \theta^2\right) \tag{5.53}$$

Next, we assume that F^2 falls to l/e at the point $x = d_1/2$, $y = 0$, which is at the surface of the waveguide. Then $a^2 = 1/2d_1^2$ and

$$g_1 = \exp\left(-\frac{1}{4}k_3^2 d_1^2 \theta^2\right) \tag{5.54}$$

Likewise, tilting of the floor or roof gives a coupling factor:

$$g_2 = \exp\left(-\frac{1}{4}k_3^2 d_2^2 \theta^2\right) \tag{5.55}$$

The loss factor for a distance z is

$$g = g_1^{N1} g_2^{N2} = \exp\left(-\frac{\pi^2\theta^2 z}{\lambda}\right) \tag{5.56}$$

where k_3 is replaced by k_0. The loss in dB is, therefore,

$$L_{\text{tilt}} = \frac{4.343\pi^2\theta^2 z}{\lambda} \tag{5.57}$$

5.3 Expected Communication Range in a Mine and Propagation Loss

Communication can be maintained between two separated individuals until the separation distance increases to a point where the signal strength is not sufficient to overcome the background electrical noise. To obtain estimates of this communication range for a mobile application involving roving miners equipped with portable personal radio transceivers, three frequency-independent loss factors should be taken into account. These factors are polarization loss – to account for likely misalignment of transmit and receive antennas; antenna-efficiency loss – to account for the non-ideal antenna installation on the portable units; and fade margin – to account for signal-cancellation effects due to destructive interference. By exercising care in the orientation and position of the portable radios in the mine tunnel cross section while communicating, these polarization and signal-fading losses can of course be reduced, thereby producing a corresponding increase in range.

Attenuation, expressed in decibels, varies linearly with distance and exhibits large losses around corners, bends, or any deviation from a straight path. Surface irregularities of the walls also contribute to path losses. Typical results of propagation losses in mine tunnels are shown in Table 5.1. These measurements show a significant difference in communication range with antenna polarization. This fact is most noticeable in mine tunnels having one dimension significantly greater than another. Antenna polarization aligned with the longest tunnel dimension will yield the greatest communication range, a property exhibited in Table 5.2 (Cawley, 1989). The waveguide propagation mode requires tunnel dimensions greater than a wavelength of the radiated signal in order to support propagation. Range is also a weak function of both receiver sensitivity and transmitted power.

The total propagation loss (between isotropic antennas) is determined using the following equation (Alfred et al., 1974):

$$L = L_0 + (A \times D) + L_C \tag{5.58}$$

Table 5.2 Propagation characteristics in a coal mine 2.33 × 4.67 m tunnel cross section

Frequency (MHz)	Polarization	Coupling loss (dB)	Attenuation (dB/33 m)	Bend loss (dB)	Approximate range in straight line (m)	One bend (m)
200	Vertical	65	15	–	160	–
415	Vertical	78	6.5	35	302	123
415	Horizontal	40	5.9	35	548	350
1,000	Vertical	65	4.3	36	558	279
1,000	Horizontal	57	2.5	38	1,160	653

where

L = total propagation loss,
L_0 = fixed (coupling) loss,
A = attenuation/33 m,
D = distance between transmitter and receiver (in 33 m), and
L_C = loss associated with a 90° bend.

Equation (5.58) is only valid for distances of $D > 33$ m.

5.4 System Description

The UHF transceivers works in the frequency range of 410–500 MHz. The maximum radiated power of the system in underground mines is restricted up to 2 W due to the safety reasons. Therefore, the radiated power of the transceivers is adjusted at two levels, namely, 1 and 2 W by setting the software algorithm and programming the microcontroller of the transceivers. Figure 5.2 illustrates the function of UHF transceivers in an underground gallery and also indicates how the electromagnetic waves travel in a narrow gallery. These UHF transceivers can also be used for communication in the shaft, incline, and longwall panel in addition to the straight gallery in underground mines. These transceivers can independently function and easy to use due to their lighter weight and less power consumption. Moreover, the system has 10 channels so that different-level mine management can communicate simultaneously in the same working place. The channels are spaced at 30 kHz frequency intervals. The impedance of the antenna is 50 ohm and power requirement of the transceiver is 7.2 V DC. The communication range of the transceivers on the surface is around 1 km. In underground gallery, communication range varied from 175 to 200 m with 1 W output power and 275 to 300 m with 2 W output power at 450 MHz frequency with a gallery dimension of 2 m height and 4 m width. The communication range increases with increase in output power and dimension of the underground gallery.

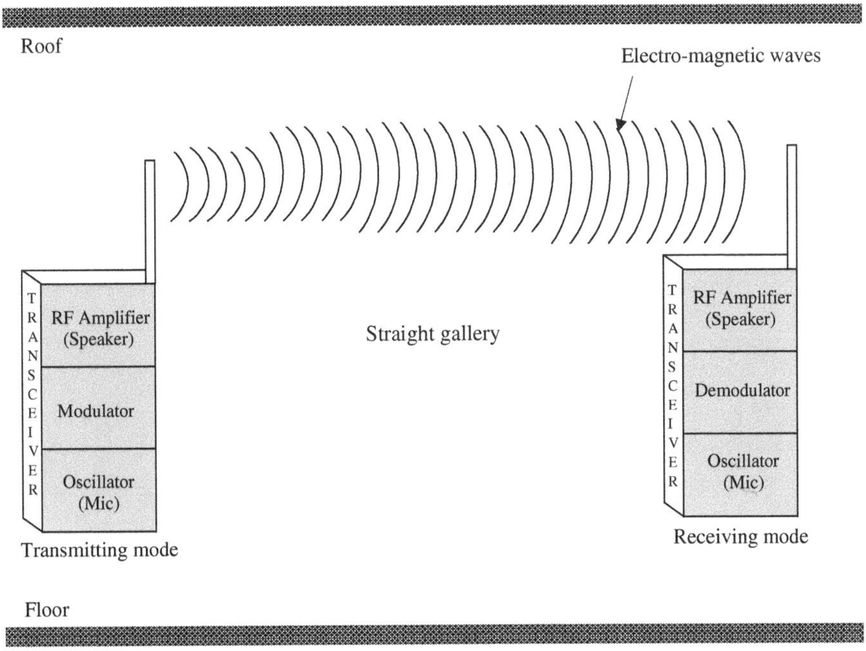

Fig. 5.2 Propagation of electromagnetic waves in underground gallery

5.5 Circuit Diagram

Figure 5.3 depicts block diagram of an UHF transceiver. The functions of different units of a transceiver are described below.

(i) *Gain control amplifier:* It is a logic level gain control that provides a step reduction in gain; which improves the operation during high input signal level conditions. Supply current is also reduced in both high gain mode as well as low gain mode. It is a mixer of low-noise amplifier (LNA) and preamplifier.
(ii) *RF band pass filter:* It selects the frequency range to be allowed for the operation, and pass the desired RF frequency. This stage acts to boost the weak signal.
(iii) *Down converter:* It is actually a frequency mixer. It mixes local oscillator (LO) and band pass filter (BPF), and also allows both high-frequency side and low-frequency side. Signals at IF Input port are mixed with signals at LO port using a double-balanced mixer.
(iv) *IF processing:* Mixer output fed to IF amplifier, which fixes tune and provide sufficient selectivity to reject adjacent channel signals.
(v) *VCO and buffer amplifier:* The job of this block is to control over signal voltage. The buffers provide load isolation to oscillator and prevent frequency pulling due to load-impedance changes.

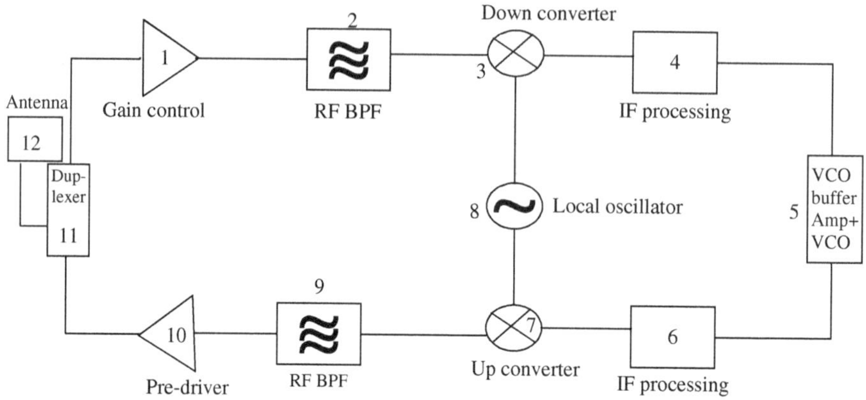

Fig. 5.3 Block diagram of UHF transceiver

(vi) *IF processing:* Mixer output fed to IF amplifier, which fixes tune and provide sufficient selectivity to reject adjacent channel signals.
(vii) *Up converter:* It is designed for low-voltage operation.
(viii) *Local oscillator:* Signals at IF input port are mixed with signals at LO port using a double-balanced mixer.
(ix) *RF band pass filter:* It selects frequency range to be allowed for the operation, and pass the desired RF frequency.
(x) *Predriver:* It drives signal to duplexer.
(xi) *Duplexer:* It combines two or more signals onto a common channel or medium to increase its transmission efficiency. It allows a transmitter operating on one frequency and a receiver operating on a different frequency to share one common antenna with a minimum interaction and degradation of the different RF signals.
(xii) *Antenna:* It receives as well as transmits the incoming RF signals.

5.6 Specification

Frequency range : 410–500 MHz
Number of channels : 10
Channel spacing : 30 kHz
Antenna impedance : 50 Ω
Power supply requirement : 7.2 V DC
Current drain (approx.) : Transmit 0.8 A (1 W)
Current drain (approx.) : Receive AF maximum 300 mA
Current drain (approx.) : Standby 85 mA

Operating temperature range : –30°C to +60°C
Output power : 1.0 and 2.0 W
External MIC impedance : 9-pin multiconnector/2.2 kΩ
Sensitivity : 0.25 μV
Squelch sensitivity : 0.25 μV
Audio output power : 0.5 W at 5% distortion with an 8 Ω load

5.7 Field Trial

Two transceivers were used for laboratory and field trials (Fig. 5.4). The transceivers were tested in the open space with frequency fixed at 450 MHz. It was found that communication range in the open space was around 1.1 km with 2 W output power and around 800 m with 1 W output power.

The experiments were also performed in different underground mines with frequencies varied from 410 to 500 MHz. Figure 5.5 shows the photographs taken during field trial of the transceiver sets. It was found that communication range in underground gallery varied from 175 to 200 m with 1 W output power and 275 to 300 m with 2 W output power and 450 MHz frequency with a gallery dimension of 2 m height and 4 m width. The communication range increases with increase in output power and dimension of the underground gallery. Further, it was found that

Fig. 5.4 View of the UHF transceiver sets

Fig. 5.5 View of communication between two miners using UHF transceivers

communication range is optimum at 450 MHz frequency. It was also observed that tub transportation does not significantly hinder communication between transceivers. The transceivers are also tested in the shaft of Chinakuri Mine of Eastern Coalfields Limited, which is the deepest coal mine of India. Clear communication could be established from pit top to pit bottom having depth of around 612 m. It was also found that two cages at same level in a shaft did not hinder communication significantly.

5.8 Capabilities

The UHF transceivers have the following capabilities:

- The transceivers can be used for straight gallery having long distance.
- The transceivers can also be used for communication in the shaft, incline, and longwall panel.
- The transceivers can further also be used in surface mines.
- The transceivers work independently and they are easy to use due to their lighter weight and less power consumption.
- The transceivers have 10 channels so that different-level mine management can communicate simultaneously in the same working place.

5.9 Conclusions

The UHF transceivers for line-of-sight communication works in the frequency range of 410–500 MHz. The maximum radiated power of the system in underground mines is restricted up to 2 W due to the safety reasons. Therefore, the radiated power of the transceivers is adjusted at two levels, namely, 1 and 2 W by setting the software algorithm and programming the microcontroller of the transceivers. These UHF transceivers can also be used for communication in the shaft, incline, and longwall panel in addition to the straight gallery in underground mines. These transceivers can work independently and easy to use due to their lighter weight and less power consumption. Moreover, the system has 10 channels so that different-level mine management can communicate simultaneously in the same working place.

References

Alfred G, Lagace RL and Strong PF (1974) Theory of the propagation of UHF radio waves in coal mine tunnels. *Information Circular*, 8745, US Bureau of Mines, Pittsburgh, PA.

Bandyopadhyay LK, Chaulya SK and Kumar S (2002) A proposed wireless communication system for underground mines. *Proceedings of the International Conference on Mineral Industry: Issues on Economics, Environment and Technology*, Mining, Geological and Metallurgical Institute of India, Kolkata, India, pp. 317–324.

Bertoni HL, Honcharenko W, Macel LR and Xia HH (1994) UHF propagation prediction for wireless personal communications. *Proceedings of the IEEE*, Vol. 82, Issue 9, 3 September 1994, pp. 1333–1359.

Cawley JC (1989) An assessment of leaky feeder radio systems in underground mines, US Department of the Interior, Bureau of Mines, Pittsburgh Research Center, Pittsburgh, PA.

Emslie A, Lagace R and Strong P (1975) Theory of the propagation of UHF radio waves in coal mine tunnels. *IEEE Transactions on Antennas and Propagation*, **23**(2): 192–205.

Holloway CL, Hill DA, Dalke RA and Hufford GA (2000) Radio wave propagation characteristics in lossy circular waveguides such as tunnels, mine shafts, and boreholes. *IEEE Transactions on Antennas and Propagation*, **48**: 1354–1366.

Kurner T, Cichon DJ and Wiesbeck W (1996) Evaluation and verification of the VHF/UHF propagation channel based on a 3-D-wave propagation model. *IEEE Transactions on Antennas and Propagation*, **44**(3): 393–404.

Kurner T, Lebherz M, Wiesbeck W and Krank W (1991) Signal characteristics of the VHF/UHF propagation channel. *Proceedings of International Symposium of Antennas and Propagation Society*, Vol. 3, 24–28 June 1991, pp. 1530–1533.

Porrat D and Cox DC (2004) UHF propagation in indoor hallways. *IEEE Transactions on Wireless Communications*, **3**(4): 1188–1198.

Rak M and Pechac P (2007) UHF propagation in caves and subterranean galleries. *IEEE Transactions on Antennas and Propagation*, **55**(4): 1134–1138.

Reed HR and Russell CM (1966) *Ultra High Frequency Propagation*. Chapman and Hall, London.

Talbi L (2001) Simulation of indoor UHF propagation using numerical technique. *Proceedings of Canadian Conference on Electrical and Computer Engineering*, Vol. 2, 13–16 May 2001, pp. 1357–1362.

Wang F, Zhou W and Li L (2007) A technique of ultra high frequency measurement. *Proceedings of IEEE 21st International Frequency Control Symposium of European Frequency and Time Forum*, 29 May–1 June 2007, pp. 820–822.

Zelley CA and Constantinou CC (1999) A three-dimensional parabolic equation applied to VHF/UHF propagation over irregular terrain. *IEEE Transactions on Antennas and Propagation*, **47**(10): 1586–1596.

Zhang C and Mao Y (2006) Effects of cross section of mine tunnel on the propagation characteristics of UHF radio wave. *Proceedings of 7th International Symposium on Antennas, Propagation & EM Theory*, 26–29 October 2006, pp. 1–5.

Zhang YP, Zheng GX and Sheng JH (2001) Radio propagation at 900 MHz in underground coal mines. *IEEE Transaction Antennas Propagation*, **49**: 757–762.

Zheng HD, Xu Z and Huo Y (2008) Research on characteristics of the loss of UHF propagation in rectangular tunnels. *Proceedings of 4th International Conference on Wireless Communications, Networking and Mobile Computing*,1 2–14 October 2008, pp. 1–4.

Chapter 6
Mine-Wide Communication

6.1 Introduction

The near impossibility of communicating by radio in mines, tunnels, and similar underground environments using conventional techniques has slowly become accepted with resignation since the first recorded attempts of the early 1920s. It is true that frequencies in the very low frequency (VLF) and low-frequency (LF) bands may be used to communicate over limited ranges through the earth itself. With this realization of the problems, the need for effective radio communication in such underground conditions has also become more urgent (Chufo, 1978; Chiba et al., 1978; Isberg et al, 1982; Davis et al., 1984, 1989; Delogne, 1991; Chow, 1992; Saindon and Chow, 1993; Bandyopadhyay et al., 2002, 2009). The inductive-loop principle, using low frequencies and a laid conductor wire, has provided a partial solution in coal mines. The first hint of a satisfactory solution to the problem came in 1956 with the publication by Monk and Winbigler (1956) regarding the use of an imperfect transmission line as a distributed aerial for communicating with railway trains in tunnels at very high frequency (VHF). This crucial innovation went largely unnoticed at the time and was certainly ignored by mining interests. In recent years, however, this technique of using the leakage fields of a specially constructed and laid transmission line has received rapidly increasing attention as the solution to communication problems in a wide range of situations. Such lines themselves are variously known as "leaky feeders" or "radiating cables." The optimum range of frequencies for use with these techniques fortuitously "encompasses the standard VHF" mobile radio wavebands and so the production of special equipment apart from the leaky cables has not usually been necessary for mine-wide communication. In order to establish mine-wide communication, keeping in view of all the prevailing underground conditions like labyrinth path, complex geological conditions, corner and bends, and intrinsic safe limit, leaky feeder cables can be used as antenna, as well as transmitting lines throughout the mine.

The first leakage-field radio communication system was developed for the Longannet mine of National Coal Board, UK, in Scotland, to meet an urgent need for two-way communication with men who patrol a 9-km long conveyor belt. Commissioned in 1970, this is believed to be the first such system installed in any mine. The total system length required was well beyond the practical range of a single

L.K. Bandyopadhyay et al., *Wireless Communication in Underground Mines*,
DOI 10.1007/978-0-387-98165-9_6, © Springer Science+Business Media, LLC 2010

base station and therefore, three base stations were used. Afterward there has been international interest in the application of leaky feeder transmission lines as a means of extending the propagation of radio waves in underground (Hill and Wait, 1976a; Isberg, 1978; Delogne and Deryck, 1980; Delonge, 1982; Dudley et al., 2006, 2007).

Most of the analytical work for leakage field radio communication has been done by earlier researchers (Delogne, 1974; Martin and Webster, 1974). In-depth investigations in the analysis of waveguide systems have also been done by the earlier researchers (Wait and Hill, 1974; Martin, 1975b; Delogne, 1976; Chufo and Vancura, 1976). There are several important considerations in the selection and application of waveguides. Cougouille (1974) has taken a single wire suspended in the mine entry to be used as two conductor transmission line under certain selected conditions in European mines. The single wire will support two types of dominant modes: monofilar and bifilar (Delogne et al., 1973; Dejyck, 1971). In the monofilar mode, the forward current is carried by the transmission line and the return current is carried by the tunnel walls or structure. In the bifilar mode, the return current is carried by the outer conductor. The advantage of the monofilar mode is that it is readily excited and is received by an antenna placed in the tunnel. It has the disadvantage of higher attenuation due to loss in surrounding rock or mine structure. An important consideration in these systems is the conversion from the monofilar or asymmetrical mode to the bifilar or symmetrical mode (Hill and Wait, 1974b). Mode conversion is accomplished either discretely, by spaced discontinuities in the cable (Seidel and Wait, 1978; Klemenschits and Bonek, 1994), or continuously through a loosely braided outer conductor (Martin, 1972) or through spaced slots in a solid external shield of a coaxial cable (Jones, 1974). The attenuation and excitation of the modes on a two-wire line have been studied both analytically (Hill and Wait, 1974a) and experimentally (Dejyck, 1971). When the transmission line is a coaxial cable, the bifilar mode propagates between the inner and outer conductors, and the monofilar mode propagates via the outer conductor and the tunnel walls. A good comparison of bifilar and monofilar transmission lines has been made by Martin (1975a) and a summary of various coaxial cable leaky feeder systems has been prepared by Lagace et al. (1974). The detailed analysis of conductors in tunnels of circular and rectangular cross section has been implemented (Hill and Wait, 1974a, b; Mahmoud and Wait, 1974; Wait and Hill, 1974, 1975a, c, 1977a; Jakubowski, 1994). It is interesting to note that Martin's design (Martin, 1972) has equalized the phase velocities between the monofilar and bifilar modes. Wait and Hill (1974) has shown that the phase velocity of the single-wire mode varies with the proximity of the transmission line to the tunnel wall. A good analysis of the effect of external structures to leaky feeder cables is given by Cree (1974).

6.2 The System

6.2.1 Basic Principle

In the simplest use of the leakage-field principle, a single length of leaky feeder is threaded through a tunnel or mine to be served and connected to a single base station in place of a normal aerial to provide two-way communication with personal or mobile sets. A leaky feeder allows signals to leak out of or into itself at a controlled rate. It effectively behaves as a long antenna that can guide radio waves around corners and bends. The communication system is featured by VHF frequency-modulated (FM) high-band operation in the 146–174 MHz with an acceptable radio frequency (RF) power output as required for an underground mine. The cable is characterized by excellent frequency response over the required spectrum to meet the specifications of mine industries. It may be fire resistant, water resistant, etc., according to the needs of the same. The exponential decay characteristic of the signal in the line (conveniently expressed in dB/km) leads immediately to some important consequences. It may be noted that the attenuation of signals along tunnels not equipped with feeders is also logarithmic, typically 0.5–1 dB/m. Transmitter powers and receiver sensitivities are seen to be almost insignificant beside the parameters of the line itself, while the total range in a simple system may be effectively increased by putting the line amplifiers at every 350 m instead of one end to automatically compensate the RF loss or other additional losses in the cable. These amplifiers operate independently in both the forward and reverse directions. The intrinsically safe VHF transceivers operating within 2 W and leaky feeder cable as transmitting media, followed by line amplifier at regular intervals would be used for communication through galleries. Due to skin depth effect phenomena, the radio waves emit (leak) in larger periphery through leaky feeder cables and with help of line amplifier, the attenuated waves can be further amplified. By placing line amplifier at regular intervals, the entire mine area can be covered. Apart from that, the signal loss in the leaky cable is directly proportional to the operating frequency. Therefore, optimum frequency would be chosen to meet the operational constraint.

6.2.2 System Description

Leaky feeder-based communication system is used for mine-wide communications. The system has two channels transceivers (at 152 and 172 MHz). The system is totally flexible to fulfill individual site requirements and offers minimal system components. It is quite easy to install and maintain. The system has the following components: (i) leaky feeder cable, (ii) head-end unit/base station, (iii) line amplifier (LA), (iv) branch units (BUs), (v) power couplers (PC), and (vi) line terminator (LT). Fig. 6.1 depicts the block diagram of the system for underground mine.

(i) *Leaky feeder cable*: Leaky feeder cable acts as an extended antenna system throughout the workings. It gives 100% coverage where installed. It is easy to install and maintain.
(ii) *Head-end unit*: The main function of the head-end equipment is to interface the leaky feeder cable with the base stations and consists of (i) base station radios, (ii) base station power supplies, (iii) surface coverage antenna, (iv) leaky feeder interface, (v) emergency broadcast bridges, and (vi) diagnostic units.
(iii) *Line amplifier*: The main function of the LA is to amplify the upstream and downstream voice signals. It is connected with the leaky feeder cable at every

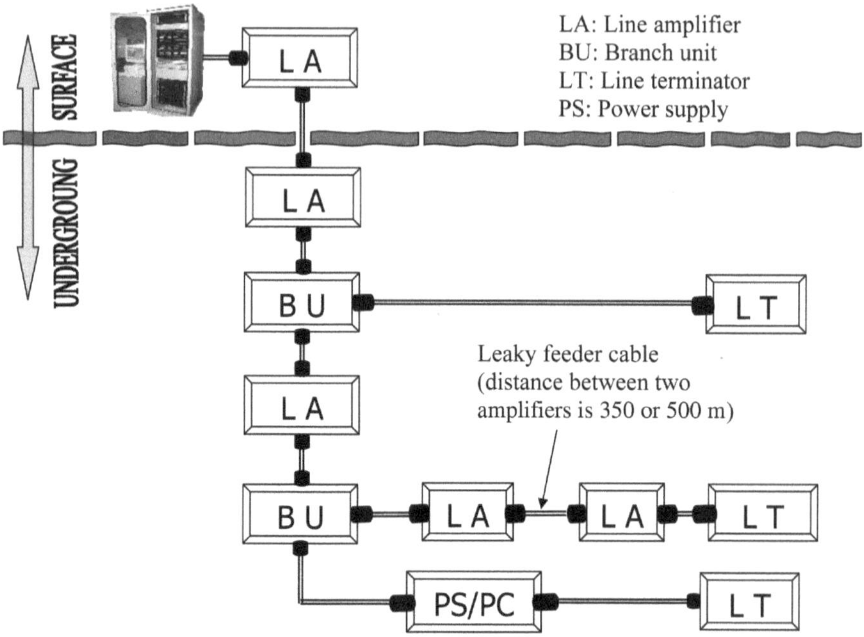

Fig. 6.1 Block diagram of the leaky feeder-based communication system

350 or 500 m intervals throughout the system to overcome cable attenuation. The LA is placed in IP68 enclosure. It is powered from the leaky feeder cable itself.

(iv) *Branch units*: The main objective of the BU is to route the voice in multiple directions wherever required. The BU is connected with the leaky feeder cable only at places where the voice or data is to be routed in the different underground mine galleries. It is placed in IP68 enclosure.

(v) *Power couplers*: Power coupler (PC) is used to reinject the DC power into the system. It is connected to the leaky feeder cable after 4–5 LAs for the same. It is enclosed in IP68 enclosure and glands.

(vi) *Line terminator*: The main objective of the LT is to provide the correct impedance match and load to the system and prevent water ingress into leaky feeder cables. Line terminators are used at each end of leaky feeder cable. The impedance of the line terminator is 75 Ω. It is also enclosed in IP68 enclosure.

6.3 Specifications of Leaky Feeder System

The specifications of the entire system are given in Table 6.1 and the specifications of the different components are given in the subsequent Tables 6.2, 6.3, 6.4, 6.5, and 6.6.

Table 6.1 Specifications of the leaky feeder-based communication system

Parameter	Value
Transmit frequency	152 and 172 MHz
Receive frequency	152.07 and 172.07 MHz
Number of channels	Two
Power supply voltage	12 V DC
Underground power distribution	Through the leaky feeder cable
System impedance	75 Ω
RF radiating medium	Leaky feeder cable
Maximum power radiated	15 mW
Radial RF coverage	20–22 m in coal mine
Head-end location	On surface
Operating temperature	−30°C to + 50°C

Table 6.2 Specifications of line amplifier

Parameter	Value
Transmit frequency	152 and 172 MHz
Receive frequency	152.07 and 172.07 MHz
Gain	16.5 dBm (typical)
Power requirement for LA	100 mA

Table 6.3 Specifications of branch unit

Parameter	Value
Impedance	75 Ω
Maximum insertion loss: main line	4.5 dBm
Maximum insertion loss: branch line	20 dBm
Gain	16.5 dBm (typical)
Power requirement for LA	100 mA

Table 6.4 Specifications of power coupler

Parameter	Value
Impedance	75 Ω
Maximum insertion loss	1 dBm
DC voltage drop	0.2 V DC

Table 6.5 Specifications of power supply unit

Parameter	Value
Input power	105–125 V AC, 1 A maximum
Output power	13.6 V DC, 1.25 A minimum
Operating temperature	0–50°C
Overload protection	Current limited, self-restoring

Table 6.6 Specifications of leaky feeder cable

Parameter	Value
Construction	
Inner conductor: plain annealed copper wire	Nominal diameter: 2.31 mm
Insulation: natural polythene thread and tube	Nominal diameter: 9.53 mm
Outer conductor: plain annealed copper wires lapped around the insulation	Nominal diameter: 10.73 mm
Sheath : Polyethylene compound black	Nominal diameter: 13.13 mm
Outer sheath : PVC yellow, 1.20 mm nominal radial	Nominal diameter: 15.53 mm
Physical Characteristics	
Minimum bending radius	100 mm
Maximum operating temperature	70°C
Electrical characteristics	
Maximum inner conductor resistance at 20°C	4.23 Ω/km
Maximum outer conductor resistance at 20°C	4.0 Ω /km
Nominal capacitance at 1 kHz	51 pF/m
Nominal characteristic impedance	75 Ω

6.4 Working Technique

There are various methods available for underground communication for the distribution of voice and data, but the most versatile and reliable wired cum wireless system is the so-called leaky feeder system. The leaky feeder cable is analogous to a surface antenna system. A good leaky feeder system will provide the basis for basic two-way voice and data applications. The leaky feeder basically contains a normal coaxial cable designed to pass RF signals from one end of the cable to the other with varying degrees of loss, depending on the quality. The purpose of the cable's outer copper shield is to keep as much signal as possible inside the cable and prevent external RF signals and/or interference from entering the cable. The leaky feeder is called so, as its external copper shield has "holes" or "gaps," allowing the RF signals to leak out evenly along its length, so they can be picked up by radios nearby. The holes in the leaky feeder cable also enable external RF signals to enter and run along the cable. In large mines, repeaters may also be used to amplify and retransmit incoming and outgoing signals to moving miners carrying portable radios. The spacing of these repeaters along the cable is governed primarily by the receiver sensitivity, the longitudinal attenuation rate of the cable, the coupling loss from the cable to the transceivers (usually VHF sets), and the transmitted power. The basic function of a leaky feeder system is to maintain a constant level of signal along a considerable length of leaky feeder cable; a solution which is both effective and reliable. The leaky feeder radio system is virtually noise free and is capable of both voice and data communications on all channels simultaneously. The system also has standard video band pass capabilities and on-board diagnostics. The system has a limited range because the high-frequency transmissions cannot pass through solid rock and limits the system to a line-of-sight application. It does, however, allow two-way mobile communication.

6.5 Advantages

Leaky feeder systems have the following advantages over the traditional antennae (Austin, 1990): (i) cannot be easily damaged, (ii) are not susceptible to corrosion, (iii) can provide consistent coverage, (iv) are fire resistant, etc.

6.6 Types of Cables

There are a variety of approaches presently in use or under evaluation as described below.

6.6.1 Long-Wire Antennas

A long-wire antenna (Cougouille, 1974; Herns, 1974; Liegeois, 1974) is an antenna that accepts power and radiates it efficiently on any frequency for which its overall length is not less than approximately half the wavelength. In general, the gain achieved with long-wire antennas is not as great as the gain obtained from the multi-element arrays. But the long-wire antenna has advantages of its own. The construction of long-wire antennas is simple, both electrically and mechanically, with no particularly critical dimensions or adjustments. The long-wire antenna will work well and give satisfactory gain and directivity over a frequency range up to twice the value for which it was cut. Another factor is that long-wire antennas have directional patterns that are sharp in both the horizontal and vertical planes. Also, they tend to concentrate the radiation at the low vertical angles.

6.6.2 Twin-Wire Feeders

A twin-wire feeder consists of two parallel conductors (Wait and Hill, 1977b) unlike coax that consists of two concentric conductors. The currents flowing in both wires run in possible directions but are equal in magnitude. As a result, the fields from them cancel out and no power is radiated or picked up. To ensure efficient operation, the spacing of the conductors is normally kept within about 0.01 times of the wavelength. The feeder exists in a variety of forms. Essentially, it is just two wires that are spaced in terms of RF of operation. In practical terms, manufactured feeders are available and consist of two wires contained within the plastic sheath that is also used as a spacer between them to keep the spacing, and hence the impedance constant. Another form, commonly called open-wire feeder, simply consists of two wires kept apart by spacers that are present at regular intervals along the feeder.

6.6.3 Delogne System

Various techniques have been used to implement the exchange of electromagnetic fields between the coaxial transmission line and the cavity of the mine entry. The Institute National des Industries Extractives (INIEX)//Delogne system uses a complete annular gap in the external conductor of the coaxial cable (Delogne, 1974). For the reduction of uncertain loss of the gap, the following circuits are routinely used:

(i) Gap is shunted by a capacitor, which reduces its impedance, while the capacitance effect is compensated at the operating frequency by an induction coil inserted in the external conductor. The bandwidth is about 20%.
(ii) Second, wide-band operation is obtained. It consists of a transformer with the windings. If the number of turns is equal then no magnetic flux would be created by the coaxial mode. But a slight difference in the number of turns is sufficient to achieve radiation. The system has been installed principally in Belgium, France, Germany, and USA.

6.6.4 Slotted Shield Cables

Slotted shield cable is designed in such a way that the outer shield of the helical coaxial cable is milled away to provide slots or apertures about 0.5×0.1 cm about every 2 cm. The cable is more expensive than other leaky cables but has performed quite well in numerous applications. The selection of the type of leaky feeder depends on the type of requirement such as whether an active or passive system is used.

6.6.5 Loose-Braided Cables

For the loose braid cable, the surface transfer impedance concept was used in order to characterize the braid, which was assumed thin with mesh dimensions small compared with the distance between the inner and the outer coaxial conductors. Considering, exclusively, the electromagnetic characteristics of the coaxial mode, it was concluded that (i) the radial decay of the outside field is very poorly influenced by the electromagnetic coverage of the outer conductor, (ii) the percentage of power that travels outside the coaxial structure is lower than the slotted cable, (iii) the characteristic impedance of the loose braid cable decreases with the increase of the mesh dimensions, and (iv) the conductor losses are higher for the loose braid cable. Calculations have been done for propagation of braided coaxial cable (Wait and Hill, 1975b; Hill and Wait, 1976b) and channel characteristics of a braided coaxial cable in mine tunnels (Mahmoud and Wait, 1976).

6.7 Drawbacks

Leaky feeder systems have the following four main drawbacks:

(i) Since the mine face moves up to 100 m in a day where coal is being extracted, the leaky feeder cable must constantly be extended. This is costly and is not always done on a timely basis.
(ii) The second issue is that the signal typically ranges up to about 20 m in radius from the feeder cable. While this provides good coverage for each tunnel it requires feeder cable to be strung in every location that requires coverage. Since this gets expensive, generally only well-traveled sections of the mine are covered.
(iii) Third, the signal is not carried over an approved Federal Communications Commission (FCC) band. Consequently, it may interfere with other equipment in the mine.
(iv) Finally, the system requires constant power. As the length of the cable increases, the signal must be amplified at intervals. These line amplifiers require power. Unfortunately in the event of an emergency communication (Stolarczyk, 1991) is shut down when the power is shut down.

6.8 Field Trial

The leaky feeder-based system was installed in Nandira Mine, Talcher Area, Mahanadi Coalfields Limited (MCL), Sambalpur, Orissa, India. A view of the installed system is depicted in the Fig. 6.2. The leaky feeder cable was laid on from control room available at pit top to different working locations of the underground mine covering total distance of about 5 km. The head-end unit/base station of the system was placed in the control room and leaky feeder was interfaced with the head-end unit as per the installation guidelines. Three line amplifiers as depicted in Figs. 6.3 and 6.4 were placed at distances 350, 700, and 1,050 m from the head-end unit, in the straight gallery to maintain the required signal strength. The first branch unit of the gallery was at 975 m. Four line amplifiers were placed to the left side from the first branch unit of the gallery at 275, 625, 975, and 1,250 m. Two line amplifiers were placed to the right side from the first branch unit of the gallery at 275 and 625 m. Branch unit as shown in Fig. 6.5 was placed near the first branch of the straight gallery at 1,050 m to route the voice in different galleries of underground mine. At each end, a line terminator was interfaced with leaky cable to provide the correct impedance match and proper load to the system and to prevent water ingress into leaky cables. Power couplers as shown in Fig. 6.4 were placed at 350 and 1,675 m for reinjecting the 12 V DC power into the system at those distances. Intrinsically safe (IS) power supply units with sufficient back up as depicted in Fig. 6.6 were

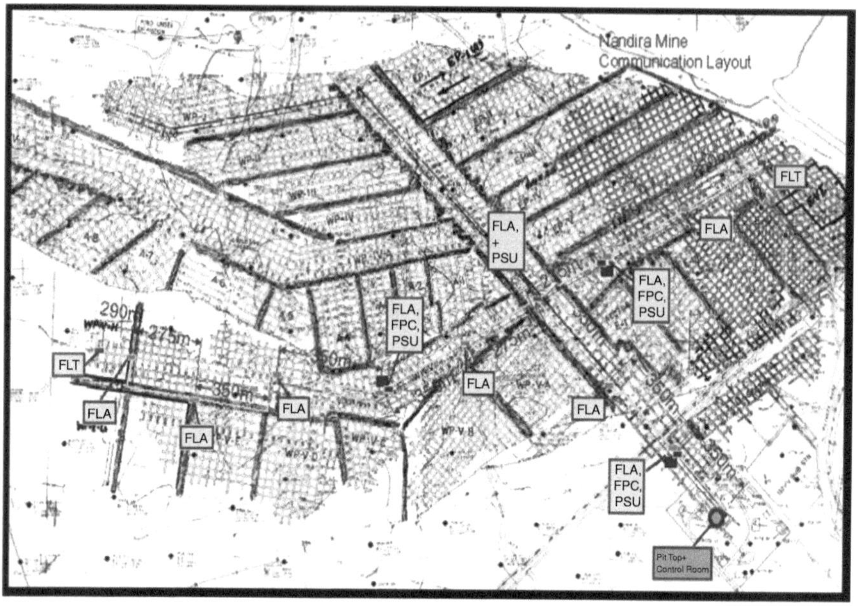

Fig. 6.2 Leaky feeder-based communication layout in the Nandira Mine, MCL

Fig. 6.3 A view of line amplifier installed in Nandira Mine of MCL

placed at 350 and 1,675 m to provide the required voltage as well as to provide back up in case of power failure. Ten IS transceivers were provided with the system for using the system by mine executives (Fig. 6.7). The system was installed in the Nandira Mine, MCL, on December 2, 2006, and since then the system is operating successfully.

Fig. 6.4 A view of line amplifier and power coupler installed in Nandira Mine of MCL

Fig. 6.5 A view of branch unit installed in Nandira Mine of MCL

6.9 Development of Passive Amplifier

6.9.1 Description

The passive amplifier for leaky feeder-based underground communication is designed to extend the coverage area of the communication beyond the limit of the leaky antenna in an underground mine at 152 MHz operating frequency. The system provides an amplifier for increasing the radiation field strength of a radiating leaky cable and enhancing the low level of radiation in the leaky cable itself to avoid use of booster for amplifying the signal. The passive amplifier for leaky feeder-based underground communication consists of leaky feeder cable of 15.53 mm diameter

Fig. 6.6 A view of IS power supply unit installed at Nandira Mine

Fig. 6.7 A view of VHF transceiver used for communication with the system

with a central conductor of 2.31 mm thickness, which is subsequently connected with a conducting bar of quarter wavelength long (49.3 cm at 152 MHz system's frequency). The conducting bar is fabricated from the copper and is concaved in such a way that it should held on the leaky feeder cable. A quarter wavelength long antenna of the same leaky feeder cable with its outer conductor removed is connected through a BNC connector.

6.9.2 Circuit Diagrams

Figures 6.8 and 6.9 illustrate the passive amplifier for leaky feeder-based underground communication. Figure 6.8 shows the perspective view of the passive amplifier for leaky feeder-based underground communication. Figure 6.9 illustrates the

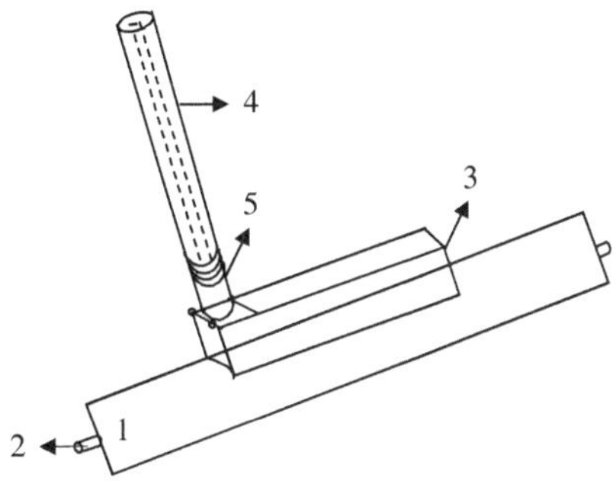

Fig. 6.8 Perspective view of the system

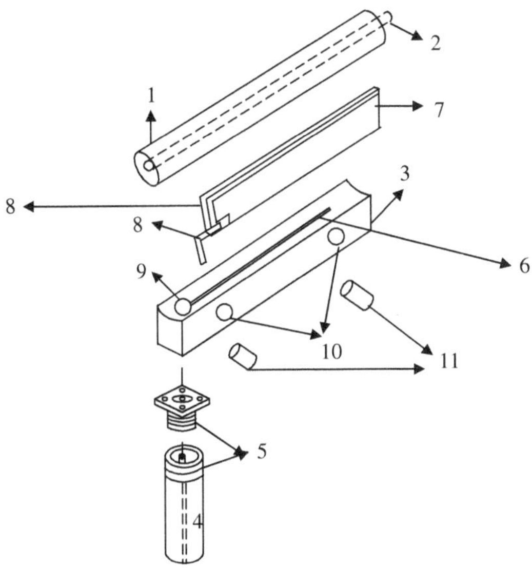

Fig. 6.9 Explored view of the system. 1, Leaky feeder cable; 2, central conductor; 3, a rectangular conducting bar; 4, antenna; 5, connector; 6, slot; 7, a printed circuit board; 8, a conductive strip of copper; 9, a hole; 10, two holes; and 11, two pins

explored view of the passive amplifier for leaky feeder-based underground communication. The details of Figs. 6.8 and 6.9 are described below.

The perspective view as depicted in Fig. 6.8 of the passive amplifier consists of leaky feeder cable (1) of 15.53 mm diameter with the central conductor (2) of 2.31 mm thickness which is subsequently connected with a conducting bar (3) which is a quarter wavelength long, i.e., 49.3 cm at 152 MHz system's frequency. The conductor bar is fabricated from the copper and is concaved in such a way that it should held on the leaky feeder cable (1). A quarter wavelength long antenna (4) of the same leaky feeder cable with its outer conductor removed is connected through a BNC connector (5). The length of the quarter wavelength antenna is 49.3 cm as the system's operating frequency is 152 MHz.

Figure 6.9 illustrates the detailed construction of the passive amplifier. The passive amplifier consists of leaky feeder cable (1) of 15.53 mm diameter with the central conductor (2) of thickness 2.31 mm. A conducting bar (3), which is a quarter wavelength long, i.e., 49.3 cm at 152 MHz system's operating frequency, is fabricated with copper and concaved to held on the leaky feeder cable (1). A slot (6) is milled along the length of the concaved conducting bar (3) in concaved side of the conducting bar. The slot (6) is dimensioned to adjust a printed circuit board (7), which is approximately a quarter wavelength long to the system's operating frequency to support a conductive strip of copper (8). A connecting wire (9) is connected to the conductive strip to attach with the antenna (4) by means of the BNC connector (5), which adjusts the length of the conductive strip (8) to the quarter wavelength of the system's operating frequency 152 MHz. The slot (6) is milled to a depth such that the printed circuit board (7) is secured in the slot and the portion of the board protruding above the concave surface will hold the conductive strip (8) within the slot (6) milled up to the proximity of central conductor (2) of the leaky feeder cable. Two holes (10) are bored through the concaved said copper conducting bar (3). The pins (11) are driven through the holes (10) to lock the printed circuit board (7) in the position. A hole (12) is bored through the conducting bar (3) at one end of the said slot (6) so that the said center conductor (2) of the leaky feeder cable (1) can be connected to the conductive strip (7). A quarter wavelength long antennas (4) of the same leaky feeder cable with its outer conductor removed are connected through a BNC connector (5). The length of the quarter wavelength antenna is 49.3 cm as the system's operating frequency is 152 MHz.

6.9.3 Specifications

The system consists of a leaky feeder cable, a concaved copper bar milled to support a conducting strip of a length of around $\lambda/4$ of the 152 MHz, and a BNC connector to be connected with the conductive strip. The diameter of the leaky cable is 15.53 mm with nominal attenuation at around 160 MHz (4.3 dB/100 m) for radiating the signals of 152 MHz. The length of the concaved copper bar is $\lambda/4$, i.e., 49.3 cm for 152 MHz signal and the thickness of the bar is 2 cm. The specifications of the different units of the system as shown in Figs. 6.8 and 6.9 are as follows:

- ► Leaky feeder cable (1): diameter – 15.53 mm
- ► Central conductor (2): diameter – 2.31 mm
- ► A rectangular conducting bar (3): length – 49.3 cm at the system's operating frequency 152 MHz, width – 2 cm, thickness – 2 cm
- ► Antenna (4): length – 49.3 cm
- ► Connector (5): BNC connector

- ▶ Slot (6): length – 49 cm, depth – 0.5 cm, width – 1.5 mm
- ▶ A printed circuit board (7): length – 49 cm, width – 1 cm, thickness – 1.5 mm
- ▶ A conductive strip of copper (8): length – 49 cm
- ▶ Two holes (10): diameter – 2 mm, depth – 0.5 cm
- ▶ Two pins (11): diameter – 2 mm, length – 1.5 cm
- ▶ One hole (12): diameter – 2 mm, depth – 1.5 cm

6.9.4 Field Installation

In case of cable fault in leaky feeder communication systems working in underground mines, it is not possible to replace the cable or place an amplifier due to physical and other constraints, and signal levels falls below the design requirement. Therefore, a simple way of enhancing the signal is desirable. Therefore, the effective way is to tap into the cable and place a simple antenna at the location of degraded signal. In some of the cases, it is required to bring coverage into areas adjacent to leaky cable coverage area. In that case also, the system can be useful.

6.9.5 Testing of Antenna

(i) Experimental setup

For measurement of radiation pattern of the antenna in the laboratory, the surrounding was kept free from all kinds of electromagnetic reflectors and absorbers, because it can affect the performance of the antenna. The antenna was placed at a smooth place. The power was fed to the antenna by a signal generator of Model – TAYLOR 68 A. The range of the frequency of signal generator used here ranges from 100 kHz to 240 MHz. It was connected with a power supply of 220 V, 50 Hz AC. The experimental setup used is illustrated in Fig. 6.10.

For the measurement of radiation pattern, a Spectrum Analyzer (Version 3.05 of Anritsu make) was used. The model of Spectrum Analyzer used was 2711D. The frequency span of Spectrum Analyzer was set at 100 kHz.

(ii) Results and discussion

At first, using the Spectrum Analyzer readings were taken without antenna. Readings were again taken at a distance of 0–4 m at an interval of 1 m with antenna. The measured values of gain with and without antenna are given in Table 6.7. From the experimental results, it was found that there was around –10 dBm gain with the use of antenna.

Measurements were again taken in different directions around the passive antenna. The distance of measurement from the passive antenna was 1 m and

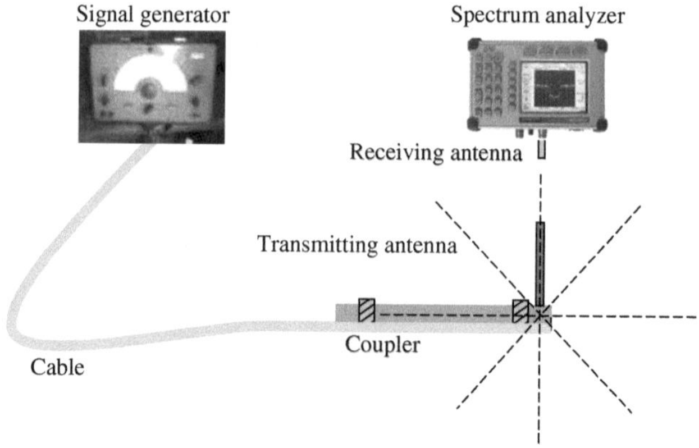

Fig. 6.10 Experimental setup

Table 6.7 Measured gain at different linear distance

Distance (m)	Gain (dBm) with antenna	Gain (dBm) without antenna
0	–47	–57
1	–54	–64
2	–58	–70
3	–66	–74
4	–70	–74

Fig. 6.11 Radiation pattern of transmitting antenna at a distance of 1 meter

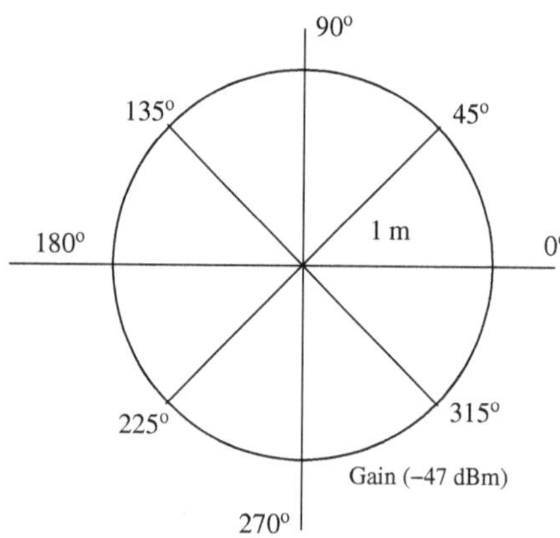

readings were taken in the directions of 0°, 45°, 90°, 135°, 180°, 225°, 270°, and 315° around the passive antenna. Using the data, a graph was plotted (Fig. 6.11) and it was found that radiation pattern was circular with a gain of –47 dBm at a distance of 1 m.

6.9.6 Capabilities

The main capabilities of the passive amplifier for leaky feeder-based underground communication are:

(i) The system is capable of extending the coverage area of the communication beyond the limit of the leaky antenna in an underground mine.
(ii) The system is capable of increasing the radiation field strength of a radiating leaky cable.
(iii) The system has provision for using the leaky cable for extending coverage area of communication in underground mine.
(iv) The system is capable of increasing the low level of the radiation in the leaky cable itself to avoid use of booster for amplifying the signal.
(v) The system provides a means for increasing radiation field strength of a leaky cable without enhancing input signal strength or use of repeater.
(vi) The system provides an amplifier to be installed in the cable without splicing.
(vii) The system is capable of increasing the level of 152 MHz signal into an underground mine.

6.10 Conclusions

Leaky feeder-based communication system is suitable for establishing data, voice, and video communication in an underground mine. The system has line amplifiers at about 350 or 500 m intervals and two channels (at 152 MHz and 172 MHz). The input supply to the system is 12 V DC and the system is intrinsically safe for use in underground mine. The wireless communication range around the leaky feeder cable is around 20–22 m in underground coal mine. To enhance the communication range in the required portion of an underground mine, a passive amplifier has been designed, which has to be coupled with the leaky feeder cable. The system provides an amplifier for increasing the radiation field strength of a radiating leaky cable and enhancing the low level of radiation in the leaky cable itself to avoid use of booster for amplifying the signal. With the passive amplifier the wireless communication range around the leaky feeder cable has increased up to 50 m.

References

Austin BA (1990) Medium frequency body loop antenna for underground use. *Proceedings of IEE Colloquium on Electrically Small Antennas*, London, UK, Vol. 3, 23 October 1990, pp. 1–5.

Bandyopadhyay LK, Chaulya SK and Kumar S (2002) A proposed wireless communication system for underground mines. *Proceedings of the International Conference on Mineral Industry: Issues on Economics, Environment and Technology*, Mining, Geological and Metallurgical Institute of India, Kolkata, India, pp. 317–324.

Bandyopadhyay LK, Chaulya SK, Mishra PK, Choure A and Baveja BM (2009) Wireless information and safety system for mines. *Journal of Scientific and Industrial Research*, **68**: 107–117.

Chiba JC, Inaba T, Kuwamoto Y, Banno O and Sato R (1978) Radio communication in tunnels. *IEEE Transactions on Microwave Theory Technique*, **26**: 439–443.

Chow S (1992) Propagation of radio waves in potash mine tunnels and its implication on radio systems performance. *Proceedings of International Conference on Computer Science*, Victoria, Australia, Vol. 2, 16–20 November 1992, pp. 890–894.

Chufo RL (1978) Medium frequency mine communications. *Proceedings of 28th IEEE Vehicular Technology Conference*, Colorado, USA, Vol. 28, 22–24 March 1978, pp. 261–266.

Chufo RL and Vancura PD (1976) The installation, operation and maintenance of an underground mine-wide radio system. *IAS Annual 1976*, pp. 1196–1202.

Cougouille R (1974) Radio transmission of speech in the French coal mines. *Preliminary Report*, Bureau of Mines Contract no. HO122061.

Cree DJ (1974) Railway radio communication using radiating coaxial cable. *Proceeding of International Conference of Radio: Roads, Tunnels and Mines*, Liege, Belgium, Vol. 2, pp. 187–199.

Davis QV, Martin DJR and Haining RW (1984) Microwave radio in mines and tunnels. *Proceedings of 34th IEEE Vehicular Technology Conference*, Pittsburgh, USA, Vol. 4, 21–23 May 1984, pp. 31–36.

Davis R, Simpson A and McGeehan JP (1989) Preliminary wireless propagation results at 1.7 GHz using a half wave dipole and a leaky feeder as the transmitting antenna. *Proceedings of Fifth International Conference on Radio Receivers and Associated Systems*, Bath, UK, 23–27 July 1990, pp. 10–14.

Dejyck L (1971) Mines et carrisres. *Bulletin Technique,* Institute National Des Industries Extractives, Liege, Belgium.

Delogne P (1974) The INIEX mine communications systems. *Proceeding of International Conference of Radio: Roads, Tunnels and Mines*. Liege, Belgium, Vol. 2, pp. 129–136.

Delogne P (1976) Basic mechanisms of tunnel propagation. *Radio Science*, **11**: 295–303.

Delonge P (1982) *Leaky Feeders and Subsurface Radio Communications*. Stevenage, UK.

Delogne P (1991) EM propagation in tunnels. *IEEE Transactions on Antennas and Propagation*, **39**(3): 401–406.

Delogne P and Deryck L (1980) Underground use of a coaxial cable with leaky sections. *IEEE Transactions on Antennas and Propagation*, **28**(6): 875–883.

Delogne P, Deryck L and Liegeois R (1973) Guided propagation of radio waves. *Proceedings of Through-the-Earth Electromagnetics Workshop*, Colorado, USA, 15–17 August, 1973, pp. 49–52.

Dudley SEM, Quinlan TJ and Walker SD (2006) 1.6 Gb/s hybrid wireless-optical leaky feeder system for tunnel applications. *Proceedings of International Conference on Railway Condition Monitoring of the Institution of Engineering and Technology*, Birmingham, UK, 29–30 November 2006, pp. 204–206.

Dudley SEM, Quinlan TJ and Walker SD (2007) 1.6 Gb/s data throughput optically-remoted leaky feeders for underground transport environments. *Proceedings of IEEE 65th Vehicular Technology Conference*, Dublin, Ireland, 22–25 April 2007, pp. 520–524.

Herns CD (1974) Practical experience gained with S.E.L. and Funke and Hunter radio systems for locomotives. *Proceeding of International Conference of Radio: Roads, Tunnels and Mines*, Liege, Belgium, Vol. 2, pp. 100–109.

Hill DA and Wait JR (1974a) Gap excitation of an axial conductor in a circular tunnel. Journal of *Applied Physics*, **45**: 4774–4777.

Hill DA and Wait JR (1974b) Excitation of monofilar and bifilar conductor modes on a transmission line in a circular tunnel. *Applied Physics*, **45**: 3402–3406.

Hill DA and Wait JR (1976a) Calculated transmission loss for a leaky feeder communication system in a circular tunnel. *Radio Science*, **11**(4): 315–321.

Hill DA and Wait JR (1976b) Propagation along a braided coaxial cable located close to a tunnel wall. *IEEE Transactions on Microwave Theory and Techniques*, **24**(7): 476–480.

Isberg RA (1978) Radio communication in subways and mines through repeater amplifiers and leaky transmission lines. *Proceedings of 28th IEEE Vehicular Technology Conference*, Colorado, USA, Vol. 28, 22–24 March 1978, pp. 248–254.

Isberg RA, Cawley JC and Chufo RL (1982) The design and implementation of a UHF radio system using distributed antennas, passive reflectors and two-way signal boosters in a room and pillar limestone mine. *Proceedings of 32nd IEEE Vehicular Technology Conference*, California, USA, Vol. 32, 23–26 May 1982, pp. 259–267.

Jakubowski RJ (1994) Results of distributed antenna and leaky feeder systems tests at 800 MHz in the Washington D.C. metro system tunnels. *Proceedings of IEEE 44th Vehicular Technology Conference*, Stockholm, Sweden, 8–10 June 1994, pp. 1113–1116.

Jones RW (1974) Performance of slotted cables for radio communication in mines. *Proceeding of Colloquium on Leaky Feeder Radio Communication System*, 21–22 April 1974, University of Surrey, Surrey, England, pp. 171–182.

Klemenschits T and Bonek E (1994) Radio coverage of road tunnels at 900 and 1,800 MHz by discrete antennas. *Proceedings of 5th IEEE International Symposium on Personal, Indoor and Mobile Radio Communications, Wireless Networks – Catching the Mobile Future*, The Hague, Netherlands, Vol. 2, 18–23 September 1994, pp. 411–415.

Lagace RL, Ginty JJ, Roetter MF and Spencer RH (1974) Leaky coaxial cable for guided wireless communications systems. *Final Report of Bureau of Mines Contract No.: H0122026*, Vol. 3, pp. 1–24.

Liegeois R (1974) Communications par radio dans les tunnels. *Proceeding of International Conference of Radio: Roads, Tunnels and Mines*, Liege, Belgium, Vol. 2, pp. 200–211.

Mahmoud SF and Wait IR (1974) Theory of wave propagation along a thin wire inside a rectangular waveguide. *Radio Science*, **9**: 417–420.

Mahmoud SF and Wait JR (1976) Calculated channel characteristics of a braided coaxial cable in a mine tunnel. *IEEE Transactions on Communication*, **24**: 82–87.

Martin DJR (1972) Transferred surface currents in braided coaxial cables. *Electron Letters*, **18**: 465–466.

Martin DJR (1975a) A general study of the leaky feeder principle. *Radio Electronics Engineering*, **45**: 206–214.

Martin DJR (1975b) Underground communications – the new role of radio. *Colliery Guardian Annual Review 1974*, pp. 99–106.

Martin DJR and Webster R (1974) The use of radio in British coal mines. *Proceeding of International Conference of Radio: Roads, Tunnels and Mines*, Liege, Belgium, Vol. 2, pp. 110–128.

Monk NM and Winbigler HS (1956) Communication with moving trains in tunnels. *IEEE Transaction on Vehicular Communication*, **7**: 21–28.

Saindon, JP and Chow S (1993) Propagation of radio waves in underground tunnels and underground radio systems. *Proceedings of IEEE Eighth International Conference on Antennas and Propagation*, Edinburgh, UK, Vol. 1, 22–24 September 1993, pp. 155–158.

Seidel D and Wait J (1978) Role of controlled mode conversion in leaky feeder mine communication systems. *IEEE Transactions on Antennas and Propagation*, **26**(5): 690–694.

Stolarczyk LG (1991) Emergency and operational low and medium frequency band radio communications system for underground mines. *IEEE Transaction on Industrial Application*, **27**: 780–790.

Wait JR and Hill DA (1974) Guided electromagnetic waves along an axial conductor in a circular tunnel. *IEEE Transaction on Antennas Propagation*, **22**: 627–630.

Wait JR and Hill DA (1975a) Electromagnetic fields of a coaxial cable with an interrupted shield located in a circular tunnel. *Journal of Applied Physics*, **46**(10): 4352–4356.

Wait JR and Hill DA (1975b) Propagation along a braided coaxial cable in a circular tunnel. *IEEE Transactions on Microwave Theory and Technique*, **23**: 401–405.

Wait JR and Hill DA (1975c) Electromagnetic fields of a dielectric coated coaxial cable with an interrupted shield. *Technical Memorandum*, U.S. Office of Telecommunications, June 1975.

Wait JR and Hill DA (1977a) Analysis of radio frequency transmission in a semicircular mine tunnel containing two axial conductors. *IEEE Transactions on Communication*, **25**: 1046–1050.

Wait JR and Hill DA (1977b) Electromagnetic fields of a dipole source in a circular tunnel containing a surface wave line. *International Journal of Electronics*, **42**(4): 377–391.

Chapter 7
Web-Based Information and Decision Support System for Mining Industry

7.1 Introduction

Mining industry is lacking in application of information technology (IT) (Das, 2001; Kumar and Guha, 2001). To achieve the future production target, coal mining industry has to improve its productivity and safety by enhancing mechanization (Reid et al., 2000; Hainsworth et al., 2001; Sen, 2001), increasing online monitoring (Bandyopadhyay et al., 2005a), implementing wireless communication systems in underground (Bandyopadhyay et al., 2005b), and adopting appropriate application-based IT system (Prasad, 1997; Chandramouli and Singh, 2003; Chaulya et al., 2008). This chapter presents a web-based information and decision support system required for coal mining industry.

7.2 Required Areas of IT Application in Coal Mines

An effective IT system aims to provide increased productivity and profitability with an eye on the miner's safety working in the mines. An integrated IT system with different modules has to be implemented in a mining area as described below.

7.2.1 *Improving Production and Productivity*

For effective mine management, managerial staff requires a complete knowledge of daily targets and achievements, output per man shift for piece rated, time rated, and overall categories, reasons for shortfall, etc. The real-time measurement of production using sensor and computer (online data feeding from dispatch conveyor or tippler where raw coal is dumped) can make mine manager proactive instead of doing production analysis and take decision at shift end or day after. Sensor as weight-o-meter will directly send data to computer. Analysis of breakdown reports obtained through feeding data from the supervisor's reports into computer, will clearly give breakup of total time lost and its item-wise cause. After going through web-based reports generated by computer, it will be easy to identify gray areas and priorities

L.K. Bandyopadhyay et al., *Wireless Communication in Underground Mines*,
DOI 10.1007/978-0-387-98165-9_7, © Springer Science+Business Media, LLC 2010

can be set for allocation of scarce resources to improve production and productivity. Further, using machine status sensors, IT system can be used for monitoring machine operation period and calculation of production in shift-wise, day-wise, month-wise manner, etc.

7.2.2 *Shift and Personnel Management*

IT will give new impetus to major functions like (a) online monitoring of production, (b) day-to-day planning, (c) resource management, and (d) face quality control. Employee personnel records can be effectively maintained at the mine level only, preventing delay in dealing with matters related to leave records, promotions, transfers, and other matters. Preparation of payrolls can also be possible at the mine level only. Performance and attendance of individuals can also be ascertained by keeping separate records of performance of each person employed for improving productivity and discipline.

7.2.3 *Reduction of Production Discrepancy*

In most of the mines, there is coal production discrepancy (13–16%), which leads to loss per tonne of coal produced. Main reason of discrepancy is coal tub, which miners fill in such a way that it looks as full, but tubs are partially loaded. Miners are paid in terms of number of tubs they fill. Using weight-o-meter, discrepancy will be reduced and mine will have to pay revenue as per their actual production in weight-o-meter. The production data will be recorded online in the system for information at different management levels.

7.2.4 *Maintenance of Equipment*

In mining industry, a number of equipment related to production and transportation are regularly maintained in order to avoid any failure or breakdown. Maintenance cost at mechanized mines (35% of operating cost of the system) goes as high as 50–60% when both direct and indirect costs are considered.

A computerized maintenance and management of equipment module has to be developed to overcome existing problems. Materials/spares have to be issued against predefined breakdown/preventive maintenance code. The modules have to be developed for (i) routine condition monitoring of major equipment using online sensors; (ii) recording frequency and duration of all breakdown events; (iii) categorizing all events by required status codes; (iv) administering defect identification and quantification; (v) linking reliability to equipment performance and duty; (vi) automating work order generation and administration; (vii) tracking engine and wheel motor hours; (viii) automating generation of service schedule; (ix) tracking fuel consumption and allocation; (x) allocating maintenance resources and priority setting; (xi) coordination of production and maintenance function; (xii) management of tire inventory and performance; (xiii) major component tracking; and (xiv) correlating maintenance performance against equipment duty.

7.2.5 Management of Inventory

Computerized system will reduce the store cost of each mine. If the material list of all the stores of a company is in a web then any mine can draw material from any store of other mine (essentially from same company) so that nonmoving item of one mine can be utilized by others. The system can have predetermined triggers so that dependence on individual decisions is minimized. Unique material code and nomenclature across the company stores will reduce inventory holdings. The study of obsolete/nonmoving items may help companies to decide on write-off. This will reduce overall costing of the inventory and remove the duplicity of items by combining part number and vendors' specifications. Different type of reports can be generated easily as per the need of mine management.

7.2.6 Environment Monitoring

In underground coal mines, generally portable devices (methanometer, CO meter, anemometer, thermometer, etc.) are used for environmental monitoring and recording huge data in record book for future use. Such measurements are infrequent and subjective to human failures, causing lapse on safety precautions.

To overcome existing problems, online sensor-based environmental monitoring system has to be integrated with the required system for continuous monitoring of vital environmental parameters in normal condition [methane (CH_4), carbon monoxide (CO), oxygen (O_2), air velocity, and temperature] and in case of sealed-off area [temperature, CH_4, CO, O_2, nitrogen oxide (NO), nitrogen dioxide (NO_2), sulfur dioxide (SO_2), and hydrogen sulfide (H_2S)]. The system provides online visual representation of trend of all monitored parameters and gives audiovisual warning signal when a particular parameter crosses respective threshold limit so that mine management can immediately take appropriate action. This would also help in enhancement of productivity and profitability of a mine, proper ventilation planning, early detection of fire, indicating occurrence of high concentration of gases, reducing response time, and minimizing dependence on human errors.

7.2.7 History of Mine

Feeding mine details and updating it regularly in database will help in taking due course of actions in already mined areas. All historical information has to be preserved in server and can easily be made available for future mine planning, preparation of environmental impact assessment and management plan reports, environmental compliance report, any R&D study in the mine, court of enquiry, statutory compliance, preparation of mine closure plan, and other necessary purposes.

7.2.8 Disaster Forecasting and Mine Safety Management

Every year disasters occur in coal mines causing death of miners, and loss of coal and property. Underground pumps are often sited at remote locations, and their operation and maintenance is very difficult. Online monitoring of pump operation and water level using sensors is highly essential for early detection of fault in the pump. A microprocessor-based multichannel intrinsically safe real-time environmental monitoring system has to be used for online continuous monitoring and providing early audiovisual warning signal to forecast different category of disasters using special sensors and techniques (Table 7.1). The system will also provide online visual representation of trend of all monitored parameters and give audiovisual warning signal when a particular parameter crosses respective threshold limit so that mine management can immediately take appropriate action.

Using computer database, all safety- and rescue-related data can be uploaded in the system for future use. Analysis of data can identify possible areas of weakness in mines safety system, and data can be used as a guideline for the decision-making process to improve mine safety performance. Online monitoring of exhaust fan speed and pressure development across fan has to be done using tachometer

Table 7.1 Techniques for disaster forecasting in underground coal mines

Category of disaster	Parameters to be monitored for forecasting	Sensor/technique
1. Explosion of flammable gas	Methane (CH_4)	CH_4 sensor
2. High concentration of toxic gas	Carbon monoxide (CO)	CO sensor
3. Mine fire due to self heating of coal	Temperature of coal strata	Temperature sensor
4. Roof fall	Movement of roof strata	Low-frequency acoustic sensor
5. Inundation	Demarcation of water-logged area, and maintaining safe distance between working area and water-logged or old working area	Updating and analysis of day-to-day survey data by ground-penetrating radar (GPR)
6. Pump management	Running status of pump and early warning of increasing water level	Pump status sensor (ON/OFF mode) and water-level indicator

and air pressure sensors, respectively, for proper ventilation in underground coal mines.

7.2.9 Statutory Requirements

All statutory-related data (acts, rules, regulations, and standards) can be fed into computer for accessing required information as and when required. Compliance to the statutory requirement can be fulfilled easily and effectively with less manpower and cost.

7.2.10 Postdisaster Management

Computerized system has to be developed to provide accident analysis, accident prone areas, probable remedies, emergency response plan, list of first aid, rescue trained personnel along with their address, telephone numbers, and place of duty during an emergency. Mobilization of manpower and resources can be done effectively to rescue the trapped miners without any delay. Sometimes during any mishap, there is misplacement of statute books, like manager's diary, supervisor's report, etc., but if these reports are filled in web then there is no chance of losing data/information. This information will help expediting the court of enquiry in case of accidents/or a disaster. A day-to-day action taken by safety managers can help in avoiding any casualty and if any thing goes wrong then the logged data can help in rectifying the problem. It will also speed up actions in providing assistance and benefits to affected people. Computerized attendance of miners can be easily done and can be used during emergency.

7.2.11 Improvement in Working by Online Record Keeping

In underground coal mines, lot of records are maintained manually for (a) explosive magazine report to prevent pilferage; (b) cap-lamp maintenance report books to maintain lamp; (c) fan record book to keep an online record of the mine resistance, pressure developed, etc.; (d) issue registers of safety items, like safety lamps, methanometers, CO meters, etc.; (e) safety wears like helmets, safety boots, etc.; and (f) ventilation records, like air measurements at each station, maintenance of fire-stopping records, stone dust barriers, stone dust quality, inventory maintenance of ventilation department like cement, stone dust, water pipes, bricks, etc. Even a ventilation officer has to maintain about 50 registers.

A computerized database for different purposes has to be prepared. Technical persons can concentrate on their technical inspection to improve upon the actual mine condition near the workings in underground. Working culture of the mine has to be improved with efficient management and improved productivity.

7.2.12 Wireless Communication in Underground Mine

A proper and reliable wireless communication system in underground mines will save machine break down time, and also help in immediate transfer of message from the vicinity of underground working area to surface for speedy rescue operation. Different types of wireless communication systems are developed, namely, personnel emergency detector, mine tracking system, leaky feeder-based system, induction-based system, very high frequency (VHF)/ultra-high-frequency (UHF) transceivers, trapped miner locator, etc., and these systems has to be integrated with the required IT system.

7.2.13 Decision-Making

An appropriate audiovisual communication link has to be established for effective, fast, and instant decision-making at different management levels and keeping history of important decisions in database for future reference. Audiovisual communication featured with integrated software would help to build a decision support system. It will also allow for storing important communication, which can be used as reference in taking the decision in similar circumstances. Important office circulars, orders, and decisions can be scanned and uploaded in intranet for wide and fast circulation and history keeping of such document.

7.2.14 Training

Computerization will help trainers for maintaining different forms. Various training modules have to be made available in required IT system for implementation in the training programs. Identification of the need for training will be easy through computerized database. With computerized training center, technical trainers will be relieved of doing unproductive work. Training facility, imparted by HRD employees through audiovisual communication, would enhance employees' expertise and morale, and enable more participation with no incremental costs. The quality of training has to be improved to provide a technically sound, motivated, alert, and safety-conscious workforce.

7.3 System Description

Required web-based information and decision support system has to be implemented especially for connecting coal mines of different regions with their respective area office (AO) and head office (HO). The system should have appropriate software and database, middleware, and hardware.

7.3.1 Software and Database

Separate database and module is essential for each category as described above. With the development of application software, networking software, and specialized software, an integrated IT system has to be made available to coal mining industry having following facilities: (i) different modules for data entry, logging, online monitoring, processing, analysis, storing, and transmitting; (ii) web page for dissemination of mining information, information exchange, video conferencing, and decision-making utilities; (iii) decision support system with mining database; and (iv) network architecture for connecting mining sites with AO and HO and providing communication strategies.

Many database software are available for data storing and maintaining, namely Oracle, SQL Server, MS Access, MS Excel, DB2, MySQL, D2K. Oracle database may be used for data storage, as it is more compatible and secure than other available databases. Initially, database can be designed on the basis of collected information. Manager/concerned authority can recognize specific data/information related to safety, productivity, maintenance of equipment, statutory requirement, environmental monitoring, and other information of a mine. The collected information of a mine can be categorized as: (i) production, (ii) mine operational, (iii) geological, (iv) geotechnical, (v) technological, (vi) mining and environmental standards, (vii) environmental status, (viii) mine planning and design, (ix) store inventory, (x) equipment maintenance, (xi) finance and accounting, (xii) store and purchase, (xiii) R&D, (xiv) policy and decision, (xv) legal, (xvi) sales, (xvii) procurement status, (xviii) safety, (xix) medical, (xx) attendance, (xxi) accident, (xxii) lease, (xxiii) HRD, (xxiv) training, and (xxv) pay role and other information.

Information can be collected online using sensors on (i) CH_4, (ii) CO, (iii) air velocity, (iv) temperature, (v) strata movement, (vi) water-level indicator, (vii) machine status, (viii) load, (ix) vibration, (x) oil temperature, (xi) bearing temperature, (xii) pressure and other sensors. The data can be received from sensors using sensor-reader software/driver, such as Data Acquisition Software (DAS), which is already developed by the authors. In this software, files can be opened for viewing graphical simulation of data in different channels. Various statistical analysis operations like maximum, minimum, average, moving average, standard deviation, etc., are also incorporated. There are alarm and warning messages when a data crosses its safe limit and helps in taking precautionary measures. The second way of getting data is the use of online web camera and headphones. Other ways of getting data are direct feeding of data by using keyboards, scanners, digitizers, etc.

All collected information can be stored in a master database having different modules. All modules can be internally connected to each other with reference and composite keys, and program codes. These modules can be managed with front-end program code and back-end tools having advanced functions, procedures, view, cursors, and triggers. These functions, keys, and tools can be used to manipulate, relate, and display data with security, and remove data redundancy. Stored information can be accessed though web. The data can be displayed in tabular, pictorial, chart, graphical, 3D formats, etc., using SQL query/commands. Retrieve can filter and display information as per the desired format of user with the help of front-end software, back-end database, web server's software, and support networking.

7.3.2 Middleware

All computer terminals placed at different places can allow access to data/information available in the database. The terminals can also be updated with new data/information placed at mine sites.

Required web-based information system can have a number of stations (mines) equipped with computer terminals having web camera, modem, and headphone, telephone, etc., facilities connected with each other using switch and all computers are connected with a common server. Specially developed software can store all data at central location at AO. Web camera facility in each station – for situation assessment purposes – which has the ability of connecting directly to any other computer in the network, can send data both in audio and video format (speed, of 100 Mbps). On each station, number of computers can be placed with web camera and headphone and all these computers can be connected with a switch. All the servers can be kept at central location at AO. Basically, two types of servers can operate (i) to keep entire database server; and (ii) to work as application server. At HO, a computer can be connected with AO with dial-up connectivity and other computers can be connected with the said computer. Server at AO can also be configured as remote access server (RAS) so that connection can be established from HO through dial-up using modem and existing infrastructure of telephone line.

For data manipulating and storing, all the data of each mining site can be stored in the server. Web server can be made responsible for collection, processing, and storage of data from all stations via a commercial program. Oracle database server can do the data storage and batch processing. The processed data can be sent through web page after execution of a code written in Visual Studio NET, which can produce a file with desired information from any station. Consecutively, a program written in Visual Basic NET can be executed, that can read experimental data and produce desired report at any location. For comparison, desired previous-day information/decisions can be available to authorized user based on the requirement through database server. Any query can be executed from any client using the application developed in Visual Basic NET, which can be made available from application server.

Complete networking system (Fig. 7.1) indicates placement of computer terminals and bidirectional feature of flow of data/information. Computers at one location can be connected with a switch using twisted-pair, category 6 cables. Switches of all stations can be connected with optic fiber cable. AO can be connected with HO through dial-up. For this, one server at AO can be configured as RAS and one modem with telephone line can be attached with it. System kept at HO can be attached with modem along with telephone line. Any time the computer at HO can dial into the servers at AO and get the required information.

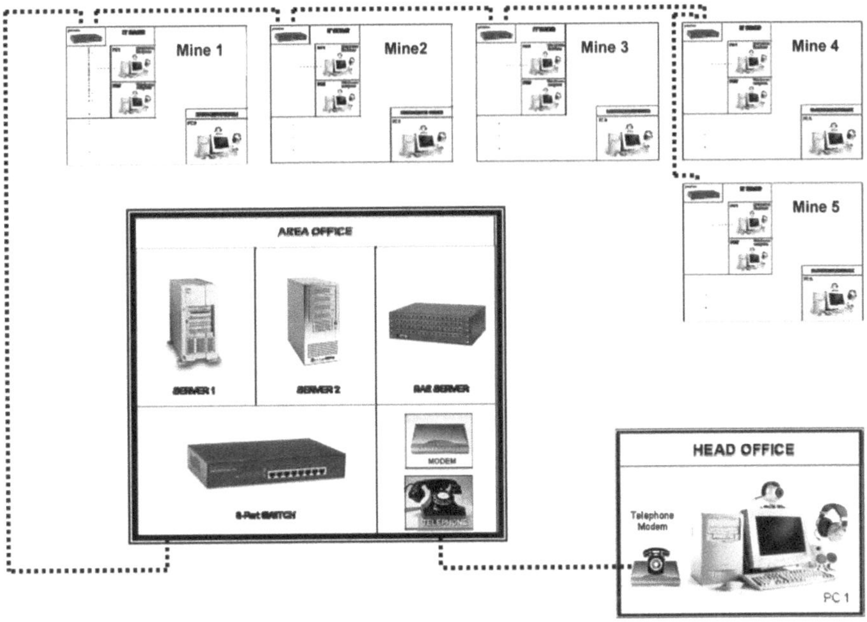

Fig. 7.1 Setup for the complete network system

7.3.3 Hardware

Schematic diagram (Fig. 7.2) of a mine indicates location of sensors, computer, and other peripherals. Different sensors can be connected to a computer in IT room using underground remote station, armored cable, junction boxes, flameproof (FLP) power supply, intrinsically safe uninterrupted power supply (UPS), and other accessories. Various parameters can be continuously monitored online using different sensors. Each mine can have one IT room and manager room, where required system can have connections. IT room can be equipped with computers, web cameras, headphones, UPS, laser printer, scanner, plotter, digitizer, mobile handset, and other accessories for calibration, testing, and maintenance. Manager room can be equipped with a computer, UPS, headphone, web camera, scanner, laser printer, and mobile handsets.

7.4 Conclusions

Adaptation of IT-based management strategy in coal mining industry will allow officials to quickly access, integrate, and display critical information for making better tactical decisions and at the same time to provide long-term information that senior management require for taking strategic decisions. Further, IT system will provide access to mining personnel on worldwide development related to mining

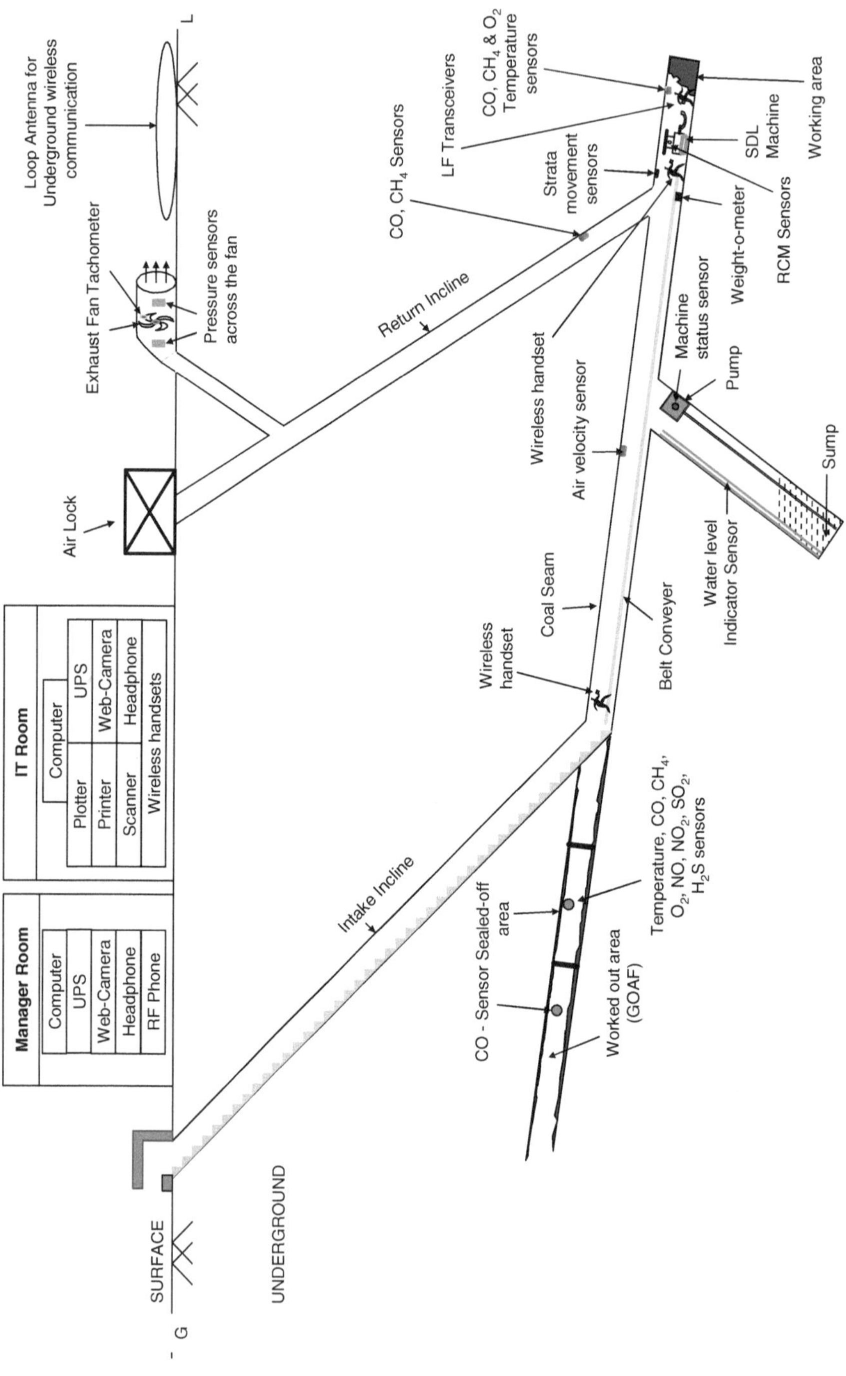

Fig. 7.2 Schematic diagram of a mine showing location of sensors and other accessories

technology, which will help them to adopt best technology for their site-specific mining conditions. With the development and implementation of required web-based information and decision support system, coal mining industry would become technologically sound, economically viable, profitable, and safer.

References

Bandyopadhyay LK, Chaulya SK, Dutta MK, Narayan A and Kumar M (2005a) Development of an environmental monitoring system for underground coal mines. *Proceedings of International Seminar on Coal Science and Technology – Emerging Global Dimensions* (eds. Singh AK and Sen K), Allied Publishers, New Delhi, India, pp. 629–635.

Bandyopadhyay LK, Kumar S, Mishra PK, Narayan A and Sinha MK (2005b) Studies on wireless communication systems for underground coal mines. *Proceedings of International Seminar on Coal Science and Technology – Emerging Global Dimensions* (eds. Singh AK and Sen K), Allied Publishers, New Delhi, India, pp. 56–64.

Chandramouli V and Singh JK (2003) Application of information technology for sustainable development of mining industry. *Proceedings of National Seminar on Role of Information Technology in the Present Scenario of Globalization* (eds. Singh JK, Lolarak R, Kalyani VK and L Kumar), Central Mining Research Institute, Dhanbad and Computer Society of India, Dhanbad, India, pp. 295–304.

Chaulya SK, Bandyopadhyay LK and Mishra PK (2008) Modernization of Indian coal mining industry: Vision 2025. *Journal of Scientific & Industrial Research*, **67**: 28–35.

Das SK (2001) Application of information technology in the mining industry. *Proceedings of Third Conference on Computer Application in Mineral Industry* (eds. Bandyopadhyay C and Sheorey PR), Oxford & IBH Publisher, New Delhi, India, pp. 35–48.

Hainsworth DW, Reid DC, McPhee RJ, Ralston JC, Corke PI and Winstanely G (2001) Automation applied to a rapid roadway development system for underground coal mining. *Proceedings of Regional Symposium in Computer Applications in the Minerals Industries*, Tampere, Finland, pp. 23–26.

Kumar VV and Guha R (2001) Computerisation of Indian mining industry – quo Vadis? *Proceedings of Third Conference on Computer Application in Mineral Industry* (eds. Bandyopadhyay C and Sheorey PR), Oxford & IBH Publisher, New Delhi, India, pp. 15–22.

Prasad SB (1997) IT strategy for mineral industry. *Proceedings of National Seminar on Information Technology in Mineral Industry,* Geological and Metallurgical Institute of India, Kolkata, India, pp. 185–192.

Reid DC, Hainsworth DW, Ralston JC and McPhee RJ (2000) Shearer guidance: a path towards longwall automation. *Proceedings of the 1st International Workshop on Robotics for Mining and Underground Applications*, Brisbane, Australia, pp. 121–126.

Sen SK (2001) Current status of mechanization and automation in mineral industry – Indian scenario. *Proceedings of International Conference on Mechanization and Automation – The Future of Mineral Industry*, Mining, Geological and Metallurgical Institute of India, Kolkata, India, pp. 7–11.

Chapter 8
ZigBee Technology: A Unique Wireless Sensor Networking Solution

8.1 Introduction

The Institute of Electrical and Electronics Engineers (IEEE) is the world's largest technical professional society for promoting development and application of electro technology and allied sciences for the benefit of humanity (www.IEEE.org). The IEEE has developed a number of standards related to IEEE 802 community (Table 8.1). The IEEE 802.1 and 802.2 standards are the basis for most of the standards. Wireless technologies for short- and medium-range communication attract attention due to their ability to support multiple applications and their cost efficiency. These technologies became synonymous with home and manufacturing automation as well as with wireless sensor networks.

The ZigBee technology is emerging as an important wireless sensor networking solution for short- and medium-range communication due to its numerous advantages, namely, unlicensed 2.4 GHz industrial, scientific, and medical (ISM) band, ultra low power (ideal for battery-operated system), operates for years on inexpensive batteries, large number of nodes/sensors, reliable and secure links between network nodes, easy deployment and configuration, low-cost system, very fast transition time, digital battery monitor facility, and smaller in size (system on chip).

ZigBee's industry covers already many applications, starting from home automation and going to military and commercial wireless sensor networks (WSN). ZigBee is based on the IEEE 802.15.4 standard (Zheng and Lee, 2004; Evans, 2003; www.zigbee.org), which supports physical and media access control (MAC) layers, as well as approved by the ZigBee Alliance network and application layers (www.en.wikipedia/wiki/ZigBee). Being a symbiosis of two organization standards, ZigBee devices are being built on open platforms. Many companies are utilizing open software and hardware for manufacturing ZigBee platforms. This technology attracts user by its simplicity, cost efficiency, and low power consumption (Bougard et al., 2005; Viswanathan and Boult, 2007) over other technologies (Table 8.2). Among many ZigBee applications, it is realized that ZigBee devices will be widely used to build smart homes, including security systems, meter reading,

L.K. Bandyopadhyay et al., *Wireless Communication in Underground Mines*,
DOI 10.1007/978-0-387-98165-9_8, © Springer Science+Business Media, LLC 2010

Table 8.1 IEEE standards

Standard	Features
IEEE 802	Local area network (LAN)/metropolitan area network (MAN)
IEEE 802.1	LAN/MAN bridging and management and remote media access control (MAC) bridging
IEEE 802.2	Logical link control (LLC) standards for connectivity
IEEE 802.3	Ethernet standards for carrier sense multiple access with collision detection (CSMA/CD)
IEEE 802.4	Token passing bus access
IEEE 802.5	Token ring access and for communication between LANs and MANs
IEEE 802.6	Information exchange between systems
IEEE 802.7	Information exchange for broadband LAN cabling
IEEE 802.8	Fiber optic connection
IEEE 802.9	Integrated services, like voice and data
IEEE 802.10	LAN/MAN security implementations
IEEE 802.11	Wireless networking – wireless fidelity (Wi-Fi)
IEEE 802.12	Demand priority access method
IEEE 802.13	Not used (Unlucky number)
IEEE 802.14	Cable television broadband communication
IEEE 802.15.1	Bluetooth
IEEE 802.15.2	Coexistence of wireless personal area networks (WPAN) with other wireless devices operating in unlicensed frequency bands such as wireless local area networks (WLAN)
IEEE 802.15.3	MAC and PHY standard for high-rate (11–55 Mbps) WPANs
IEEE 802.15.4	Wireless sensor/control networks – ZigBee
IEEE 802.16	Wireless networking – world interoperability for microwave access (WiMAX)
IEEE 802.16d	Fixed WiMAX
IEEE 802.16e	Mobile WiMAX

remote commanders, industrial monitoring, and others (Callaway et al., 2002; Bai et al., 2007; Lin et al., 2007; Tsang et al, 2007; Pinedo and Garcia, 2008). ZigBee devices are good fit for both military and commercials. ZigBee devices can work years on the same battery and support 100–250 kbps transmissions. ZigBee competes with Bluetooth, ultra wide band, and 802.11 techniques to prove its importance. None of competing technologies so far could prove its uniqueness. The trend is observed when vendors trying to build chip set supporting several competing technologies. Some of the essential features of ZigBee are (i) over the air data (data rates of 250, 40, and 20 kbps), (ii) star or peer-to-peer operation, (iii) allocation of guaranteed time slots (GTSs), (iv) low power consumption (few milli-Amperes), (v) carrier sense multiple access with collision avoidance (CSMA-CA) channel access, (vi) link-quality indication (LQI), and (vii) fully acknowledged protocol (IEEE Standard 802.15.4-2003) for transfer reliability (www.zigbee.org). The ZigBee technology is briefly enumerated subsequently.

Table 8.2 Comparison of emerging wireless technologies

Parameters	Technologies Wi-Fi	ZigBee	WiMAX
Application	Wireless LAN, Internet	Sensor networks	Metro area broadband Internet connectivity
Typical range	100 m	70–100 m	50 km
Frequency range (GHz)	2.4	2.4	2–11
Data rate (Mbps)	108–600	250×10^{-3}	75
Modulation	Direct sequence spread spectrum (DSSS)	DSSS	Quadrature amplitude modulation (QAM)
Network	IP and P2P	Mesh	IP
IT network connectivity	Yes	No	Infrastructure
Network topology	Infrastructure (ad hoc also possible)	Ad hoc	Request/grant
Access protocol	CSMA-CA	CSMA-CA	Request/grant
Key attributes	Wider bandwidth and more flexibility	Low cost and less consumption of power	Throughput and wide coverage

8.2 ZigBee Technology

8.2.1 ZigBee Stack

The IEEE 802.15.4 standard is a simple packet data protocol for light-weight wireless networks and specifies the media access control (MAC) and physical (PHY) networking layers (Fig. 8.1). ZigBee technology takes full advantage of the IEEE 802.15.4 standard and adds the logical network, security, and application software (Ergen, 2004).

The focus of the ZigBee standard permits monitoring over 10–100 m, with multiple nodes per network. The IEEE 802.15.4 standard uses two PHYs. In case of 2.4 GHz PHY band the data rate is 250 kbps and for 868/915 MHz band the data rate is 20–40 kbps. Channels used for 2.4 GHz and 868/915 MHz bands are 16 and 11, respectively. The CSMA-CA and slotted CSMA-CA are used for channel accessing in IEEE 802.15.4 standard. Information related to property and ranges of IEEE 802.15.4 are given in Table 8.3.

8.2.2 The Network Layer

Like all IEEE 802 standards, the IEEE 802.15.4 standard encompasses only those layers which are within the up-data-link-layer (DLL). Higher-layer protocols are at the discretion of individual applications. In traditional wired networks, the network

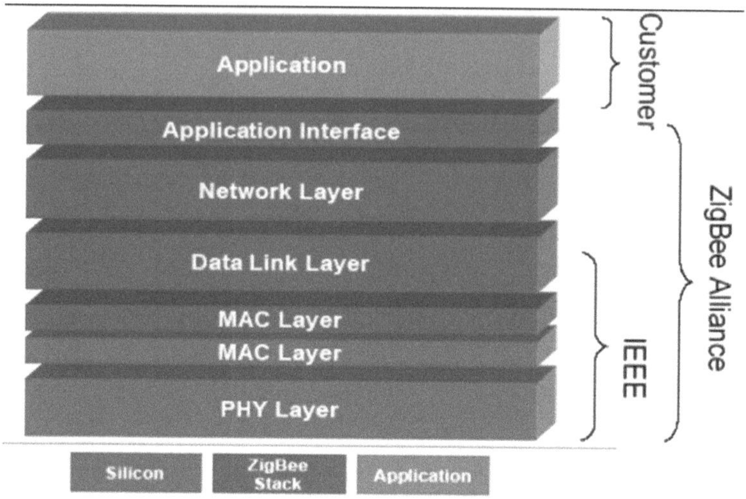

Fig. 8.1 ZigBee stack

Table 8.3 ZigBee technology at a glance

Property	Range
Raw data rate	(i) 868 MHz: 20 kbps (ii) 915 MHz: 40 kbps (iii) 2.4 GHz: 250 kbps
Range	10–100 m
Latency	Down to 15 ms
Channels	(i) 868/915 MHz: 11 channels (ii) 2.4 GHz: 16 channels
Frequency band	Two physical layers: 868/915 MHz and 2.4 GHz
Addressing	Short 8 bit or 64 bit IEEE
Channel access	CSMA-CA and slotted CSMA-CA
Temperature	Industrial temperature range: –40 to +81°C

layer is responsible for topology construction and maintenance, as well as naming and binding services, which incorporate the necessary tasks of addressing, routing, and security. In fact, it is important for any network layer implementation built on the already energy conscious IEEE 802.15.4 draft standard to be equally conservative. Network layers built on the standard are expected to be self-organizing and self-maintaining to minimize total cost to the user. The IEEE 802.15.4 draft standard supports multiple network topologies, including both star and peer-to-peer networks (Fig. 8.2). The topology is an application design choice (Kolla et al., 2003; Al-karaki and Kamal, 2004; Li et al., 2008); some applications, such as PC peripherals, may require the low-latency connection of the star network, while others, such as perimeter security, may require the large-area coverage of peer-to-peer network. ZigBee networks use three devices, namely (i) the network coordinator, (ii) full

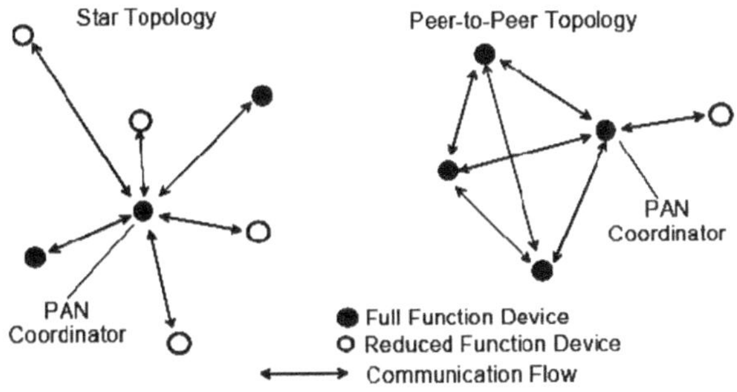

Fig. 8.2 Star and peer-to-peer network

function device, and (iii) reduced function device. The working principles (Darji, 2004) of these three types of devices are given below:

(i) *Network coordinator (Coordinator)*: It maintains overall network knowledge. It is the most sophisticated of all the three types and requires higher memory and high computing power.
(ii) *Full function device (Router)*: It supports all 802.15.4 functions and features specified by the standard. It can function as a network coordinator. Additional memory and computing power make it ideal for network router functions or it could be used in network-edge devices.
(iii) *Reduced function device (End device)*: It carries limited functionality to complexity and lower cost. It is generally found in network-edge device.

8.2.3 The Data Link Layer

The IEEE 802 standard splits the data link layer (DLL) into two sublayers namely, the medium access control (MAC) and logical link control (LLC). The LLC is standardized in 802.2 and is common among the 802 standards, such as 802.3, 802.11, and 802.15.1, while the MAC sublayer is closer to the hardware and may vary with the physical layer implementation. Figure 8.3 shows the inclusion method of IEEE 802.15.4 into the International Organization for Standardization (ISO) open systems interconnection (OSI) reference model. The IEEE 802.15.4 MAC provides services to an IEEE 802.2 type (I) LLC through the service-specific convergence sublayer (SSCS), or a proprietary LLC can access the MAC services directly without going through the SSCS. The SSCS ensures compatibility between different LLC sublayers and allows the MAC to be accessed through a single set of access points. Using this model, the 802.15.4 MAC allows more complex network topologies. The features of the IEEE 802.15.4 MAC are (i) acknowledged frame delivery, (ii) channel access mechanism, (iii) frame validation, (iv) GTS management, and (v) beacon management.

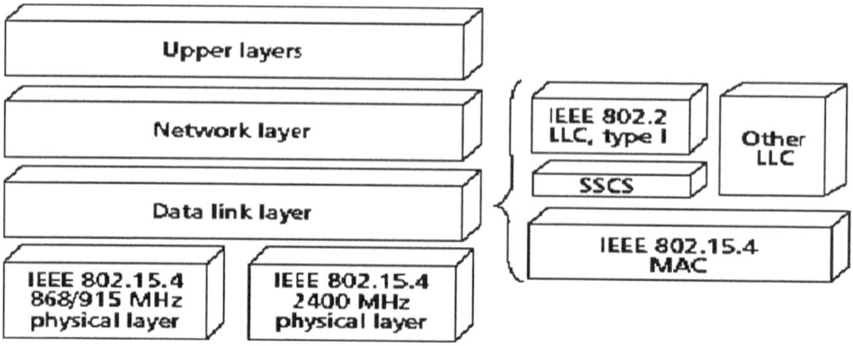

Fig. 8.3 IEEE 802.15.4 in the ISO-OSI layered network model

The MAC sublayer provides two services to higher layers that can be accessed through two service access points (SAPs). The MAC data service is accessed through the MAC common part sublayer (MCPS-SAP) and the MAC management services are accessed through the MAC layer management entity (MLME-SAP). These two services provide an interface between the SSCS or another LLC and the PHY layer. The MAC management service has 26 primitives compared to Bluetooth (IEEE 802.15.1), which has about 131 primitives and 32 events. The 802.15.4 MAC is of very low complexity, very suitable for its intended low-end applications, making it at the cost of a smaller feature set than 802.15.1. A new standard for communication technology in control is IEEE 802.15.4 and a nonprofit consortium called the ZigBee Alliance has developed protocols using the physical layer of IEEE 802.15.4 (Fig. 8.1).

8.2.4 The General MAC Frame Structure

The MAC frame structure is kept very flexible to accommodate the needs of different applications and network topologies while maintaining a simple protocol. The general format of a MAC frame is shown in Fig. 8.4. The MAC frame is called the MAC protocol data unit (MPDU) and is composed of MAC header (MHR), MAC service data unit (MSDU), and MAC footer (MFR). The first field of the MAC header is the frame control field. It indicates the type of MAC frame being used.

The size of the address field may vary between 0 and 20 bytes. For instance, a data frame may contain both source and destination information, while the return acknowledgment frame does not contain any address information at all. On the other hand, a beacon frame may only contain source address information. In addition, short 8-bit device addresses or 64-bit IEEE device addresses may be used. This flexible structure helps to increase the efficiency of the protocol by keeping the packets short. The payload field is variable in length; however, the complete MAC frame may not exceed 127 bytes in length. The data contained in the payload is

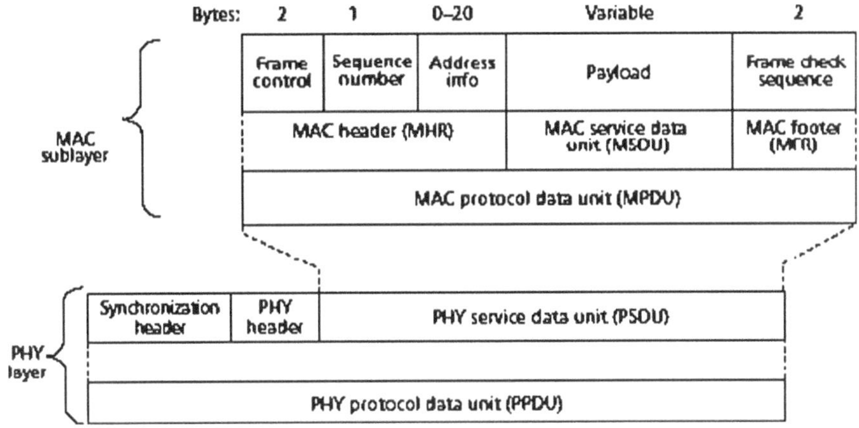

Fig. 8.4 General MAC frame format

dependent on the frame type. Other fields in a MAC frame are the sequence number and frame check sequence (FCS). The sequence number in the MAC header matches the acknowledgment frame with the previous transmission. The transaction is considered successful only when the acknowledgment frame contains the same sequence number as the previously transmitted frame. The FCS helps to verify the integrity of the MAC frame. The FCS in an IEEE 802.15.4 MAC frame is a 16-bit International Telecommunication Union – Telecommunication (ITU-T) Standardization Sector cyclic redundancy check (Lu et al., 2004).

Superframe structure: The low-rate wireless personal area network (LR-WPAN) standard allows the optional use of a superframe structure. The coordinator defines the format of the superframe. The superframe is bounded by network beacons, is sent by the coordinator (Fig. 8.5) and is divided into 16 equally sized slots (Kinney, 2003). The beacon frame is transmitted in the first slot of each superframe. If a coordinator does not wish to use a superframe structure it may turn off the beacon transmissions. The beacons are used to synchronize the attached devices, to identify

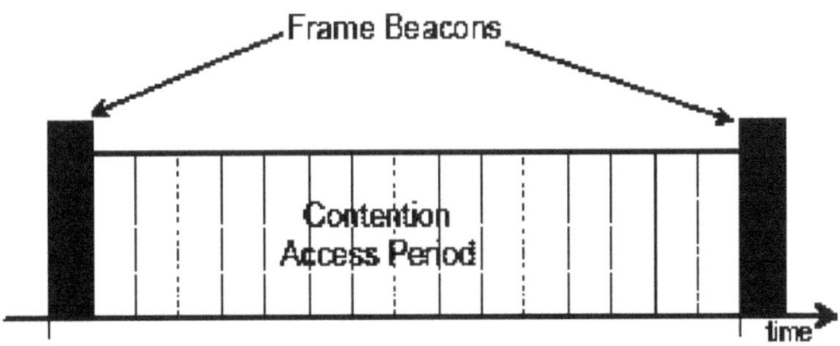

Fig. 8.5 Superframe structure

the PAN, and to describe the structure of the superframes. Any device wishing to communicate during the contention access period (CAP) between two beacons shall compete with other devices using a slotted CSMA-CA mechanism. All transactions shall be completed by the time of the next network beacon.

For low-latency applications or applications requiring specific data bandwidth, the PAN coordinator may dedicate portions of the active superframe to that application. These portions are called guaranteed time slots (GTSs). The GTSs comprise the contention-free period (CFP), which always appears at the end of the active superframe starting at a slot boundary immediately following the CAP, as shown in Fig. 8.6. The PAN coordinator may allocate up to seven of these GTSs and a GTS may occupy more than one slot period. However, a sufficient portion of the CAP shall remain for contention-based access of other networked devices or new devices wishing to join the network. All contention-based transactions shall complete before the CFP begins. Also, each device transmitting in a GTS shall ensure that its transaction is complete before the time of the next GTS or the end of the CFP (Kinney, 2003).

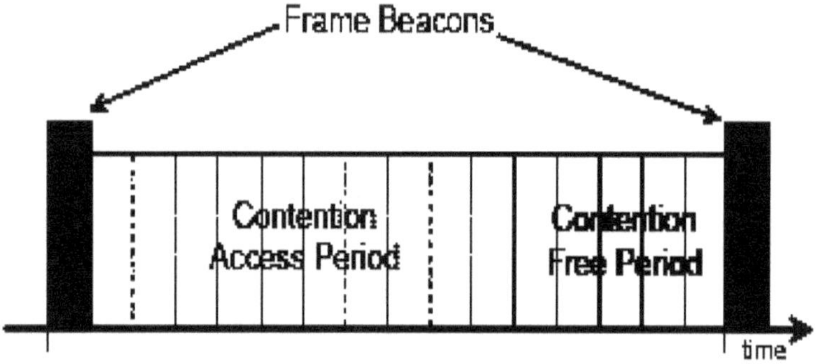

Fig. 8.6 Contention access period and contention-free period

8.2.5 MAC Protocol Overview in IEEE 802.15.4

Depending on network configuration, an LR-WPAN may use one of two channel access mechanisms. In a beacon-enabled network with superframes, a slotted CSMA-CA mechanism is used. In networks without beacons, unslotted or standard CSMA-CA is used. When a device wishes to transmit data to PAN in a nonbeacon-enabled network, it first checks if another device is currently transmitting on the same channel. If so, it may back off for a random period or indicate a transmission failure if unsuccessful after some retries. Acknowledgment frames confirming a previous transmission do not use the CSMA mechanism since they are sent immediately following the previous packet. The PAN coordinator announces the superframe structure to PAN devices periodically through beacon frames. By changing active and inactive portion via parameter superframe order (SO) and beacon order (BO), WPAN can operate under low duty cycle to conserve energy (Lu et al., 2004).

8.2.5.1 Contention Access Period

In a beacon-enabled network, any device wishing to transmit during the contention access period (CAP), enables its receiver and waits for the beginning of the next time slot and then determines if another device is currently transmitting in the same slot. If another device is already transmitting in the slot, the device backs off for a random number of slots and determines if channel is clear or indicates a transmission failure after some retries.

Three variables are maintained at each device for a channel access: number of back off (NB), contention window (CW), and back-off exponent (BE). NB is the number of times the CSMA-CA backs off while attempting the current transmission, and is reset to 0 for each new data transmission. CW is contention window length, which is reset to 2 either for a new data transmission or when the channel is found to be busy. BE is back-off exponent, which is related to back-off periods a device should wait before attempting carrier sensing. In addition, in a beacon-enabled network, acknowledgment frames do not use CSMA. An important function of the MAC is confirming successful reception of a received frame. Successful reception and validation of a data or MAC command frame is confirmed with an acknowledgment. If the receiving device is unable to handle the incoming message for any reason, the receipt is not acknowledged. The frame control field indicates whether or not an acknowledgment is expected. The acknowledgment frame is sent immediately after successful validation of the received frame. Beacon frames sent by a PAN coordinator and acknowledgment frames are never acknowledged. In CSMA-CA, a lot of energy is generally consumed by long back-off period during high traffic to avoid collisions. As IEEE 802.15.4 supports a battery life extension (BLE) mode, back-off exponent is limited to 0–2. This reduces the period of ideal listening in low offered traffic applications. A network device can put its radio to sleep to conserve energy immediately after the reception of acknowledgement packet if there is no more data to be sent or received (Kinney, 2003).

8.2.5.2 Contention-Free Period

As mentioned earlier, the IEEE 802.15.4 standard allows the optional use of contention-free period (CFP) for devices that required dedicated bandwidth to achieve low latencies. Device requiring dedicated bandwidth and low latencies transmission can be assigned GTS in CFP by PAN coordinator. When the device wishes to transmit the frame during GTS, it first checks a list on the beacon frame to see whether it has been allocated a valid GTS. If a valid GTS is found, the device enables its receiver at a time prior to start of the GTS and transmits the data during the GTS period. The MAC of the PAN coordinator ensures that its receiver is enabled for all allocated GTS. All contention-based actions must be completed before the CFP begins (Kinney, 2004).

8.2.5.3 Synchronization

The PAN coordinator transmits beacon frames periodically to announce the superframe structure in a PAN. Devices need to synchronize with coordinator by receiving and decoding the beacon frames before any data transmission. There are two methods of synchronization: tracking and nontracking. With tracking, the device receives the first beacon, gets current superframe structure, knows when to activate its receiver for the next beacon, and keeps track of it. To transmit a frame, the device can enable its receiver just a little earlier before the beacon arrival. With nontracking, the device attempts to acquire the beacon only once. The device needs to enable its receiver and searches for a specific period until it receives a beacon from its associated coordinator, when it attempts to transmit a frame.

8.2.6 The Physical Layer

IEEE 802.15.4 offers two PHYs options that combine with the MAC to enable a broad range of networking applications. Both PHYs are based on direct sequence spread spectrum (DSSS) methods that result in low-cost digital integrated circuit (IC) implementation, and both share the same basic packet structure for low duty cycle, low-power operation. The fundamental difference between the two PHYs is the frequency band. The 2.4 GHz PHY specifies operation in the 2.4 GHz industrial, scientific, and medical (ISM) band, which has nearly worldwide availability, while the 868/915 MHz PHY specifies operation in the 868 MHz band in Europe and 915 MHz ISM band in the United States. While mobility between countries is not anticipated for most home networking applications, the international availability of the 2.4 GHz band does offer advantages in terms of larger markets and lower manufacturing costs. On the other hand, the 868 and 915 MHz bands offer an alternative to the growing congestion and other interference like microwave ovens, etc., associated with the 2.4 GHz band and longer range for a given link budget due to lower propagation losses.

Twenty-seven frequency channels are available across the three bands given in Fig. 8.7 and Table 8.4. The 868/915 MHz PHY supports a single channel between 868.0 and 868.6 MHz, and 10 channels between 902.0 and 928.0 MHz. Due to the regional support for these two bands, it is unlikely that a single network would ever use all 11 channels. However, the two bands are considered close enough in terms of frequency, therefore (even though they are not identical) the same can be used for both bases to reduce the cost for both, lowering manufacturing costs. The 2.4 GHz PHY supports 16 channels between 2.4 and 2.4835 GHz with channel spacing (5 MHz) aimed at easing transmit and receive filter requirements. The IEEE 802.15.4 standard includes the necessary hooks to implement dynamic channel selection, although the specific selection algorithm is left to the network layer. The MAC layer includes a scan function that steps through the list of supported channels in search of a beacon, while the PHY layers contain several lower-level functions, such as receiver energy detection, link-quality indication, and channel

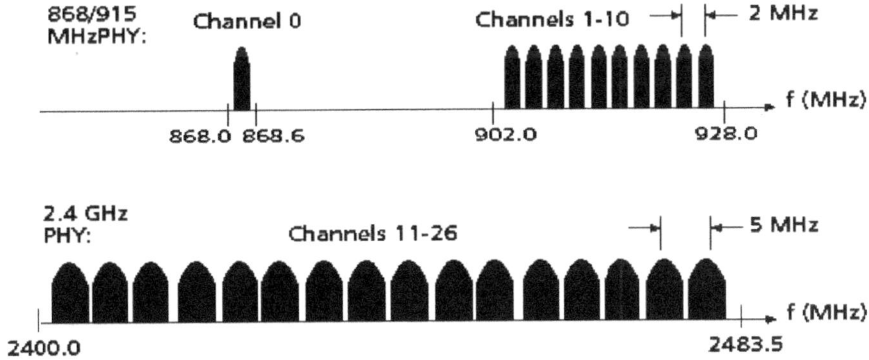

Fig. 8.7 IEEE 802.15.4 channel frequencies

Table 8.4 IEEE 802.15.4 channel frequencies

Channel number	Channel center frequency (Hz)
$K = 0$	868.3
$K = 1, 2, 10$	$906 + 2\,(K-1)$
$K = 11, 12,\ldots,26$	$2405 + 5\,(K-11)$

switching, which enable channel assessment and frequency agility. The network to establish its initial operating channel and to change channels in response to a prolonged outage uses these functions (Callaway et al., 2002).

The packet structure: To maintain a common simple interface with the MAC, both PHY layers share a single packet structure as shown in Fig. 8.8. Each packet, or PHY protocol data unit (PPDU), contains a synchronization header (preamble plus start of packet delimiter), a PHY header to indicate the packet length, and the payload, or PHY service data unit (PSDU). The 32-bit preamble is designed for acquisition of symbol and chip timing, and in some cases may be used for coarse frequency adjustment. Channel equalization is not required for PHY due to the combination of small coverage area and relatively low chip rates. Within the PHY header, 7 bits are used to specify the length of the payload (in bytes).

This supports packets of length 0–127 bytes, although due to MAC layer overhead, zero-length packets will not occur in practice. Typical packet sizes for home applications such as monitoring and control of security, lighting, air conditioning, and other appliances are expected to be in the order of 30–60 bytes, while more demanding applications such as interactive games and computer peripherals, or multihop applications with more address overhead, may require larger packet sizes. Adjusting for the transmission rates in each band, the maximum packet durations are 4.25 ms for the 2.4 GHz band, 26.6 ms for the 915 MHz band, and 53.2 ms for the 868 MHz band (Darji, 2004).

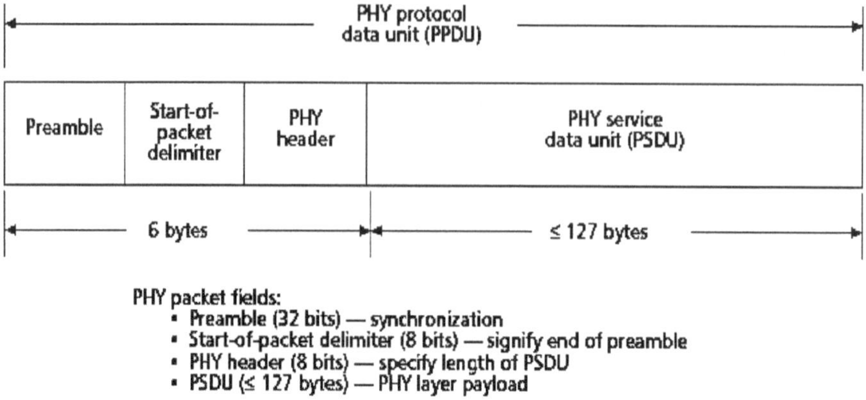

Fig. 8.8 IEEE 802.15.4 physical layer packet structure

8.2.7 Sensitivity and Range

IEEE 802.15.4 currently specifies receiver sensitivities of –85 dBm for the 2.4 GHz PHY and –92 dBm for the 868/915 MHz PHY. These values include sufficient margin to cover manufacturing tolerances as well as to permit very low cost implementation approaches. In each case, the best device may be of the order of 10 dBm better than the specification. Naturally, the achievable range will be a function of the receiver sensitivity as well as transmit power. The standard specifies that each device shall be capable of transmitting at least 1 mW, but depending on the application needs, the actual transmit power may be lower or higher (within regular limits). Typical devices (1 mW) are expected to cover a 10–20 m range. However, with good sensitivity and moderate increase in transmit power, a star network topology can provide complete home coverage.

8.2.8 Security

When security of MAC layer frames is desired, ZigBee uses MAC layer security to secure MAC command, beacon, and acknowledgment frames. ZigBee may secure messages transmitted over a single hop using secured MAC data frames, but for multihop messaging, ZigBee relies upon upper layers for security. The MAC layer uses the advanced encryption standard (AES) as its core cryptographic algorithm and describes a variety of security suites that use the AES algorithm. These suites can protect the confidentiality, integrity, and authenticity of MAC frames. The MAC layer does the security processing, but the upper layers, which set up the keys and determine the security levels to use, control this processing. When the MAC layer transmits (receives) a frame with security enabled, it looks at the destination (source) of the frame, retrieves the key associated with that destination (source), and then uses this key to process the frame according to the security suite designated for the key being used. Each key is associated with a single security suite and the

MAC frame header has a bit that specifies whether security for a frame is enabled or disabled.

8.2.9 Limitations

All wireless technologies are different from each other due to their structural differences. Due to this all the technologies have some specific limitations. The limitations of ZigBee technology are as follows:

(i) Application requiring long range without using router.
(ii) Mobile applications.
(iii) Streaming data only, video-data transmission is not possible.
(iv) The ZigBee and IEEE 802.15.4 protocols are designed for battery-powered applications. In practice, not all devices in a network can be battery powered, particularly those that need to be switched on all the time (active mode), such as routers and co-coordinators. Such device is permanently connected to the mains supply, e.g., a ceiling lamp or an electric radiator (Howitt and Gutierrez, 2003).

8.3 Specialty of ZigBee Technology

There are a multitude of standards like Bluetooth and Wi-Fi that address mid-to-high data rates for voice, PC LANs, video, etc. However, a wireless network standard is not available that meets the unique needs of sensors and control devices. Sensors and controls do not need high bandwidth but they do need low latency and very low energy consumption for long battery lives and for large device arrays.

The ZigBee Alliance is not pushing a technology, rather it provides a standardized base, set of solutions for sensor, and control systems. Following are some specialties in ZigBee technology (www.freescale.com/zigbee):

(i) The physical layer was designed to accommodate the need for low cost yet allowing for high levels of integration. The use of direct sequence allows the analog circuitry to be very simple and very tolerant toward inexpensive implementations.
(ii) The MAC layer has been designed to allow multiple topologies without complexity. The power management operation does not require multiple modes of operation. The MAC allows a reduced functionality device (RFD) that need not have flash nor large amounts of ROM or RAM. The MAC was designed to handle large numbers of devices without requiring them to be "parked" (Stein and Kibitzes, 2002).
(iii) The network layer has been designed to allow the network to spatially grow without requiring high-power transmitters. The network layer can also handle large amounts of nodes with relatively low latencies.

(iv) In contrast to Bluetooth, which has many different modes and states depending upon latency and power requirements, ZigBee/IEEE 802.15.4 has two major modes states active (transmit/receive) and sleep. The application software needs to focus on the application, not on which power mode is optimum for each aspect of operation (Pahlavan and Krishnamurthy, 2006).
(v) ZigBee's use of the IEEE 802.15.4 PHY and MAC allows networks to handle any number of devices. This attribute is critical for massive sensor arrays and control networks (Stein and Kibitzes, 2002).
(vi) Zigbee-compliant RFID is a wireless link to uniquely identify objects or people.
(vii) Zigbee-compliant RFID is a means of storing and retrieving data through electromagnetic transmission to an RF-compatible integrated circuit. It is now being seen as a radical means of enhancing data-handling processes. Zigbee-compliant RF readers and tags use a defined radio frequency and protocol to transmit and receive data.

8.4 Comparison of ZigBee with Other Wireless Technologies

ZigBee and other wireless technologies are different solutions for various application areas (Akyildiz et al., 2007; Bai et al., 2007). The differences are from their approach to their desired application. Bluetooth has addressed a voice application by embodying a fast-frequency hopping system with a master–slave protocol (Held, 2001). ZigBee has addressed sensors, controls, and other short-message applications by embodying a direct sequence system with a star or peer-to-peer protocols, whether GPRS/GSM/LxTT/CDMA are used in wide area for data and voice. Again Wi-Fi technology is best suited for web, e-mail, and for video. Minor changes to ZigBee or other technologies would not change their inherent behavior or characteristics. The different behaviors come from architectural differences. Table 8.5 gives the comparison between ZigBee and other wireless technologies. In comparison to other wireless technologies, ZigBee devices can quickly attach, exchange information, detach, and then go to deep sleep to achieve a very long battery life (Howitt and Guterrez, 2003; Petrova et al., 2006; Shuaib et al., 2006; Jo et al., 2007; www.maxstream.net/wireless/zigbee.php).

The comparative studies reveal that there are several advantages of the ZigBee technology. The no-contact nature of this technology is the significant advantage common among all types of RFID systems. All ZigBee tags can be read despite extreme environmental factors such as snow, fog, ice, paint, and other visually and environmentally challenging conditions:

- ZigBee-compliant tags placed on objects to be tracked
    - Tag holds unique identification for the object
    - Tag moves along with the object

Table 8.5 Comparison of ZigBee technology with GSM, Wi-Fi, and Bluetooth

Market name standard	GPRS/GSM 1×RTT/CDMA	Wi-Fi™ 802.11b WLAN	Bluetooth™ 802.15.1 WPAN	ZigBee™ 802.15.4 LR-WPAN
Application focus	Wide area voice and data	Web, e-mail, video	Cable replacement	Monitoring and control
System resources	16 MB+	1 MB+	250 kB+	4–32 kB
Battery life (days)	1–7	1.5–5	1–7	100–1,000+
Network size	1	32	7	255/65,000
Band width (kbps)	64–128+	11,000+	720	20–250
Transmission range (m)	1,000+	1–100	1–10+	1–100+
Success metrics	Reach, quality	Speed, flexibility	Cost, convenience	Reliability, power, cost

- ZigBee-compliant readers placed at strategic locations
  - Readers are placed in form of network
  - Readers detect tags in vicinity and generate events
- ZigBee-compliant middleware deployed in an organization
  - Middleware manages readers
  - Middleware processes events collected from readers
- ZigBee-compliant RFID applications
  - Application implements the business processes and workflow
  - When RFID-enabled, uses services and interface provided by middleware
- Access control and security
  - Rule-based tracking of secure items
  - Access-rights-based denial of access
- Transport and logistics
  - Automated traffic and parking management
  - Planning and tracking of movement and delivery of goods
- Supply chain visibility
  - Easy inventory management
  - Increased accuracy
  - Reduced human effort

- Manufacturing process enhancement
  - Just-in-time manufacturing process
  - Tracking of manufacturing assembly process
- Visitor identification through ZigBee-compliant RFID
  - RFID-enabled passports
  - Quick and accurate identification at ports
- Evacuation management during emergencies
  - Regular tracking of personnel at critical places
  - Predetermined evacuation procedures
- New-born and infant protection systems
  - Tracking at hospitals, day-care centers, schools, parks, etc.
- Anti-counterfeiting of products
  - Scan and lookup of product tags for legitimacy check
  - Secure authentication to avoid cheating of the system
- Securing the supply chain
  - Avoid and capture pilferage
  - Non-intrusive inspection
- Food tracking during transport and at storage
  - Improved visibility and security of food containers
  - Assuring quality and maintenance of food
- Sterilization process for medical instruments
  - Example: operation equipment
  - RFID-based automated tracking and verification
- Regular servicing updates for building security systems
  - Example: fire extinguisher
  - RFID-based automated tracking and reminders
- Conformance of storage requirements for medicines
  - Example: refrigeration requirements
  - RFID-based reminders if medicines are not shelved back in time

- Valuable items are secured behind self-closing rolling shutters
- Users identify to the systems through RFID cards
- Access rights, associated to RFID card, are predefined with the system
- Access to valuables is based on time-based access rights
- Integration with backend policy system possible through network
- Cartons of perishable product items are tagged at supplier location
- Tags uniquely identify the product items
- Information (variable selling price, expiry date, storage conditions, etc.) about the product is hosted on server.

8.5 Features of ZigBee

ZigBee is poised to become the global control/sensor network standard. It has been designed to provide some specific features. The specific features of ZigBee are as follows (Ergen, 2004):

(i) Low power consumption and simple to implement.
(ii) Bluetooth has many different modes and states depending upon latency and power requirements such as sniff, peak, hold, park, active, etc. ZigBee/IEEE 802.15.4 has active (transmit/receive) or sleep mode.
(iii) ZigBee devices will be more economical than its predecessors, saving megawatts of power at its full deployment.
(iv) Low-cost device and easy to install and maintain. ZigBee device allows batteries to last for years using primary cells without any charges. ZigBee's simplicity allows for inherent configuration, and redundancy of network devices provides low maintenance.
(v) High density of nodes per network is offered. ZigBee's protocol stack is estimated to be about one-fourth of Bluetooth's or 802.11.5. The IEEE 804.15.4 PHY adopted by ZigBee has been designed for 868 MHz band in Europe, the 915 MHz band in North America, Australia, etc., and the 2.4 GHz band has been accepted in almost all countries of the world (Keshav, 2004).
(vi) Reliability is high.
(vii) Support for a large number of nodes.
(viii) Security is high.
(ix) Product interoperability.
(x) Vendor independence.
(xi) The 128-bit AES encryption provides secure connection between devices.
(xii) ZigBee-based products can access up to 16 separate, 5 MHz channels in the 2.4 GHz band, several of which do not overlap with US and European versions of IEEE 802.11 or Wi-Fi (www.tutorial-reports.com/wireless/ZigBee).

8.6 Application Areas

ZigBee protocol meets the unique needs of sensors and control devices and it has also been developed with the emphasis on low-cost battery-powered application. It is well suited to a wide range of building automation, industrial, medical and residential control, and monitoring applications and consumer electronics. Examples include the following fields:

(i) Mining applications: (a) Miners' tracking, (b) prevention of fatal accident between men and vehicles, (c) prevention of vehicle collisions, (d) route tracing for opencast transport vehicles, (e) providing warning signals when miner entering the unsafe area, (f) monitoring underground gases, (g) proximity warning, (h) message communication, etc.
(ii) Remote sensing: Water/sewage-level monitoring and temperature sensing.
(iii) Building automation: Security, light, thermostat, and air-conditioning control.
(iv) Industrial and commercial: Monitor, control, and automation links.
(v) Health care: Patient monitoring, data logger, and remote diagnosis.
(vi) Memory tagging: Automotive service record, maintenance logging, and inventory control/tracing.
(vii) Device and service discovery.
(viii) Optional acknowledge service.
(ix) Messaging and optional response.
(x) Remote metering.
(xi) Active RFID/asset tracking.
(xii) Wireless smoke and carbon monoxide detectors.
(xiii) HVAC control.
(xiv) Glass-break detectors.
(xv) Standing water sensors.
(xvi) Loud-sound detectors.

8.7 Manufacturers of ZigBee Devices

Various companies manufactured ZigBee devices as per the IEEE 802.15.4 standard. Table 8.6 includes names of some companies with their product details.

8.8 Conclusions

ZigBee technology is very important to automation and remote control application. ZigBee provides low-cost and low-power-consumption devices for those type of equipment that does not need high data transfer rate but long battery life of several months. With the standardization of MAC and PHY almost complete, now the

Table 8.6 Manufacturers of ZigBee devices and their product details

Company name	Name of device	Specification
Jennic (www.jennic.com)	JN513× second-generation microcontrollers	(i) 32-bit on-chip RISC coprocessors. (ii) The memory has been increased to 192 K ROM and 96 K RAM. (iii) Modem architecture that provides 97 dBm receiver sensitivity. (iv) It is more robust in noisier environments with greater tolerance to carrier offset.
Freescale (www.free-scale.com)	ZigBee-compliant transceivers: (a) 1319× (b) 1320× (c) 1321×	(i) One-stop shop solution. (ii) High-sensitivity radio solution. (iii) Adjustable Tx output power. (iv) Multiple power-down modes for extreme battery life.
Ubec (www.ubec.com)	UZ2400	(i) –95 dBm sensitivity. (ii) Differential RF input and output. (iii) Integrated 20 MHz and 100 kHz real-time clock output. (iv) Digital VCO and filter calibration. (v) Support SMA sleep mode.
Ember (www.ember.com)	Em 260 coprocessor	(i) Isolates timing of critical ZigBee functions (routing, etc.) from the application processor. (ii) Prevents resource conflicts between ZigBee stack and user application by separating ZigBee stack processing. (iii) Flash based for easy upgrades.
Atmel (www.atmel.com)	AVR 8-bit RISC microcontroller	(i) Flash memory densities ranging from 1 kb to 256 kb. (ii) With the help of its fast core and feature rich peripherals AVR will be able to offer both wireless communication and handle main application.
Oki Semi-conductor (www.okisemi.com)	ML 7065	(i) Suitable for both RFD and FFD device. (ii) DSSS with 2 M chips and 250 kbps maximum data transmission rate. (iii) Offset quadrature phase shift keying modulation/demodulation function.
Silicon Laboratories (www.silabs.com)	C8051F52× and C8051F3× MCU	(i) All devices are designed to operate across the full automotive temperature (–40 to 125°C) and voltage (2.7–5.25 V) range. (ii) The C8051F52× family is the first to combine a ±0.5% integrated precession internal oscillator with 8 kb flash, 25 MIPs, 12-bit ADC, 16-bit timers, SPI, UART, and 6/16 bit I/O lines in small 10-pin QFN.
Microchip (www.microchip.com)	MRF2J40: 2.4 GHz, IEEE 802.15.4 transceiver	(i) Supports MiWi and ZigBee protocol. (ii) 4-wire SPI interface. (iii) Inline/standalone encryption. (iv) Automatic MAC retransmission. (v) On-chip Tx/Rx switch.

Table 8.6 (continued)

Company name	Name of device	Specification
Texas Instruments (www.ti.com)	RF transceivers: CC2430, CC2431	(i) Three different flash versions: CC2430 F32/64/128. (ii) 32/64/128 kb of flash memories is used. (iii) The CC2430 combines the performance with the industry standard such as 8051 MCU, 32/64/128 kb flash memories, 8 kb RAM, and many other powerful features.
Integration Associates (www.integretion.com)	IA OEM-DAUB1 2400	(i) Provides a compliant 802.15.4 interface. (ii) 2.4 GHz operation. (iii) Integrated antenna. (iv) Wireless range over 30 m. (v) USB 1.1 interfaces. (vi) Various low-power sleep modes.
Maxtream (www.maxstream.net)	Xbee OEM RF module	(i) ISM 2.4 GHz operating frequency. (ii) 1 mw (0 dBm) power output (up to 100 m range). (iii) Industrial (–40 to 85°C) rating. (iv) Advance networking and low-power modes supported.
Helicomm (www.helicomm.com)	IP-link 1220 embedded module	(i) It includes an IEEE 802.15.4-compliant radio, an 8051-based microcontroller, programmable I/O, ZigBee-compliant IP-NET networking software, and a flexible antenna. (ii) It comes in a variety of short, medium, and long-range solutions.
Rajkamal Bar-Scan System Pvt. Ltd. (www. rajkamalbarscan.com)	AeroScout's T2 and T3	(i) LF channel: 125 kHz. (ii) Operating temperature: –20 to 60°C. (iii) Input voltage: 12 V DC.
Crossbow (www.xbow.com)	Imote2	(i) Development kit for Imote2-based wireless sensor networks. (ii) PXA271 XScale® Processor at 13–416 MHz. (iii) Wireless MMX DSP Coprocessor, 256 kB SRAM, 32 MB FLASH, 32 MB SDRAM. (iv) Integrated 802.15.4 radio.
Shenzhen Techno. Co. Ltd. (www.soczigbee.com)	Zigbee WS-MA/MB	(i) WS-M CC2430/31 ZigBee IEEE 802.15.4. (ii) 2.4 GHz operation. (iii) Integrated antenna. (iv) Operating voltage: 2.7–3 V.

focus is given to protocol layers and application profiles. The ZigBee Alliance is taking the lead in this effort. In parallel, several semiconductor manufacturers are expected to announce integrated circuit support for implementation of ZigBee stack. The focus of 802.15.4 development was on maintaining simplicity by concentrating on the essential requirements that provide a successful standard. The standard is targeting the residential market by taking advantage of lower cost and low power consumption. Now it is expected that several users of wireless technologies will shift to the ZigBee standard solution due to the expected lower cost and improved performance.

References

Akyildiz IF, Melodiaa T and Chowdhury RK (2007) A survey on wireless multimedia sensor networks. *Computer Networks*, **51**(4): 921–960.

Al-Karaki JN and Kamal AE (2004) Routing techniques in wireless sensor networks: a survey. *IEEE Wireless Communications*, **11**(6): 6–28.

Bai M, Zhao X, Hou ZG and Tan M (2007) A wireless sensor network used in coal mines. *Proceedings of IEEE International Conference on Networking, Sensing and Control*, London, UK, 15–17 April 2007, pp. 319–323.

Bougard B, Catthoor F, Daly DC, Chandrakasan A and Dehaene W (2005) Energy efficiency of the IEEE 802.15.4 standard in dense wireless microsensor networks: modeling and improvement perspectives. *Proceedings of Design, Automation and Test in Europe*, Munich, Germany, 7–11 March 2005, Vol. 1, No. 1, pp. 196–201.

Callaway E, Gorday P and Hester L (2002) Home networking with IEEE 802.15.4: a developing standard for low-rate wireless personal area networks. *IEEE Communication Magazine*, **40**: 70–77.

Callaway E, Gorday P, Hester L, Gutierrez JA, Naeve M, Heile B and Bahl V (2002) Home Networking with IEEE 802.15.4: a developing standard for low-rate wireless personal area networks. *IEEE Communications Magazine*, **40**(8): 70–77.

Darji AD (2004) *ZigBee-wireless sensor networking for automation*. M. Tech. Thesis, Electronic Systems Group, Electrical and Electronics Department, Indian Institute of Technology, Mumbai, India.

Ergen SC (2004) *ZigBee/IEEE 802.15.4 Summary*. www.pages.cs.wisc.edu.

Evans PC (2003) ZigBee wireless standard. *IEE Review*, **49**(3): 28–31.

Held G (2001) *Data over wireless networks Bluetooth, WAP and Wireless LANs*. Tata McGraw-Hill Publishing Company Limited, New Delhi.

Howitt I and Gutierrez JA (2003) IEEE 802.15.4 low rate – wireless personal area network coexistence issues. *Proceedings of IEEE Wireless Communications and Networking Conference*, Illinois, USA, 16–20 March 2003, Vol. 3, pp. 1481–1486.

Jo WC, Ho YH, Chang YJ and Dan KS (2007) Analysis of throughput in a ZigBee network under the presence of WLAN interference. *Proceedings of International Symposium on Communications and Information Technologies*, Sydney, Australia, 17–19 October 2007, pp. 1166–1170.

Keshav S (2004) *An Engineering Approach to Computer Networking ATM Networks, the Internet, and the Telephone Network*. Pearson Education, India.

Kinney P (2003) *ZigBee Technology: Wireless Control that Simply Works*. www.centronsolutions.com.

Kolla S, Border D and Mayer E (2003) Field bus networks for control system implementations. *Proceedings of IEEE Electrical Insulation Conference and Electrical Manufacturing and Coil Winding Technology Conference*, Southempton, UK, 23–25 September 2003, Vol. 3, 493–498.

Li X, Fang K, Gu J and Zhang L (2008) An improved ZigBee routing strategy for monitoring system. *Proceedings of First International Workshop on Intelligent Networks and Intelligent Systems*, Beijing, China, 1–3 November 2008, pp. 255–258.

Lin S, Liu J and Fang Y (2007) ZigBee based wireless sensor networks and its applications in industrial. *Proceedings of IEEE International Conference on Automation and Logistics*, Shandong, China, 18–21 August 2007, pp. 1979–1983.

Lu G, Krishnamachari B and Raghavendra CS (2004) Performance evolution of the IEEE802.15.4 MAC for low-rate low-power wireless networks. *Proceedings of the 23rd IEEE International Conference on Performance Computing and Communications*, Arizona, USA, 15–17 April 2004, Vol. 4, 701–706.

Pahlavan K and Krishnamurthy P (2006) *Principles of Wireless Networks*. Prentice-Hall of India Private Limited, New Delhi.

Petrova M, Riihijärvi J, Mähönen P and Labella S (2006) Performance study of IEEE 802.15.4 using measurements and simulations.*Proceedings of IEEE Wireless Communications and Networking Conference*, Las Vegas, USA, 16–20 April 2006, pp. 487–492.

Pinedo FED and Garcia MJA (2008) An experimental analysis of ZigBee networks. *Proceedings of 33rd IEEE Conference on Local Computer Networks*, Montreal, Canada, 14–17 October 2008, pp. 723–729.

Shuaib K, Boulmalf M, Sallabi F and Lakas A (2006) Co-existence of ZigBee and WLAN, a performance study. *Proceedings of Wireless Telecommunications Symposium*, California, USA, 6–9 April 2006, pp. 1–6.

Stein G and Kibitzes K (2002) Concept for an architecture of a wireless building automation. *Proceedings of Sixth IEEE Africon Conference*, George, South Africa, 2–4 October 2002, Vol. 1, 139–142.

Tsang KF, Lee WC, Lam KL, Tung HY and Kai X (2007) An integrated ZigBee automation system: an energy saving solution. *Proceedings of 14th International Conference on Mechatronics and Machine Vision in Practice*, Xiamen, China, 4–6 December 2007, pp. 252–258.

Viswanathan A and Boult TE (2007) Power conservation in ZigBee networks using temporal control. *Proceedings of 2nd International Symposium on Wireless Pervasive Computing*, Puerto Rico, 5–7 February 2007, pp. 327–331.

Zheng J and Lee MJ (2004) Will IEEE 802.15.4 make ubiquitous networking a reality? a discussion on a potential low power, low bit rate standard. *IEEE Communications Magazine*, **27**(6): 23–29.

Websites

www.atmel.com
www.ember.com
www.en.wikipedia.org/wiki/ZigBee: ZigBee-Wikipedia, the free encyclopedia
www.freescale.com/zigbee: ZigBee Technology from Freescale
www.helicomm.com
www.IEEE.org
www.integretion.com
www.jennic.com
www.maxstream.net/wireless/zigbee.php
www.microchip.com
www.okisemi.com
www.rajkamalbarscan.com
www.silabs.com
www.soczigbee.com
www.ti.com
www.tutorial-reports.com/wireless/zigbee
www.ubec.com
www.xbow.com
www.zigbee.org: ZigBee Alliance

Chapter 9
Wireless Information and Safety System for Mines

9.1 Introduction

In case of disaster in an underground mine, it is very difficult for mine management to identify actual person trapped, their number, and exact location. Therefore, identification and coding of miners is a vital need for underground mine management in case of disaster as well as normal operating conditions. Mining industry is generally capital intensive; cost of maintenance (35% of operating cost of system) at mechanized mines goes as high as 50–60% when both direct and indirect costs are taken into account (Sen, 2001). Sometimes, it constitutes 30% of total production cost. In today's globally competitive market scenario, efforts to reduce production cost have awaken mining industry for automation and optimum utilization of equipment by increasing its availability and performance (Kumar and Guha, 2001; Bandyopadhyay et al., 2001, 2008a). Therefore, continuous monitoring of equipment location and their operation with respect to dynamic working places is necessary to make the underground mines viable, competitive, and profitable. Considering the importance of wireless tracking and monitoring devices for mines, authors have developed a "wireless information and safety system (WISS)" which will help in overcoming the day-to-day mining problems as well as the problems during disaster conditions.

9.2 Function

The system is used for tracking and monitoring miners and equipment in underground mines using ZigBee-enabled active RFID devices, which form a wireless network among themselves and other static and mobile ZigBee devices placed at strategic locations (Bandyopadhyay et al., 2008b, c, d, e; Kumari et al., 2008). It is a low-power data communication and wireless LAN system designed to perform radio transmission of digital data signals. The same active RFID device is programmed to act as a tag (end device), router, or coordinator that enables them to form a mesh network. Tags (end devices) are assigned to the miners or moveable equipment. Routers are placed in strategic locations in an underground mine along the desired

L.K. Bandyopadhyay et al., *Wireless Communication in Underground Mines*,
DOI 10.1007/978-0-387-98165-9_9, © Springer Science+Business Media, LLC 2010

path for wireless communication by forming dynamic mesh network. Coordinator is placed in the pit top control room to collect information from all the routers.

Core system component, ZigBee-compliant active RFID device, can be programmed to act as tag (end device), router, or coordinator using resident application-specific embedded software that enables them to form an IEEE 802.15.4-based mesh network (Ergen, 2004). It uses a unified wireless mesh-networking infrastructure to locate, trace, and manage mobile assets and people as well as monitor different environmental conditions using sensors (Callaway et al., 2002; Howitt and Gutierrez, 2003; Pahlavan and Krishnamurthy, 2006). ZigBee devices have numerous advantages (Held, 2001; Stein and Kibitzes, 2002; Kinney, 2003; Darji, 2004), some of these are as follows: (i) unlicensed 2.4 GHz industrial, scientific, and medical (ISM) band; (ii) ultra-low-power (ideal for battery-operated system) requirement; (iii) operates for years on inexpensive batteries; (iv) large number of nodes/sensors; (v) reliable and secure links between network nodes; (vi) easy deployment and configuration; (vii) low-cost system; (viii) very fast transition time; (ix) digital battery monitor facility; and (x) smaller in size (system on chip).

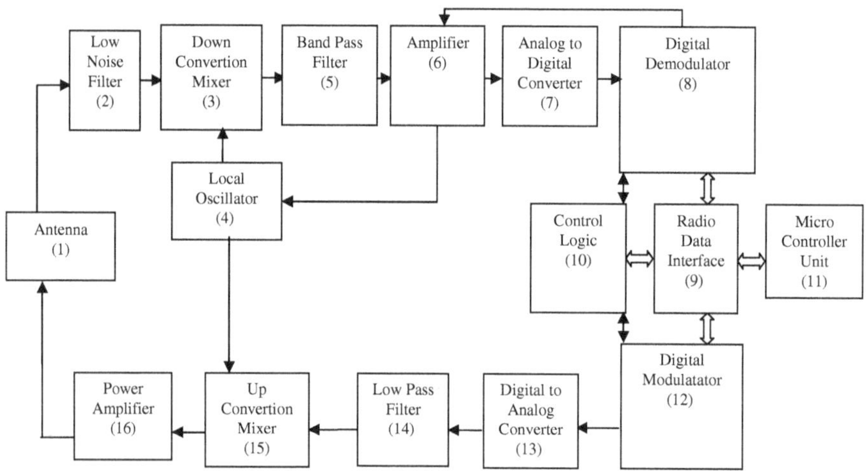

Fig. 9.1 Block diagram of ZigBee transceiver

In ZigBee transceiver (Fig. 9.1), incoming 2.4 GHz RF signal is picked up by antenna (1) and passed to low-noise filter (2) through duplexer. Low-noise filter (2) eliminates DC offset and noise problem and amplifies radio frequency (RF) signal from antenna to a suitable level before feeding to down-conversion mixer (3), which mixes amplified signal with a high-frequency signal generated by local oscillator (4). Output of down-conversion mixer (3) is then fed to a band pass filter (5), and signal thus obtained is down converted in quardrature to 2 MHz intermediate frequency (IF). Down-conversion mixer (3) converts received RF frequency to IF, which is filtered and amplified by a band pass filter (5), and then autotuned by phase-locked loop. Output of band pass filter (5) is fed to amplifier (6), which controls over gain by inbuilt automatic gain control (AGC). Automatic frequency control (AFC) of amplifier (6) controls frequency of local oscillator (4), which is also inbuilt in amplifier (6). AGC output is sampled with 4 MHz sampling rate at analog-to-digital converter (7) for further digitization. The digitized signal is fed to digital demodulator (8), where channel filtering and demodulation are performed in digital domain. Signal obtained from digital demodulator (8) is an efficient digitized actual signal containing data, which is fed to radio data interface (9) and control logic (10). Radio data interface (9) interfaces signal to microcontroller unit (11). Control logic (10) controls signal of digital demodulator (8) and digital modulator (12). Transmission of IF signal coming from radio data interface (9) is modulated by digital modulator (12), and fed to digital-to-analog converter (13) for conversion of digital signal. Output of digital-to-analog converter (13) is fed to low pass filter (14), where signal is filtered and again passed to up-conversion mixer (15), which up converts filtered signal directly to RF by a single sideband modulator using local oscillator (4). The up-conversion mixer (15) is designed for low-voltage operation and is ideal for use in portable consumer equipment. Up-conversion mixer (15) operates with IF input frequencies (40–500 MHz), and up converts to output frequencies as high as 2.5 GHz. Output of up-conversion mixer (15) is then fed to power amplifier (16), which power ups the signal. Powered-up signal is then passed to antenna (1) through duplexer, which combines two or more signals onto a common channel or medium to increase its transmission efficiency. It allows a transmitter to operate on one frequency and a receiver on a different frequency to share one common antenna with minimum interaction and degradation of different RF signals. Then, RF signal is transmitted to antenna (1) for transmission. Specifications of RF part of the ZigBee device are given in Table 9.1. It uses direct sequence spread spectrum (DSSS) for modulation at 2.4 GHz frequency. It uses operating voltage of 2.7–3.6 V.

Table 9.1 Specifications of RF part of the system

Parameter	Values
1. Operating ambient temperature range	–40 to 85°C
2. Operating supply voltage range	2.7–3.6 V
3. Microcontroller	High-performance and low-power 8051
• Type	microcontroller core
• Programmable flash memory	128 KB
• RAM	8 KB
• System clock frequency	32 MHz
• Time delay	10 ns
4. Radio part	
• RF frequency range	2.4–2.41 GHz (ISM)
• Radio bit rate	250 kbps
• Radio chip rate	2.0 M Chips/s
• Receiver sensitivity	–92 dBm
• Transmitter spurious emission	–43 dBm
• Receiver spurious emissions	–75 dBm
5. Modulation	Direct sequence spread spectrum (DSSS)
6. Wireless standard compliance	ETSI EN 300 (Europe), FCC CFR47 (USA), and ARIB STD-T66 (Japan)

9.3 The System

WISS is consists of five subdevices, namely (1) coordinator, (2) router/end device, (3) carbon-monoxide-monitoring device, (4) methane-monitoring device, and (5) message device. Further, each device has two parts, namely (i) tracking/monitoring/messaging, and (ii) power supply unit.

9.3.1 Coordinator

9.3.1.1 Circuit Details

Tracking and Monitoring Unit

The block diagram of coordinator is depicted in Fig. 9.2, which consists of external cap-lamp battery, internal battery supply, voltage regulator, voltage inverter, MAX232, embedded radio, analog-to-digital converter input, and light emitting diode. The external cap-lamp battery and internal battery is connected to

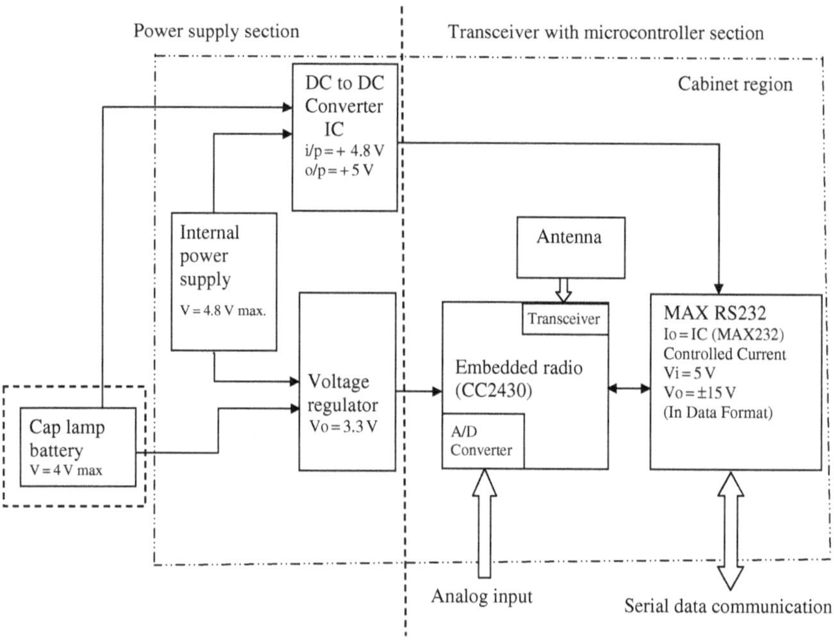

Fig. 9.2 Block diagram of coordinator circuit. Key: Analog input description: $V = 3.6$ V, $I = 4$ mA maximum, C and L negligible. Serial data communication description: $V = \pm 15$ V, $I =$ MAX232 IC controlled, C and L negligible

voltage regulator. The said voltage regulator through safety arrangement is connected to an embedded radio. The said embedded radio is connected to programming port, analog-to-digital converter input, light emitting diode, and MAX232 serial interface.

The cap-lamp battery is connected with the voltage regulator to provide the regulated 3.3 V power supply to the circuit. The internal battery supply is also connected with the RFID device through turnover switch so that the device can be operated using external battery supply or internal battery supply. The output after the safety arrangement is feed to programmable port, analog-to-digital converter input, and light emitting diode for programming, sending digital signal to external device, taking analog input from external device, and indicating the data transmission operation. Serial interface MAX232 gets its power supply from internal or external power supply.

Figure 9.3 shows the circuit diagram of the RFID device. A three pin jack (CON3) is provided to connect external cap-lamp battery. Power supply to the

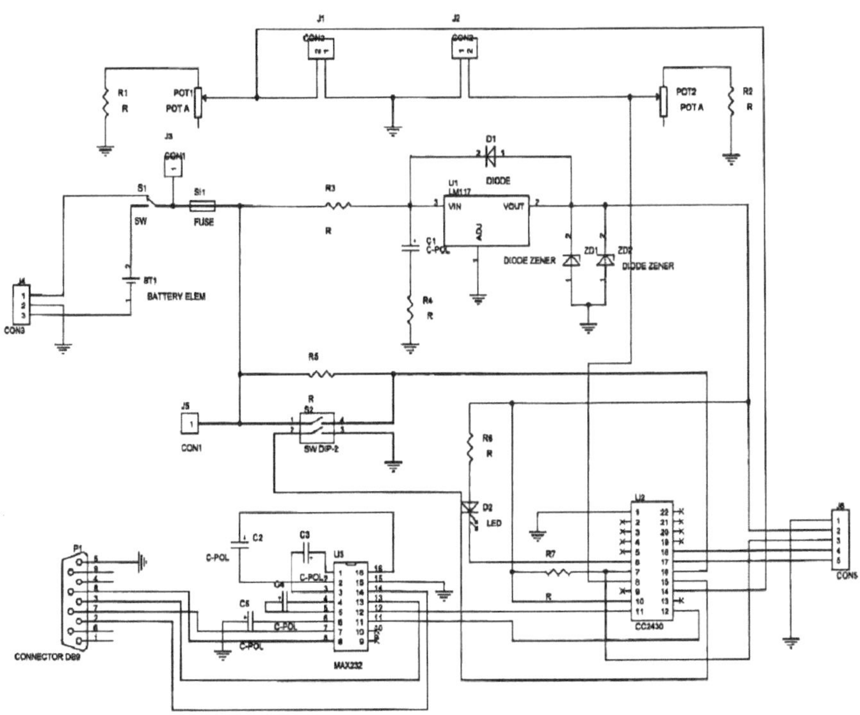

Fig. 9.3 Circuit diagram of RFID part of coordinator. Key: J1 and J2: ADC input port (V = 3.6 V, I = 4 mA maximum, C and L negligible), J4: Cap-lamp battery supply 4.0 V (max), J6: Programming port, VDD=2.0–3.6 V, J3 and J5: Voltage measuring point via voltmeter (L and C of voltmeter are negligible and R is very high, MΩ); BT1: Four AA size rechargeable batteries, 4.8 V. P1: Serial port for connection with computer: this port will be used when the system is placed in safe zone only

RFID is given through the voltage regulator LM1117, 3.3 V. For safety reasons, two Zener diodes and a diode are connected with the voltage regulator. LED (D2) via resistance (R6) is connected for indicating transmission in the RFID chip. S2 (SW DIP-2) is a change over switch between voltage inverter circuit output and internal battery supply. Jack (J1) and jack (J2) is for external sensor connection. POT1 and POT2 are given to provide voltage to RFID chip which will be in RFID chip's range. Jack (J6) is for programming. MAX232 is used for serial interfacing between RFID chip and computer. Figures 9.4, 9.5, 9.6, and 9.7 illustrate the PCB layout for top, bottom, and component (top and bottom) layers, respectively. Table 9.2 gives the list of components used along with their specification.

Power Supply Unit

Power supply to the circuit is supplied from two different sources: external source (i.e., cap-lamp battery) and internal source (i.e., four AA size rechargeable batteries). A switch is used for interchanging the power supply between internal and external power supplies. The required voltage of 5 V for MAX232 is provided by the DC-to-DC converter (LM2577) of the power supply unit. Figure 9.8 shows the circuit diagram of the regulated power supply to the device. Two pin jack JR3 (CON2) is used to connect cap-lamp battery and internal battery.

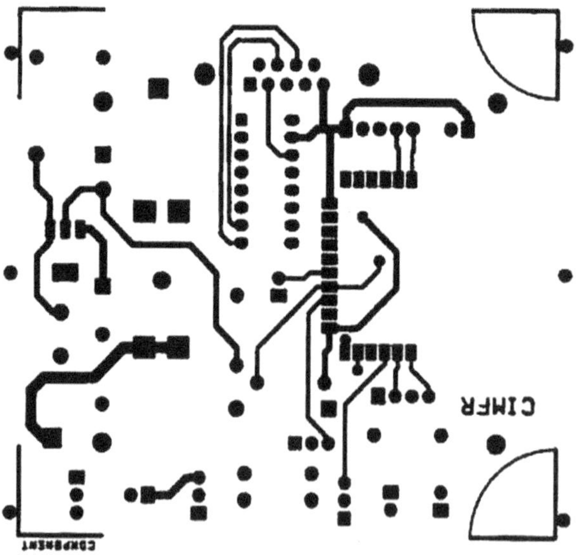

Fig. 9.4 PCB top layer of RFID part of coordinator

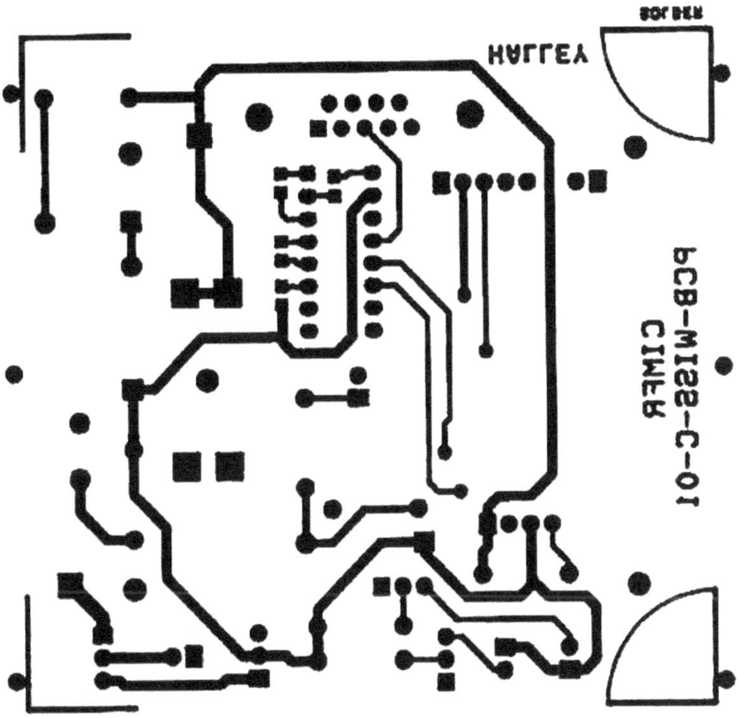

Fig. 9.5 PCB bottom layer of RFID part of coordinator

(a) Details of cap-lamp battery

Battery type: Intrinsically safe lithium-ion cap-lamp battery
Operating voltage: 3.7 V
Maximum voltage: 4.0 V
Current: 0.4 A
Ampere-hour: 4 Ah

(b) Internal power supply

Battery type: Four AA size rechargeable alkaline batteries
Voltage: 4×1.2 V = 4.8 V
Ampere-hour: 700 mAh each

Voltage regulator (U2) is used for 5 V voltage output. Considering the safety purpose, fuse (S/1) and diode (D1) are used. LED1 is used to check transmission in the circuit. Capacitors (C6, C7, C8, and C9) are used to remove any type of ripples in the circuit. Figures 9.9, 9.10, 9.11, and 9.12 show the PCB layouts for top, bottom, and component layouts (top and bottom), respectively. Table 9.3 gives the list of components used in the device. Figures 9.14 and 9.15 illustrate the wiring diagram and enclosure diagram of the device.

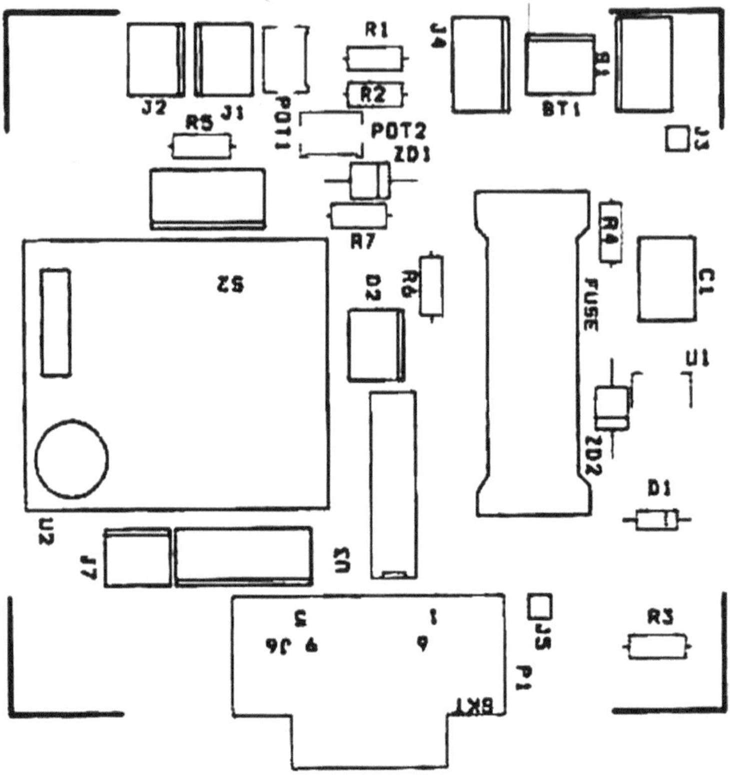

Fig. 9.6 Component layout (*top*) of RFID part of coordinator

9.3.1.2 Current

Transmitting Mode

(a) From cap-lamp: In this mode, the maximum current consumption by the circuit is 400 mA. It is measured by keeping an ampere meter in series between cap-lamp and RFID device in transmitting mode. As per the IEC60079-11, 2007 standard (Fig. 12.7 as given in Chapter 12 – Resistive circuits), the minimum ignition current at 4.0 V is greater than 3 A. Therefore, the current consumption of the RFID device powered with cap-lamp (maximum voltage 4.0 V) is much lesser than the minimum ignition current as stipulated in the standard.

(b) From internal battery: In this mode, the maximum current drawn by the circuit is 250 mA. It is measured by keeping an ampere meter in series between internal battery and RFID device in transmitting mode. As per the IEC60079-11, 2007 standard (Fig. 12.7 as given in Chapter 12 – Resistive circuits), the minimum ignition current at 4.8 V is greater than 3 A. Therefore, the current consumption of the RFID device powered with internal battery (maximum voltage 4.8 V) is much lesser than the minimum ignition current as stipulated in the standard.

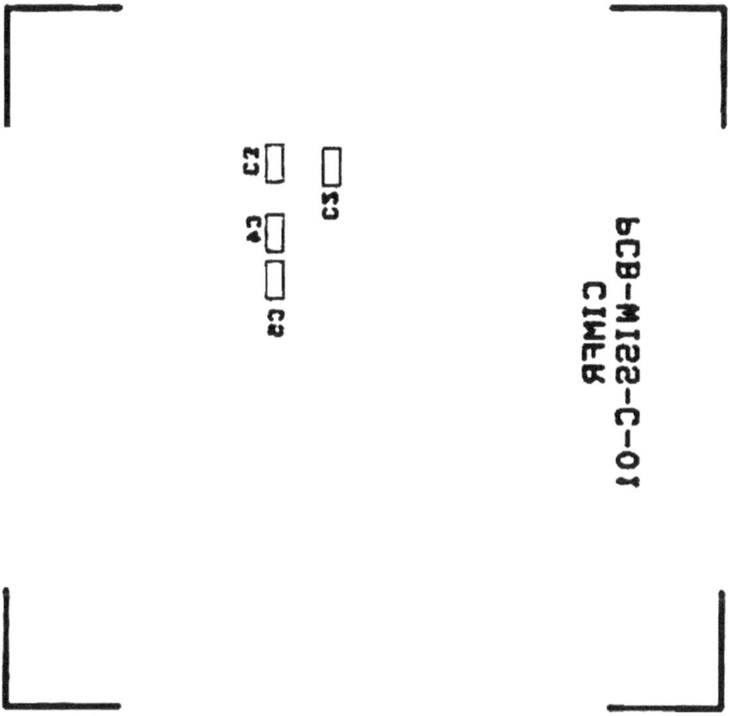

Fig. 9.7 Component layout (*bottom*) of RFID part of coordinator

Table 9.2 Component list of RFID part of coordinator

Name	Designation	Value	Tolerance
Resistors	R1	1 kΩ/0.25 W, MFR	1%
	R2	1 kΩ/0.25 W, MFR	1%
	R3	6.8 Ω/5 W, MFR	1%
	R4	82 Ω/0.5 W, MFR	1%
	R5	10 kΩ/0.25 W, MFR	1%
	R6	470 Ω/0.25 W, MFR	1%
	R7	43 kΩ/0.25 W, MFR	1%
Variable resistors	POT1	10 kΩ	
	POT2	10 kΩ	
Light emitting diode	D2 (LED)	3 mm Green LED $I = 7$ mA (max)	
Capacitors	C1	0.33 μF (ceramic), 100 V	10%
	C2	1 μF (SMD), 16 V	10%
	C3	1 μF (SMD), 16 V	10%
	C4	1 μF (SMD), 16 V	10%
	C5	1 μF (SMD), 16 V	10%
Switches	S1	On–off–on switch	
	S2	DIP-2 switch	

Table 9.2 (continued)

Name	Designation	Value	Tolerance
Zener diode	ZD1	3.6 V/5 W	
	ZD2	3.6 V/5 W	
Diode	D1	1N4007	
Voltage regulator	U1	LM1117	
Embedded radio	U2	CC2430, RFID chip	
Serial interface chip	U3	MAX232 IC	
Fuse	S/1	500 mA	250 V, 35 A, breaking capacity
Antenna	Antenna	Dielectric rod antenna	
Connectors	P1	9 pin serial port connector (this part will be used only when the system is placed in safe zone)	
	J1 & J2	2 pin jack for ADC input, (3.6 V, 4 mA, *C* and *L* negligible)	
	J4	3 pin jack for cap-lamp battery connection (4 V, 0.4 Amax)	
	J6	5 pin jack for programming to CC2430, VDD = 2.36 V	
	BT1	Connector for 4 rechargeable batteries of 1.2 V each	

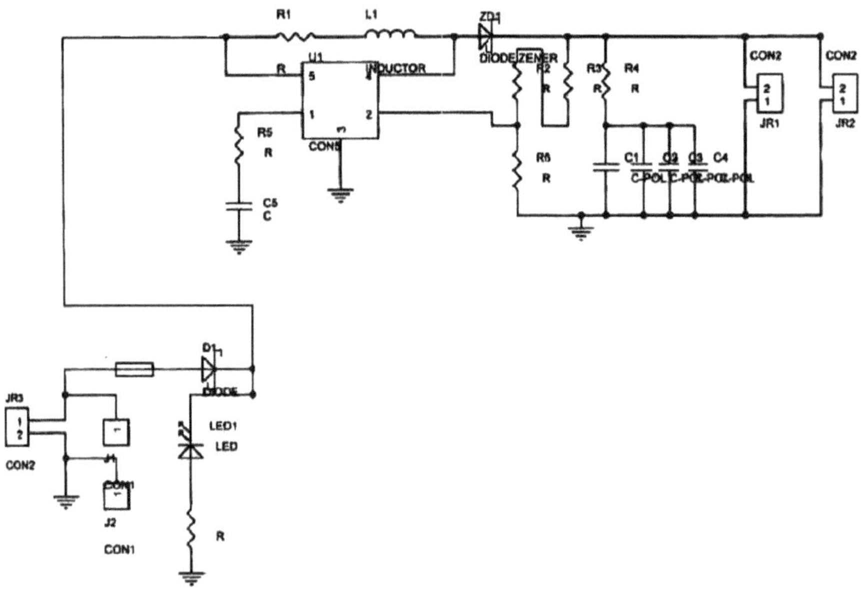

Fig. 9.8 Circuit diagram of power supply part of coordinator. Key: J1: voltage measuring point via voltmeter, J2: voltage measuring point via voltmeter (*L* and *C* of voltmeter are negligible and *R* is very high in MΩ), JR3: cap-lamp battery connection with maximum voltage of 4.0 V, 400 mA (max), JR2: output terminal, +5 V, JR1: output measuring point (voltmeter) (*L* and *C* of voltmeter are negligible and *R* is very high in MΩ)

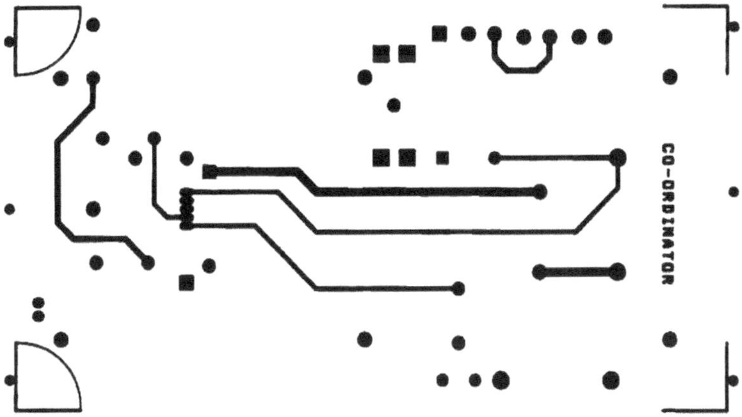

Fig. 9.9 PCB top layer of power supply part of coordinator

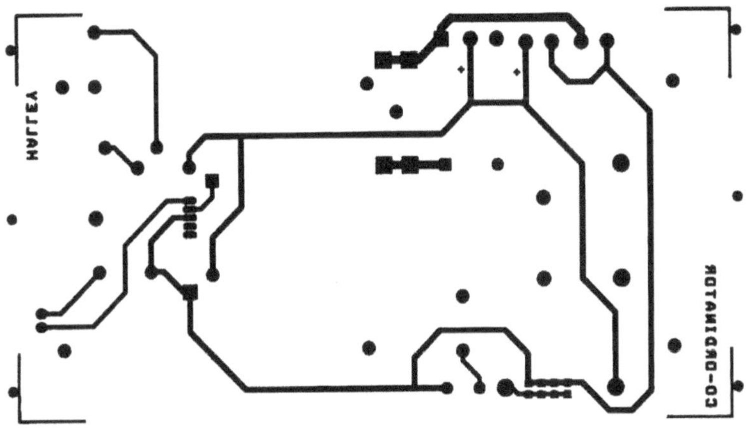

Fig. 9.10 PCB bottom layer of power supply part of coordinator

Receiving Mode

(a) From cap-lamp: In this mode, the maximum current consumption by the circuit is 400 mA. It is measured by keeping an ampere meter in series between cap-lamp and RFID device in transmitting mode. As per the IEC60079-11, 2007 standard (Fig. 12.7 as given in Chapter 12 – Resistive circuits), the minimum ignition current at 4.0 V is greater than 3 A. Therefore, the current consumption of the RFID device powered with cap-lamp (maximum voltage 4.0 V) is much lesser than the minimum ignition current as stipulated in the standard.

(b) From internal battery: In this mode, the maximum current drawn by the circuit is 250 mA. It is measured by keeping an ampere meter in series between the internal battery and RFID device in receiving mode. As per the IEC60079-11, 2007

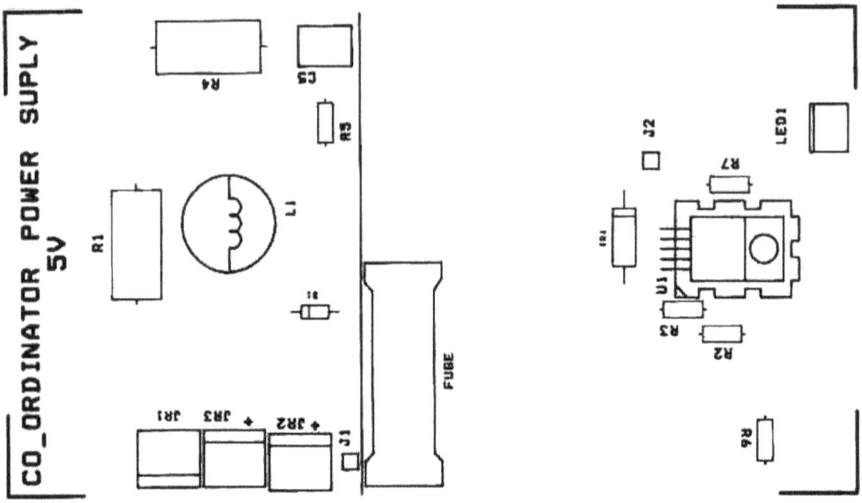

Fig. 9.11 Component layout (*top*) of power supply part of coordinator

Fig. 9.12 Component layout (*bottom*) of power supply part of coordinator

standard (Fig. 12.7 as given in Chapter 12 – Resistive circuits), the minimum ignition current at 4.8 V is greater than 3 A. Therefore, the current consumption of the RFID device powered with internal battery (maximum voltage 4.8 V) is much lesser than the minimum ignition current as stipulated in the standard.

Table 9.3 Component list for power supply part of coordinator

Name	Designation	Value	Tolerance
Resistor	R1	12 Ω/2 W, MFR	1%
	R2	2.7 kΩ/0.25 W, MFR	1%
	R3	390 Ω/0.25 W, MFR	1%
	R4	22 Ω/2 W, MFR	1%
	R5	2.2 kΩ/0.25 W, MFR	1%
	R6	1 kΩ/0.25 W, MFR	1%
	R7	470 Ω/0.25 W, MFR	1%
Light emitting diode	LED1	3 mm red LED I = 7 mA (max)	
Capacitor	C1	1 μF (SMD)/16 V	10%
	C2	1 μF (SMD)/16 V	10%
	C3	1 μF (SMD)/16 V	10%
	C4	1 μF (SMD)/16 V	10%
	C5	0.33 μF (ceramic)/100 V	10%
Schkotty diode	ZD1	1N5818	
	D1	1N5818	
Voltage regulator	U1	LM2577 ADJ	
Inductor	L1	100 μH	10%
Fuse	FUSE1	500 mA	250 V, 35 A, breaking capacity
Connector	JR3	2 pin jack for cap-lamp battery connection (4 V, 0.4 A max)	
	JR2	2 pin jack for taking output voltage (5 V)	
	JR1	2 pin jack for measuring output voltage via voltmeter	

9.3.1.3 Operating Voltage

The RFID chip has wide operating voltage range from 2.7 to 3.6 V. The coordinator operates in the following two voltages:

(i) With cap-lamp: The voltage provided to the regulator is 4.0 V (maximum).
(ii) With internal battery: The voltage provided to the regulator is 4.8 V.

9.3.1.4 Power

In case of cap-lamp battery, the maximum power drawn by the circuit is (maximum cap-lamp battery voltage) × (maximum current drawn by the circuit), i.e., = 4.0 V × 400 mA = 1.6 W. In case of internal battery the maximum power drawn by the circuit is (internal battery voltage) × (maximum current drawn by the circuit), i.e., = 4.8 V × 250 mA = 1.2 W. All the components used in the circuit are well within the IS allowed maximum power rating.

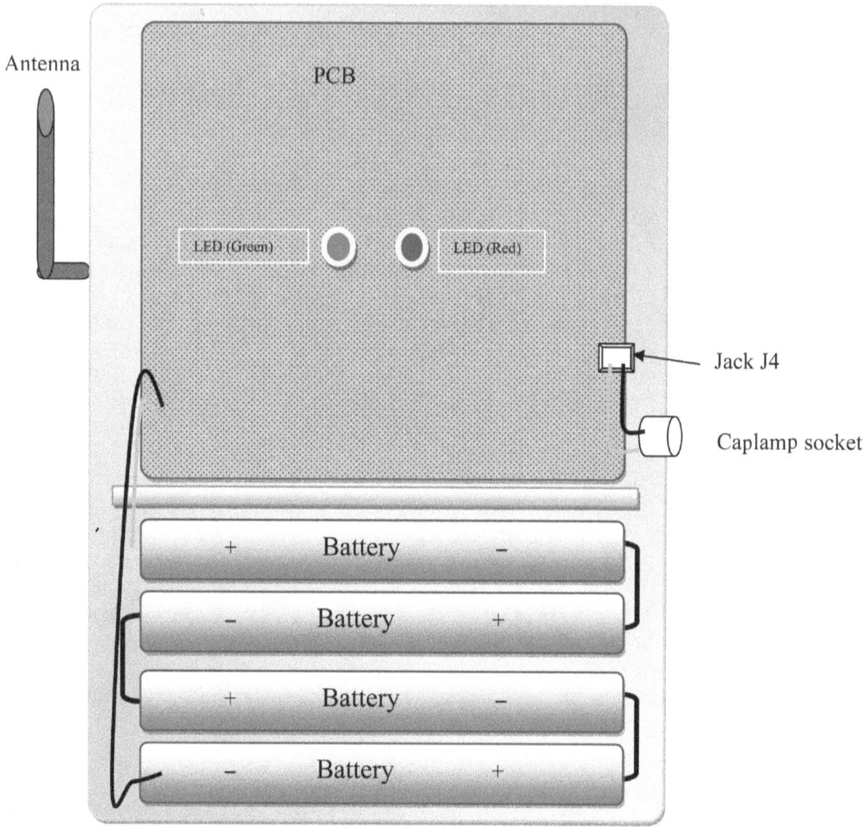

Fig. 9.13 Wiring diagram of coordinator

9.3.1.5 Inductance

Inductance of the circuit is 28.8 nH for antenna and 100 μH inductor is used in DC-to-DC converter circuit part. So, inductance (Li) = 100.0288 μH. Inductance to ADC (Lc), UART pins, and conducting wires is almost negligible. As per the IEC60079-11, 2007 standard (Fig. 12.8 as given in Chapter 12 – Inductive circuit) the maximum allowable inductance (Lo) is more than 6 mH at 0.4 A. Therefore, Lo is >> Li + Lc.

9.3.1.6 Capacitance

Capacitance of the circuit (Ci) is 8.66 μF. Capacitance to ADC and UART pins, and conducting wires (Cc) are almost negligible. As per the IEC60079-11, 2007 standard (Fig.12.6 as given in Chapter 12 – Group I capacitive circuit), the maximum allowed capacitance (Co) at 4.8 V is >1000 μF. Thus, Co is >> Ci + Cc.

Fig. 9.14 View of coordinator along with enclosure

9.3.1.7 Short Circuit Current

(i) With cap-lamp battery: The maximum current that can flow through a circuit is limited by a current-limiting resistor of 4.2 Ω. The maximum current through circuit is 4.0 V/ 4.2 Ω = 0.95 A. As per the IEC60079-11, 2007 standard (Fig. 12.7 as given in Chapter 12 – Resistive circuits), the minimum ignition current at 4.0 V is more than 3 A. Therefore, the maximum current that can flow in the circuit is much lesser than the allowable current as mentioned in the standard.

(ii) With internal battery: The maximum current that can flow through the circuit is limited by a current-limiting resistor of 4.2 Ω. The maximum current through circuit is 4.8 V/ 4.2 Ω = 1.14 A. As per the IEC60079-11, 2007 standard (Fig. 12.7 as given in Chapter 12 – Resistive circuits), the minimum ignition current at 4.8 V is more than 3 A. Therefore, the maximum current that can flow in the circuit is much lesser than the allowable current as mentioned in the standard.

9.3.1.8 Safety Analysis

Each component used in the device is provided by its own power rating. The current across the resistor is calculated taking the shortage route to ground by the formula V/R. Then power rating of the component is calculated by the formula I^2R. This power rating is compared with the component power rating which should be less in regard to intrinsic safety. Ratio ($W1/W2$) of the given power rating of the component to the calculated power rating should be greater then 1.5 as per the IEC60079-11, 2007 standard. Tables 9.4 and 9.5 present the safety analysis for the RFID unit and power unit of the coordinator, respectively. Ratio ($W1/W2$) is greater than 1.5 for all the safety components.

9.3.1.9 Antenna Characteristics

The characteristics of the used dielectric rod antenna operating at 2.4–2.41 GHz frequency are given in Table 9.6. Figure 9.15 shows the omnidirectional radiation pattern of the antenna. Figure 9.16 illustrates the plot for voltage standing wave ratio (VSWR) and return loss of the used antenna.

9.3.1.10 Radiated Power

Radiated power of the device using dielectric antenna in frequency range of 2.4–2.41 GHz is given in Table 9.7. As per the measurement it was observed that the radiated power was very less in the frequency range 2.4–2.41 GHz. The maximum value of 803.526 μW was measured at 3 m distance from the antenna, kept in vertical mounting position. The radiated power of the RFID devices is very low,

Table 9.4 Details of safety analysis for RFID unit of coordinator

Component designation	Component location	Value	Maximum rating used by circuit (W2)	Component rating (W1)	$W1/W2$
R1	Power supply part	1 kΩ/0.25 W	23.04 mW	0.25 W	10.85
R2	Power supply part	1 kΩ/0.25 W	10.89 mW	0.25 W	10.85
R3	Power supply part	6.8 Ω/5 W	3.3 W	5 W	1.51
R4	Power supply part	82 Ω/0.5 W	254.22 mW	0.5 W	1.96
R5	Power supply part	10 kΩ/0.25 W	2.30 mW	0.25 W	108.69
R6	Microcontroller part	470 Ω/0.25 W	48.99 mW	0.25 W	5.10
R7	Microcontroller part	43 kΩ/0.25 W	0.529 mW	0.25 W	472.58
ZD1	Power supply part	3.6 V/5 W	0.6336 W	5 W	7.89
ZD2	Power supply part	3.6 V/5 W	0.6336 W	5 W	7.89
S/1	Power supply part	500 mA	400 mA		

Table 9.5 Details of safety analysis for power unit of coordinator

Component designation	Component location	Value	Maximum rating used by circuit (*W2*)	Component rating (*W1*)	*W1*/*W2*
R1	Power supply part	12 Ω/2 W	1.306 W	2 W	1.53
R2	Power supply part	2.7 kΩ/0.25 W	4.0 mW	0.25 W	62.5
R3	Power supply part	390 Ω/0.25 W	0.578 mW	0.25 W	432.5
R4	Power supply part	22 Ω/2 W	475.773 mW	2 W	4.2
R5	Power supply part	2.2 kΩ/0.25 W	11.336 mW	0.25 W	22.0
R6	Power supply part	1 k Ω/0.25 W	1.48 mW	0.25 W	168.9
R13	Power supply part	470 Ω/0.25 W	53.108 mW	0.25 W	4.7
FUSE1	Power supply part	500 mA	400 mA		

Table 9.6 Antenna characteristic of coordinator

Parameters	Value
Operating frequency	2.4–2.41 GHz
Gain	18 dBi
Radiation pattern	Omnidirectional
Polarization	Circular
Impedance	50 Ω
Voltage standing wave ratio (VSWR):	1.9949 at 2.405040 GHz
Return loss	–9.5728 dB at 2.405040 GHz
Antenna type	Dielectric rod antenna

compared to the permissible value (100 mW) given in ETSI EN 300 (Europe) standard.

9.3.1.11 Threshold Power

Maximum voltage in the circuit supplied from internal battery is 4.8 V. Maximum current in the circuit is 400 mA. Threshold power = voltage × current = 4.0 V × 400 mA = 1.6 W. As per the IEC 60079-0, 2007 standard for gas group I, the recommended threshold power is 6 W. The threshold power of the RFID device is very much lesser than the recommended standard limit, since it is a very low power device.

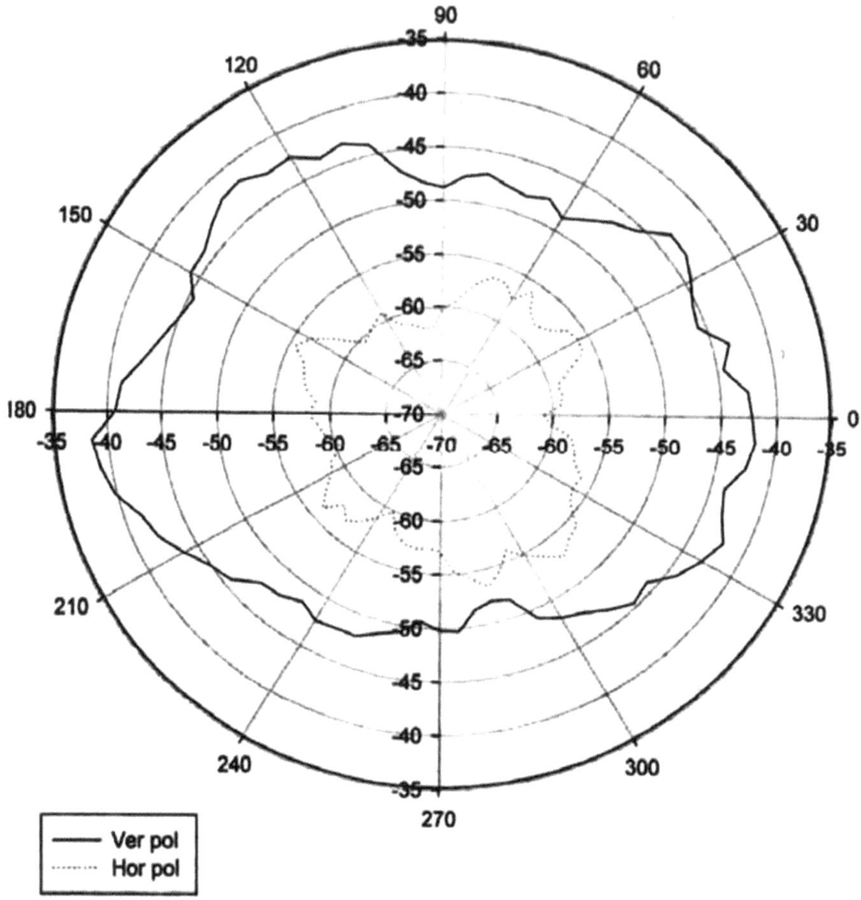

Fig. 9.15 Radiation pattern of dielectric antenna

9.3.2 Router/End device

9.3.2.1 Circuit Details

Tracking and Monitoring Unit

The block diagram of router/end device is depicted in Fig. 9.17, which consists of external cap-lamp battery, light emitting diode, voltage regulator, turnover switch, internal battery supply, safety arrangement, embedded radio, programming port, universal asynchronous receiver–transmitter, analog-to-digital converter input, buzzer, and message switch, and light emitting diode. The external cap-lamp battery is connected to light emitting diode and voltage regulator. The voltage regulator output is connected to turnover switch. The turnover switch is also connected to internal battery supply, thus power to the circuit can be given by either cap-lamp or

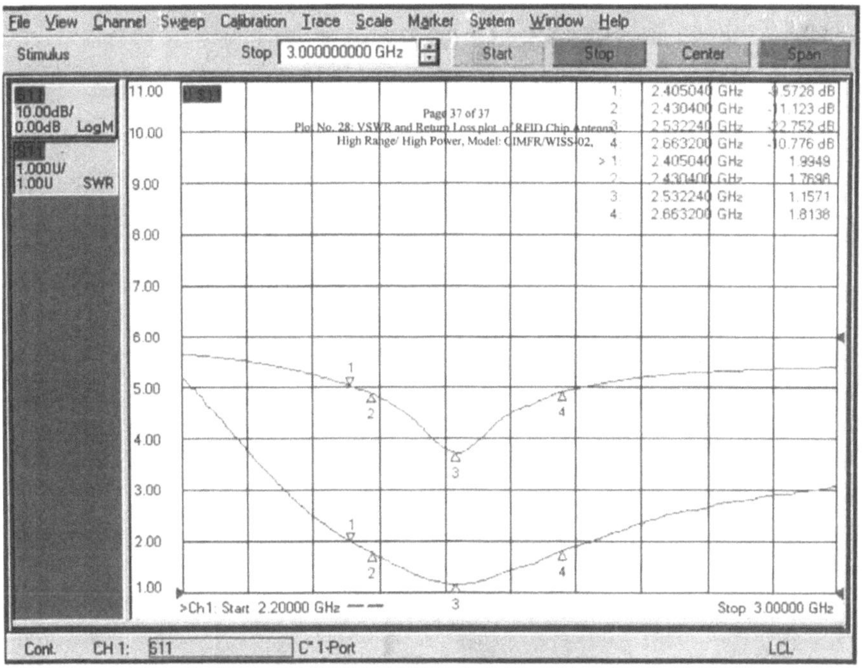

Fig. 9.16 Plot of VSWR and return loss of dielectric antenna

Table 9.7 Radiated power of dielectric antenna

Frequency (GHz)	Distance (m)	Device mounting	Power radiated (dBm)	Power radiated (μW)
2.405340000	3	Horizontal	–46.84	20.7014
2.405340000		Vertical	–30.95	803.526

internal battery via turnover switch. The output after turnover switch is connected to programming port and embedded radio. The embedded radio is connected to programming port, universal asynchronous receiver–transmitter output, analog-to-digital converter input, buzzer, message switch, and light emitting diode.

The external cap-lamp battery is connected with the light emitting diode to indicate the connection of the power supply with the caplamp battery. The cap-lamp battery is also connected with the voltage regulator to provide the regulated 3.3 V power supply to the circuit. The internal battery supply is also connected with the RFID device through turnover switch so that the device can be operated using external battery supply or internal battery supply. The output after the turnover switch is feed to programmable port, universal asynchronous receiver–transmitter output, analog-to-digital converter input, buzzer, message switch, and light emitting diode for programming, sending digital signal to external device, taking analog input from external device, generating alarm, transmitting message, and indicating the data transmission operation, respectively.

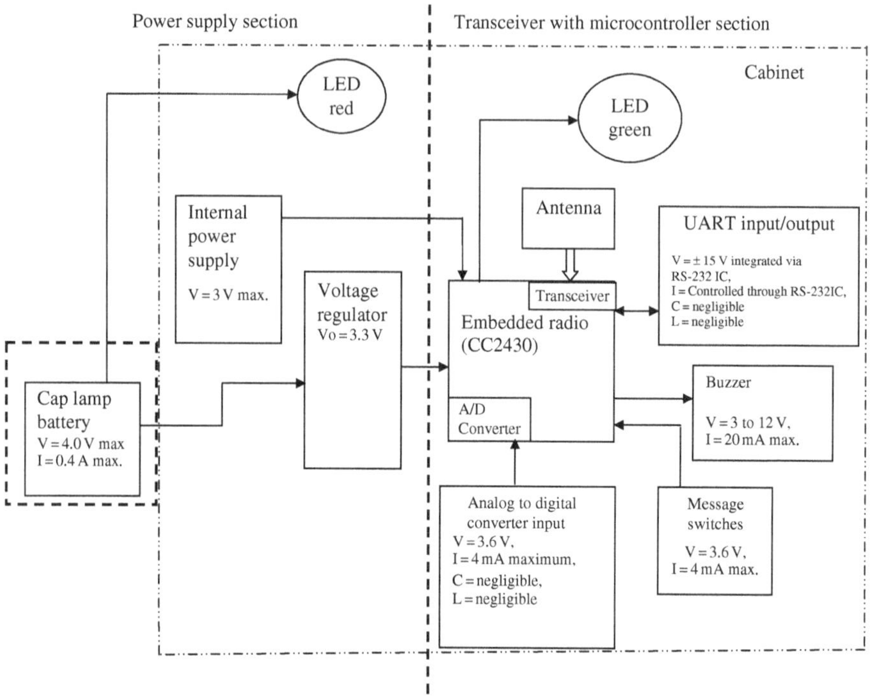

Fig. 9.17 Block diagram of router/end device circuit

Figure 9.18 shows the circuit diagram of the RFID device. A two pin jack (CN1) is provided to connect external cap-lamp battery. LED (LED2) via resistance (R8) is connected to this jack. The jack power is connected to voltage regulator (MAX604 CSA+) via fuse (F1) and resistance (R9). The battery (BT1) is connected via fuse (S/1) to another endpoint of the on–off–on switch (S5). The output from all stages of the switch (S5) is common and connected via resistance (R10) to RFID chip, Zener diodes (ZD1, ZD2, and ZD3) are in parallel with the final supply after S5. The cathode point of the Zener diode is grounded. The capacitor (C1) is connected via resistance (R3) as a filter in parallel with the Zener diode. The embedded radio chip (ER) is connected to LED (LED1) via resistance (R4) and power supply through the resistance (R10). The reset switch (S1) is connected between CC2430 embedded radio (ER) and ground, rest of message switches (S2, S3, and S4) are connected between the embedded radio chip (ER) with power supply through resistances (R5, R6, and R7), respectively, and grounded. Thus, one point of the each message switches (SX) is grounded. The buzzer (SU1) is connected to the embedded radio chip (ER) via resistance (R2) and diode (D1). The programming port (J1) is connected with the embedded radio chip (ER) and power supply. The UART output jack (J2) and ADC in jack (J3) are connected with the embedded radio chip (ER) and grounded. The output of the voltage regulator is connected to one endpoint of on–off–on switch (S5). The various pins of embedded radio chip (ER) are shown in Fig. 9.19 and

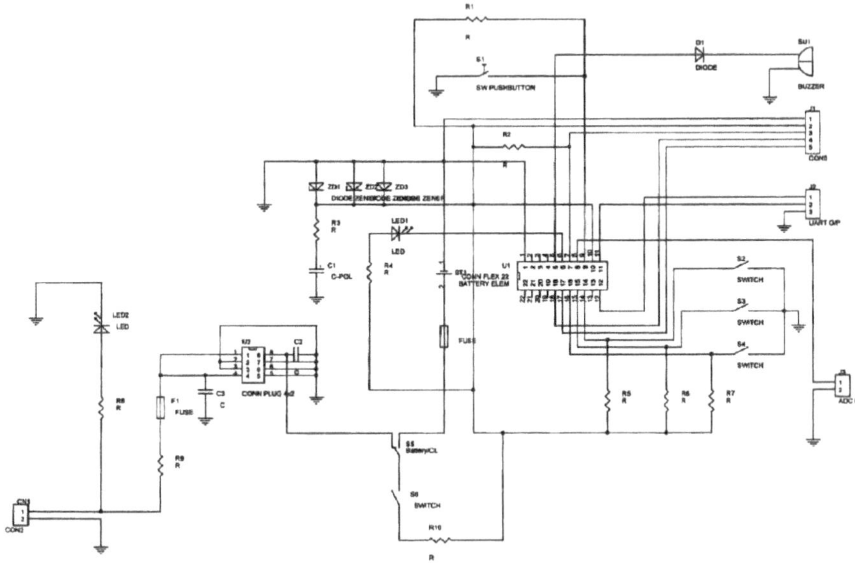

Fig. 9.18 Circuit diagram of router/end device. Key: CN1: Cap-lamp battery supply 4.0 V, 0.4 A maximum; J1: Programming port, VDD = 2.0 to 3.6 V; J2: UART Output $V = \pm 15$ V integrated via RS-232 IC, I = Controlled through RS-232 IC, C = negligible, L = negligible; J3: ADC Input $V = 3.6$ V, $I = 4$ mA maximum, C = negligible, L = negligible, analog signal

details of pin used in the circuit are given in Table 9.8. Figures 9.20, 9.21, 9.22, and 9.23 illustrate the PCB layout for top, bottom, and component (top and bottom) layers, respectively. Table 9.9 gives the list of components used in the circuit along with its value.

Power Supply Unit

Power supply in the circuit is supplied from two different sources: External source (i.e., cap-lamp battery) and internal source (i.e., two AA size batteries).

(a) Details of cap-lamp battery:

Battery type: Intrinsically safe lithium-ion cap-lamp battery
Operating voltage: 3.7 V
Maximum voltage: 4.0 V
Limited maximum current: 0.4 A
Ampere-hour: 4 Ah

(b) Internal power supply

Battery type: Two AA size non-rechargeable batteries
Voltage: 1.5 V each
Ampere-hour: 1500 mAh each

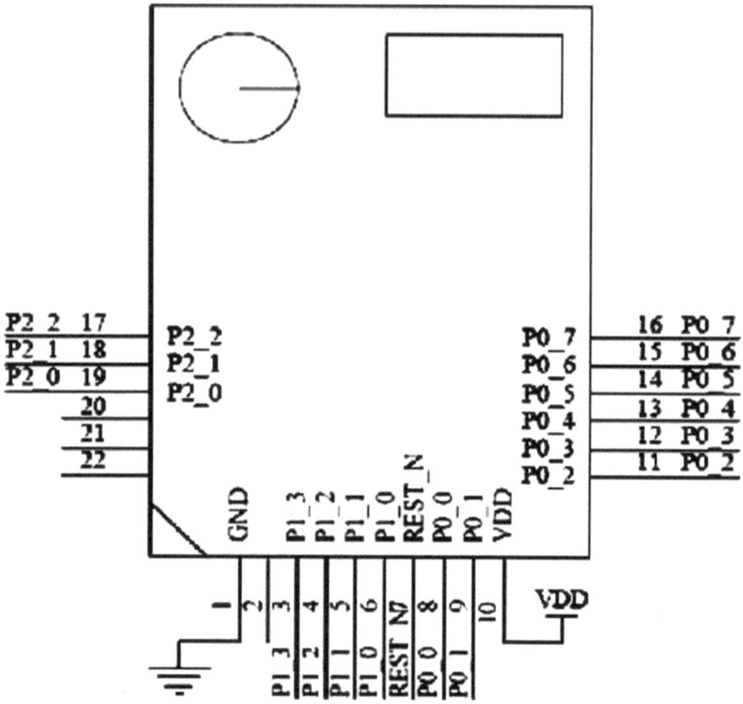

Fig. 9.19 Pin diagram of CC2430 RFID chip (22 pins)

Table 9.8 Details of CC2430 RFID chip pins

Pin no.	Pin definition	Connection	Input parameter	
			Voltage	Current
1	Ground	Ground		
2	Null	Not used		
3	P1_3	Not used		
4	P1_2	Not used		
5	P1_1	Buzzer input	3–12 V	20 mA (max)
6	P1_0	Programming pins through diode	4 V	
7	Reset			
8	P0_0	12 bit ADC	3.6 V(max)	
9	P0_1			
10	VDD	VCC	2.7–3.6 V	
11	P0_2	UART RXD	± 15 V (max)	IC (RS-232)
12	P0_3	UART TXD	± 15 V (max)	IC (RS-232)
13	P0_4	Not used		
14	P0_5	Message	3.6 V (max)	4 mA (max)
15	P0_6	Message	3.6 V (max)	4 mA (max)
16	P0_7	Message	3.6 V (max)	4 mA (max)

Table 9.8 (continued)

Pin no.	Pin definition	Connection	Input parameter	
			Voltage	Current
17	P2_2	Programming	3 V	4 mA (max)
18	P2_1	Programming	3 V	4 mA (max)
19	P2_0	Not used		
20	Null	Not used		
21	Null	Not used		
22	Null	Not used		

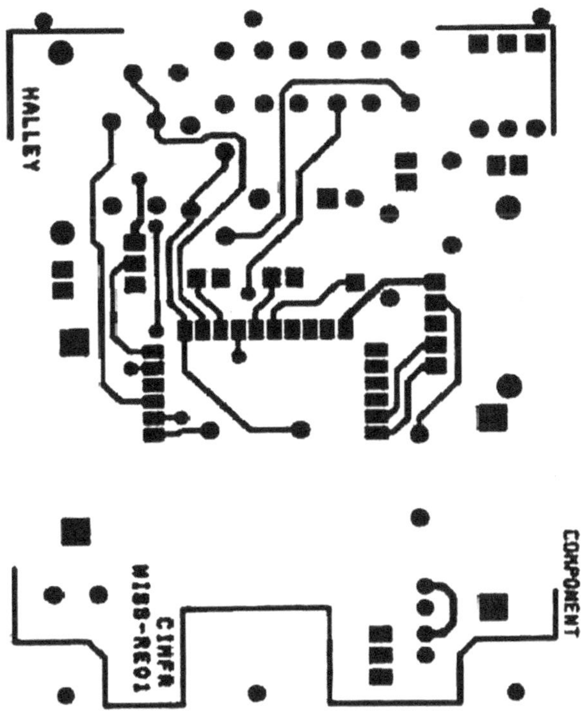

Fig. 9.20 PCB top layer of router/end device

A switch is used for interchanging the power supply between internal and external supplies. External power supply arrangement has additional circuitry using high-quality voltage regulator in order to regulate the power supply considering the safety factor in account. A fuse is connected with battery to provide safety by limiting the excess current in the circuit. A current-limiting resistor is also used for limiting the current in the circuit. Figures 9.24 and 9.25 show the wiring and enclosure diagram of the complete device.

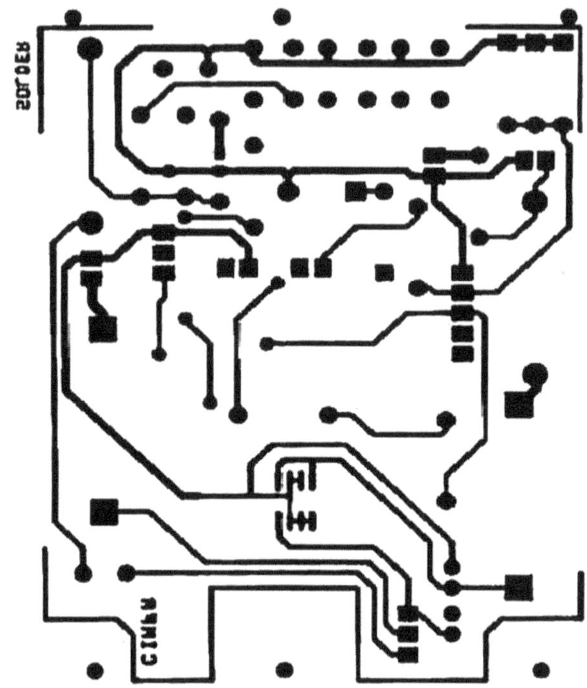

Fig. 9.21 PCB bottom layer of router/end device

9.3.2.2 Current

Transmitting Mode

(a) From cap-lamp: In this mode, the maximum current consumption by the circuit is 44 mA. It is measured by keeping an ampere meter in series between cap-lamp and RFID device in transmitting mode. As per the IEC60079-11, 2007 standard (Fig. 12.7 as given in Chapter 12 – Resistive circuits), the minimum ignition current at 4.0 V is greater than 3 A. Therefore, the current consumption of the RFID device powered with cap-lamp (maximum voltage 4.0 V) is much lesser than the minimum ignition current as stipulated in the standard.

(b) From internal battery: In this mode, the maximum current drawn by the circuit is 39 mA. It is measured by keeping an ampere meter in series between internal battery and RFID device in transmitting mode. As per the IEC60079-11, 2007 standard (Fig. 12.7 as given in Chapter 12 – Resistive circuits), the minimum ignition current at 3.0 V is greater than 3 A. Therefore, the current consumption of the RFID device powered with internal battery (maximum voltage 3.0 V) is much lesser than the minimum ignition current as stipulated in the standard.

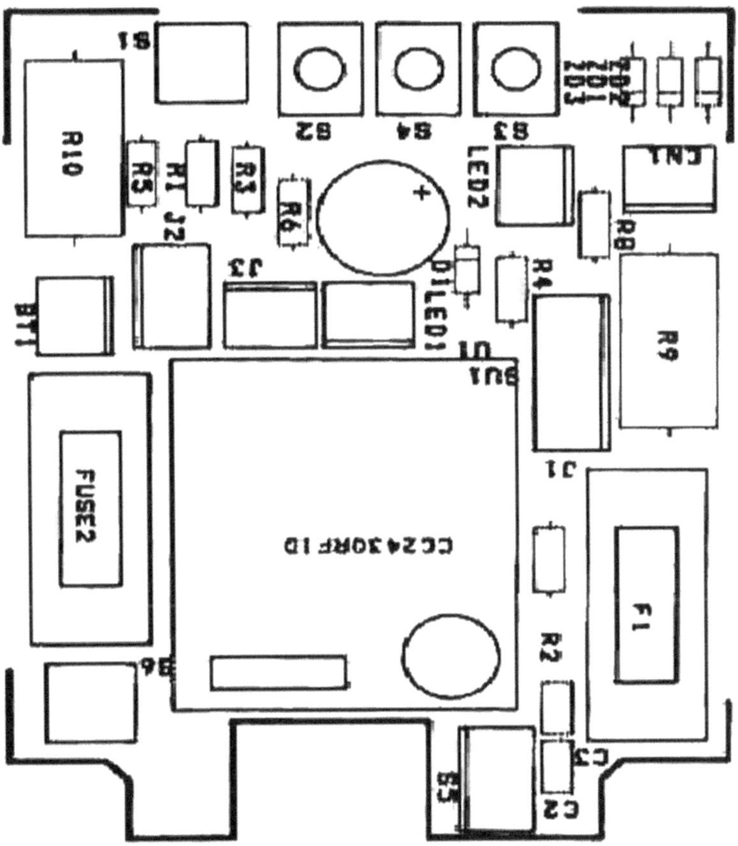

Fig. 9.22 Component layout (*top*) of router/end device

Receiving Mode

(a) From cap-lamp: In this mode, the maximum current drawn by the circuit is 23 mA. It is measured by keeping an ampere meter in series between the cap-lamp and RFID device in receiving mode. As per the0 IEC60079-11, 2007 standard (Fig. 12.7 as given in Chapter 12 – Resistive circuits), the minimum ignition current at 4.0 V is greater than 3 A. Therefore, the current consumption of the RFID device powered with cap-lamp (maximum voltage 4.0 V) is much lesser than the minimum ignition current as stipulated in the standard.

(b) From internal battery: In this mode, the maximum current drawn by the circuit is 15 mA. It is measured by keeping an ampere meter in series between the internal battery and RFID device in receiving mode. As per the IEC60079-11, 2007 standard (Fig. 12.7 as given in Chapter 12 – Resistive circuits), the minimum ignition current at 3.0 V is greater than 3 A. Therefore, the current consumption of the RFID device powered with internal battery (maximum voltage 3.0 V) is much lesser than the minimum ignition current as stipulated in the standard.

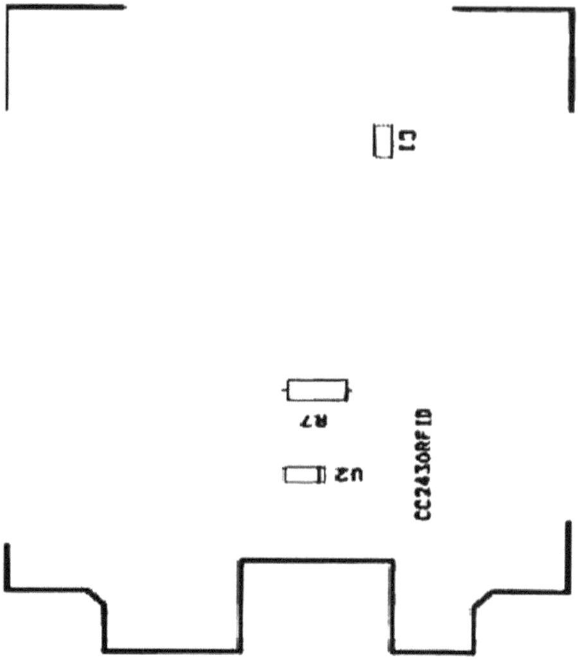

Fig. 9.23 Component layout (*bottom*) of router/end device

9.3.2.3 Operating Voltage

The RFID chip has wide operating voltage range from 2.7 to 3.6 V. In the developed RFID device, it operates in the following two voltages.

(i) With cap-lamp: The operating voltage provided to the RFID chip after voltage regulation is 3.3 V.
(ii) With internal battery: The operating voltage provided to the RFID chip is 3.0 V.

9.3.2.4 Power

In case of cap-lamp battery the maximum power drawn by the circuit is (maximum cap-lamp battery voltage) × (maximum current drawn by the circuit), i.e., power = 4.0 V × 44 mA = 176 mW. In case of internal battery the maximum power drawn by the circuit is (internal battery voltage) × (maximum current drawn by the circuit), i.e., power = 3.0 V × 39 mA = 108 mW. All the components used in the circuit are well within the IS allowed maximum power rating. Calculation of power of each component is given separately.

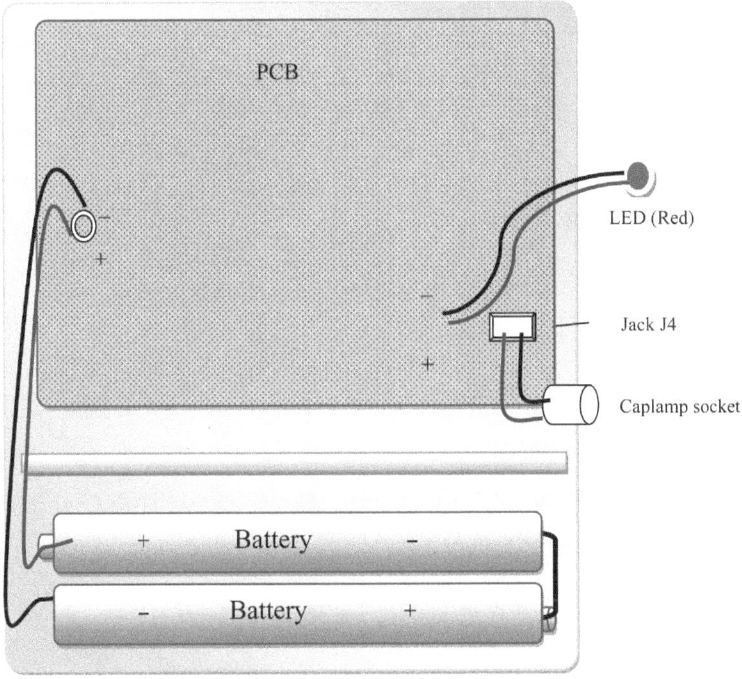

Fig. 9.24 Wiring diagram of router/end device

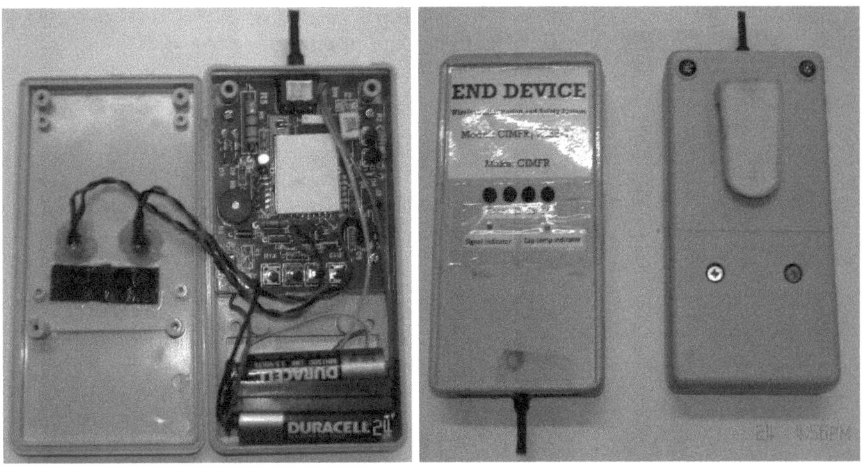

Fig. 9.25 View of router/end device along with enclosure

9.3.2.5 Inductance

Inductance of the circuit (Li) is 28.8 nH. Inductance to ADC, UART pins, and conducting wires (Lc) is almost negligible. All the inductors are located in antenna parts

Table 9.9 Component list for router/end device

Name	Designation	Value	Tolerance
Resistors	R1	43 kΩ/0.25 W, MFR	1%
	R2	43 kΩ/0.25 W, MFR	1%
	R3	72 Ω/0.25 W, MFR	1%
	R4	1 kΩ/0.25 W, MFR	1%
	R5	43 kΩ/0.25 W, MFR	1%
	R6	43 kΩ/0.25 W, MFR	1%
	R7	43 kΩ/0.25 W, MFR	1%
	R8	470 Ω/0.25 W, MFR	1%
	R9	10 Ω/5 W, MFR	1%
	R10	10 Ω/2 W, MFR	1%
Light emitting diode	LED1	3 mm Green LED, I = 7 mA (max)	
	LED2	3 mm Red LED, I = 7 mA (max)	
Capacitor	C1	1 μF (SMD), 16 V	10%
	C2	0.01 μF (ceramic), 50 V	10%
	C3	0.01 μF (ceramic), 50 V	10%
Zener diode	ZD1	3.6 V/1 W	
	ZD2	3.6 V/1 W	
	ZD3	3.6 V/1 W	
Diode	D1	1N4007	
Switches	S1 to S4 & S6	Push to 'on' button switch	
	S5	On-Off-On switch	
Embedded radio	U1	CC2430, RFID Chip	
Voltage regulator	U2	MAX 604 CSA+ 3.3 V, 200 mA max, 1.8 W	
Buzzer	SU1	Magnetic field effect buzzer, 3–12 V	
Fuses	F1	100 mA	250 V, 35 A, Breaking Capacity
	FUSE2	80 mA	250 V, 35 A Breaking Capacity
Connector	J1	5 pin jack for programming, VDD = 2.0–3.6 V	
	J2	3 pin jack for UART output V = ± 15 integrated via RS-232 IC, I=Controlled through RS-232IC, C = negligible, L = negligible	
	J3	2 pin jack for ADC input, V = 3.6 V, I = 4 mA maximum, C = negligible, L = negligible, analog signal.	
	CN1	2 pin jack for caplamp battery (4 V/0.4 A)	
	BT1	Connector for 2 AA size batteries of 1.5 V each	

of CC2430 chip so current in the inductor is limited by CC2430 chip. Pin current, i.e., 4 mA which is well within the limit of gas group I inductive circuit. As per the IEC60079-11, 2007 standard (Fig. 12.8 as given in Chapter 12 – Inductive circuit) the maximum allowable inductance (Lo) is around more than 300 mH at 4 mA. Therefore, Lo is >> Li + Lc.

9.3.2.6 Capacitance

Capacitance of the circuit (Ci) is 1.02 μF. Capacitance to ADC and UART pins, and conducting wires (Cc) are almost negligible. As per the IEC60079-11, 2007 standard (Fig. 12.6 as given in Chapter 12 – group I capacitive circuit), the maximum allowable capacitance (Co) at 4 V is >1000 μF. Thus, Co is >> Ci + Cc.

9.3.2.7 Short Circuit Current

(i) With cap-lamp battery: The maximum current that can flow through a circuit is limited by a current-limiting resistor (CLR) of 10 Ω. The maximum current through circuit is 4.0 V/10 Ω = 0.4 A. As per the IEC60079-11, 2007 standard (Fig. 12.7 as given in Chapter 12 – Resistive circuits), the minimum ignition current at 4.0 V is more than 3 A. Therefore, the maximum current that can flow in the circuit is much lesser than the allowable current as mentioned in the standard.

(ii) With internal battery: The maximum current that can flow through the circuit is limited by a current-limiting resistor of 10 Ω. The maximum current through circuit is 3.0 V/10 Ω = 0.3 A. As per the IEC60079-11, 2007 standard (Fig. 12.7 as given in Chapter 12 – Resistive circuits), the minimum ignition current at 3.0 V is more than 3 A. Therefore, the maximum current that can flow in the circuit is much lesser than the allowable current as mentioned in the standard.

9.3.2.8 Safety Analysis

Each component which is used in the device is provided by its own power rating. The current across the resistor is calculated taking the shortage route to ground by the formula V/R. Then power rating of the component is calculated by the formula I^2R. This power rating is compared with the component power rating which should be less in regard to intrinsic safety. Ratio ($W1/W2$) of the given power rating of the component to the calculated power rating should be greater then 1.5 as per the IEC60079-11, 2007 standard. Table 9.10 gives the analysis of each safety component used in the device. Ratio ($W1/W2$) for each component is more than 1.5.

9.3.2.9 Antenna Characteristics

The characteristic of the chip antenna operating at 2.4 GHz frequency is given in Table 9.11. It has linear polarization. Figure 9.26 depicts the omnidirection radiation pattern of the chip antenna. Figure 9.27 illustrates the plot for voltage standing wave ratio (VSWR) and return loss of the used antenna.

Table 9.10 Details of safety analysis for router/end device

Component designation	Component location	Value	Maximum rating used by circuit ($W2$)	Component rating ($W1$)	$W1/W2$
R1	Microcontroller part	43 kΩ / 0.25 W	0.372 mW	0.25 W	672
R2	Microcontroller part	43 kΩ / 0.25 W	0.372 mW	0.25 W	672
R3	Power supply part	72 Ω / 0.25 W	136.104 mW	0.25 W	1.836
R4	Power supply part	1 kΩ / 0.25 W	15.21 mW	0.25 W	16.436
R5	Microcontroller part	43 kΩ / 0.25 W	0.372 mW	0.25 W	672
R6	Microcontroller part	43 kΩ / 0.25 W	0.372 mW	0.25 W	672
R7	Microcontroller part	43 kΩ / 0.25 W	0.372 mW	0.25 W	672
R8	Power supply part	470 kΩ / 0.25 W	34.053 mW	0.25 W	7.34
R9	Power supply part	10 Ω / 5 W	0.4 W	5 W	12.5
R10	Power supply part	10 Ω / 2 W	0.9 W	2 W	2.22
ZD1	Power supply part	3.6 V/ 1 W	0.072 W	1 W	13.88
ZD2	Power supply part	3.6 V/ 1 W	0.072 W	1 W	13.88
ZD3	Power supply part	3.3 V/ 1 W	0.072 W	1 W	13.88
F1	Power supply part	100 mA	44 mA		
FUSE2	Power supply part	80 mA	39 mA		

Table 9.11 Characteristics of chip antenna

Parameters	Value
Operating frequency	2.4–2.41 GHz
Radiation pattern	Omnidirectional
Gain	0.2 dBi
Polarization	Linear
Impedance	50 Ω
Voltage standing wave ratio (VSWR):	2.5232 at 2.4050 GHz
Return loss	–7.2642 dB at 2.4050 GHz
Antenna type	Chip antenna

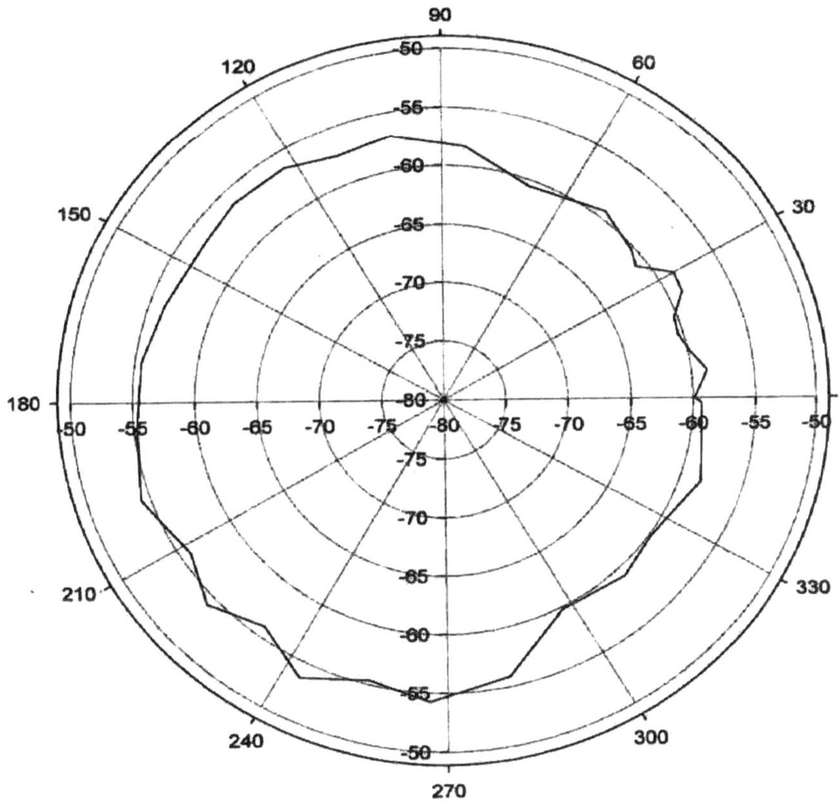

Fig. 9.26 Radiation pattern of chip antenna

9.3.2.10 Radiated Power

Radiated power of the device using chip antenna in frequency range of 2.4–2.41 GHz is given in Table 9.12. As per the measurement, it was observed that the radiated power was very less in the frequency range 2.4–2.41 GHz. The maximum value of 1.329852 μW was measured at 1 m distance from the antenna kept in vertical mounting position. It was gradually decreased with increase in distance from the antenna. The radiated power of the RFID devices is very low compared to the permissible value (100 mW) recommended in ETSI EN 300 (Europe) standard.

9.3.2.11 Threshold Power

Maximum voltage in the circuit supplied from cap-lamp is 4.0 V. Maximum current in the circuit is 44 mA. Threshold power = voltage × current = 4.0 V × 44 mA = 176 mW. As per the IEC 60079-0, 2007 standard for gas group I, the recommended

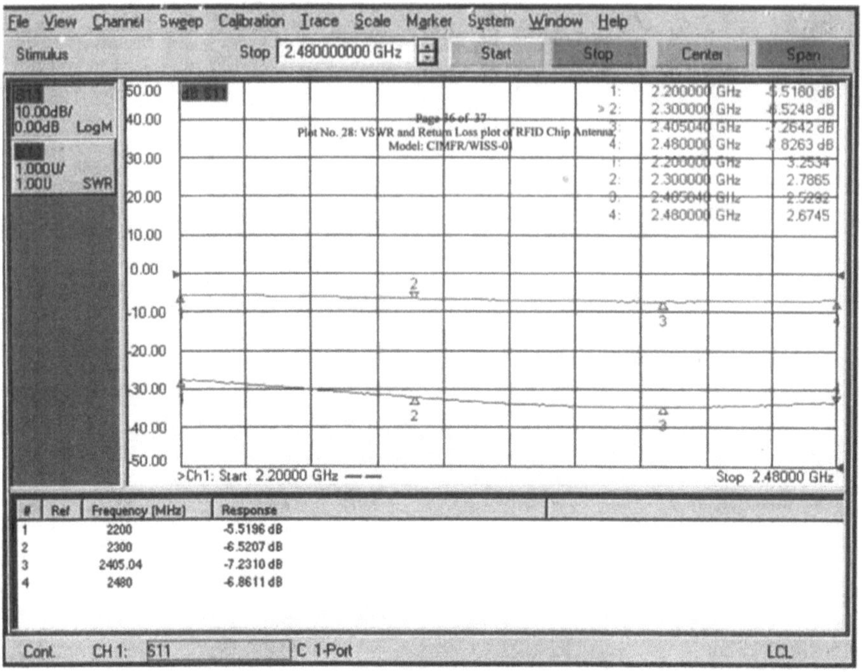

Fig. 9.27 Plot of VSWR and return loss of chip antenna

Table 9.12 Radiated power of the chip antenna

Sl. No.	Frequency (GHz)	Device mounting	Power radiated (dBm)	Power radiated (μW)
1	2.405272000	Horizontal	–63.18	0.480839
	2.405272000	Vertical	–58.77	1.329852
2	2.404528000	Horizontal	–76.74	0.021183
	2.404528000	Vertical	–71.47	0.071285
3	2.404584000	Horizontal	–72.59	0.050807
	2.405272000	Vertical	–71.35	0.073282

threshold power is 6 W. The threshold power of the RFID device is very much lesser than the recommended standard limit, since it is a very low power device.

9.3.3 Carbon-Monoxide-Monitoring Device

9.3.3.1 Circuits Detail

Power Supply Unit

Power supply in the circuit is supplied from two different sources: external source (i.e., cap-lamp battery) and internal source (i.e., four AA size batteries). A switch

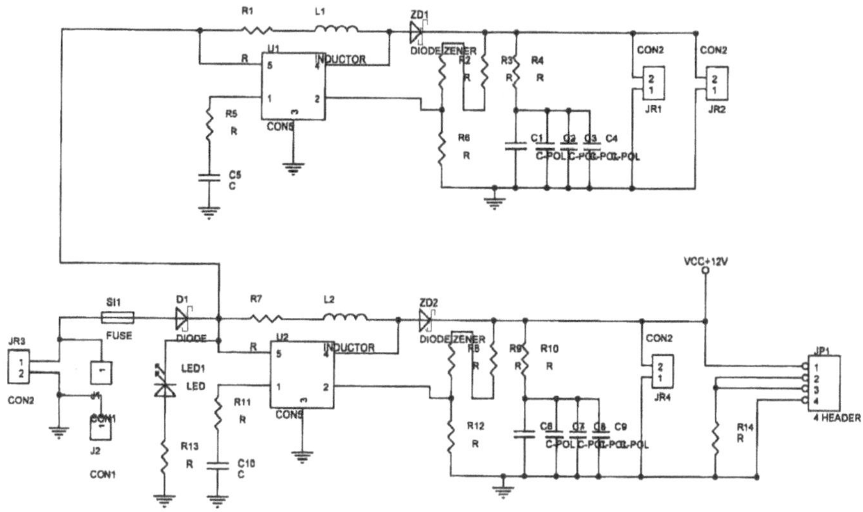

Fig. 9.28 Circuit diagram of power supply part of CO-monitoring device. Key: J1 & J2: Voltage measuring point via voltmeter (*L* and *C* of voltmeter are negligible and *R* is very high in MΩ); JR3: Cap-lamp battery connection with maximum voltage of 4.0 V, 400 mA (max); JR2: Output terminal, +5 V; JR1 & JR4: Output measuring point (Voltmeter) (*L* and *C* of voltmeter are negligible and *R* is very high in MΩ)

is used for interchanging the power supply between internal and external supplies. External power supply arrangement has additional circuitry using high-quality voltage regulator in order to regulate the power supply considering the safety factor in account. A fuse is connected with battery to provide safety by limiting the excess current in the circuit. A current-limiting resistor is also used for limiting the current in the circuit.

The device requirement of regulated voltage of 5 and 12 V is provided by the DC-to-DC converter (LM2577). Figure 9.28 shows the circuit diagram of the regulated power supply to the device. Two pin jack JR3 (CON2) is used to connect cap-lamp battery and internal battery. Voltage regulator (U1 and U2) is used for 5 and 12 V voltage output. Considering the safety purpose, fuse (S/1) and diode (D1) are used. LED (LED1) is used to check transmission in the circuit. Capacitors (C1, C2, C3, and C4) are used to remove any type of ripples in +5 V circuit. Capacitors (C6, C7, C8, and C9) are used to remove any type of ripples in +12 V circuit. Figures 9.29, 9.30, 9.31, and 9.32 show the PCB layout for top, bottom, and component (top and bottom) layers, respectively. List of components used along with its specification is given in Table 9.13.

(a) Details of cap-lamp battery:

Battery type: Intrinsically safe lithium-ion cap-lamp battery
Operating voltage: 3.7 V

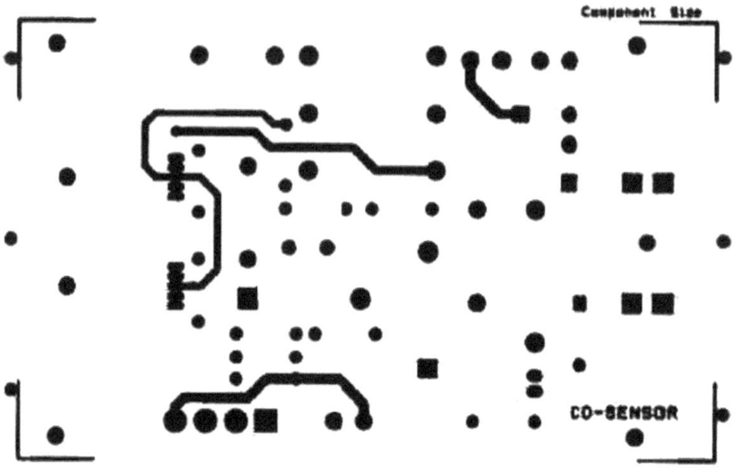

Fig. 9.29 PCB (*top*) layer of power supply part of CO-monitoring device

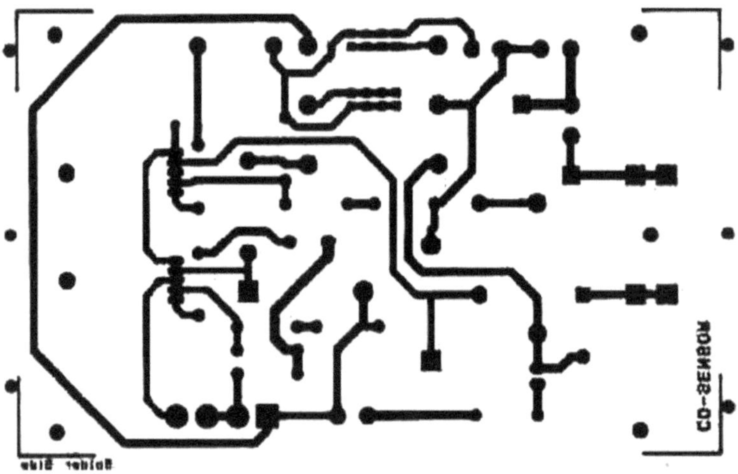

Fig. 9.30 PCB (*bottom*) layer of power supply part of CO-monitoring device

Maximum voltage: 4.0 V
Limited maximum current: 0.4 A
Ampere-hour: 4 Ah

(b) Internal power supply

Battery type: Four AA size rechargeable alkaline batteries
Voltage: $4 \times 1.2\ \text{V} = 4.8\ \text{V}$
Ampere-hour: 700 mAh each

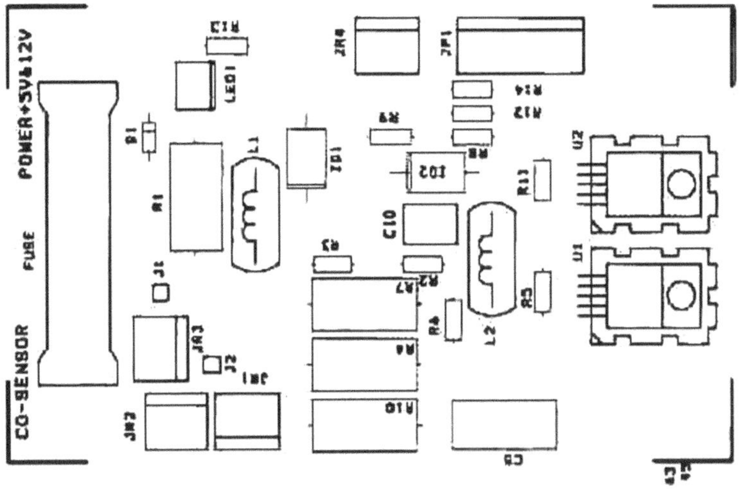

Fig. 9.31 PCB component layout (*top*) of power supply part of CO-monitoring device

Fig. 9.32 PCB component (*bottom*) layout of power supply part of CO-monitoring device

Gas-Monitoring Unit

The block diagram of "carbon-monoxide-monitoring device" is depicted in Fig. 9.33 which consists of external cap-lamp battery, light emitting diode, voltage regulator, turnover switch, internal battery supply, safety arrangement, embedded radio, programming port, universal asynchronous receiver–transmitter, analog-to-digital converter input, buzzer, and light emitting diode. The external cap-lamp battery is connected to light emitting diode and voltage regulator. The voltage regulator output is connected to turnover switch. The turnover switch is also connected to

Table 9.13 Component list for power supply part of CO-monitoring device

Name	Designation	Value	Tolerance
Resistors	R1	12 Ω/2 W, MFR	1%
	R2	2.7 kΩ/0.25 W, MFR	1%
	R3	390 Ω/0.25 W, MFR	1%
	R4	22 Ω/2 W, MFR	1%
	R5	2.2 kΩ/0.25 W, MFR	1%
	R6	1 kΩ/0.25 W, MFR	1%
	R7	12/2 W, MFR	1%
	R8	8.2 kΩ/0.25 W, MFR	1%
	R9	560 Ω/0.25 W, MFR	1%
	R10	47 Ω/5 W, Wire Wound	10%
	R11	2.2 kΩ/0.25 W, MFR	1%
	R12	1 kΩ/0.25 W, MFR	1%
	R13	470 Ω/0.25 W, MFR	1%
	R14	250 Ω/1 W, MFR	1%
Inductor	L1	100 μH	
	L2	100 μH	
Light emitting diode	LED1	3 mm red LED $I = 7$ mA (max)	
Capacitors	C1	22 μF/10 V, SMD	10%
	C2	22 μF/10 V, SMD	10%
	C3	22 μF/10 V, SMD	10%
	C4	22 μF/10 V, SMD	10%
	C5	0.33 μF/100 V, ceramic	10%
	C6	10 μF/16 V, SMD	10%
	C7	10 μF/16 V, SMD	10%
	C8	10 μF/16 V, SMD	10%
	C9	10 μF/16 V, SMD	10%
	C10	0.33 μF/100 V, ceramic	10%
Schottky diode	ZD1	1 N 5818	
	ZD2	1 N 5818	
	D1	1 N 5818	
Voltage regulator	U1	LM2577 ADJ	
	U2	LM2577 ADJ	
Fuse	FUSE	250 mA	250 V, 35 A
Connector	JR1	2 pin jack for measuring output voltage via voltmeter	
	JR2	2 pin jack for taking output voltage (5 V)	
	JR3	2 pin jack for cap-lamp battery connection (4 V) 0.4 A)	
	JR4	2 pin jack for measuring output voltage via voltmeter (12)	
	JP1	4 pin connector for sensor (3E) 3E/F,CiTiceLs)	

internal battery supply, thus power to the circuit can be given by either cap-lamp or internal battery via turnover switch. The output after turnover switch is connected to programming port and embedded radio. The embedded radio is connected to programming port, universal asynchronous receiver–transmitter output, analog-to-digital converter input and light emitting diode. The diagram also consists

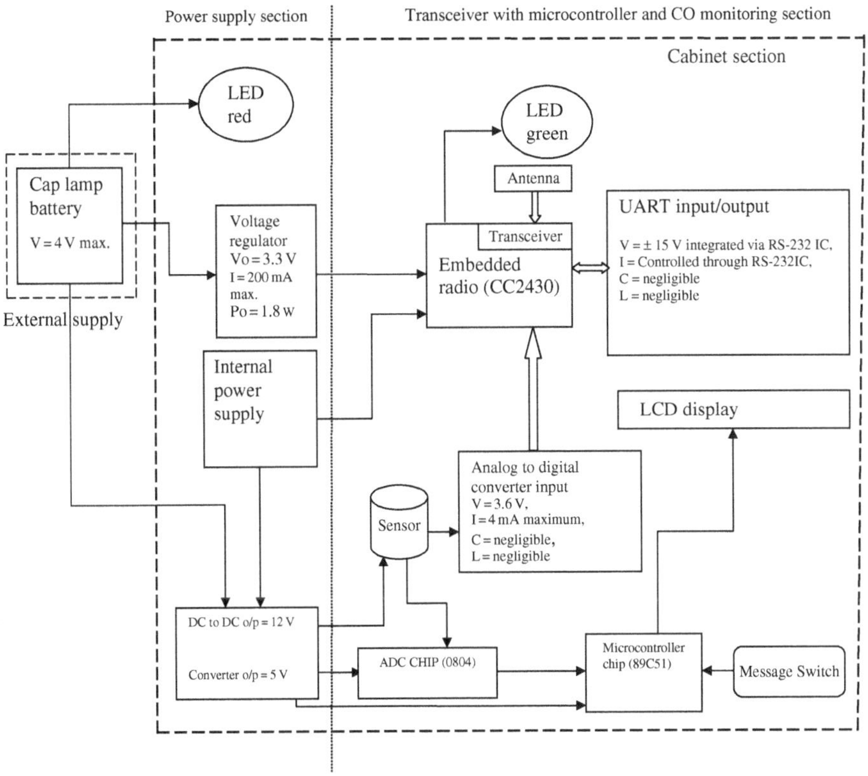

Fig. 9.33 Block diagram of CO-monitoring device

of sensor ADC chip, AT89C51 microcontroller, and LCD. The sensor is connected to the ADC pin and ADC chip to give its data. Subsequently, ADC chip is connected to the microcontroller AT89C51 for processing the data. The measured value is displayed on the LCD which is connected to the microcontroller chip.

The external cap-lamp battery is connected with the light emitting diode to indicate the connection of the power supply with the cap-lamp battery. The cap-lamp battery is also connected with the voltage regulator to provide the regulated 3.3 V power supply to the circuit. The internal battery supply is also connected with the RFID device through turnover switch so that the device can be operated using external battery supply or internal battery supply. The output after the turnover switch is feed to programmable port, universal asynchronous receiver–transmitter output, analog-to-digital converter input, buzzer, and light emitting diode for programming, sending digital signal to external device, taking analog input from external device, generating alarm, and indicating the data transmission operation, respectively.

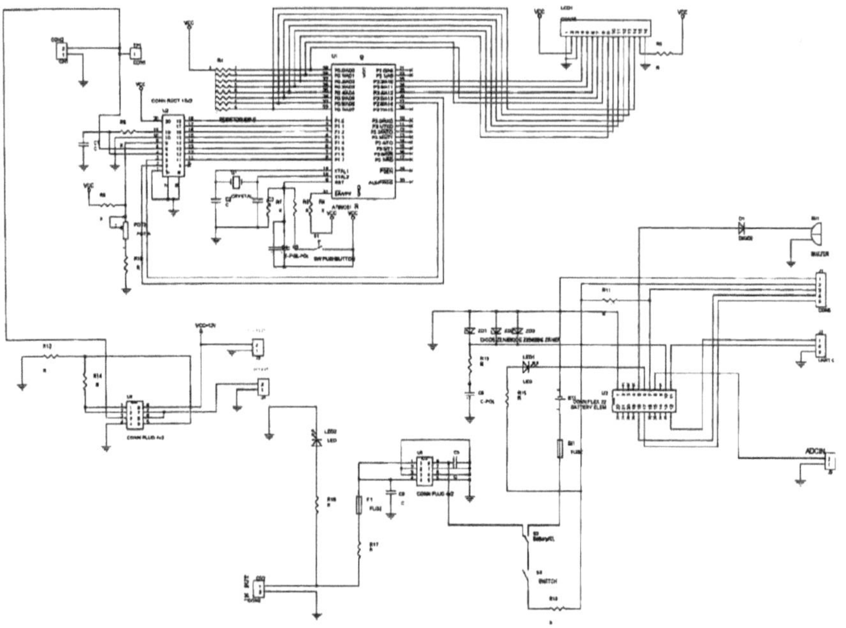

Fig. 9.34 Circuit diagram of RFID and sensor unit of CO-monitoring device. Key: CN2: Cap-lamp battery supply 4.0 V, 0.4 A (maximum); BT1: Four AA size rechargeable batteries, 3.6 V; J1: Programming port, VDD = 2.0 to 3.6 V; J2: UART Output $V = \pm$ 15 V integrated via RS-232 IC, I = controlled through RS-232 IC, C = negligible, L = negligible; J5: ADC Input V = 3.6 V, I = 4 mA maximum, C = negligible, L = negligible, analog signal

Figure 9.34 shows the circuit diagram of the RFID unit and gas-sensing unit. A two pin jack (CN1) is provided to connect external cap-lamp battery. LED (LED2) via resistance (R8) is connected to this jack. The jack power is connected to voltage regulator (MAX604 CSA+) via fuse (F1) and resistance (R9). The battery (BT1) is connected via fuse (S/1) to another endpoint of the on–off–on switch (S5). The output from all stages of the switch (S5) is common and connected via resistance (R10) to RFID chip, Zener diodes (ZD1, ZD2, and ZD3) are in parallel with the final supply after switch (S5). The cathode point of the Zener diodes (ZD1, ZD2, and ZD3) is grounded. The capacitor (C1) is connected via resistance (R3) as a filter in parallel with the Zener diodes (ZD1, ZD2, and ZD3). The embedded radio chip (ER) is connected to LED (LED1) via resistance (R4) and power supply through the resistance (R10). The reset switch (S1) is connected between CC2430 embedded radio chip (ER) and ground. The buzzer (SU1) is connected to the embedded radio chip (ER) via resistance (R2) and diode (D1). The programming port (J1) is connected with the embedded radio chip (ER) and power supply. The UART output jack (J2) and ADC in jack (J3) are connected with the embedded radio chip (ER) and grounded. The output of the voltage regulator is connected to one endpoint of on–off–on switch (S5).

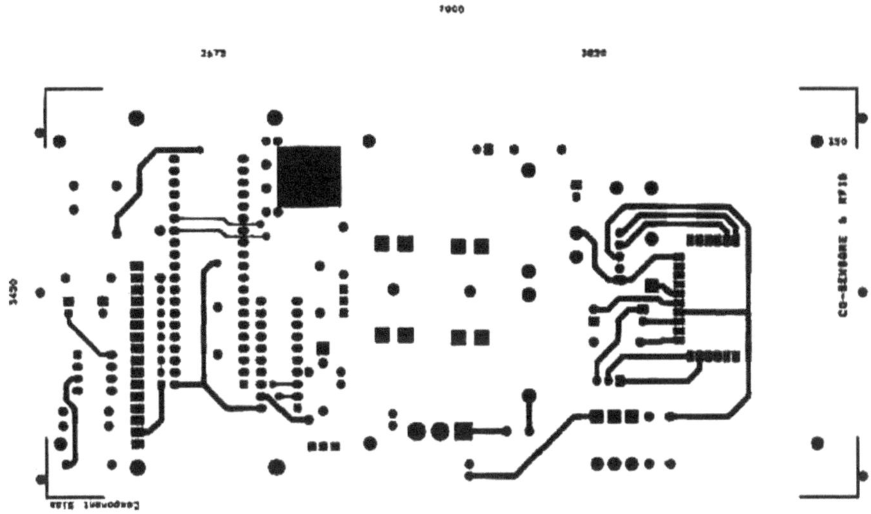

Fig. 9.35 PCB (*top*) layout of RFID and sensor unit of CO-monitoring device

The ADCIN (J5) of the RFID unit is connected to the TP1 (CON1) of the sensing unit to get analog voltage from the CO-sensing device through which it is processed by the RFID device and wirelessly transmitted. Regulated voltage of +12 V VCC is given to the jack (J3), which powers the chip (U4) and +5 V VCC is given to power the ADC chip (U2) and microcontroller chip (U1) from the regulated power supply unit. CO sensor is attached to two pin jack (J4). The analog input from the sensor goes to the ADC chip (U2) which converts the data into the digital output and is fed to port 1 of the microcontroller (U1), which displays the sensor data on the LCD (LCD1) attached to port 2. Figures 9.35, 9.36, 9.37, and 9.38 show the PCB layout for top, bottom, and component (top and bottom) layers, respectively. List of components used along with their specification is given in Table 9.14. Figure 9.39 shows the enclosure diagram of the whole device.

9.3.3.2 Safety Analysis

Each component used in the device is provided by its own power rating. The current across the resistor is calculated taking the shortage route to ground by the formula V/R. Then power rating of the component is calculated by the formula I^2R. This power rating is compared with the component power rating which should be less in regard to intrinsic safety. Ratio ($W1/W2$) of the given power rating of the component to the calculated power rating should be greater then 1.5 as per the IEC60079-11, 2007 standard. Table 9.15 gives the safety analysis of the power supply unit and Table 9.16 gives the safety analysis of RFID and sensor unit. Ratio ($W1/W2$) is greater than 1.5 for all the safety components used in the device.

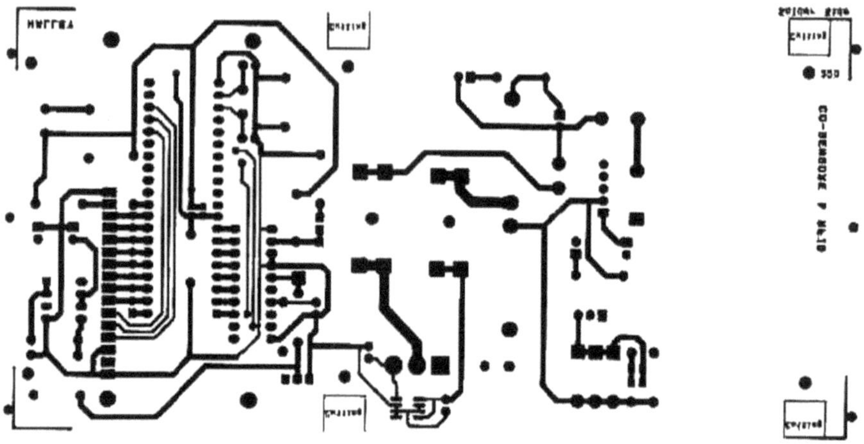

Fig. 9.36 PCB (*bottom*) layout of RFID and sensor unit of CO-monitoring device

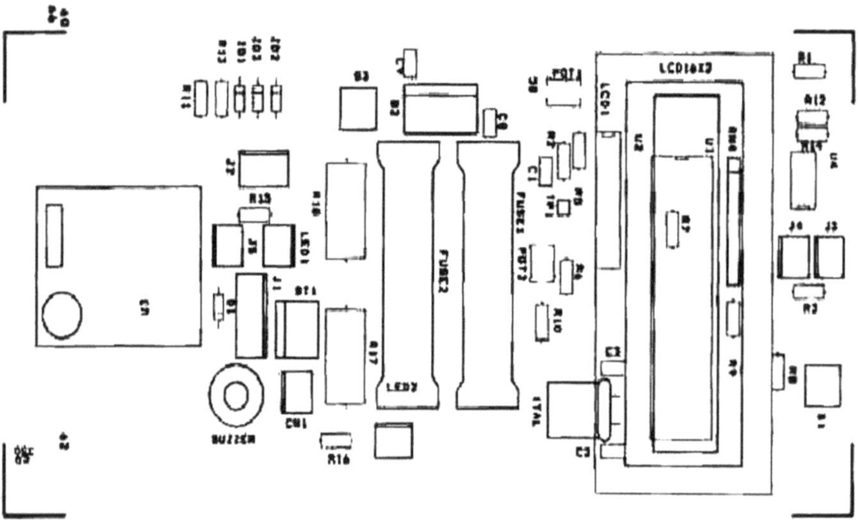

Fig. 9.37 PCB component (*top*) layout of RFID and sensor unit of CO-monitoring device

9.3.4 *Methane-Monitoring Device*

9.3.4.1 Circuit Details

Power Supply Unit

Power supply in the circuit is supplied from two different sources: external source (i.e., cap-lamp battery) and internal source (i.e. four AA size batteries). A switch is used for interchanging the power supply between internal and external supplies.

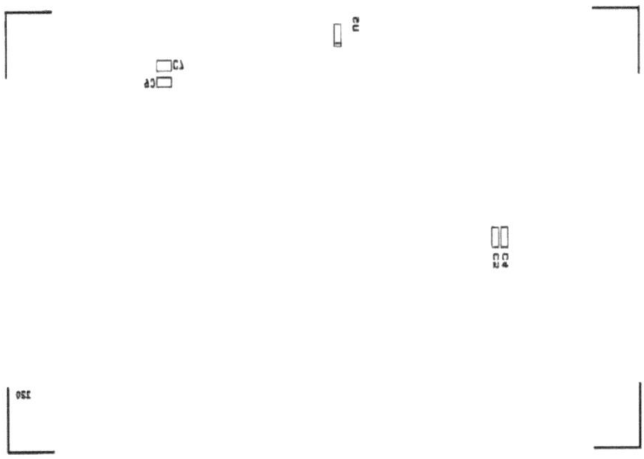

Fig. 9.38 PCB component (*bottom*) layout of RFID and sensor unit of CO-monitoring device

Table 9.14 Component list of RFID and sensor unit of CO-monitoring device

Name	Designation	Value	Tolerance
Resistors	R3	3.3 kΩ/0.25 W, MFR	1%
	R5	10 kΩ/0.25 W, MFR	1%
	R6	2.2 kΩ/0.25 W, MFR	1%
	R7	8.2 kΩ/0.25 W, MFR	1%
	R8	100 kΩ/0.25 W, MFR	1%
	R9	33 kΩ/5 W, MFR	1%
	R10	470 Ω/0.25 W, MFR	1%
	R11	43 kΩ/0.25 W, MFR	1%
	R12	10 kΩ/0.25 W, MFR	1%
	R13	72 Ω/0.25 W, MFR	1%
	R14	10 kΩ/0.25 W, MFR	1%
	R15	470 Ω/0.25 W, MFR	1%
	R16	470 Ω/0.25 W, MFR	1%
	R17	10 Ω/5 W, Wire Wound	10%
	R18	10 Ω/2 W, MFR	1%
Variable resistor	POT2	3.3 kΩ/0.25 W, MFR	
8-pin resistor network	RN4 (2–9)	10 kΩ/0.25	
Capacitor	C1	56 pF/50 V, ceramic	10%
	C2	33 pF/50 V, ceramic	10%
	C3	33 pF/50 V, ceramic	10%
	C4	4.7 μF/16 V, SMD	10%
	C5	4.7 μF/16 V, SMD	10%
	C6	1 μF /16 V, SMD	10%
	C8	0.01 μF/50 V, ceramic	10%
	C9	0.01 μF/50 V, ceramic	10%
Light emitting diode	LED1	3 mm Green LED, $I = 7$	
	LED2	3 mm Red LED, $I = 7$ mA (max)	
Zener diode	ZD1	3.6 V / 1 W	
	ZD2	3.6 V / 1 W	
	ZD3	V / 1 W	
Diode	D1	1N4007	
Switches	S1, S3	Push to on button switch	
	S2	On–off–on switch	

Table 9.14 (continued)

Name	Designation	Value	Tolerance
Microcontroller	U1	AT 89C51	
ADC chip	U2	ADC 0804	
Embedded radio	U3	CC2430, RFID chip	
Amplifier	U4	LM358	
Voltage regulator	U5	MAX 604 CSA+3.3 V, 200	
Buzzer	SU1	Magnetic field effect buzzer	
LCD	LCD1	16 × 2 character	
Crystal	XTAL		
Fuse	FUSE1	100 mA	250 V, 35 A, breaking capacity
	FUSE2	80 mA	250 V, 35 A, breaking capacity
Connector	J1	5 pin jack for programming, VDD = 2.0–3.6 V	
	J2	3 pin jack for UART output $V = \pm$ 15 integrated via RS-232 IC, I = controlled through RS-232 IC, C = negligible, L = negligible	
	J3	2 pin jack for 12 V supply	
	J4	2 pin jack for connecting RFID unit for transmitting	
	J5	2 pin jack for ADC input, $V = 3.6$ V, $I = 4$ mA maximum, C = negligible, L = negligible, analog signal	
	CN2	2 pin jack for cap-lamp battery (4 V/ 0.4 A)	
	BT1	Connector for 3 AA size rechargeable batteries of 1.2 V each	

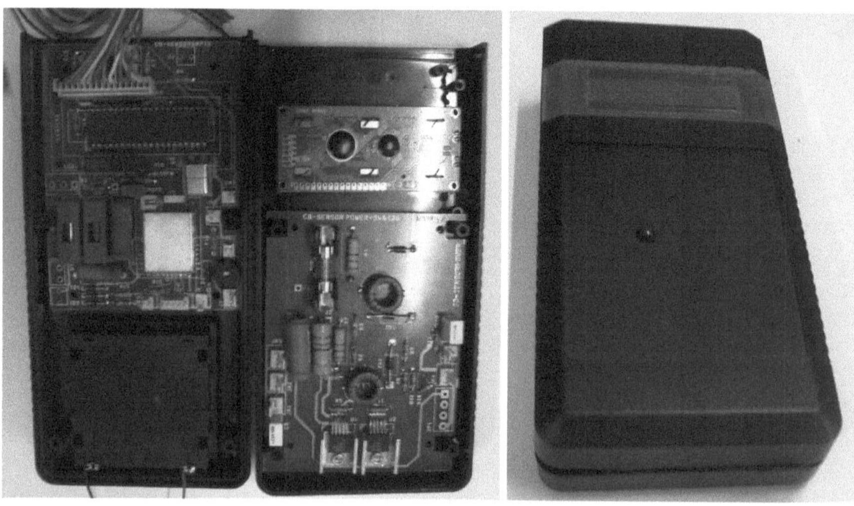

Fig. 9.39 View of carbon dioxide monitoring device along with enclosure

Table 9.15 Details of safety analysis for power supply unit of CO-monitoring device

Component designation	Component location	Value	Maximum rating used by circuit (*W*2)	Component rating (*W*1)	*W*1/*W*2
R1	Power supply part	12 Ω/2 W	259.51 mW	2 W	7.7
R2	Power supply part	2.7 kΩ/0.25 W	4 mW	0.25 W	62.5
R3	Power supply part	390 Ω/0.25 W	0.578 mW	0.25 W	432.5
R4	Power supply part	22 Ω/2 W	475.77 mW	2 W	4.20
R5	Power supply part	2.2 kΩ/0.25 W	11.33 mW	0.25 W	22.0
R6	Power supply part	1 kΩ/0.25 W	1.48 mW	0.25 W	168.9
R7	Power supply part	12 Ω/2 W	496.40 mW	2 W	4.0
R8	Power supply part	8.2 kΩ/0.25 W	12.20 mW	0.25 W	20.49
R9	Power supply part	560 kΩ/0.25 W	0.83 mW	0.25 W	301.2
R10	Power supply part	47 Ω/5 W	1.9 W	5 W	2.63
R11	Power supply part	2.2 kΩ/0.25 W	65.34 mW	0.25 W	3.82
R12	Power supply part	1 kΩ/0.25 W	0.83 mW	0.25 W	301.2
R13	Power supply part	470 Ω/0.25 W	49.02 mW	0.25 W	5.0
R14	Power supply part	250 Ω/1 W	0.57	1	1.75
FUSE	Power supply part	250 mA			

External power supply arrangement has additional circuitry using high-quality voltage regulator in order to regulate the power supply considering the safety factor in account. A fuse is connected with battery to provide safety by limiting the excess current in the circuit. A current-limiting resistor is also used for limiting the current in the circuit.

The required voltage of +5 V for MAX232 is provided by the DC-to-DC converter (LM2577) of the power supply unit. Figure 9.41 shows the circuit diagram of the regulated power supply to the device. Two pin jack JR1 (CON2) is used to connect cap-lamp battery and internal battery. Voltage regulator (U1) is used for 5 V voltage output. Considering the safety purpose, fuse (FUSE1) and diode (D1) is used. LED (LED1) is used to check transmission in the circuit. Capacitor (C1) is used to remove any type of ripples in the circuit.

Table 9.16 Details of safety analysis of RFID and sensor unit for CO-monitoring device

Component designation	Component location	Value	Maximum rating used by circuit (*W*2)	Component rating (*W*1)	*W*1/*W*2
R3	CO- sensing unit	3.3 kΩ/0.25 W	7.57 mW	0.25 W	33.0
R5	CO-sensing unit	10 kΩ/0.25 W	2.5 mW	0.25 W	100
R6	CO-sensing unit	2.2 kΩ/0.25 W	7.69 mW	0.25 W	32.50
R7	CO-sensing unit	8.2 kΩ/0.25 W	3.048 mW	0.25 W	82
R8	CO-sensing unit	100 Ω/0.25 W	36 μW	0.25 W	6944.44
R9	CO-sensing unit	33 kΩ/0.25 W	0.75 mW	0.25 W	333.33
R10	CO-sensing unit	470 Ω/0.25 W	1.64 mW	0.25 W	152.43
R12	CO-sensing unit	10 kΩ/0.25 W	3.6 mW	0.25 W	69.44
R14	CO-sensing unit	10 kΩ/0.25 W	3.6mW	0.25 W	69.44
R11	Microcontroller part	43 kΩ/0.25 W	0.372 mW	0.25 W	672.0
R13	Power supply part	72 Ω/0.25 W	136.104 mW	0.25 W	1.83
R15	Power supply part	470 Ω/0.25 W	31.29 mW	0.25 W	7.98
R16	Power supply part	470 Ω/0.25 W	34.03 mW	0.25 W	7.34
R17	Power supply part	10 Ω/5 W	0.4 W	5 W	12.5
R18	Power supply part	10 Ω/2 W	0.4 W	2 W	5
RN4	CO- sensing unit	10 kΩ/0.25 W	2.5 mW	0.25 W	100
ZD1	Power supply part	3.6 V/1 W	0.072 W	1 W	13.88
ZD2	Power supply part	3.6 V/1 W	0.072 W	1 W	13.88
ZD3	Power supply part	3.3 V/1 W	0.072 W	1 W	13.88
FUSE1	Power supply part	100 mA			
FUSE2	Power supply part	80 mA			

(a) Details of cap-lamp battery:

Battery type: Intrinsically safe lithium-ion cap-lamp battery
Operating voltage: 3.7 V
Maximum voltage: 4.0 V
Limited maximum current: 0.35 A
Ampere-hour: 4 Ah

(b) Internal power supply:

Battery type: Rechargeable alkaline battery
Voltage: $4 \times 1.2\text{ V} = 4.8\text{ V}$
Ampere-hour: 700 mAh each

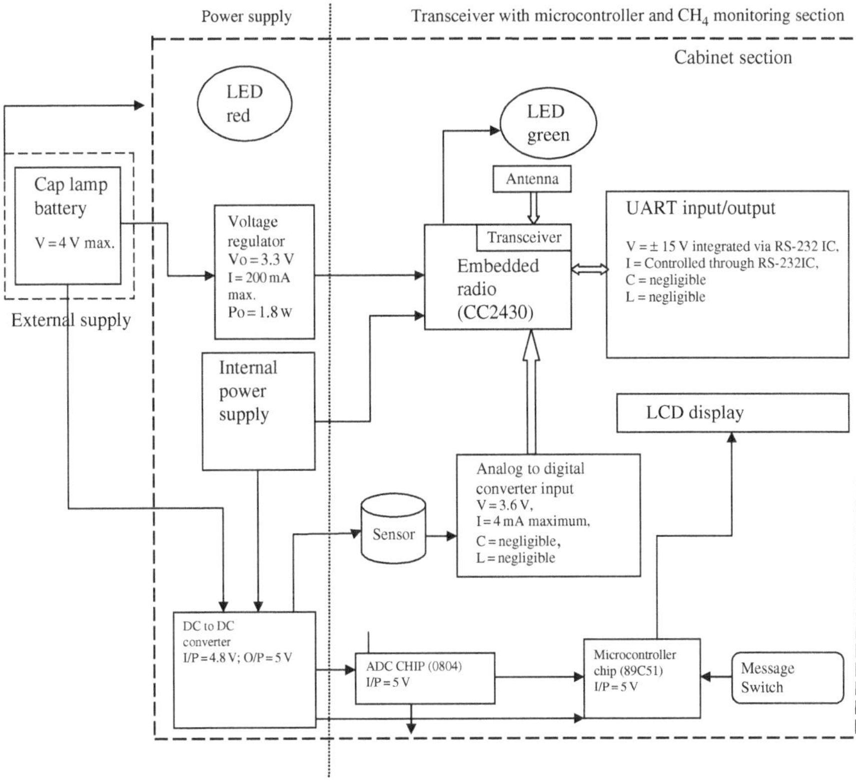

Fig. 9.40 Block diagram of CH_4-monitoring device

Gas-Monitoring Unit

The block diagram of "methane-monitoring device" is depicted in Fig. 9.40, which consists of external cap-lamp battery, light emitting diode, voltage regulator, turnover switch, internal battery supply, safety arrangement, embedded radio, programming port, universal asynchronous receiver–transmitter, analog-to-digital converter input, buzzer, and light emitting diode. The external cap-lamp battery is connected to light emitting diode and voltage regulator. The voltage regulator output is connected to turnover switch. The turnover switch is also connected to internal battery supply, thus power to the circuit can be given by either cap-lamp or internal battery via turnover switch. The output after turnover switch is connected to programming port and embedded radio. The embedded radio is connected to programming port, universal asynchronous receiver–transmitter output, analog-to-digital converter input, buzzer, and light emitting diode.

The diagram also consists of sensor ADC chip, AT89C51 microcontroller, and LCD. The sensor is connected to the ADC pin and ADC chip to give its data. Subsequently, ADC chip is connected to the microcontroller AT89C51 for processing the data. The measured value is displayed on the LCD which is connected to the microcontroller chip.

The external cap-lamp battery is connected with the light emitting diode to indicate the connection of the power supply. The cap-lamp battery is also connected with the voltage regulator to provide the regulated 3.3 V power supply to the circuit. The internal battery supply is also connected with the RFID device through turnover switch so that the device can be operated using external battery supply or internal battery supply. The output after the turnover switch is feed to programmable port, universal asynchronous receiver–transmitter output, analog-to-digital converter input, buzzer, and light emitting diode for programming, sending digital signal to external device, taking analog input from external device, generating alarm, and indicating the data transmission operation, respectively.

Figure 9.41 shows the circuit diagram of the RFID device. A two pin jack (CN1) is provided to connect external cap-lamp battery. LED (LED3) via resistance (R10) is connected to this jack. The jack power is connected to voltage regulator (MAX604 CSA+) via fuse (FUSE3) and resistance (R11). The battery (BT1) is connected via fuse (FUSE2) to another endpoint of the on–off–on switch (S1). The output from all stages of the switch (S2) is common and connected via resistance (R12) to RFID

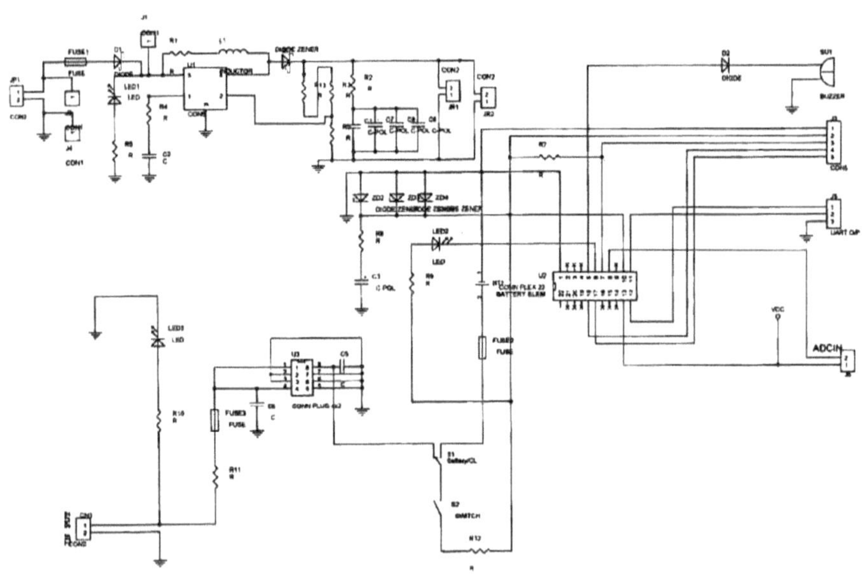

Fig. 9.41 Circuit diagram of power and RFID unit of CH_4-monitoring device. Key: J6: ADC input port ($V = 3.6$ V, $I = 4$ mA maximum, C and L negligible); JP1 and CN1: Cap-lamp battery supply 4.0 V (maximum); J3: Programming port, VDD=2.0 to 3.6 V; JR1 and J1: Voltage measuring point via voltmeter (L and C of voltmeter are negligible and R is very high, MΩ); BT1: Four AA size rechargeable batteries, 4.8 V; J5: UART Output $V = \pm$ 15 V integrated via RS-232 IC, I = Controlled through RS-232 IC, C = negligible, L = negligible; JR2: Output terminal, 5 V

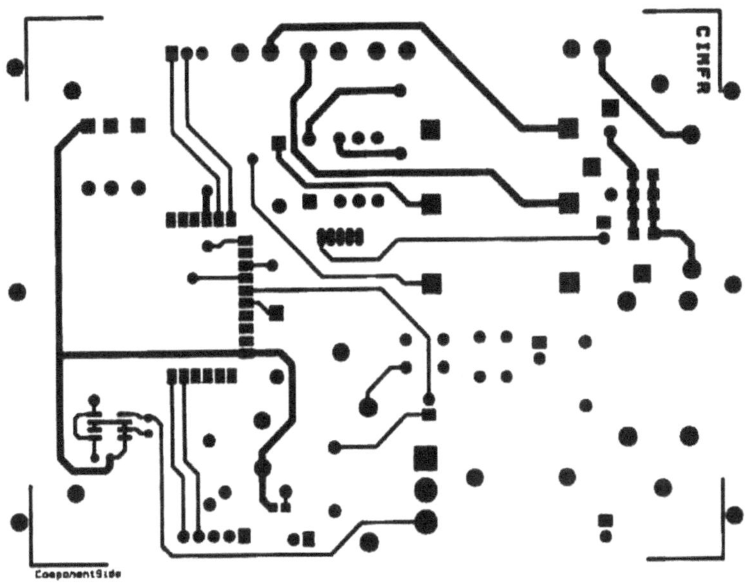

Fig. 9.42 PCB (*top*) layout of power and RFID unit of CH_4-monitoring device

chip, Zener diodes (ZD2, ZD3, and ZD4) are in parallel with the final supply after S2. The cathode point of the Zener diodes (ZD2, ZD3, and ZD4) is grounded. The capacitor (C3) is connected via resistance (R8) as a filter in parallel with the Zener diodes (ZD2, ZD3, and ZD4). The embedded radio chip (ER) is connected to LED (LED2) via resistance (R9) and power supply through the resistance (R12). The reset switch (S2) is connected between CC2430 embedded radio chip (ER) and ground. The buzzer (SU1) is connected to the embedded radio chip (ER) diode (D2). The programming port (J5) is connected with the embedded radio chip (ER) and power supply. The UART output jack (J5) and ADC in jack (J6) are connected with the embedded radio chip (ER) and grounded. The output of the voltage regulator is connected to one endpoint of on–off–on switch (S1). Figures 9.42, 9.43, 9.44, and 9.45 show the PCB layout of top, bottom, and component (top and bottom) layers, respectively. Table 9.17 gives the list of component of the RFID and power units of the device.

Figure 9.46 shows the circuit diagram of the methane-sensing unit. A 5 V regulated voltage is provided to the jack (J4). Voltage regulator (U1) is used to get the output voltage of 3.3 V. Two Zener diodes (ZD1 and ZD2) are used for safety. Jack (J3) is connected to the ADC pin of the RFID part. Amplifiers (U2) and (U6) are used for compensating the voltage at sensor output and ADC pin of the RFID unit, respectively. Sensor is connected to the jack (J6) whose analog output goes to the ADC chip (U3) from where the digital data are processed by the microcontroller chip (U4) and the data are displayed on the LCD. Figures 9.47, 9.48, and 9.49 depict the PCB layout of top, bottom, and component layers, respectively.

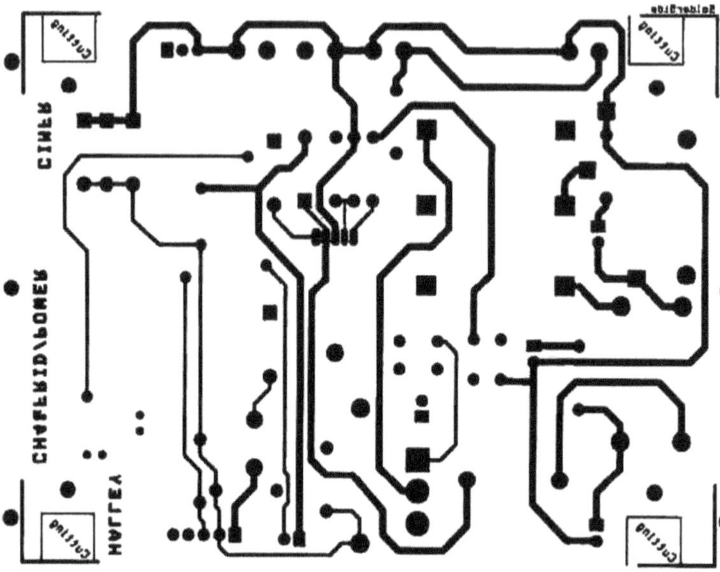

Fig. 9.43 PCB (*bottom*) layout of power and RFID unit of CH_4-monitoring device

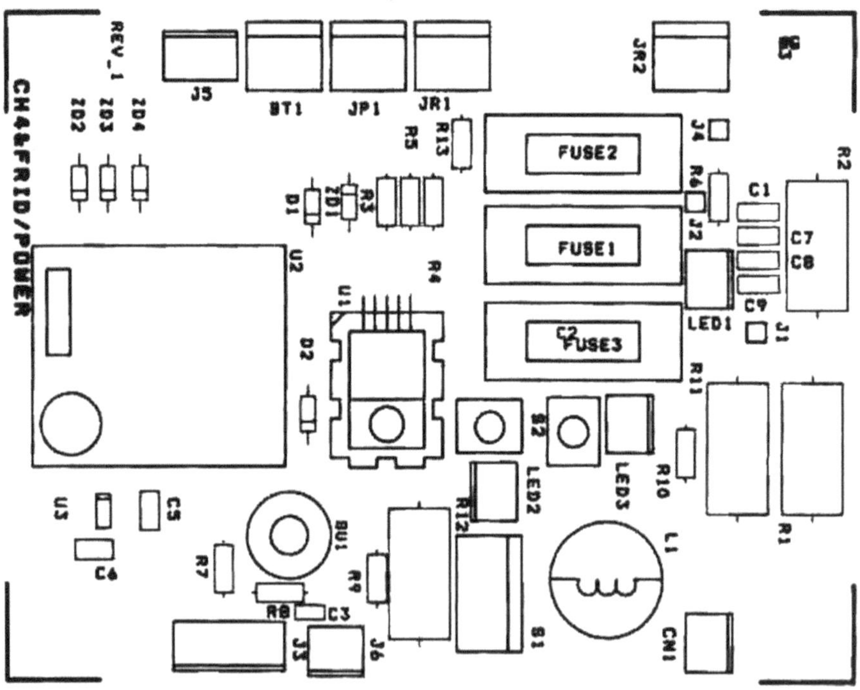

Fig. 9.44 PCB component (*top*) layout of power and RFID unit of CH_4-monitoring device

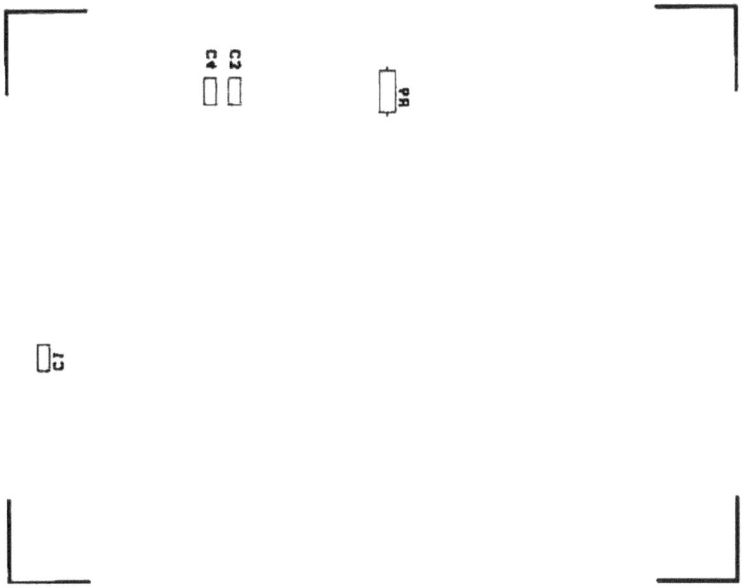

Fig. 9.45 PCB component (*bottom*) layout of power and RFID unit of CH_4-monitoring device

Table 9.17 Component list of RFID and power unit of CH_4-monitoring device

Name	Designation	Value	Tolerance
Power Supply Part			
Resistors	R1	12 Ω/2 W, MFR	1%
	R2	22 Ω/2 W, MFR	1%
	R3	390 Ω/0.25 W, MFR	1%
	R4	2.2 kΩ/0.25 W, MFR	1%
	R5	1 kΩ/0.25 W, MFR	1%
	R6	470 Ω/0.25 W, MFR	1%
	R13	2.7 kΩ/0.25 W, MFR	1%
Capacitor	C1	22 μF/10 V, SMD	10%
	C2	0.33 μF/100 V, ceramic	10%
	C7	22 μF/10 V, SMD	
	C8	22 μF/10 V, SMD	
	C9	22 μF/10 V, SMD	
Light emitting diode	LED1	3 mm red LED, $I = 7$ mA (max)	
Schottky diode	D1	1N5818	
	ZD1	1N5818	
Voltage regulator	U1	LM 2577 ADJ	
Inductor	L1	100 μH	
Fuse	FUSE1	400 mA	250 V, 35 A, breaking capacity
Connector	JP1	2 pin jack for cap-lamp battery connection. (4 V max, 0.4 A max)	
	JR2	2 pin jack for taking output voltage (5 V)	
	JR1	2 pin jack for measuring output voltage via voltmeter	

Table 9.17 (continued)

Name	Designation	Value	Tolerance
RFID Part			
Resistors	R7	43 kΩ/0.25 W, MFR	1%
	R8	72 Ω/0.25 W, MFR	1%
	R9	470 Ω/0.25 W, MFR	1%
	R10	470 Ω/0.25 W, MFR	1%
	R11	10 Ω/5 W, Wire Wound	10%
	R12	10 Ω/2 W, MFR	1%
Light emitting diode	LED2	3 mm green LED, $I = 7$ mA (max)	
	LED3	3 mm red LED, $I = 7$ mA (max)	
Capacitor	C3	1 μF/50 V, SMD	10%
	C5	0.01 μF/50 V, ceramic	10%
	C6	0.01 μF/50 V, ceramic	10%
Zener diode	ZD2	3.6 V/1	
	ZD3	3.6 V/1 W	
	ZD4	3.6 V/1 W	
Diode	D2	1N4007	
Switches	S2	Push to "on" button switch	
	S1	On–off–on switch	
Embedded radio	U2	CC2430, RFID chip	
Voltage regulator	U3	MAX 604 CSA+V, 3.3 V, 200 mA max, 1.8 W	
Buzzer	SU1	Magnetic field effect buzzer 3 – 12 V	
Fuse	FUSE3	100 mA	250 V, 35 A,breaking capacity
	FUSE2	80 mA	250 V, 35 A, breaking capacity
Connectors	J3	5 pin jack for programming, VDD = 2.0–3.6 V	
	J5	3 pin jack for UART output $V = \pm 15$ integrated via RS-232 IC, I=controlled through RS-232 IC, $C =$ negligible, $L =$ negligible	
	J6	2 pin jack for ADC input, $V = 3.6$ V, $I = 4$ mA maximum, $C =$ negligible, $L =$ negligible, analog signal	
	CN1	2 pin jack for cap-lamp battery (4 V/0.4 A)	
	BT1	Connector for 3 AA size rechargeable batteries of 1.2 V each	

Table 9.18 gives the list of components used in the sensor unit of the device. Figure 9.50 illustrates the enclosure diagram of the whole device.

Figures 9.51 and 9.52 illustrate the graphs depicting voltage versus concentration of methane and carbon monoxide, respectively, separately for both the devices. Wiring diagram of both the devices are almost similar, which is depicted in Fig. 9.53.

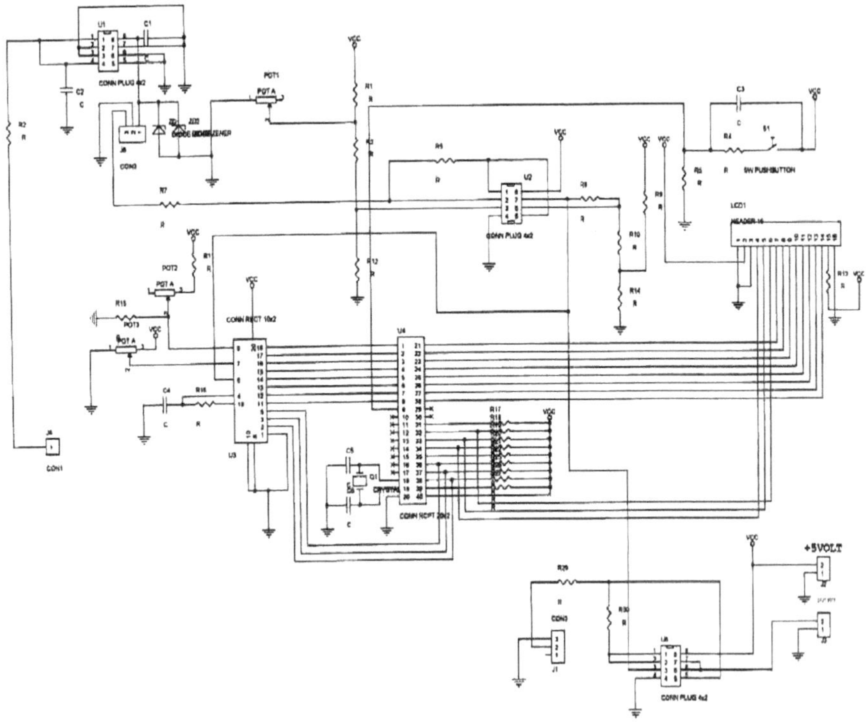

Fig. 9.46 Circuit diagram of sensor unit of CH_4-monitoring device. Key: J2: 2 pin jack for connecting +5 V supply; J3: 2 pin jack for connecting RFID unit; J4: 2 pin jack for connecting +5 V supply; J6: 3-point for connecting the CH_4 sensor

9.3.4.2 Safety Analysis

Each component used in the device is provided by its own power rating. The current across the resistor is calculated taking the shortage route to ground by the formula V/R. Then power rating of the component is calculated by the formula I^2R. This power rating is compared with the component power rating which should be less in regard to intrinsic safety. Ratio ($W1/W2$) of the given power rating of the component to the calculated power rating should be greater than 1.5 as per the IEC60079-11, 2007 standard. Table 9.19 gives the safety analysis of the power supply unit, and Tables 9.20 and 9.21 provide the safety analysis of RFID and sensor unit, respectively. Ratio ($W1/W2$) is greater than 1.5 for all the safety components used in the device.

9.3.5 Message Device

Two types of message devices have been developed using CC2430 RFID chips having 48 and 22 pins. CC2430 RFID chip having 48 pins is manufactured by Texas

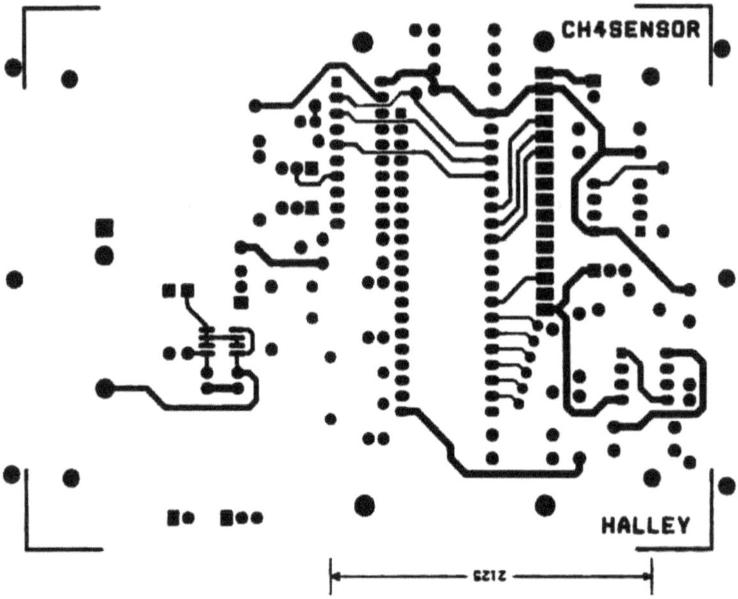

Fig. 9.47 PCB (*top*) layout of sensor unit of CH_4-monitoring device

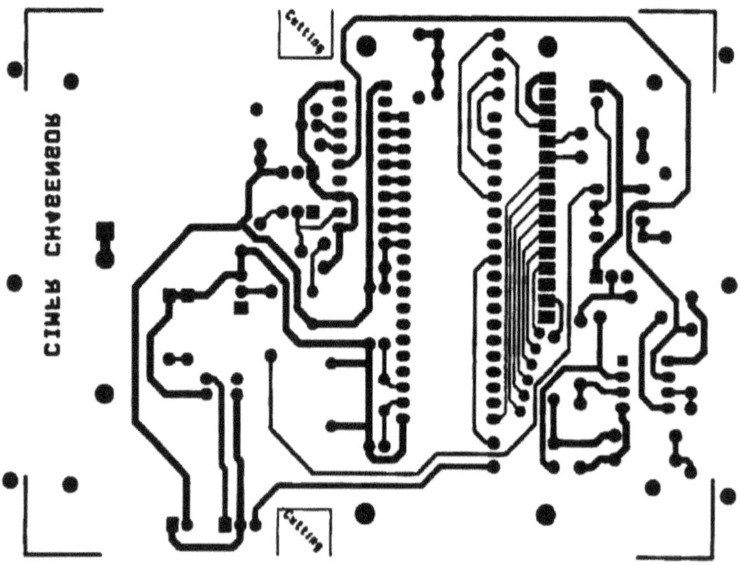

Fig. 9.48 PCB (*bottom*) layout of sensor unit of CH_4-monitoring device

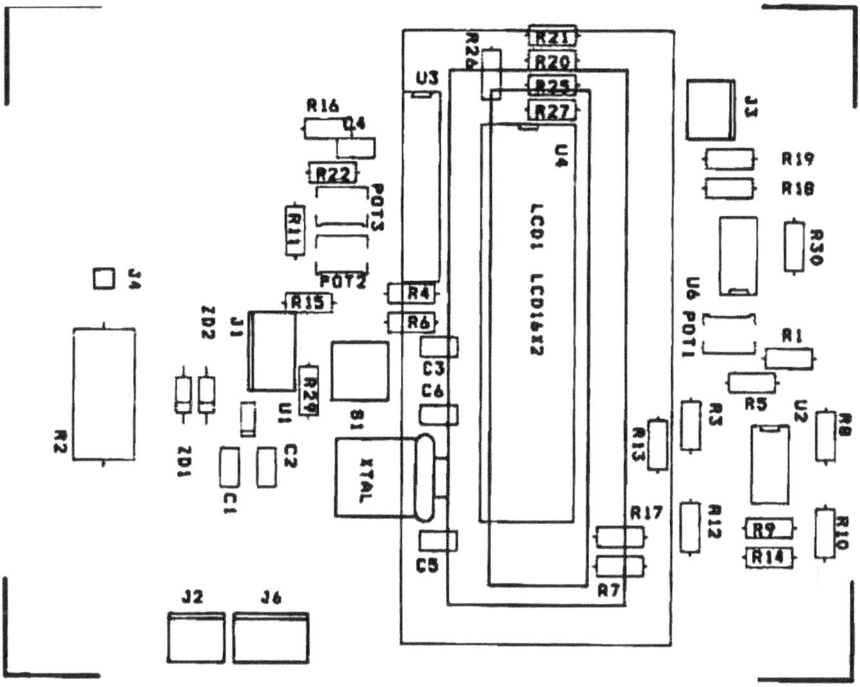

Fig. 9.49 PCB component layout of sensor unit of CH_4-monitoring device

Table 9.18 Component list of sensor unit of CH_4-monitoring device

Name	Designation	Value	Tolerance
Resistors	R1	1 kΩ/0.25 W, MFR	1%
	R2	10 Ω/5 W, Wire Wound	10%
	R3	100 kΩ/0.25 W, MFR	1%
	R4	100 Ω/0.25 W, MFR	1%
	R5	100 kΩ/0.25 W, MFR	1%
	R6	8.2 kΩ/0.25 W, MFR	1%
	R7	100 kΩ/0.25 W, MFR	1%
	R8	100 kΩ/0.25 W, MFR	1%
	R9	1 kΩ/0.25 W, MFR	1%
	R10	1 kΩ/0.25 W, MFR	1%
	R11	2.2 kΩ/0.25 W, MFR	1%
	R12	100 kΩ/0.25 W, MFR	1%
	R13	3.3 kΩ/0.25 W, MFR	1%
	R14	100 Ω/0.25 W, MFR	1%
	R15	470 Ω/0.25 W, MFR	1%
	R16	10 kΩ/0.25 W, MFR	1%
	R17	33 kΩ/0.25 W, MFR	1%
	R18	10 kΩ/0.25 W, MFR	1%
	R19	10 kΩ/0.25 W, MFR	1%
	R20	10 kΩ/0.25 W, MFR	1%

Table 9.18 (continued)

Name	Designation	Value	Tolerance
R21	10 kΩ/0.25 W, MFR	1%	
R22	10 kΩ/0.25 W, MFR	1%	
R25	10 kΩ/0.25 W, MFR	1%	
R26	10 kΩ/0.25 W, MFR	1%	
R27	10 kΩ/0.25 W, MFR	1%	
R29	10 kΩ/0.25 W, MFR	1%	
	R30	10 kΩ/0.25 W, MFR	1%
Variable resistors	POT1	1 kΩ	
	POT2	1 kΩ	
	POT3	100 kΩ	
	J1	1 kΩ	
Capacitors	C1	0.01 μF/50 V, ceramic	10%
	C2	0.01 μF/50 V, ceramic	10%
	C3	10 μF/16 V, SMD	10%
	C4	56 Pf/50 V, ceramic	10%
	C5	33 pF/50 V, ceramic	10%
	C6	33 pF/50 V, ceramic	10%
Zener diode	ZD1	3.6 V/1 W	
	ZD2	3.6 V/1 W	
Switch	S1	Switch push button “on” switch	
Voltage regulator	U1	MAX 604CSA+	
ADC chip	U3	ADC 0804	
Microcontroller	U4	AT89C51	
LCD	LCD1	16 × 2 character	
Crystals	XTAL	11.0592 MHz	
Sensor	4P–90		
Amplifiers	U6	LM358	
	U2	LM358	
Connectors	J2	2 pin jack for 5 V power supply	
	J3	2 pin jack for connecting sensor unit to RFID unit	
	J4	1 pin jack for 5 V power supply	
	J6	3-point for connecting the CH_4 sensor	

Instruments, USA (www.ti.com) and CC2430 RFID chip having 22 pins is manufactured by Shenzhen Technology Co. Ltd., China (www.soczigbee.com). Circuit diagram of both the message devices is described separately.

9.3.5.1 Message Device Using 22 Pins RFID Chip

Power supply and RFID unit of the message device is similar to that described in Section 9.3.4.1. Figure 9.54 shows the circuit diagram of the message device. Microcontroller (U2) is interfaced to serial interface chip (U3), 4 dual input and gate chip (U5), buffer chip (U1), keypads, and LCD. When any key is pressed interrupt is send to microcontroller (U2) and interrupt subroutine is called. Identification of key

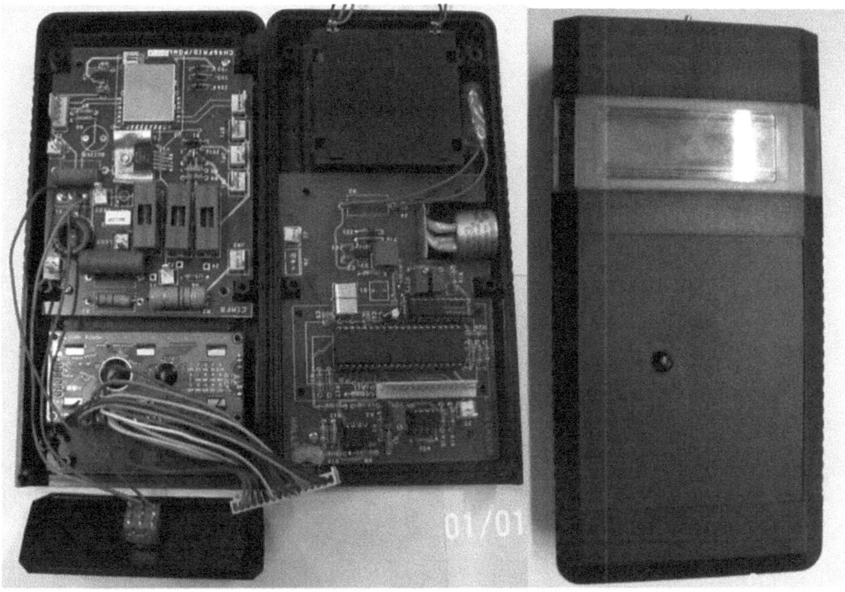

Fig. 9.50 View of methane-monitoring device along with enclosure

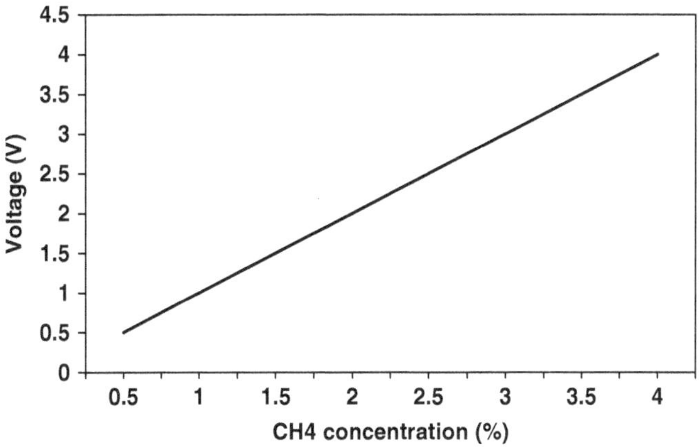

Fig. 9.51 CH_4 concentration versus sensor output voltage

number in program is done and subsequently it is displayed on the LCD. Proper check for appropriate delay is done by the microcontroller chip (U4). Any key is forwarded only after the delay time is completed. The text message which is displayed on the LCD is subsequently fed to the UART port of the embedded radio for transmitting through the serial interface chip (U3). Figures 9.55, 9.56, and 9.57 illustrate the PCB layout of top, bottom, and component layer, respectively. Table 9.22 gives the list of the components used in the message unit along with their values.

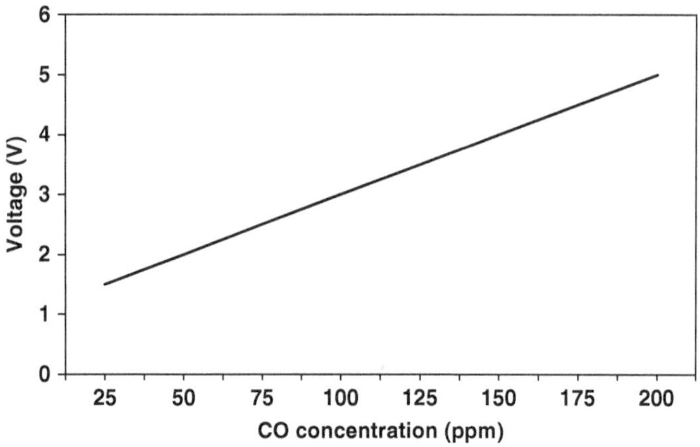

Fig. 9.52 CO concentration versus sensor output voltage

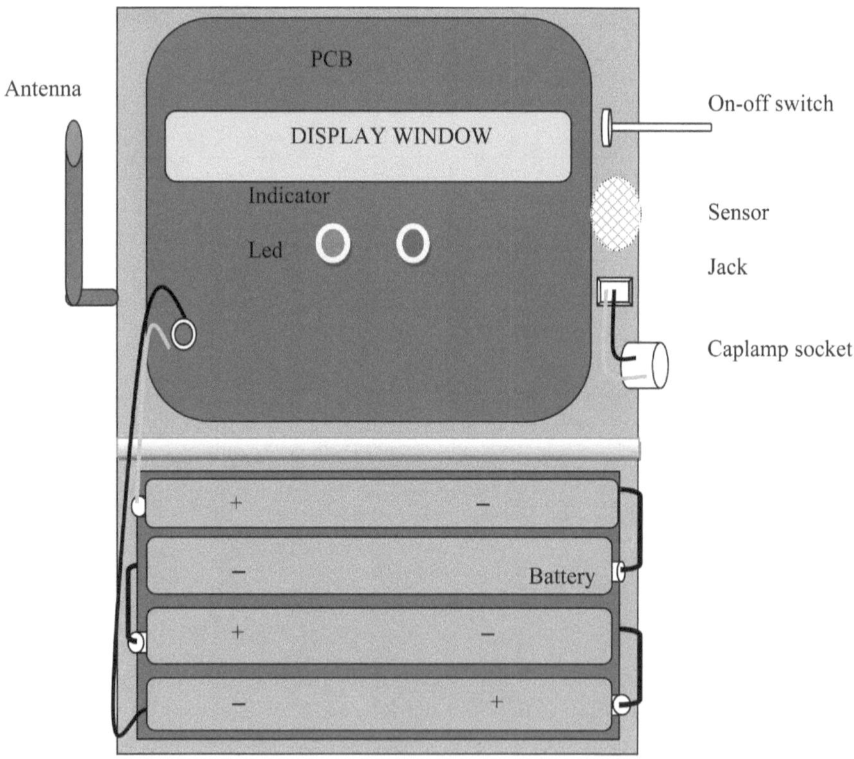

Fig. 9.53 Wiring diagram of wireless sensor device for mines

Table 9.19 Details of safety analysis of power supply unit for CH_4-monitoring device

Component designation	Component location	Value	Maximum rating used by circuit (*W*2)	Component rating (*W*1)	*W*1/*W*2
R1	Power supply unit	12 Ω / 2 W	259.51 mW	2 W	7.70
R2	Power supply unit	22 Ω/2 W	475.77 mW	2 W	4.20
R3	Power supply unit	390 kΩ/0.25 W	0.58 mW	0.25 W	431
R4	Power supply unit	2.2 kΩ/0.25 W	11.33 mW	0.25 W	22.06
R5	Power supply unit	1 kΩ/0.25 W	1.48 mW	0.25 W	168.9
R6	Power supply unit	470 Ω/0.25 W	49.02 mW	0.25 W	5.09
R13	Power supply unit	2.7 Ω/0.25 W	4.01 mW	0.25 W	62.34
FUSE1	Power supply part	400 mA			

Table 9.20 Details of safety analysis of RFID unit for CH_4-monitoring device

Component designation	Component location	Value	Maximum rating used by circuit (*W*2)	Component rating (*W*1)	*W*1/*W*2
R7	Microcontroller part	43 kΩ/0.25 W	0.372 mW	0.25 W	672.0
R8	Power supply part	72 Ω/0.25 W	136.104 mW	0.25 W	1.83
R9	Power supply part	470 Ω/0.25 W	31.29 mW	0.25 W	7.98
R10	Power supply part	470 Ω/0.25 W	34.03 mW	0.25 W	7.34
R11	Power supply part	10 Ω/5 W	0.4 W	5 W	12.5
R12	Power supply part	10 Ω/2 W	0.4 W	2 W	5
ZD2	Power supply part	3.6 V/1 W	0.072 W	1 W	13.88
ZD3	Power supply part	3.6 V/1 W	0.072 W	1 W	13.88
ZD4	Power supply part	3.3 V/1 W	0.072 W	1 W	13.88
FUSE3	Power supply part	100 mA			
FUSE2	Power supply part	80 mA			

Table 9.21 Details of safety analysis of sensor unit for CH_4-monitoring device

Component designation	Component location	Value	Maximum rating used by circuit (*W*2)	Component rating (*W*1)	*W*1/*W*2
R1	CH_4-sensing unit	1 kΩ/0.25 W	0.618 μW	0.25 W	
R2	CH_4-sensing unit	10 Ω/5 W	25 mW	5 W	200
R3	CH_4-sensing unit	100 kΩ/0.25 W	61.8 mW	0.25 W	4.0
R4	CH_4-sensing unit	100 Ω/0.25 W	36 μW	0.25 W	6,944.44
R5	CH_4-sensing unit	100 kΩ/0.25 W	62.5 μW	0.25 W	4,000
R6	CH_4-sensing unit	8.2 kΩ/0.25 W	2.95 mW	0.25 W	84.74
R7	CH_4-sensing unit	100 kΩ/0.25 W	62.5 μW	0.25 W	4,000
R8	CH_4-sensing unit	100 kΩ/0.25 W	0.24 mW	0.25 W	1,041
R9	CH_4-sensing unit	1 kΩ/0.25 W	20.61 mW	0.25 W	12.1
R10	CH_4-sensing unit	1 kΩ/0.25 W	2.44 μW	0.25 W	102,459
R11	CH_4-sensing unit	2.2 kΩ/0.25 W	7.69 mW	0.25 W	32.5
R12	CH_4-sensing unit	100 kΩ/0.25 W	61.8 mW	0.25 W	4.04
R13	CH_4-sensing unit	3.3 kΩ/0.25 W	7.42 mW	0.25 W	33.69
R14	CH_4-sensing unit	100 Ω/0.25 W	2.06 mW	0.25 W	121.35
R15	CH_4-sensing unit	470 Ω/0.25 W	1.64 mW	0.25 W	152.4
R16	CH_4-sensing unit	10 kΩ/0.25 W	2.5 mW	0.25 W	100
R17	CH_4-sensing unit	33 kΩ/0.25 W	0.74 mW	0.25 W	337.83
R18	CH_4-sensing unit	10 kΩ/0.25 W	2.5 mW	0.25 W	100
R19	CH_4-sensing unit	10 kΩ/0.25 W	2.5 mW	0.25 W	100
R20	CH_4-sensing unit	10 kΩ/0.25 W	2.5 mW	0.25 W	100
R21	CH_4-sensing unit	10 kΩ/0.25 W	2.5 mW	0.25 W	100
R22	CH_4-sensing unit	10 kΩ/0.25 W	2.5 mW	0.25 W	100
R25	CH_4-sensing unit	10 kΩ/0.25 W	2.5 mW	0.25 W	100
R26	CH_4-sensing unit	10 kΩ/0.25 W	2.5 mW	0.25 W	100
R27	CH_4-sensing unit	10 kΩ/0.25 W	2.5 mW	0.25 W	100
R29	CH_4-sensing unit	10 kΩ/0.25 W	2.5 mW	0.25 W	100
R30	CH_4-sensing unit	10 kΩ/0.25 W	2.5 mW	0.25 W	100
ZD1	CH_4-sensing unit	3.6 V/1 W	0.504 W	1 W	1.98
ZD2	CH_4-sensing unit	3.6 V/1 W	0.504 W	1 W	1.98

9.3.5.2 Safety Analysis

Each component used in the device is provided by its own power rating. The current across the resistor is calculated taking the shortage route to ground by the formula V/R. Then power rating of the component is calculated by the formula I^2R. This power rating is compared with the component power rating which should be less in regard to intrinsic safety. Ratio ($W1/W2$) of the given power rating of the component to the calculated power rating is taken, which should be greater then 1.5 as per the IEC60079-11, 2007 standard. Table 9.23 gives the safety analysis of message device. Ratio ($W1/W2$) is greater than 1.5 for all the safety components used in the device.

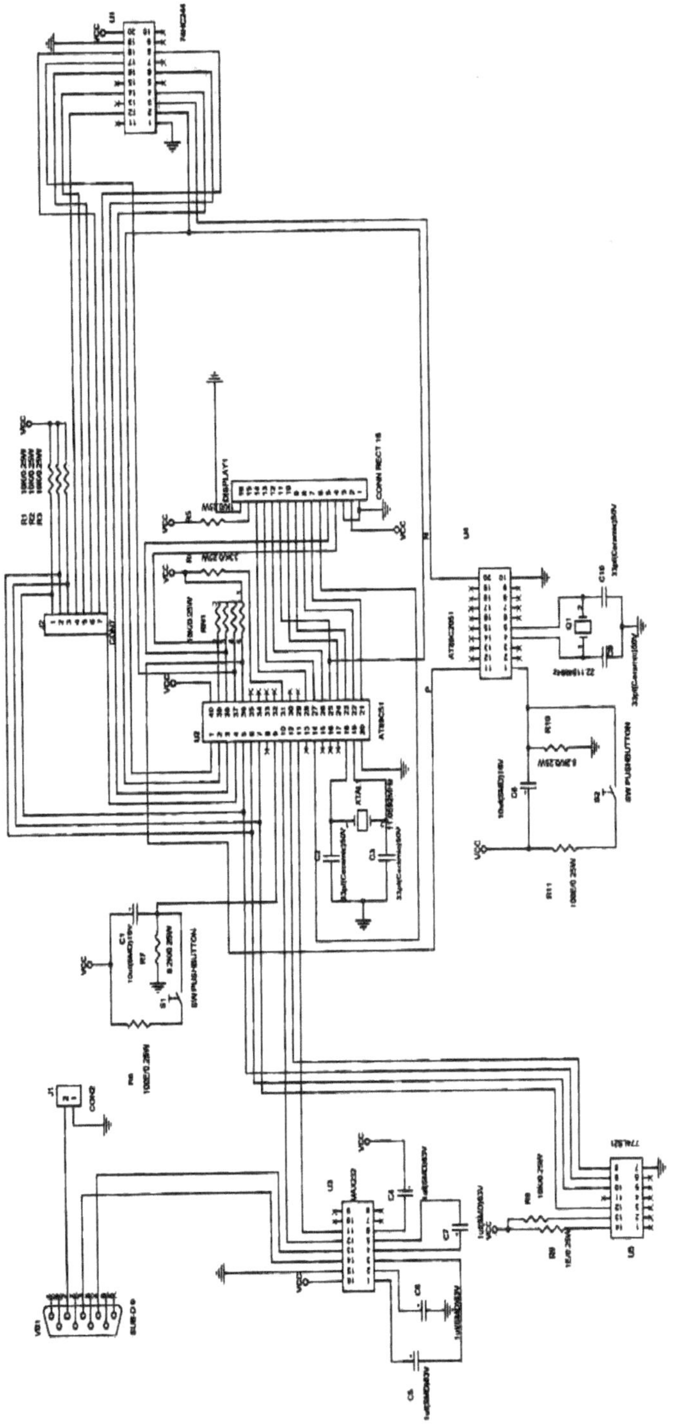

Fig. 9.54 Circuit diagram of wireless message device for mines

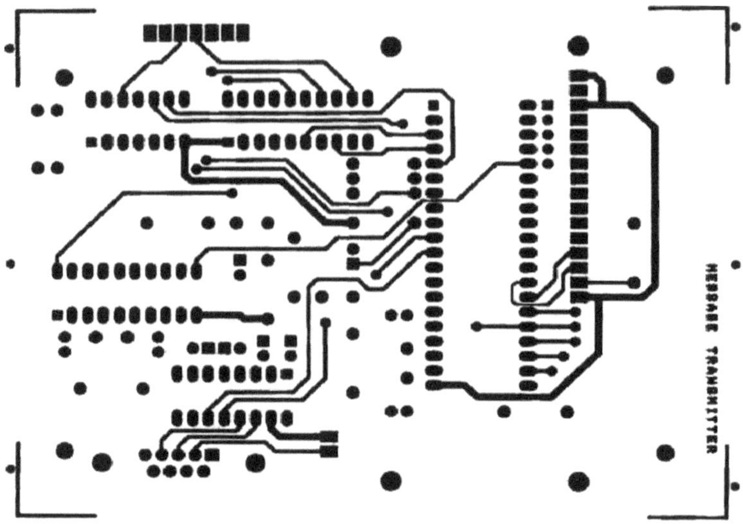

Fig. 9.55 PCB (*top layer*) layout of wireless message device for mines

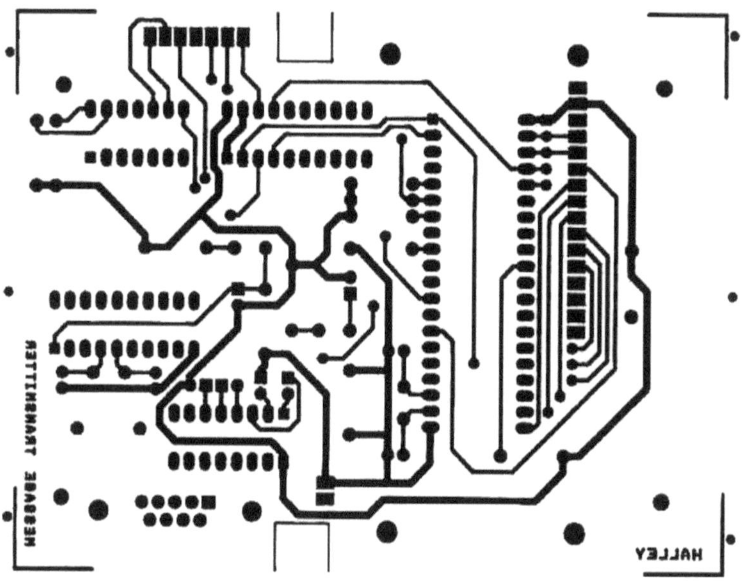

Fig. 9.56 PCB (*bottom layer*) layout of wireless message device for mines

9.3.5.3 Message Device Using 42 Pins RFID Chip

Figure 9.58 illustrates the circuit diagram of microcontroller unit of message device. Figures 9.59 and 9.60 show the circuit details of the charging unit and power supply unit of the message device, respectively. The whole device gets it power supply

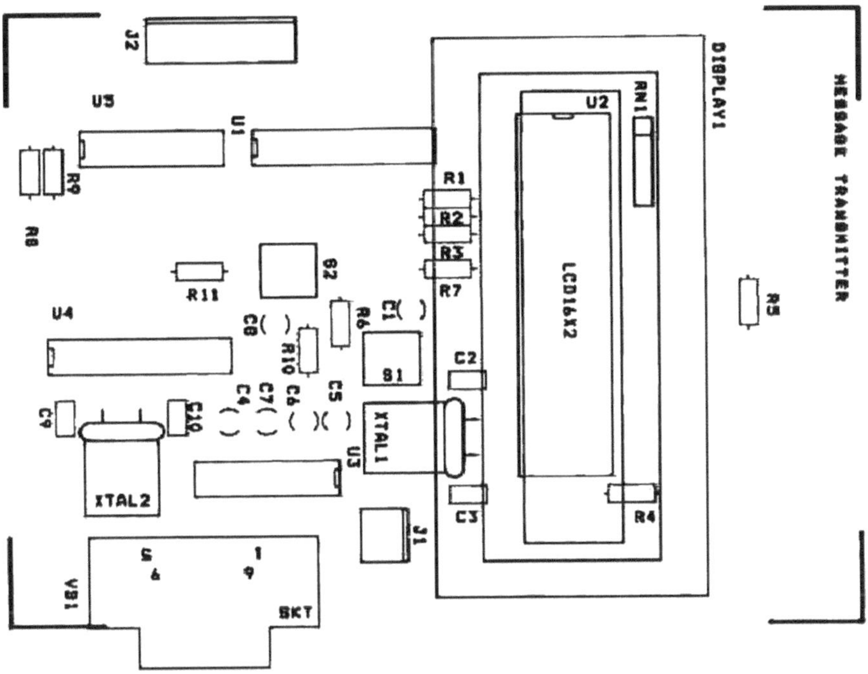

Fig. 9.57 PCB (*bottom layer*) layout of wireless message device for mines

Table 9.22 Component list of message device

Name	Designation	Value	Tolerance
Resistors	R1	10 kΩ/0.25 W, MFR	1%
	RN2	10 kΩ/0.25 W, MFR	1%
	R2	10 kΩ/0.25 W, MFR	1%
	R3	10 kΩ/0.25 W, MFR	1%
	R4	33 kΩ/0.25 W, MFR	1%
	R5	1 kΩ/0.25 W, MFR	1%
	R6	100 Ω/0.25 W, MFR	1%
	R7	8.2 kΩ/0.25 W, MFR	1%
	R8	10 kΩ/0.25 W, MFR	1%
	R10	8.2 kΩ/0.25 W, MFR	1%
	R11	100 Ω/0.25 W, MFR	1%
Capacitors	C1	10 μF/16 V, ceramic	10%
	C2	33 pF/50 V, ceramic	10%
	C3	33 pF/ 50 V, ceramic	10%
	C4	1 μF/16 V, SMD	10%
	C5	1 μF/16 V, SMD	10%
	C6	1 μF/16 V, SMD	10%
	C7	1 μF/16 V, SMD	10%
	C8	10 μF/16 V, SMD	10%
	C9	33 pF/50 V, ceramic	10%
	C10	33 pF/50 V, ceramic	10%

Table 9.22 (continued)

Name	Designation	Value	Tolerance
Switches	S1 & S2	Push to on button switch	
Buffer	U1	74HC244	
Microcontrollers	U2	AT89C51	
	U4	AT89C2051	
Serial Interface	U3	MAX232	
To provide interrupt	U5	74LS21	
Crystal	XTAL1	11.0592 MHz	
	XTAL2	22.1184 MHz	
LCD	LCD1	16 × 2 character	
Connectors	VB1	9-pin serial port connector to computer	
	J1	2 pin jack for connecting RFID unit for transmitting	
	J2	7 pin jack for programming	

Table 9.23 Details of safety analysis of message device

Component designation	Component location	Value	Maximum rating used by circuit ($W2$)	Component rating ($W1$)	$W1/W2$
R1	Power supply part	10 kΩ/0.25 W	2.5 mW	0.25 W	100
RN1	Power supply part	10 kΩ/0.25 W	2.5 mW	0.25 W	100
R2	Power supply part	10 kΩ/0.25 W	2.5 mW	0.25 W	100
R3	Power supply part	10 kΩ/0.25 W	2.5 mW	0.25 W	100
R4	Power supply part	33 kΩ/0.25 W	0.74 mW	0.25 W	337.8
R5	Power supply part	1 kΩ/0.25 W	25 mW	0.25 W	10
R6	Power supply part	100 Ω/0.25 W	36.24 mW	0.25 W	6.89
R7	Power supply part	8.2 kΩ/0.25 W	2.97 mW	0.25 W	84
R8	Power supply part	10 kΩ/0.25 W	2.5 mW	0.25 W	100
R10	Power supply part	8.2 kΩ/0.25 W	3.04 mW	0.25 W	82
R11	Power supply part	100 Ω/0.25 W	36 μW	0.25 W	6944

from two sources: external battery, i.e., cap-lamp and internal source, i.e., AA size rechargeable battery. VIN point in Fig. 9.60 gets 5 V, which is regulated to 3.2 V for the microcontroller unit (Fig. 9.58). Figure 9.61 shows the circuit details of the LCD connection for displaying message to be sent. Figures 9.62, 9.63, and

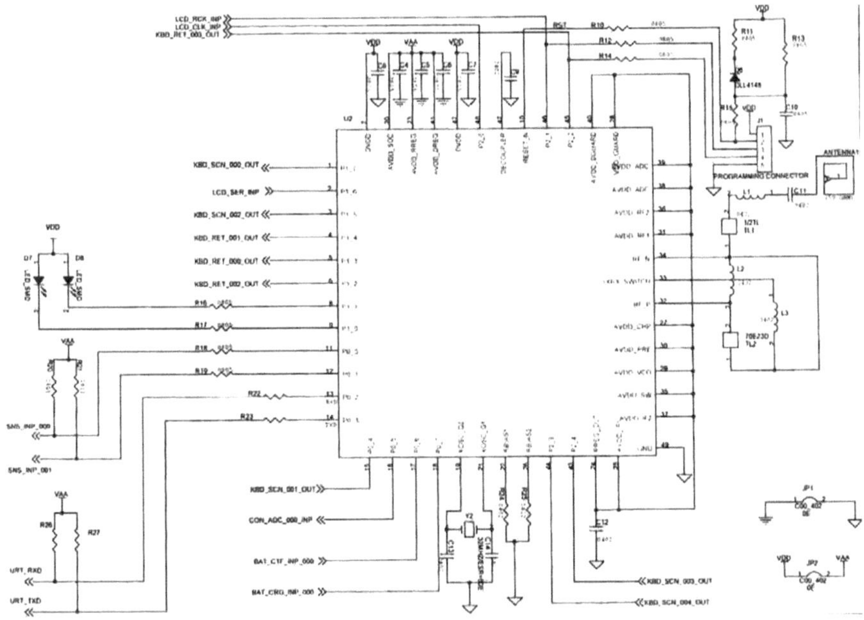

Fig. 9.58 Circuit diagram of microcontroller unit of message device

9.64 illustrate the circuit details of the keypad connection for the message device. Figure 9.65 depicts the view of the whole device along with its enclosure diagram. Table 9.24 gives the list of the components used in the message unit along with their values. Figure 9.65 illustrates the view of PCB and enclosure of the message device.

9.3.5.4 Radiated Power

Radiated power of the device using PCB dipole antenna in frequency range 2.4–2.41 GHz is given in Table 9.25. As per the measurement it was observed that the radiated power was very less in the frequency range 2.4–2.41 GHz. The maximum value of 34.43 μW was measured at 1 m distance from the antenna kept in vertical mounting position. It was gradually decreased with increase in distance from the antenna.

9.4 Principle and Operating Procedure

9.4.1 Tracking and Monitoring System

With the help of tracking and monitoring system, it will be easier to locate each miner and equipment during disaster situation as well as during normal working

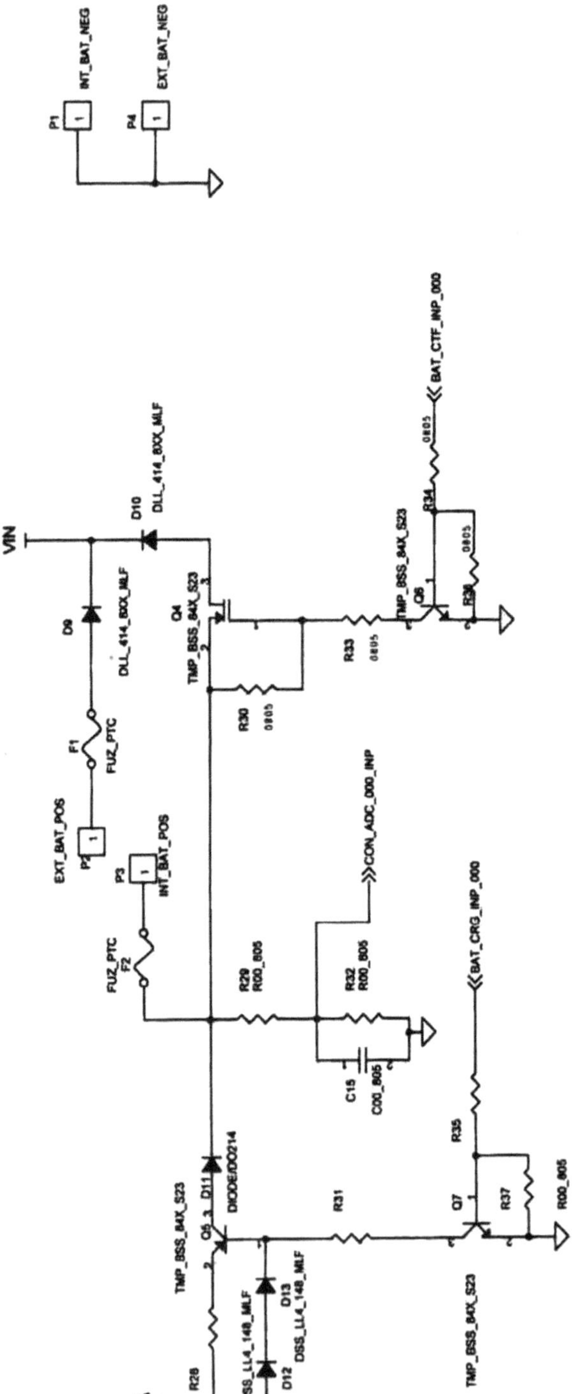

Fig. 9.59 Circuit diagram of battery charging unit of message device

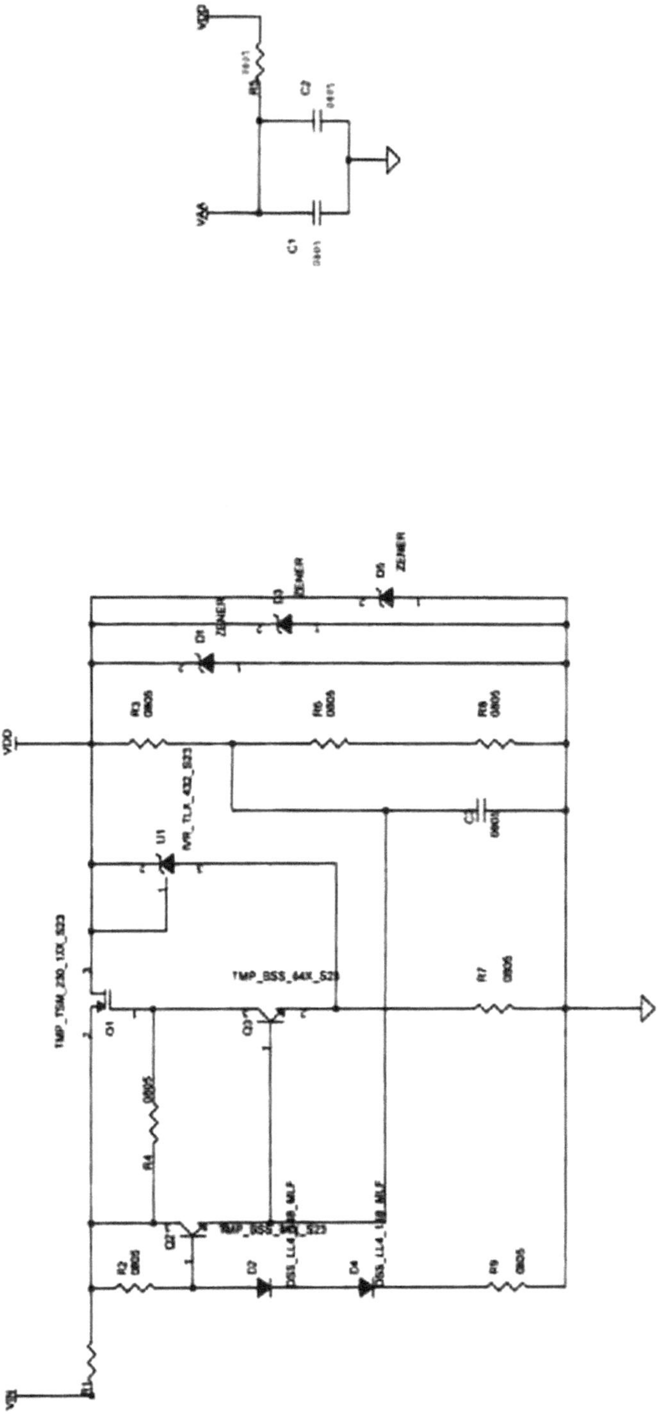

Fig. 9.60 Circuit diagram of power supply unit of message device

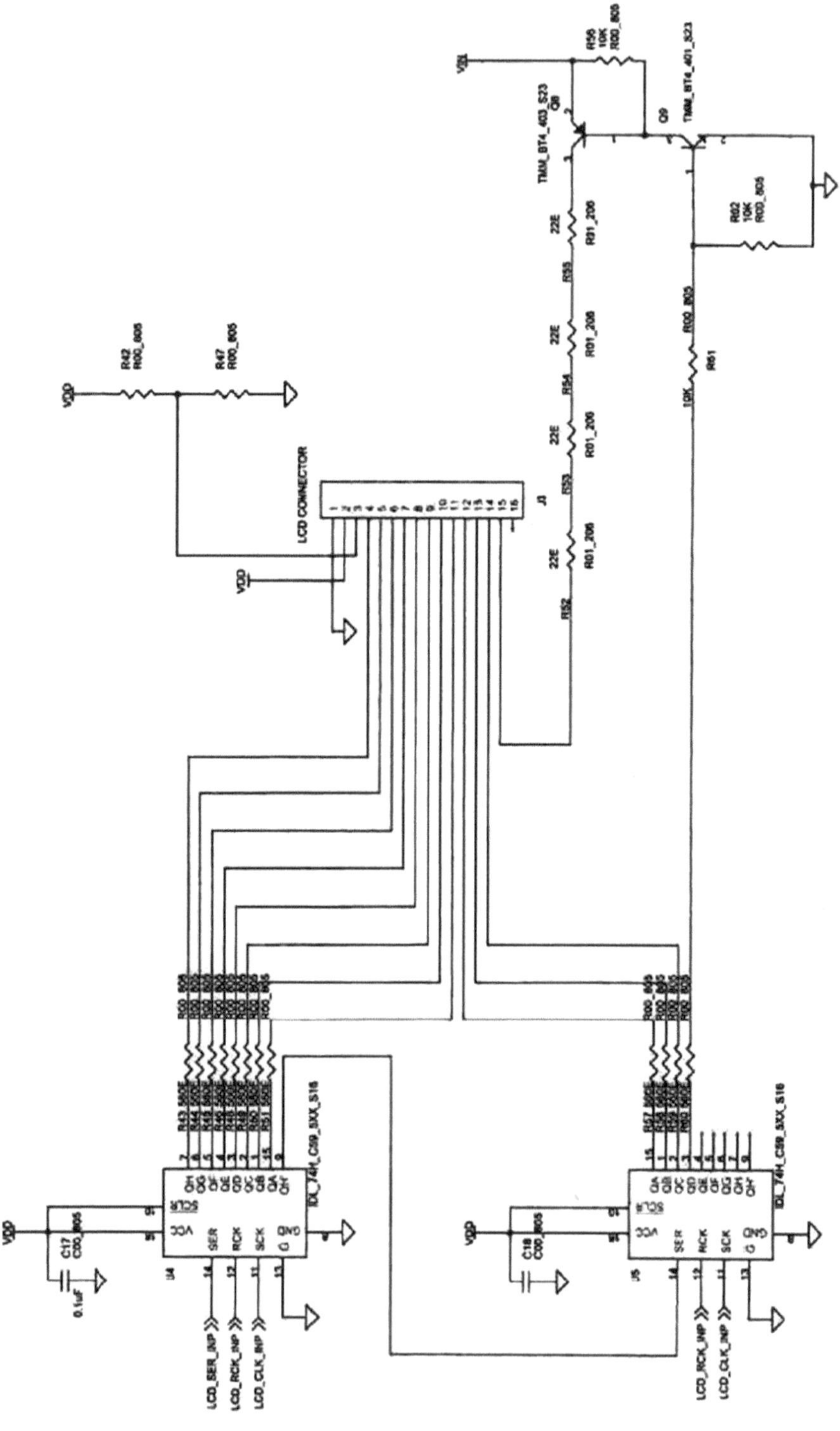

Fig. 9.61 Circuit diagram of LCD interface to the message device

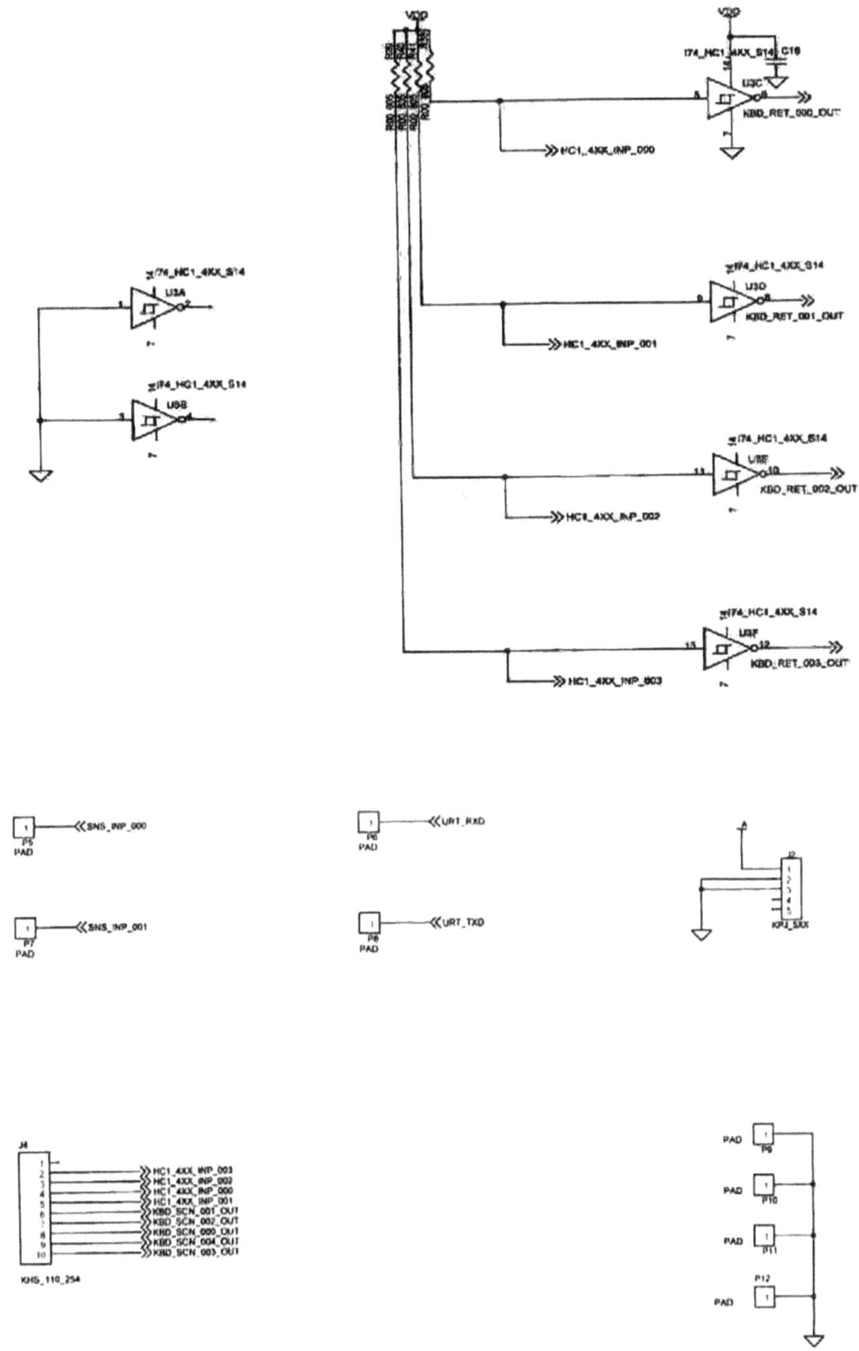

Fig. 9.62 Circuit diagram for the keypad connection to the microcontroller

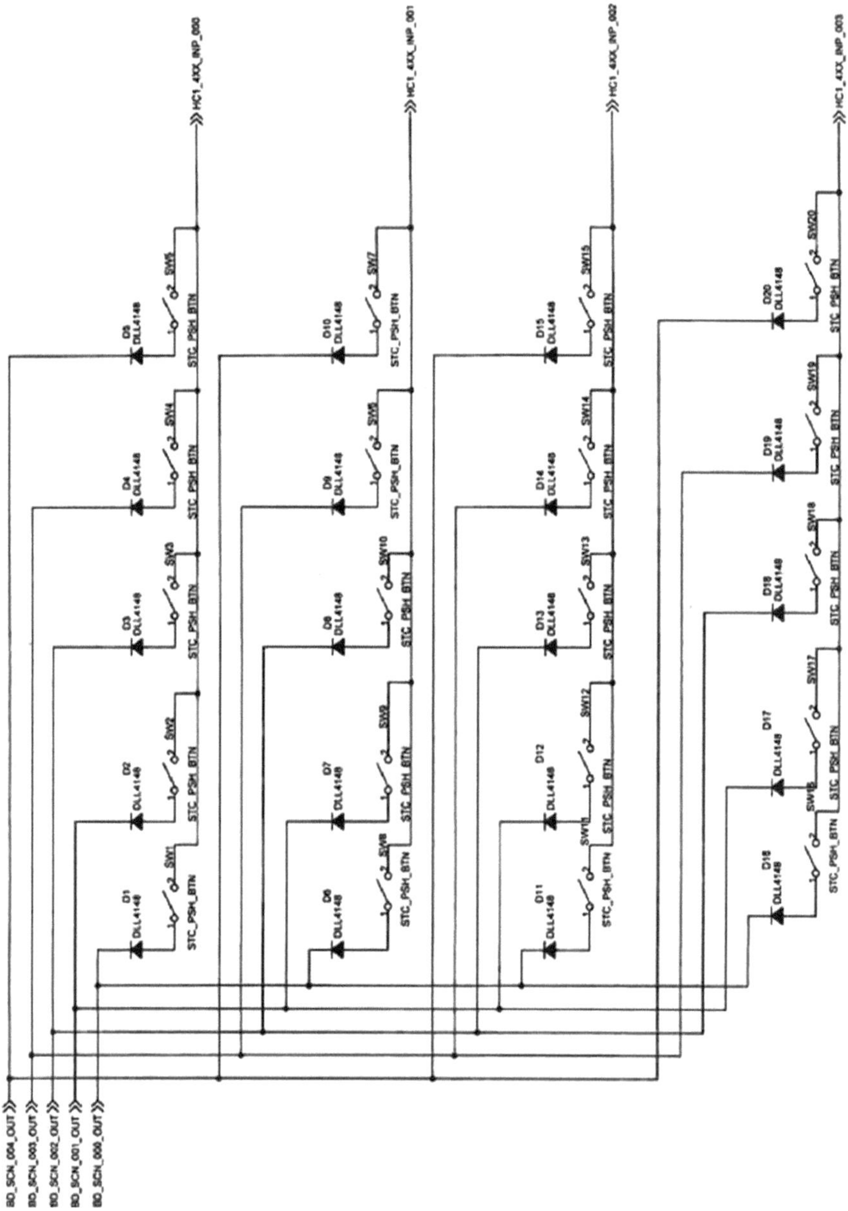

Fig. 9.63 Circuit diagram of keypads of message device

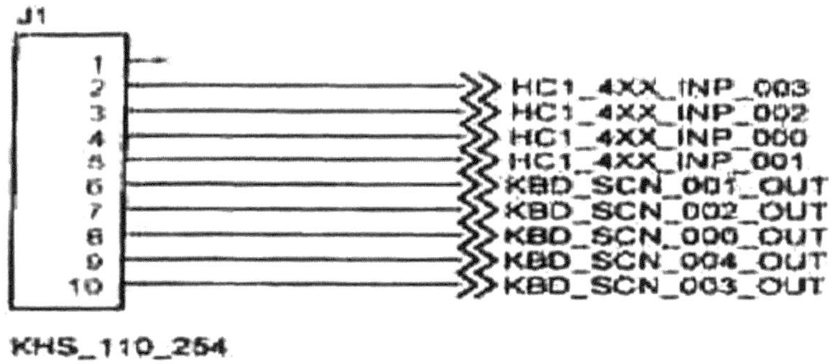

Fig. 9.64 Circuit diagram of keypad connection of message device

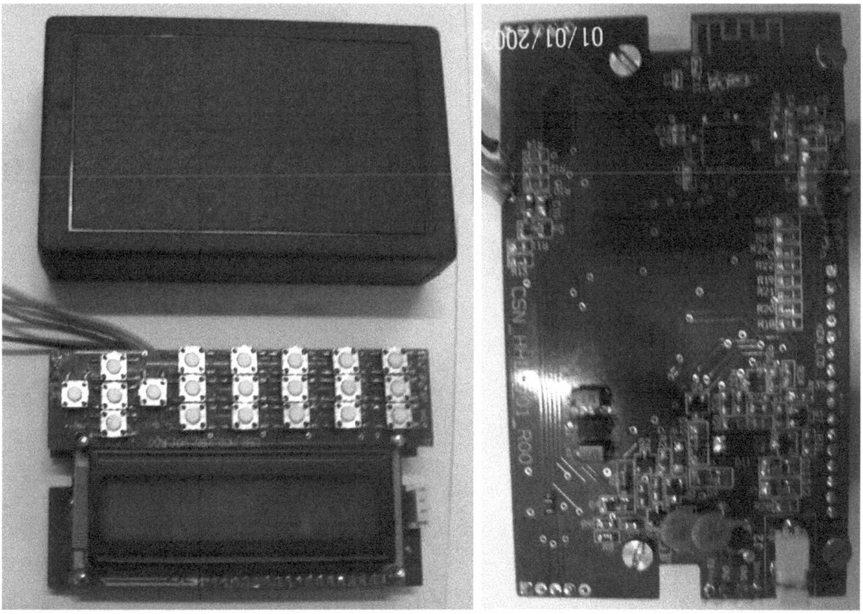

Fig. 9.65 View of message device along with enclosure

periods for productivity measurement and employees' safety. This will improve output per man shift (OMS) as well as safety of miners. This will also reduce idle time of equipment operating without going physically inside underground mine. Core hardware components of tracking and monitoring system are active RFID devices (end devices) attached to mobile workers, vehicles, and equipment in mining and tunnel operations. Each device transmits/receives messages to/from neighboring devices having ZigBee-compliant network interfaces and autonomously form network among themselves and with other static devices (routers) placed at

Table 9.24 List of components of message device

Name	Designation	Value	Tolerance
Integrated Circuits:			
Regulator, TL-432, sot-23	U1		
IC: CC2430, 48 qfn	U2		
IC:74HC14, sot-23	U3		
IC: 74HC595, sot-16	U4,U5		
Transistors:			
BJT,PNP,MMBT4403, sot-23	Q1,Q5,Q8		
TRANSISTOR BSS_84, sot-23	Q4		
BJT, NPN, MMBT4401, sot-23 / BC848	Q2,Q3,Q7,Q6,Q9		
Diodes:			
Small signal, LL4148, mlf	D2,D4,D6,D9, D10,D12,D13		
Small signal, 1N4007,SMA	D11		
Zener, 5V6, sot-23, SMA	D1		
Capacitors:			
Discrete, metal oxide, 0805	C3	2.2 μF/5 V	5%
Discrete, metal oxide, 0402	C9,C12	1 μF/5 V	5%
Discrete, metal oxide, 0805	C4,C5,C6,C7,C8, C10,C15,C16	1 μF/5 V	5%
Discrete, multilayer ceramic, 0805	C1,C2,C17,C18	0.1 μF/5 V	5%
Discrete, multilayer ceramic, 0402	C13	33 pF/5 V	5%
Discrete, multilayer ceramic, 0805	C14	27 pF/5 V	5%,
Inductors:	L2	6.8 nH, 0402	
	L3	22 nH, 0402	
	L1	1.8 nH, 0402	
Resistors:			
Discrete, metal oxide, 1206	R5, JP1, JP2	0 E/30 mW	5%
Discrete, metal oxide, 0805	R1	680 E/30 W	5%
Discrete, metal oxide, 0805	R7,R31,R43,R44,R45, R46,R48,R49,R50,R51, R57,R58,R59,R60,	560 E/30 W	5%
Discrete, metal oxide, 0805	R4,R9,R11,R30,R33, R34,R35	1 kΩ/30 mW	5%
Discrete, metal oxide, 0805	R19,R10,R12,R14,R18, R22,R23,	100 E/30 mW	5%,
Discrete, metal oxide, 0805	R16,R17	220 E/30 mW	5%
Discrete, metal oxide, 0805	R2,R3,R20,R21,R24, R25,R26R27,R36,R37, R38,R39,R40,R41,R42, R47,R56,R61,R62	10 kΩ/30 mW	5%
Discrete, metal oxide, 0805	R6,R8,R15	22 kΩ/30 W	5%,
Discrete, metal oxide, 0805	R32	33 kΩ/30 W	5%
Discrete, metal oxide, 0805	R28,R52,R53,R54,R55	22 E/30 mW	5%
Discrete, metal oxide, 0805	R29	100 kΩ/30 W	5%
XTAL 32 MHz	Y2	esr <60 E	
FUSE RESETABLE	F1,F2	0.5 A/30 V	
LED red, 1206	D7		
LED green, 1206	D8		
Tact switch, 6×6 mm	SW1 to SW20		

Table 9.25 Radiated power of PCB dipole antenna

Sl. No	Frequency (GHz)	Distance	Device mounting	Power radiated (dBm)	Power radiated (μW)
1	2.405480000	1 m	Horizontal	–45.26	29.78
	2.404480000		Vertical	–44.63	34.43
2	2.404408000	2 m	Horizontal	–68.43	0.143
	2.404624000		Vertical	–59.67	1.07
3	2.404528000	3 m	Horizontal	–68.63	0.137
	2.405524000		Vertical	–68.87	0.129

strategic locations. The location of tagged personnel/equipment/vehicles is determined in terms of those static routers. Each device communicates data in multihop to a remote computer at a control station in pit top.

9.4.2 Reduction in Fatal Accident

By implementing miner safety application, risk of fatal accidents will be greatly reduced. This will also reduce loss of life, financial loss of mine, as well as machine downtime.

RFID reader device is installed inside the cabs of mine vehicles. Ultra-long-range (ULR) RFID tags worn by miners are seen by reader units, and transmission signal triggers a light inside cab to warn driver to reduce his speed drastically. Reader unit consists of an RFID reader, a Wiegand interface card, latch relay, reset switch, and a flashing strobe; all housed in a PVC box and is placed inside vehicle cab. A quarter-wavelength whip antenna is mag-mounted to the roof of cab and cabled to reader. Power is supplied from standard battery of vehicle. All miners wear an RFID tag and when reader sees tag, transmission signal triggers the flashing strobe. Once pilot slowed down and passed miner(s) safely, he pushes reset button to stop flashing strobe.

9.4.3 Collision Prevention System

In mechanized mine environment, locos (vehicles transporting coal/mineral) may collide with each other, resulting often in serious injuries to drivers and sometimes loss of lives. For mines, this in turn also leads to downtime and lost productivity. By installing RFID collision prevention system, risk of accidents and machine downtime will be greatly reduced and productivity improved. In ZigBee-compliant RFID collision prevention system, each loco is fitted with two asset tags (one each at front and back) and an RFID reader unit, consisting of an ULR RFID reader, a Wiegand interface, antenna, and a timer relay. Reader unit is interfaced with loco controller to control speed (average, 40 km/h). When reader detects an asset tag signal from another loco in the vicinity at a range of approximately 50 m, power supply from power supply unit (PSU) to controller is reduced (50%), thus automatically reducing speed, at which loco is traveling. This in itself is also a warning to loco driver that another loco is approaching or in the same area. When RFID tag is no longer seen by reader, power supply is automatically increased (100%) and activates controller to run at full speed again.

9.4.4 Efficiency and Productivity Monitoring System

RFID technology and devices can be used effectively for monitoring and measuring productivity through an application that allows for identification of location and tracking of movement of a mine's fleet. By using this automatic (and accurate) means of collecting data over a period of time, it is possible to determine how efficiently and productively mine site is being managed. It is also an indication of whether mines' assets (fleet) are managed cost-effectively. In other words, to optimize productivity and asset management, mine management may decide that it needs to either increase or decrease number of vehicles and/or number of hours worked per shift. Costs can also be calculated more accurately and measured against productivity of a particular mine site. This will also improve output per man shift, which leads to efficient productivity.

At an opencast mine, each vehicle is fitted with an end device. Routers are mounted at specific locations around mine site to mark routes followed by vehicles. When a vehicle fitted with data collection unit moves past a "route marker," first and last transmission from tag is recorded in database of the unit. Data are downloaded and analyzed to determine each vehicle route. As all records are time stamped, it is also possible to determine time taken for each vehicle to complete the route. At present, system based on GPS technology (Karmel, 2007; Stauffer and Crystal, 2008) is popular for tracking and monitoring of vehicles in opencast mines. But it is costly and not specifically developed for opencast mine applications like analysis of shovel-dumper performance. Even application software of developed technologies is different from GPS system, which specifically incorporates module for optimizing shovel-dumper performance.

9.4.5 Monitoring Miners' Unsafe Practice and Warning System

Generally, underground miners are unknowingly going to the danger area, like, unsupported area where chances of roof fall are frequent, near blasting area, gaseous area, etc. These unsafe practices cause several accidents in underground mine. Therefore, a continuous monitoring and warning system must be implemented in the underground mines to avoid such common accidents. People working in the underground mine will be safe and an audiovisual warning will be provided while particular miner entering in an unsafe area. Thus, frequent accidents will be minimized.

Miners are continuously tracked using RFID tag (end device) provided to each worker. Unsafe practice of concerned person is analyzed by computerized software, and based on the information, mine management gives warning to defaulter. Further, an audiovisual warning system is incorporated in each tag to automatically warn particular miner entering into unsafe areas.

9.4.6 Message Communication System

Keypad-based message device is provided to the mine management for establishing message communication from underground to the surface computer. Message can be sent from any location in the wireless network. Four precoded messages can be sent to the surface computer from any of the end devices by pressing the switches attached with the end device.

9.5 System Installation Procedure

Installation procedures for underground mine with shaft entrance and with incline entrance (Fig. 9.66) and opencast mine (Fig. 9.67) are depicted. ZigBee transceivers/devices (end devices, routers, and coordinator) are housed in hard and tough structure to sustain tough mining conditions. Coordinator (C) is connected with computer (PC) to using RS232 cable in surface control room. Routers (R) are hanged on poles on surface, or hanged from roof/side of shaft (SH)/incline (I)/gallery (G) so that it should not obstruct movement of man and machinery. Distance between two routers (R) may vary (40–80 m) in straight gallery (G)/road (TR)/shaft (SH)/incline (I), where line-of-sight range is more. In case of small and labyrinth gallery (G), distance between routers (R) should be kept at lesser distance so that line-of-sight distance is maintained among routers (R), which should be placed in such a way that in case of disaster in a particular portion of mine, communication can be established by alternative routes automatically. Following above guidelines, routers (R) are placed on required portion of mine, where monitoring and tracking of miners and moveable equipment are needed. These routers (R) form wireless network with coordinator (C). Required environmental monitoring sensors are attached to particular routers, where gas monitoring is essential. End devices (E) are assigned to miners and moveable equipment. End devices (E) transmit signal to the respective routers (R), which receive signal and transmit to next routers (R) and subsequently data are transmitted to coordinator (C) through intermediate routers by multihop transmission mechanism. Finally, coordinator (C) sends data to computer (PC). Same data are processed, analyzed, and stored in computer (PC) using WSN software, which controls and commands all operations performed by total network.

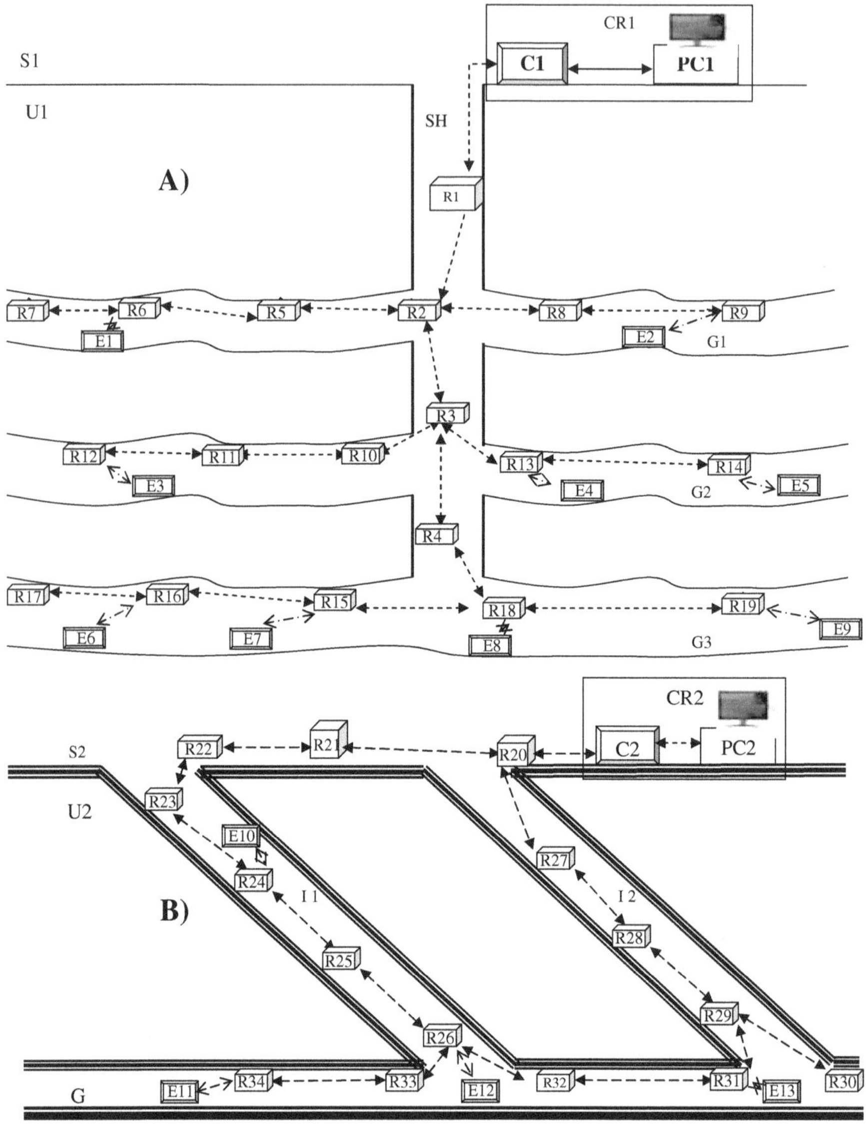

Fig. 9.66 Layout of ZigBee devices in an underground mine having: (**A**) shaft entrance; and (**B**) incline entrance

Wireless sensor network (Fig. 9.66A) in an underground mine having shaft entrance consists of a personnel computer (PC1), coordinator (C1), router (R1–R19) and end devices (E1–E9). PC1 is connected to coordinator (C1) using RS232 cable in surface (S1) control room (CR1). Routers (R1–R4) are wirelessly connected with coordinator (C1) at a distance 60 m apart in an underground (U1) shaft (SH). Routers (R5–R7) placed in left and routers (R8–R9) placed in right

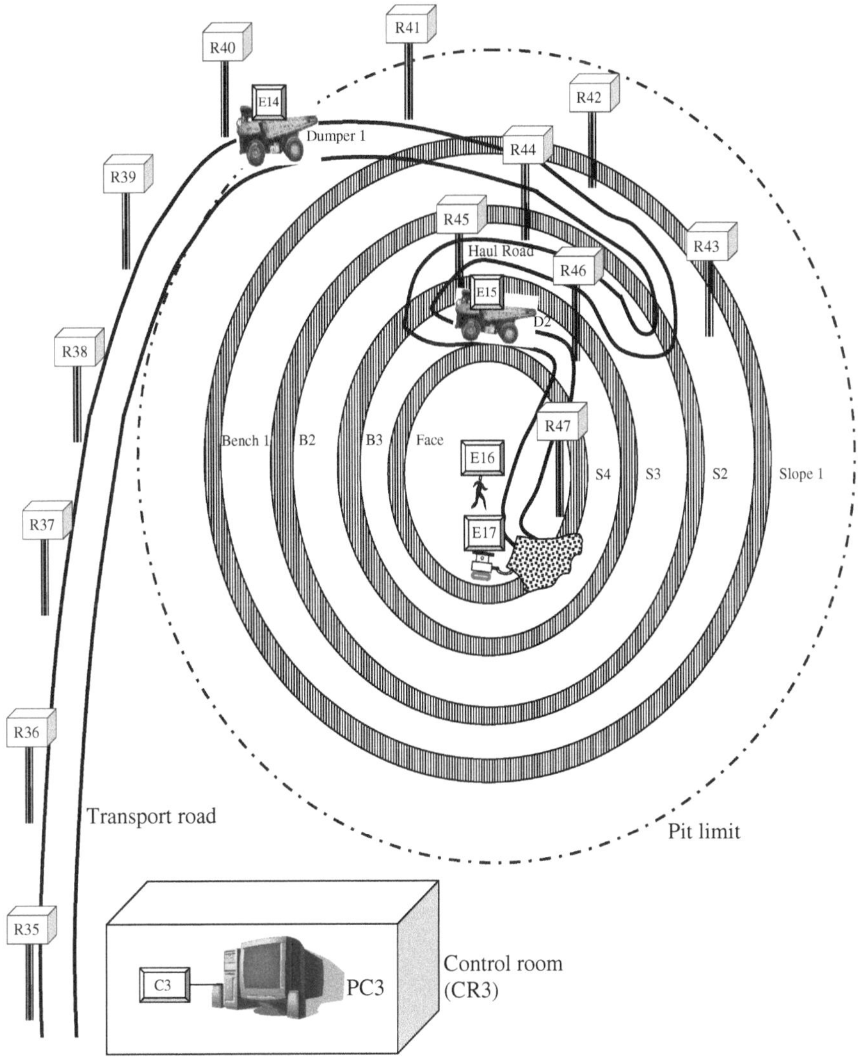

Fig. 9.67 Layout of ZigBee devices in an opencast mine

side of first galley (G1), at 50 m apart are wirelessly connected to router (R2). End devices (E1 and E2) attached with miners/moveable equipment are wirelessly communicated with routers (R6 and R9), respectively. Routers (R10–R12) placed in left and routers (R13 and R14) placed in right side of second galley (G2) are wirelessly connected to router (R3). End devices (E3, E4, and E5) attached with miners/moveable equipment is wirelessly communicated with routers (R12, R13, and R14), respectively. Routers (R15–R17) placed in left and routers (R18 and R19) placed in right side of third galley (G3) are wirelessly connected to router (R4). End devices (E6–E9) attached with miners/moveable equipment are wirelessly communicated with routers (R16, R15, R18, and R19), respectively.

Wireless sensor network in an underground mine having incline entrance (Fig. 9.66B) consists of a personnel computer (PC2), coordinator (C2), router (R20–R34), and end devices (E10–E13). PC2 is connected to coordinator (C2) using RS232 cable in surface (S2) control room (CR2). Routers (R20, R21, and R22) are wirelessly connected to coordinator (C2) at 80 m apart in surface (S). Routers (R23–R26) placed in intake incline (I1) at 50 m are wirelessly connected to router (R22). Routers (R27–R29) placed in return incline (I2) are wirelessly connected to router (R20). Routers (R30–R34) placed in gallery (G4) at 50 m are wirelessly connected to routers (R26 and R29). End device (E10–E13) fitted with miners/moveable equipment is wirelessly communicated with routers (R24, R34, R26, and R31), respectively.

Wireless sensor network in an opencast mine (Fig. 9.67) consists of a personnel computer (PC3), coordinator (C3), router (R35R47), and end devices (E14 and E17). PC3 is connected to coordinator (C3) using RS232 cable in surface control room (CR3). Routers (R35 to R47) placed along transport road and haul road of an opencast mine at an interval of around 80 m distance are wirelessly connected with coordinator (C3). End devices (E14 and E15) fitted with dumpers (D1 and D2) are wirelessly communicated with routers (R40 and R45), respectively.

9.6 Capabilities of the System

The main capabilities of the wireless information and safety system for mines are as follows:

- The system tracks and monitor miners and equipment in underground mine using ZigBee-enabled active RFID devices forming a dynamic wireless network among themselves and other static and mobile ZigBee devices placed at strategic locations.
- The system identifies the miners entering in underground mine to keep the track of the miners and maintains computerized attendance.
- The system monitors equipment locations and their operation to improve productivity and reduce fatal collision accident.
- The system locates and tracks the miners in case of disaster for speedy rescue operation.
- The system monitors miners' unsafe practice and provides warning to the respective miner.
- The system monitors environmental parameters in underground mine in real-time.
- The system sends coded message from underground mine to the surface computer.
- The system enables message communication from underground mine.

- The system automatically forms alternative network among the undisturbed and reachable routers in case of disaster in particular area in underground mine so that communication of the whole mine does not get disturbed in the whole mine.
- The system monitors movement of dumpers in opencast mine, which ultimately helps in optimum shovel-dumper performance and improving productivity.
- The system provides a low powered, intrinsically safe, easy to install, and cost-effective wireless information and safety mechanism for underground and surface mines.

9.7 Performance Analysis of RFID Devices in Underground Mines

ZigBee technology is a new technique for wireless sensor networking in underground mine. At present no significant field experiments have been carried out to evaluate the performance of ZigBee technology in underground mine. Therefore, a detailed study has been carried out to evaluate performance of ZigBee network in underground mine. Experiments were carried out for: (i) measuring maximum operating distance (ODmax) among coordinator (C), router (R), and end device (E); (ii) analyzing data communication capability of a router to remote coordinator; (iii) self-healing and self-forming capability of a mesh network; and (iv) data communication through an L-shaped or S-shaped routing path. Experiments were carried out in two ways, namely (i) experiments for evaluating packet delivery ratio and (ii) experiments for evaluating beacon rate.

9.7.1 Experiments for Evaluating Packet Delivery Ratio

Packet delivery ratio is defined as the number of data packet received by destination nodes divided by number of data packets transmitted by source nodes. In context of this definition, packet delivery ratio of ZigBee network topology can be defined as the ratio of actual number of packet received by coordinator to total number of packets sent by end device. Depending upon packet delivery ratio, efficiency of ZigBee network topology, position of routers, and maximum operating distance among end device, router, and coordinator were determined.

During experiments, various parameters were measured, namely (i) constant parameters and (ii) variable parameters. Packet injection rate, packet size, and total number of packets sent by end device were treated as constant parameters, while minimum operating distance, data communication from end device to a remote coordinator, and self-healing and self-forming capability of mesh network treated as variable parameters. Information regarding variable parameters and constant parameters are given in Table 9.26. The packet injection rate and total number of packets sent by end devices were kept at 300 ms and 100, respectively, for all the experiments. In order to evaluate the performance of ZigBee network, the experiments

Table 9.26 Experimental parameters and their specifications

Parameters	Specification
Constant parameter of ZigBee network	(i) Packet injection rate: 300 ms (ii) Total number of packets sent by end device: 100 (iii) User-defined packet size: 16 bytes
Variable parameter of ZigBee network	(i) Minimum operating distance (ii) Data communication from end device to a remote coordinator (iii) Self-healing and self-forming capability of mesh network (iv) Data communication through L-shaped or S-shaped routing path in underground mines

Table 9.27 Environmental parameters of the underground mine

Environmental parameters in mine	Value
Temperature	26–40°C
Humidity	80–90%
Air velocity	0.5–1.5 m/s
Methane gas concentration	0–1%

were divided again in two categories, namely (i) studies in underground mine scenario and (ii) studies in normal environment.

9.7.1.1 Studies in Underground Mine Scenario

Experiments were carried out at Bagdigi Colliery of Bharat Coking Coal Limited, Dhanbad, India. At the time of experiments in the underground mine, temperature and other parameters related to environment, i.e., humidity, air velocity, and methane gas concentration were measured. The value of temperature varied from 26 to 40°C, while methane gas concentration ranged from 0 to 1% (Table 9.27). Experiments in underground mines were carried out in different modes by changing the variable parameters while keeping the constant parameters fixed.

Experimental Procedure

Mode-1: Experiments were carried out twice to determine the maximum operating distance between router and coordinator to determine the optimum placement distance of routers.

(a) Distance between router and coordinator was increased gradually from 40 to 60 m and 100 m for finding out the maximum operating distance considering packet loss detection (Fig. 9.68).

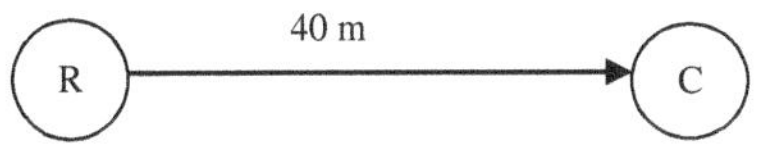

Fig. 9.68 Communication between router and coordinator

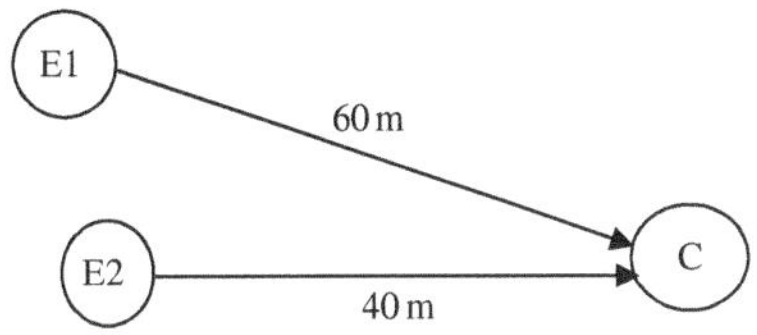

Fig. 9.69 Communication among two end devices and coordinator

(b) Communication among one coordinator and two routers was carried out simultaneously by rising the intermediate distance gradually from 60 to 80 m and 100 m (Fig. 9.69).

***Mode-2*:** Experiments were carried out to test data communication from end device to a remote coordinator in multihop topology by placing one router (Fig. 9.70b) and two routers, respectively (Fig. 9.70c), between coordinator and end device.

***Mode-3*:** Experiments were carried out to test the data transmission from a tag to a remote coordinator in multihop even if the routing path was not straight, i.e., the routing path was L-shaped (Fig. 9.71a) or S-shaped (Fig. 9.71b), which are very common in underground mines.

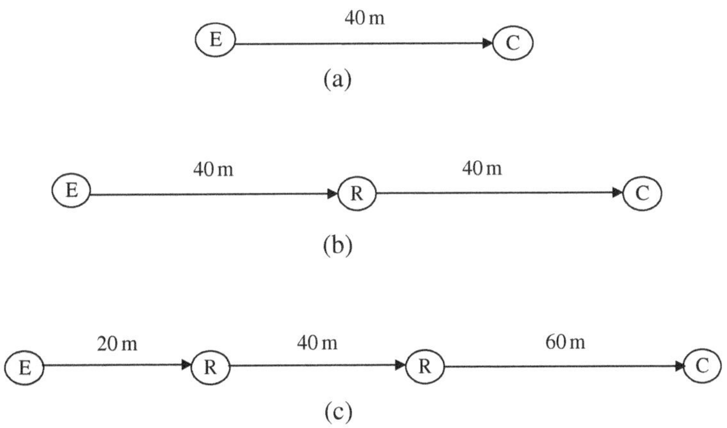

Fig. 9.70 Communication between end device and coordinator via router

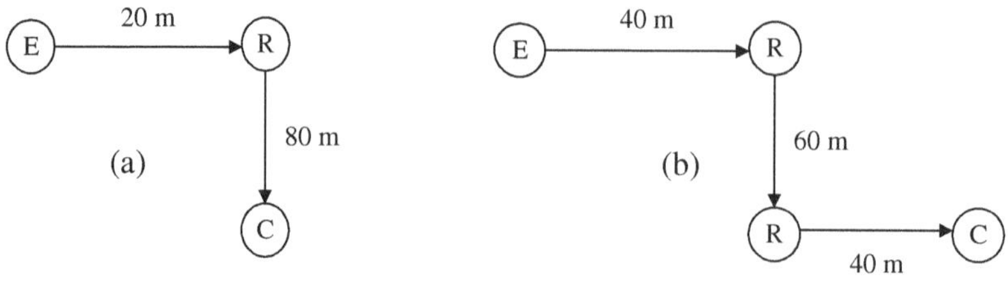

Fig. 9.71 Communication between end device and coordinator via router in bends and tunnels

Table 9.28 Variation of packet delivery ratio with increasing intermediate distance

Intermediate distance between C and E (m)	Packet received at coordinator	Packet delivery ratio (%)
40	99	99
60	99	99
80	95	95
100	86	86

Results

***Mode-1*:** (a) Number of packets received by coordinator from a single end device by gradually increasing intermediate distance is given in Table 9.28. The packet delivery ratio was 99% at an intermediate distance of 40 m and it was gradually decreased to 86% when the intermediate distance was 100 m.

(b) A single coordinator simultaneously received packets from two end devices at a time. At intermediate distance of 60 m, the packet delivery ratio was 97%, whereas at intermediate distance of 100 m, the packet delivery ratio was 92% (Table 9.29).

***Mode-2*:** Absence of routers between coordinator and end device is called single-hop system, whereas presence of one router and two routers will increase the hop number from single to two (2) and three (3), respectively. In case of single-hop system, the packet delivery ratio was 97%, while increasing hops number by two

Table 9.29 Variation of packet delivery ratio with increasing intermediate distance

Intermediate distance between C and E (m)	Packets received at the coordinator (from 2 devices)	Packet delivery ratio (%)
60	194	97
80	190	95
100	183	92

Table 9.30 Variation of packet delivery ratio with increasing number of hops

Number of hops between E and C	Packet received at the coordinator (from 2 devices)	Packet delivery ratio (%)
1	97	97
2	87	87
3	52	52

Table 9.31 Variation of packet delivery ratio with increasing number of hops

Number of hops between E and C	Packet delivery ratio (%)
1 router (Fig. 9.70b)	97
2 routers (Fig. 9.70c)	98

and three there was a decrement in packet delivery ratio to 87 and 52%, respectively (Table 9.30).

***Mode-3*:** Packets were transmitted by end device to coordinator through L-shaped or S-shaped topology. In L-shape topology, there was one router in between end device and coordinator, while in case of S-shape topology there were two routers. Packet delivery ratio was 97% for the case of L-shape topology and 98% for S-shape topology (Table 9.31).

Discussion

It was observed that increment of intermediate distance between coordinator and end device decreased packet delivery ratio. Therefore, it can be concluded that the intermediate distance between coordinator and end device is inversely proportional to the packet delivery ratio. Again it was observed that lesser number of hop routers between end device and coordinator increased the packet delivery ratio. Therefore, the relation between number of hops and packet delivery ratio is also inversely proportional. In underground mines near bends and tunnels, packets were transmitted in either L-shape topology or S-shape topology and these types of topologies did not hamper the rate of packet delivery ratio.

9.7.1.2 Studies in Normal Environment (Outside Mine)

Experiments were carried out in normal environment in different modes for comparative study with experiments in underground mine to analyze the performance of ZigBee network. Temperature, air velocity, and humidity were measured during experiments in normal environment. Temperature was varied from 30 to 32°C, whereas humidity and air velocities were varied from 24 to 38% and 1 to 3 m/s, respectively (Table 9.32).

Table 9.32 Environmental parameters and their values

Environmental parameters	Value
Temperature	30–32°C
Humidity	24–38%
Air velocity	1–3 m/s

Experimental Procedure

***Mode-1*:** Experiments were carried out twice to determine the maximum operating distance between coordinator and end device.

(a) Distance between router and coordinator was increased gradually from 40 to 60 m and 100 m (Fig. 9.68). When distance was increased more than 100 m, a significant packet loss was detected.
(b) Data communication to a single coordinator from two routers was carried out simultaneously by varying the intermediate distance steadily from 60 to 80 m and 100 m (Fig. 9.69).

***Mode-2*:** Experiment was carried out to test data communication from end devices to a remote coordinator in multihop by positioning one or two routers between coordinator and end device (Fig. 9.70).

***Mode-3*:** Experiment was carried out to test the data transmission from a tag to a remote coordinator in multihop even if the routing path was not straight, i.e., the routing path was L-shaped (Fig. 9.71a) or S-shaped (Fig. 9.71b).

***Mode-4*:** Experiment was carried out to study the automatic self-healing networking feature of ZigBee devices (Fig. 9.72).

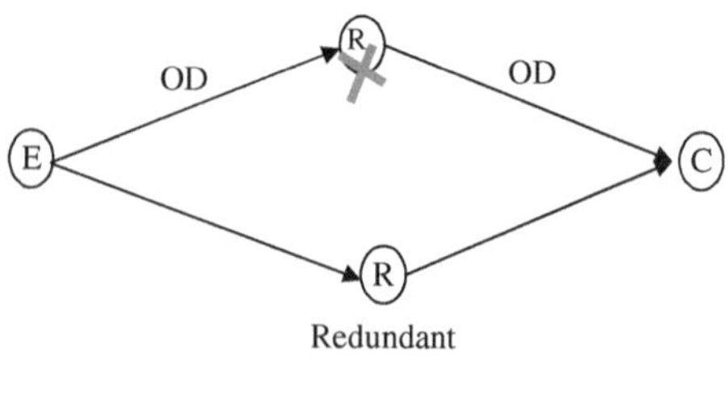

Fig. 9.72 Communication between end device and coordinator via redundant router

Table 9.33 Variation of packet delivery ratio with increasing intermediate distance

Intermediate distance between C and E (m)	Packet received at coordinator	Packet delivery ratio (%)
40	100	100
60	100	100
80	100	100
100	95	95

Results

***Mode-1*:** (a) Number of packets received by the coordinator from a single end device by gradually increasing intermediate distance are given in Table 9.33. Packet delivery ratio was decreased with the increment of intermediate distance from 40 to 100 m.

(b) The single coordinator simultaneously received packets from two end devices at a time. Packet delivery ratio decreased from 100 to 97% when intermediate distance increased from 60 to 100 m (Table 9.34).

***Mode-2*:** For a single-hop system the packet delivery ratio was 100%, whereas presence of one and two routers between coordinator and end device reduced the packet delivery ratio to 99 and 98%, respectively (Table 9.35).

***Mode-3*:** Packets were received by the coordinator through L-shape topology near bends in the gallery. Using one router, packet delivery ratio was 99%, whereas with two routers packet delivery ratio was 97% (Table 9.36).

Table 9.34 Variation of packet delivery ratio with increasing intermediate distance

Intermediate distance between C and E (m)	Packets received at the coordinator	Packet delivery ratio (%)
60	100	100
80	99	99
100	97	97

Table 9.35 Variation of packet delivery ratio with increasing number of hops

Number of hops between E and C	Packets received at the coordinator	Packet delivery ratio (%)
1	100	100
2	99	99
3	98	98

Table 9.36 Variation of packet delivery ratio with increasing number of hops

Number of hops between E and C	Packet delivery ratio (%)
1 router (Fig. 9.70b)	99
2 routers (Fig. 9.70c)	97

Table 9.37 Packet delivery ratio in multihop mode

Number of hops between E and C	Packets received at the coordinator	Packet delivery ratio (%)
E–R1–C	100	100
E–R2–C	97	97

***Mode-4*:** Packets were received through R1 from end device to coordinator when R2 was at switch-off state. If both the routers were kept at switch-on state then packets were transmitted through R1. Packets were transmitted to coordinator through R2 when R1 was kept at switch-off state. Therefore, it can be concluded that change in routing path is determined as soon as any route path failure is detected (Table 9.37).

Discussion

It can be concluded from the above results that the packet delivery ratio reduces if routers are introduced in between end device and coordinator. It has been also observed that the positions of routers between end device and coordinator affect the packet delivery ratio. It can be also concluded that packet delivery ratio is always better in normal environment (outside of mine) rather than in underground mines. Further, outside of mine the L-shaped or S-shaped topology, as described above, does not hamper the packet delivery ratio and redundant routing path is detected when conventional route path is failed.

9.7.2 Experiments for Evaluating Beacon Rate

Probability of getting beacon is defined as the number of beacons received by the destination nodes divided by the number of beacons transmitted by the source nodes, and depending upon its performance, ZigBee network can be analyzed. In order to achieve the tracking performance of ZigBee technology, the effect of beaconing rate of tags was evaluated. Keeping the beaconing rate at 2 and 5 seconds, experiments were carried out twice by increasing the number of tags from 1 to 10.

9.7.2.1 Experimental Procedure

***Mode-1*:** It is observed that for a single end device η (probability of getting beacon) at 2 second beacon interval was 0.90 while η at 5 second beacon interval was 0.95. Taking 50 packets in 1 second, the total number of beacons that was expected to reach the coordinator in 2 seconds was (50×0.9) 45. Similarly, taking 20 packets in 1 second, the number of beacons expected to reach at coordinator in 5 seconds was (20×0.95) 19.

***Mode-2*:** It is observed that for three end devices η at 2 second beacon interval was 0.87, while η at 5 second beacon interval was 0.91. The number of packets expected to reach the coordinator for 2 seconds was (50×0.87) 44. Again the number packets expected to reach in 5 seconds toward the coordinator was (20×0.91) 18.

9.7.2.2 Results

(a) For the first case, high beaconing rate (2 second), coordinator was getting 45 packets with (1.00–0.90) 0.1 beacon loss probability. At lower beaconing rate (5 second), coordinator was getting 19 packets with (1.00–0.95) 0.05 beacon loss probability.
(b) For the second case at high beaconing rate (2 second), coordinator was getting 44 packets with (1.00–0.87) 0.13 beacon loss probability. At lower beaconing rate (5 second), coordinator was getting 19 packets with (1.00–0.91) 0.09 beacon loss probability.

9.7.2.3 Performance Analysis

The probability of getting beacons in different scenarios is shown in Fig. 9.73. The graph shows that the probability of getting beacons decreases very slightly for high beaconing rate. Enhancement in beaconing rate increases chance of beacon loss due to congestion caused by multiple beacons in the system. As the beaconing rate decreases, the probability of getting beacons increases.

9.7.2.4 Discussion

Higher beacon loss probability does not imply low number of received beacons. In a specified time period, if the beaconing rate is high then total number of beacons received by the tag is high, even if some beacons get lost due to congestion. Beacon received from a tag provides the up-to-date location information of the tag to the system. Therefore, it is also expected that if the frequency of beacon is high that implies system is getting location updates frequently and that will eventually improve the tracking performance. The beaconing rate can also be adjusted

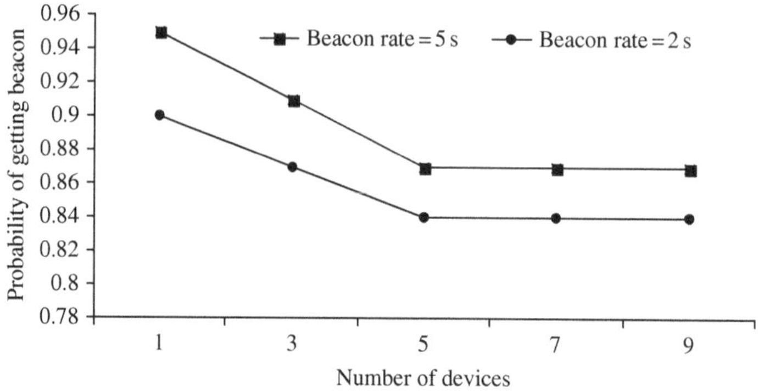

Fig. 9.73 Variation of probability of getting beacon with increasing number of devices

depending on the frequency of movement of tagged object and on the period within which the object is in the range of a particular router. High beaconing rate is required where movement of tag is very frequent. But high beacon rate increases the congestion in the system. On the other hand, in mine scenario, the walking speed (assume approximately 7.2 km/hr) of the miners is not very fast and they do not change their position very frequently. So, speed of a tagged miner is (7,200/3,600) 2 m/s. Let us assume that the distance between two routers in miners' tracking scenario is 40 m. So, a miner is staying under one router for [(40 m)/2 m/s] 20 seconds before switching to another router. In these 20 seconds it can emit 10 beacons with 2 second beaconing rate or 4 beacons with 5 second beaconing rate. But all those beacons will contain same location information. Therefore, keeping into account the beacon loss probability at the coordinator from a tag is approximately (10×0.9) 9 or (4×0.95) 4 (with 5 second beaconing rate). Even with 50% beacon loss probability, coordinator will receive at least 2 beacons at 5 second beaconing rate. Thus, further increase in number of beacons received at coordinator will not be able to produce any further improvement in tracking performance. Therefore, this analysis indicates that beacon loss has no significant effect on tracking performance.

9.8 Conclusions

Tracking and monitoring of miners and mining equipment are basic needs in underground mines. Introduction of RFID system in underground mines is a viable and cost-effective system. It can be concluded from the experiments of evaluating packet delivery ratio that it is better to keep the intermediate distance at 40–50 m between end device and coordinator. Some additional routers need to be placed in the multi-hop system to combat against sudden failure of some routers and to ensure reliable data communication. For the case of evaluating beacon rate, it also may be concluded that loss of beacon is not that much alarming as it appears intuitively. It may be further concluded that reduction in range of end device and optimum placement of routers in between end devices and coordinator can significantly improve tracking performance.

To overcome day-to-day problem faced by mine management, installation of WISS is a vital need for mining industry. With the help of central processing unit at pit top, it will be possible to keep track of miners and machines moving in underground. It will also be possible to keep record of time when respective miner is going inside the mine and coming back. Implementation of system will also help mine management to keep record of attendance and to identify persons who are delaying to start his scheduled duty and/or coming back early. In case of disaster, system will help in identifying trapped miner along with their location and numbers, and this will improve safety of miners.

References

Bandyopadhyay LK, Chaulya SK and Mishra PK (2008a) Development of tracking and monitoring system based on RFID tags for disaster management in underground mines. *Proceedings of 17th International Conference on Automation in Mining* (eds. Foganek B and Miskiewicz K), 21st World Mining Congress, Cracow, Poland.

Bandyopadhyay LK, Chaulya SK and Mishra PK (2008b) Wireless communication system for underground mines. *Proceedings of Conference on Emerging Trends in Mining and Allied Industries* (eds. Mishra MK and Sahu HB), National Institute of Technology, Rourkela, India, 2–3 February, pp. 571–577.

Bandyopadhyay LK, Chaulya SK, Mishra PK, Choure A and Baveja BM (2008c) Wireless information and safety system for mines. *Journal of Scientific & Industrial Research*, **68**: 107–117.

Bandyopadhyay LK, Chaulya SK, Mishra PK and Choure A (2008d) Wireless information and safety system for underground mines. *Proceedings of the International Union of Radio Science (URSI) General Assembly*, Chicago, USA, August 9–16, 2008.

Bandyopadhyay LK, Mishra PK, Chaulya SK, Choure A, Kumari S and Gautam S (2008e) RFID based wireless networking system for underground mines. *Proceedings of National Seminar on Frontiers in Electronics, Communication, Instrumentation and Information Technology* (ed. Kumar V), Indian School of Mines University, Dhanbad, 13–15 October, 2008, pp. 220–225.

Bandyopadhyay LK, Narayan A and Kumar A (2001) Mining automation – requirements and worldwide implementations. *Indian Mining Engineering*, **41**: 29–33.

Callaway E, Gorday P and Hester L (2002) Home networking with IEEE 802.15.4: a developing standard for low-rate wireless personal area networks. *IEEE Communication Magazine*, **40:** 70–77.

Darji AD (2004) *ZigBee-Wireless Sensor Networking for Automation*, M Tech Thesis, Electronic Systems Group, Electrical and Electronics Department, Indian Institute of Technology, Bombay, India.

Ergen SC (2004) *ZigBee/IEEE 802.15.4 Summary*, www.pages.cs.wisc.edu.

Held G (2001) *Data Over Wireless Networks Bluetooth, WAP and Wireless LANs*. Tata McGraw-Hill Publishing Company Limited, New Delhi, India.

Howitt I and Gutierrez JA (2003). IEEE 802.15.4 low rate-wireless personal area network coexistence issues. *Proceedings of IEEE Conference on Wireless Communications and Networking*, Los Angeles, USA, 20–23 March 2003, Vol. 3, 1481–1486.

Karmel CR (2007), Positioning system using packet radio to determine position and to obtain information relative to a position. *US Patent 7313401*, 25 December 2007.

Kinney P (2003) *ZigBee Technology: Wireless Control that Simply Works*. www.centronsolutions.com.

Kumar VV and Guha R (2001) Computerisation of Indian mining industry – quo Vadis? *Proceedings of Third Conference on Computer Application in Mineral Industry* (eds. Bandyopadhyay C and Sheorey PR), Oxford & IBH Publisher, New Delhi, India, pp. 15–22.

Kumari S, Jha V, Mahato B, Kumar B, Bandyopadhyay LK, Chaulya SK and Mishra PK (2008) Performance of RFID devices in underground mines. *Proceedings of National Seminar on Policies, Statutes and Legislation in Mines*, Central Institute of Mining and Fuel Research, Dhanbad, India, 20–21 December 2008, pp. 244–253.

Pahlavan K and Krishnamurthy P (2006) *Principles of Wireless Networks*. Prentice-Hall of India Private Limited, New Delhi, India.

Sen SK (2001) Current status of mechanization and automation in mineral industry – Indian scenario. *Proceedings of International Conference on Mechanization and Automation – The Future of Mineral Industry*, Mining, Geological and Metallurgical Institute of India, Kolkata, India, pp. 7–11.

Stauffer JHS and Crystal H (2008) GPS tool and equipment tracking system. *US Patent 2008030322*, 7 February 2008.

Stein G and Kibitzes K (2002) Concept for an architecture of a wireless building automation. *Proceedings of 6th IEEE Africon Conference*, George, South Africa, Vol. 1, pp. 139–142.

Chapter 10
Programming of RFID Devices

10.1 Introduction

Embedded programming is a device software which plays an integral part in the electronics device. Embedded software's principal role is not information technology, but rather the interaction with the physical world. It is written for machines that are not, first and foremost, computers. Embedded software is "built in" to the electronics in cars, telephones, audio equipment, robots, appliances, toys, security systems, pacemakers, televisions, digital watches, airplanes, missiles, process control, RFID devices, and monitoring systems, for example. Embedded software is usually written for special-purpose hardware. Application-specific embedded software has been developed for programming RFID devices to function as coordinator, router, end device, and various other purposes (Bandyopadhyay et al., 2007).

10.2 Function of RFID Devices

RFID is a wireless link to uniquely identify objects or people. It is sometimes called dedicated short-range communication (DSRC). RFID systems include electronic devices called transponders or tags, and reader to communicate with the tags. These systems communicate through radio signals that carry data either unidirectionally or bidirectionally. As shown in Fig. 10.1, when a transponder enters a read zone, its data are captured by the reader and can then be transferred through standard interfaces to a host computer, printer, or programmable logic controller for storage or action.

RFID is a means for storing and retrieving data through electromagnetic transmission to an RF-compatible integrated circuit. At present, it is viewed as a radical means of enhancing data-handling processes. An RF reader can read data emitted from RF tags. RF readers and tags use a defined radio frequency and protocol to transmit and receive data. RF tags are categorized as either passive or active.

L.K. Bandyopadhyay et al., *Wireless Communication in Underground Mines*,
DOI 10.1007/978-0-387-98165-9_10, © Springer Science+Business Media, LLC 2010

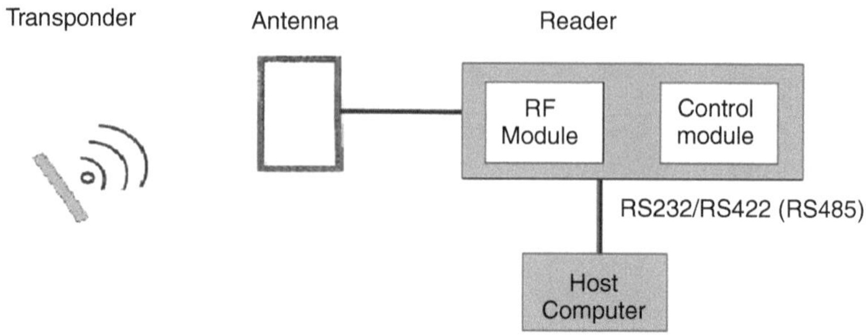

Fig. 10.1 Function of RFID device

Passive RF tags operate without a battery. They draw their power from the radio wave transmitted by reader and get energized. Then they reflect RF signal transmitted to them from a reader and add information by modulating the reflected signal. Their read ranges are very limited.

Active tags contain both a radio transceiver and a button-cell battery to power the transceiver. Since each tag has an onboard radio, active tags have larger ranges than passive tags. Active tags are ideally suited for the identification of high-unit-value products moving through a tough assembly process. They also offer the durability essential for permanent identification of captive product carriers.

All RF tags can be read despite extreme environmental factors such as snow, fog, ice, paint, and other visually and environmentally challenging conditions. Different companies manufacture RFID devices for various applications. For underground mining applications, CC2430-based RFID devices (Texas Instruments, USA) are used and programmed to form a dynamic wireless mesh network. Technical aspects of CC2430-based RFID device and programming details are elaborated subsequently.

10.3 CC2430 Chip

The CC2430 is a single-chip, IEEE 802.15.4-compliant, and ZigBee™ SOC (system on chip) RF transceiver with integrated microcontroller. It provides a highly integrated, flexible low-cost solution for applications using the worldwide unlicensed 2.4-GHz frequency ISM band. The CC2430ZDK (ZigBee development kit) is a powerful tool for developing complete ZigBee™ applications. The hardware contains an integrated PCB antenna the IEEE 802.15.4-compliant RF transceiver CC2430 with necessary support components, joystick, buttons, and LEDs that can be used for different purposes. The CC2430 is highly suited for systems where ultra-low power consumption is required. This is ensured by various operating modes. Short transition times between operating modes further ensure low power consumption. The key features of CC2430 chip are as follows (www.ti.com):

(a) RF layout

- 2.4-GHz IEEE 802.15.4-compliant RF transceiver (CC2420 radio core).
- Excellent receiver sensitivity and robustness to interferers.
- Very few external components. Only a single crystal needed for mesh network systems.
- RoHS-compliant 7 × 7 mm QLP48 package.

(b) Low power

- Low current consumption (RX: 27 mA, TX: 27 mA, microcontroller running at 32 MHz).
- Only 0.5 μA current consumption in power down mode, where external interrupts or the RTC can wake up the system.
- 0.3 μA current consumption in standby mode, where external interrupts can wake up the system.
- Very fast transition times from low-power modes to active mode enables ultra-low average power consumption in low duty cycle systems.
- Wide supply voltage range (2.0–3.6 V).

(c) Microcontroller

- High-performance and low-power 8051 microcontroller core.
- 32, 64, or 128 KB in-system programmable flash.
- 8 KB RAM, 4 KB with data retention in all power modes.
- Powerful DMA functionality.
- Watchdog timer.
- One IEEE 802.15.4 MAC timer, one general 16-bit timer, and two 8-bit timers.
- Hardware debug support.

(d) Peripherals

- CSMA/CA hardware support.
- Digital RSSI/LQI support.
- Battery monitor and temperature sensor.
- 12-bit ADC with up to eight inputs and configurable resolution.
- AES security coprocessor.
- Two powerful USARTs with support for several serial protocols.
- 21 general I/O pins, 2 with 20 mA sink/source capability.

(e) Development tools

- Powerful and flexible development tools available.

10.4 CC2430 Modules

Texas Instrument, USA manufactures different modules, namely evaluation board (CC2430EB), evaluation module (CC2430EM), and development board (CC2430DB) for programming and networking of RFID devices. CC2430EB includes a digital signal controller, RS-232 interface, user LEDs, user push-button switches, and various other components (Fig. 10.2). CC2430EM (Fig. 10.3) is used for receiving and transmitting data from routers/end devices. CC2430EB integrated with CC2430EM is used as a coordinator of a wireless network (Fig. 10.4). CC2430DB (Fig. 10.5) is used as a router/end device/sensor/messaging device. The basic components of CC2430EB are

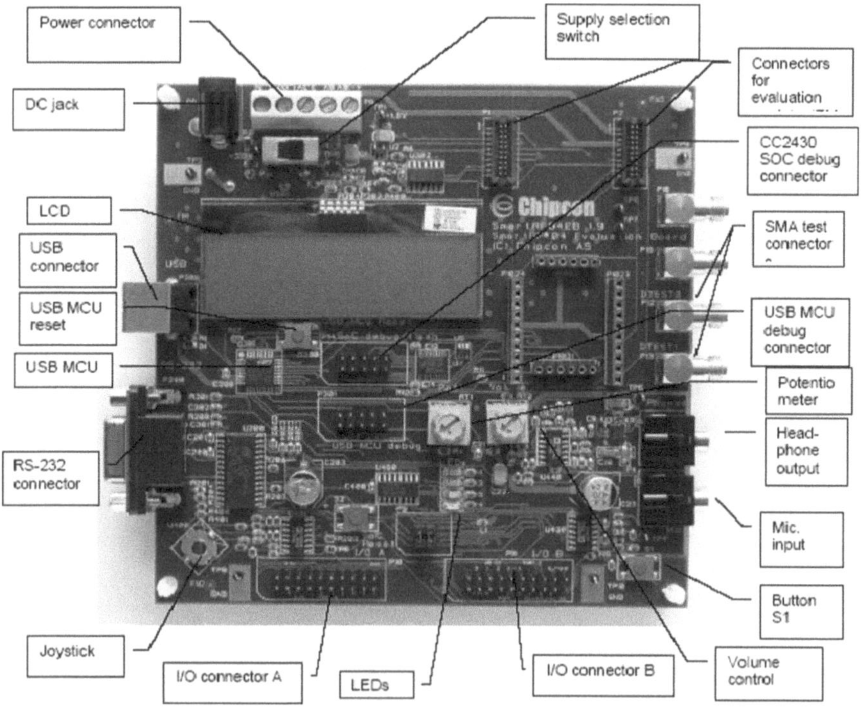

Fig. 10.2 CC2430EB evaluation board (www.ti.com)

(i) *USB Interface*: The USB interface is used as interface to a PC and for programming and debugging using the PC debugging tools and programmers. The CC2430EB can be bus-powered from the USB interface.
(ii) *RS-232 interface*: The RS-232 can be used by custom applications for communication with other devices. The RS-232 interface utilizes a voltage translation device so that the RS-232 port is compatible with bipolar RS-232 levels. Note

Fig. 10.3 CC2430EM evaluation module (www.ti.com)

Fig. 10.4 CC2430EB integrated with CC2430EM (www.ti.com)

that this RS-232 level converter contains a charge-pump power supply that generates electric noise.

(iii) *User interface*: The CC2430ZDK (ZigBee development kit) includes a joystick and a push button as user input devices, and four LEDs and a 2×16 character LCD display as user output devices. The display and user interface

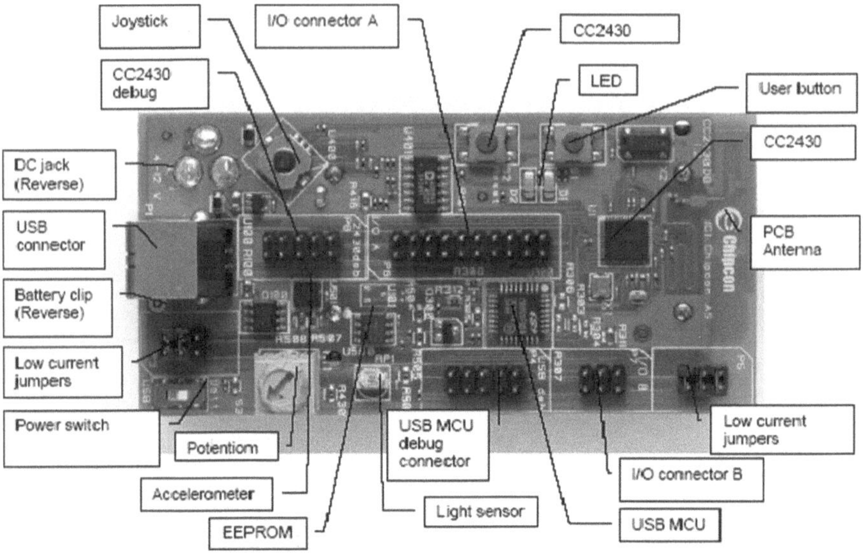

Fig. 10.5 CC2430DB development board (www.ti.com)

is controlled by the application example program loaded in the CC2430. The factory firmware uses the joystick, push button, and LCD display to implement a menu system used to control the packet error rate handling.

10.5 Programming of RFID Devices

RFID devices (coordinator, routers, and end devices) are programmed based on IEEE 802.15.4 mesh network. It uses a unified wireless mesh network infrastructure to locate, trace, and manage mobile assets. The core system component is a small, universal, battery-powered RF device, designed specifically for low-power, low-voltage RF applications in 2.4 GHz unlicensed ISM band. The devices can be programmed to act as end device and router that enables them to form an IEEE 802.15.4-based wireless mesh network (http://www.pages.cs.wise.edu).

10.5.1 Debugging Using USB Interface

The USB interface of CC2430DB is most commonly used to control CC2430 on-chip in-circuit emulator while developing embedded software. The USB interface supplies power to the board, so there is no need for additional DC power or batteries. The USB interface can also be used to program CC2430 in circuit using Chipcon programming software. To use the USB interface, simply connect the USB cable to the board and start the IAR Embedded Workbench software supplied with the kit. Follow the user guide for the IAR tools for instructions how to compile and download the code (http://iar.com).

10.5.2 Debugging with CC2430 Debug Connector

It is possible to use CC2430DB with the Chipcon packet sniffer or other application that requires USB interface for communication with PC and debug with an emulator simultaneously. In these cases, the CC2430 SOC debug connector can be used for connecting the ICE. The SmartRF04EB can be used as emulator interface with a cable from P14 "SOC debug flash" on SmartRF04EB to P8 "SOC debug connector" on CC2430DB. The jumpers between pin 1–2, 3–4, and 4–5 on header P5 must be removed.

10.6 Network Creation

A ZigBee system consists of several components. The most basic is the device. A device can be a full-function device (FFD), i.e., coordinator and router or reduced-function device (RFD), i.e., end device (Fig. 10.6). A network shall include at least one FFD, operating as the personal area network (PAN) coordinator. An FFD can talk to RFDs or FFDs, while an RFD can only talk to an FFD.

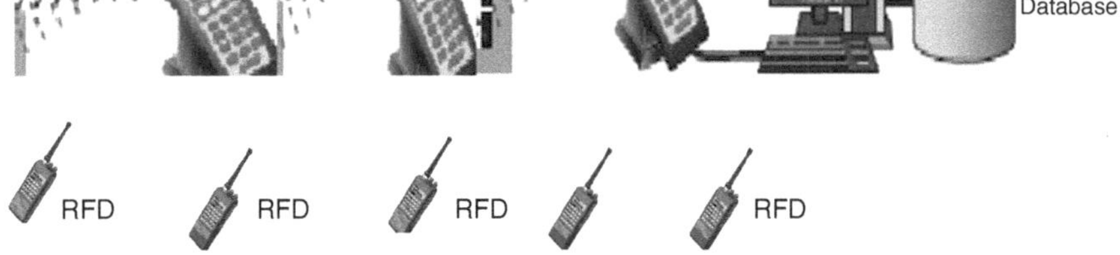

Fig. 10.6 Network structure of RFID devices

10.7 Network Topologies

ZigBee supports three types of topologies: star topology, peer-to-peer topology (mesh), and cluster tree (Fig. 10.7).

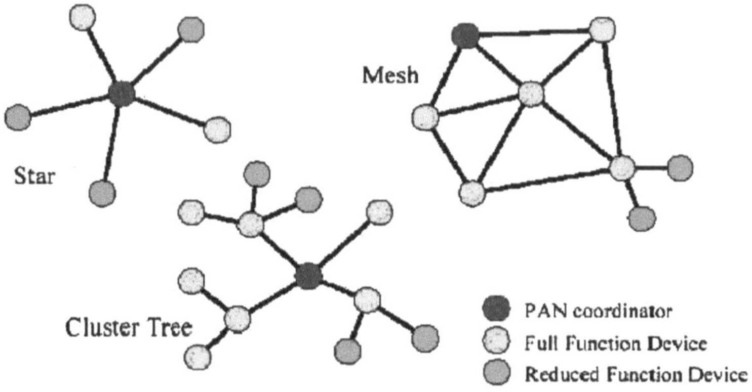

Fig. 10.7 Topology models

10.7.1 Star Topology

In the star topology, the communication is established between devices and a single central controller, called PAN coordinator. The PAN coordinator may be mainly powered while the devices will most likely be battery powered.

After an FFD is activated for the first time, it may establish its own network and become the PAN coordinator. Each star network chooses a PAN identifier, which is not currently used by any other network within the radio sphere of influence. This allows each star network to operate independently.

10.7.2 Peer-to-Peer Topology

In peer-to-peer topology, there is also one PAN coordinator. In contrast to star topology, any device can communicate with any other device as long as they are in range of one another. A peer-to-peer network can be ad hoc, self-organizing, and self-healing. Applications such as industrial control and monitoring, wireless sensor networks, asset and inventory tracking would benefit from such a topology. It also allows multiple hops to route messages from any device to any other device in the network. It can provide reliability by multipath routing.

10.7.3 Cluster-tree Topology

Cluster-tree network is a special case of a peer-to-peer network in which most devices are FFDs and an RFD may connect to a cluster-tree network as a leaf node at the end of a branch. Any of the FFD can act as a coordinator and provide synchronization services to other devices and coordinators. Only one of these coordinators however is the PAN coordinator. Cluster-tree network can be formed in two different ways, namely (i) dynamic routing network and (ii) static routing network. RFID devices based on CC2430 chip can be programmed to form dynamic and static networks using Z-stack library (http://ti.com) and TI-MAC library (http://focus.ti.com/docs/toolsw/folders/print/timac.html; http://www.ti-mac.com).

In Z-stack library, network routing is written based on C skip algorithm and address range is assigned to supporting device, which is the main drawback of this network. In C skip algorithm, as the level increases in the network, the number of accessibility of the address range assigned to the device decreases (Ahn et al., 2008). Finally, at some level, the accessibility of the address range is zero, which means no communication after that level.

10.8 Network Layers

The physical layer is the top level in the seven-level open system interconnection (OSI) model (Fig. 10.8) of computer networking as well as in five-layer TCP/IP reference model. It performs services requested by the data link layer.

Fig. 10.8 Network layers

OSI Model	
7	Application layer
6	Presentation layer
5	Session layer
4	Transport layer
3	Network layer
2	Data link layer
	a. LLC sublayer
	b. MAC sublayer
1	Physical layer

The physical layer is the most basic network layer, providing only means of transmitting raw bits rather than packets over a physical data link connecting network nodes. Neither packet headers nor trailers are consequently added to data by the physical layer. The bit stream may be grouped into code words or symbols, and converted to a physical signal, which is then transmitted through a physical transmission medium.

10.8.1 Application (Layer 7)

This layer supports application and end-user processes. Communication partners are identified, quality of service is identified, user authentication and privacy are considered, and any constraints on data syntax are identified. Everything at this layer is application specific. This layer provides application services for file transfers, e-mail, and other network software services. Telnet and FTP are applications that exist entirely in the application level. Tiered application architectures are part of this layer.

10.8.2 Presentation (Layer 6)

This layer provides independence from differences in data representation (e.g., encryption) by translating from application to network format, and vice versa. The presentation layer works to transform data into the form that the application layer can accept. This layer formats and encrypts data to be sent across a network, providing freedom from compatibility problems. It is sometimes called the syntax layer.

10.8.3 Session (Layer 5)

This layer establishes, manages, and terminates connections between applications. The session layer sets up, coordinates, and terminates conversations, exchanges, and dialogues between the applications at each end. It deals with session and connection coordination.

10.8.4 Transport (Layer 4)

This layer provides transparent transfer of data between end systems, or hosts, and is responsible for end-to-end error recovery and flow control. It ensures complete data transfer.

10.8.5 Network (Layer 3)

This layer provides switching and routing technologies, creating logical paths, known as virtual circuits, for transmitting data from node to node. Routing and forwarding are functions of this layer, as well as addressing, internetworking, error handling, congestion control, and packet sequencing.

10.8.6 Data Link (Layer 2)

At this layer, data packets are encoded and decoded into bits. It furnishes transmission protocol knowledge and management, and handles errors in the physical layer, flow control, and frame synchronization. The data link layer is divided into two sublayers: The media access control (MAC) layer and the logical link control (LLC) layer. The MAC sublayer controls how a computer on the network gains access to the data and permission to transmit it. The LLC layer controls frame synchronization, flow control, and error checking.

The MAC data communication protocol sublayer is a part of the data link layer specified in the seven-layer OSI model (layer 2). It provides addressing and channel access control mechanisms that make it possible for several terminals or network nodes to communicate within a multipoint network. A MAC protocol is not required in full-duplex point-to-point communication. In single-channel point-to-point communications full-duplex can be emulated. This emulation can be considered a MAC layer.

The MAC sublayer acts as an interface between the LLC sublayer and the network's physical layer and has a unique serial number assigned to each network. The MAC layer provides an addressing mechanism called physical address or MAC address.

10.8.7 Physical (Layer 1)

This layer conveys the bit stream – electrical impulse, light, or radio signal – through the network at the electrical and mechanical level. It provides the hardware means of sending and receiving data on a carrier, including defining cables, cards, and physical aspects. Fast Ethernet, RS-232, and ATM are protocols with physical layer components.

10.9 Wireless Application Protocol

Wireless application protocol (WAP) is an open international standard for applications that use wireless communication. Its principal application is to enable access to the Internet from a mobile phone or PDA. The physical signaling sublayer (PSS) is the portion of the physical layer that:

- Interfaces with the MAC, which is a part of the DLL; and
- Performs character encoding, transmission, reception, and decoding that further performs mandatory isolation functions as shown in Fig. 10.9.

An unassociated device shall initiate the association procedure by sending an associate request command to the coordinator of an existing PAN. If the association

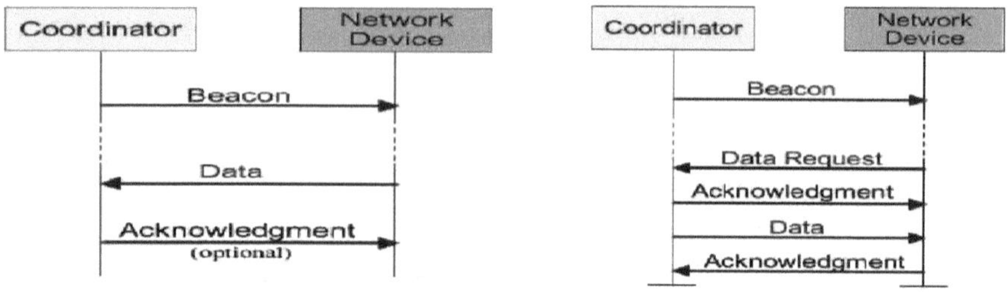

Fig. 10.9 Process of data sending by PSS

request command is received correctly, the coordinator shall send an acknowledgment. This acknowledgment however does not mean that the device has associated. The coordinator needs time to determine whether the current sources available on a PAN are sufficient to allow another device to associate. This decision should be made within a ResponseWaitTime symbols.

If already associated, remove all information. If sufficient resources are available, the coordinator shall allocate a short address to the device and generate an association response command containing the new address and a status indicating the successful association. If there are not enough resources, the coordinator shall generate an association response command containing a status indicating failure. This response is sent to the device using indirect transmission (pending, request).

On the other side, the device, after getting the acknowledgment frame, waits for the response for a ResponseWaitTime symbols. It either checks the beacons in the beacon network or extracts the association response command from the coordinator after a ResponseWaitTime symbols. On reception of association response command, the device shall send an acknowledgment. If the association is successful, store the address of the coordinator with which it has associated. General MAC frame format is shown in Fig. 10.10. The 16-bit data control field and acknowledgment frame format are given in Tables 10.1 and 10.2, respectively.

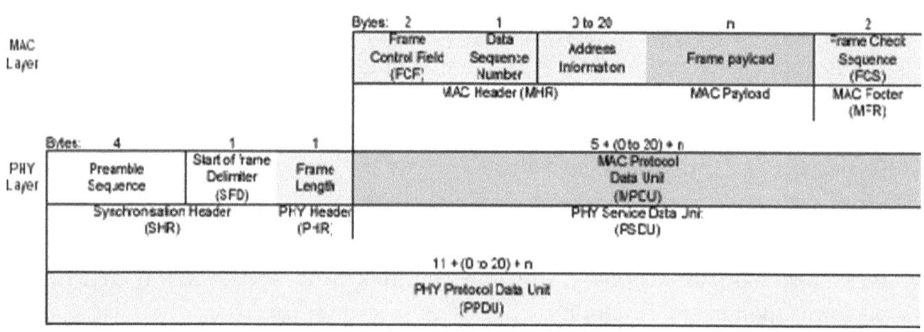

Fig. 10.10 General MAC frame format

Table 10.1 16-bit Data frame control field

Bits: 0–2	3	4	5	6	7–9	10–11	12–13	14–15
Frame type	Security enabled	Frame pending	Acknowledge request	Intra PAN	Reserved	Destination address mode	Reserved	Source address mode

Table 10.2 Acknowledge frame format

Bytes 4	1	1	2	1	2
Preamble sequence	Start of frame delimiter (SFD)	Frame length	Frame control field (FCF)	Data sequence number (DSN)	Frame check sequence (FCS)
Synchronization header (SHR)		PHY header (PHR)	MAC header (MHR)		MAC footer (MFR)

10.10 Medium Access Control

10.10.1 MAC Application

MAC application contains basic protocols and functions for the user to test and verify the 802.15.4 MAC operation. The application receives commands either automatically or from a key press and performs basic procedure such as:

- Powering up and setting device type: coordinator or device (with or without beacon enabled),
- Joining/forming the network,
- Association/disassociation,
- Scanning,
- Transmitting/receiving data, and
- Beacon support.

10.10.2 MAC Operation

The MAC performs the following procedures:

- Processing of operating system abstraction layer (OSAL) events and messages from API functions, received command frames, transmit frame status, and timers
- Action functions for MAC procedures such as scan, association, disassociation, network start, PAN Id, conflict, indirect data, polling, and portions of beacon processing
- PIB (PAN information base) management

- Building and parsing frames
- Radio management
- MAC timer management
- Frame reception and transmission including CSMA
- Frame acknowledgment handling
- Encryption and decryption
- Power management

OSAL is an operating system abstraction layer. It provides event handling, message passing, timers, memory, allocation, and other services.

HAL is a hardware abstraction layer. It provides a platform-independent interface to hardware services such as GPIOs, timers, and universal array register transfers (UARTs). 802.15.4 MAC operation is performed under C compiler.

10.10.3 MAC Application Design

MAC application (Fig. 10.11) contains four major components: main, initialization, system event processing, and callback processing.

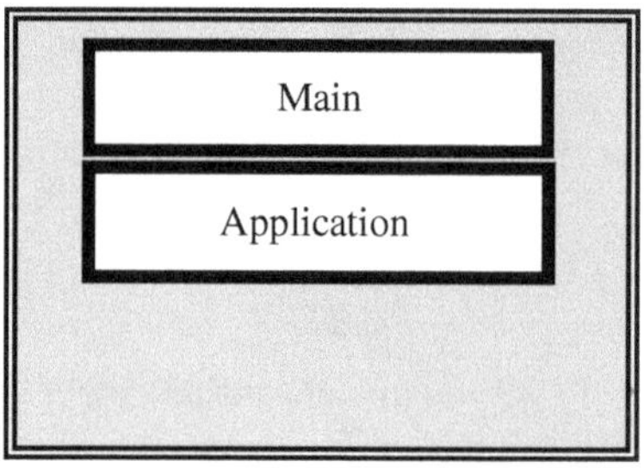

Fig. 10.11 Layers in MAC application design

10.10.3.1 Main

Main is the starting point of the system. This is where hardware, MAC, and HAL initialization are performed. This is also where OSAL is initialized and tasks are added in the scheduler. OSAL timer and HAL keys are configured and started. HAL drivers must be initialized at this stage for the system to work properly. Other initializations such as radio, interrupt, and MAC also need to be performed here. After all system initializations, OSAL_start_system is called to start the tasks.

10.10.3.2 Application

The application module contains files that perform tasks for which the application is designed. This is where the tasks can be initialized and added. Under MAC application, there are three main tasks: MAC, HAL, and application task. During the initializing process, Init function and event handling function from each task will be initialized and added in the OSAL task scheduler. Under the application module the functioning tasks are

Init: This is where the initialization of the application takes place and MAC is initialized as device or coordinator. The MAC is also reset here. Application is registered with the HAL keys so that the key callbacks can be properly handled.

Process event: This is where the application messages are handled. Application events that are distributed to the application through OSAL will be parsed and processed in two categories. All the messages will be retrieved one by one until there is nothing left, and then the routine will return the remaining events to the task manager. All other events will be handled one by one.

Callback event: This callback function sends MAC events to the application. The application must implement this function. A typical implementation of this function would allocate an OSAL message, copy the event parameters to the message, and send the message to the application's OSAL event handler. This function may be executed from task or interrupt context and therefore must be reentrant.

Coordinator startup: This routine will set up the device as a PAN coordinator. As the coordinator, MAC_ON status is true in order for the coordinator to receive incoming messages. Depending on the beacon order and super frame order, coordinator will be set up with beacon enabled or not.

Device startup: This routine will set up the device as a device. Depending on the beacon order and super frame order, coordinator will be set up with beacon enabled or not.

10.10.4 Programming with TI-MAC Library

10.10.4.1 Static Routing Network Platform

Static routing network is a fix network according to the site or map. If the network structure is changed according to site-specific conditions, then program of all the devices has to be changed, because every fix device sends the data to his predefined parent and child devices as shown in Fig. 10.12. Further, in case of failure of one router, the subsequent network formed with the particular router will not function. This is the main drawback of the static network.

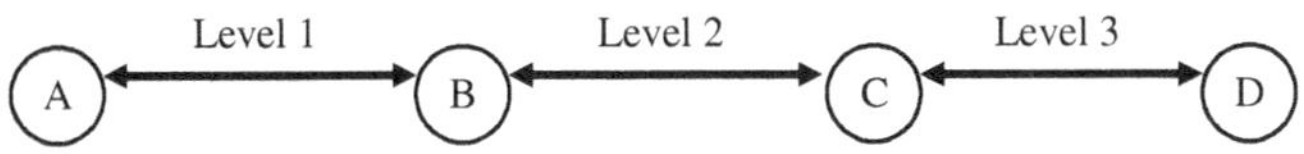

Fig. 10.12 Static routing network

To program the RFID devices for static routing network, first finalize the area network and accordingly design the network branches. Assign all the address levels to the coordinator and each router as leveled in Fig. 10.13, which is static, i.e., once we have assigned, the address of the devices cannot be changed. The address of each end device will change dynamically, according to the accessing router level.

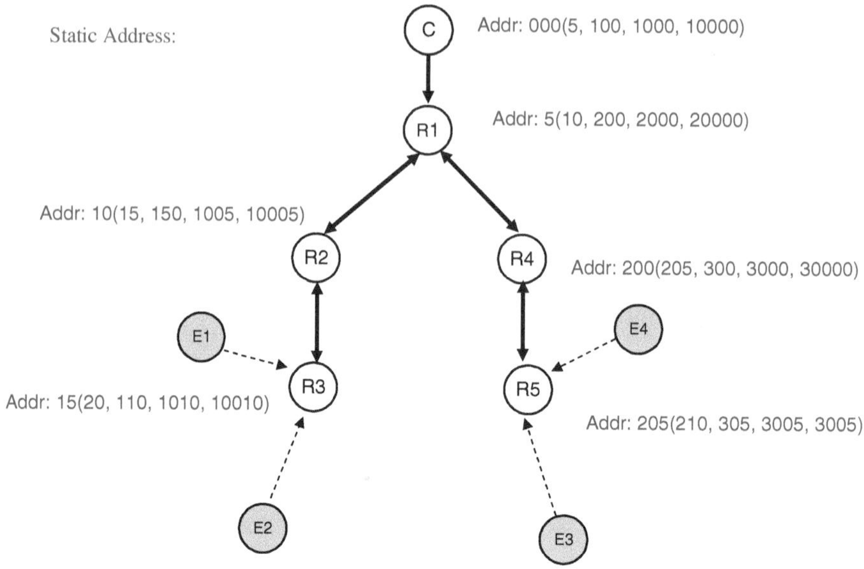

Fig. 10.13 Static routing network platform

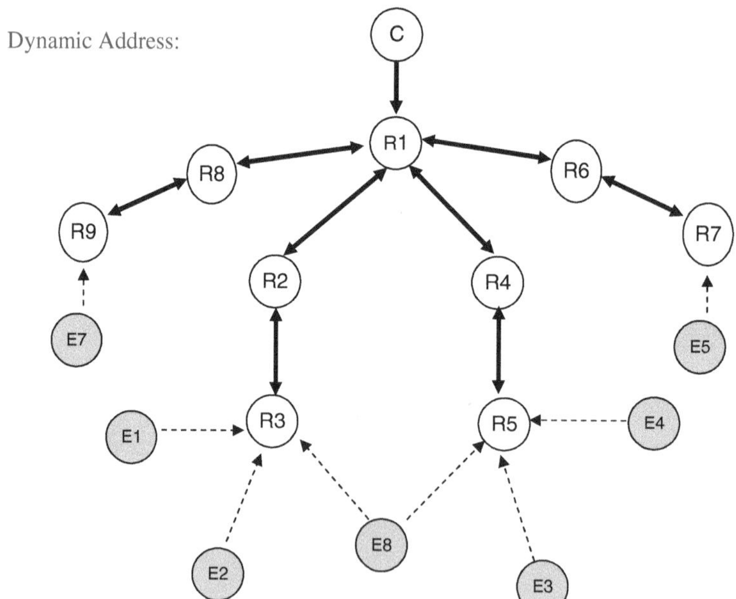

Fig. 10.14 Dynamic routing network platform

10.10.4.2 Dynamic Routing Network Platform

Dynamic routing network is an auto-configured, self-organizing, self-healing, and a multipath routing network (Gang et al., 2006; Lee et al., 2006; Yen and Tsai, 2008; http://reserach.microsoft.com/mesh). This routing network is configured according to its neighboring device (Fig. 10.14). If one device is lost/damaged inside the dynamic routing network, the remaining network will be automatically configured with the neighboring RFID devices within the range according to the topology structure. In this routing network, every device has a unique program. The network structure can be changed according to the site-specific conditions without changing the programs of RFID devices. Any type of area network and branches can be developed with the dynamic routing network.

10.11 Programming and Functionality of RFID Devices

10.11.1 Wireless Network Program

Wireless network program (WNP) has been developed to use RFID devices as coordinator, router, and end device for tracking of miners and moveable equipment in both opencast and underground mines. The WNP is used to form a dynamic wireless mesh network among the RFID devices to be placed at strategic locations of a mine. Dynamic routing network is an auto-configured, self-organizing, self-healing, and multipath routing network. This routing network is configured according to their neighboring device. If one device is lost or damaged inside a dynamic routing network, remaining network will be automatically configured according to topologies' structure. In this routing network, every device has a unique program. When network structure according to site is changed, devices automatically form wireless mesh network without changing device positions. Further, when any device fails, then remaining devices automatically form wireless mesh network with the nearest devices. Moreover, while programming devices for dynamic routing network, there is no need to declare parent and child devices of a particular device, it is assigned dynamically, thus overcomes the limitations of static routing networking.

The WNP is especially designed for programming RFID devices to perform as: (i) coordinator, (ii) router, (iii) end device, (iv) proximity warning device, (v) router with sensor device, (vi) end device with data logger, and (vi) messaging device.

10.11.2 Function of Different Devices

The same RFID device can be used for various purposes by changing hardware-specific embedded software, i.e., WNP. The function of various RFID devices is described below.

(i) *Coordinator:* It is the most capable device. It bridges to other network, and forms the root of the network tree. It is able to store information about network. It is a FFD. It is directly connected to RS-232 serial port of a computer where data interface application software is loaded.

(ii) *Router:* It is also a FFD. It forwards data packets along networks. A router is connected to at least two networks through mesh network. Routers are located at gateways, the places where two or more networks connect. Routers use headers and forwarding tables to determine the best path for forwarding packets, and they use protocols such as internet control message protocol (ICMP) to communicate with each other and configure the best route between any two devices. A router may create or maintain a table of the available routes and their conditions, and use this information along with distance and cost algorithms to determine the best route for a given packet. Typically, a packet may travel through a number of network points with routers before arriving at its destination. Routing is a function associated with network layer in a standard model of network programming, OSI mode.

(iii) *End device:* It is a RFD. It sends data to the coordinator or routers. It communicates to router and coordinator, but not to the other end devices. Basically it is a leaf node of a network tree. The device is assigned to miner or vehicle for performing tracking operation in an underground mine.

(iv) *Proximity warning device:* The proximity device is used to prevent vehicle-to-vehicle collisions and vehicle-to-man collisions. As per the current practice, it is very difficult to control vehicles' accidents. Generally, the vehicles are running in normal zigzag route in underground mine, so in case of reverse or bend there is no communication to other nearest vehicle, which results in an accident. It is a device to transmit signal, when it receives RF signal from other device and then generates alert message or beep.

(v) *Router with sensor device:* This device measures physical parameters and converts them into an analog signal that can be read by an observer or by an instrument. Many types of sensors are used like temperature sensor, barometric pressure sensor, humidity sensor, air-velocity sensor, gas sensor (oxygen, methane, carbon monoxide, and others), and routine condition-monitoring sensors for machines. Any of these sensors is connected to an RFID device, and device sends data to co-coordinator via routers. A device can use multiple sensors with different sensor identifications.

(vi) *End device with data logger:* This device is attached with a required sensor for monitoring particular parameter along gallery and working faces in an underground mine. The device is carried by the respective supervisor and moved along the desired path to continuously measure the particular parameter using the attached sensor. The device continuously monitors the parameter and stores data in the microcontroller. After completing monitoring, stored data are downloaded in a computer using RS-232 interface.

(vii) *Messaging device:* This device is used for sending or receiving text messages. This is a type of end device having keypad facility for sending any text message to another messaging device. Text message can also be sent from surface computer to any messaging device located within the wireless network.

10.11.3 Programming Procedure

WNP has been developed for programming RFID devices consisting of 8051 microcontroller chip manufactured by Texas Instrument, USA (http://.ti.com). The program is compiled and debugged in an IAR Embedded Workbench (www.iar.com) and TI-make Library using "C" language. Programming procedures along with program codes for different devices are given in Appendix A for different devices, namely (i) coordinator, (ii) router, (iii) proximity device, (iv) end device, (v) router with sensor, (vi) end device with data logger, and (vii) messaging device.

10.12 Conclusions

WNP has been developed to form dynamic wireless mesh network. Dynamic routing network is an auto-configuring, self-organizing, self-healing, and a multipath routing network. This routing network is configured according to its neighboring device. If one device is lost/damaged inside the dynamic routing network, the remaining network will be automatically configured according to the topologies' structure. Thus, WNP solves the redundancy problems in a large network.

References

Ahn S, Cho J and An S (2008) Slotted beacon scheduling using ZigBee Cskip mechanism. *Proceedings of Second International Conference on Sensor Technologies and Applications*, Cap Esterel, France, 25–31 August 2008, pp. 103–108.

Bandyopadhyay LK, Mishra PK, Chaulya SK, Mahato B and Kumar M (2007) Wireless sensor network in underground mines. *Proceedings of National Seminar on Advancement in Electronics & Communication Technology*, Bhopal, India, 2–3 December 2007.

Gang D, Sahinoglu Z, Orlik P, Jinyun Z and Bhargava B (2006) Tree-based data broadcast in IEEE 802.15.4 and ZigBee networks. *IEEE Transactions on Mobile Computing*, **5**(11): 1561–1574.

Lee KK, Kim SH, Choi YS and Park HS (2006) A mesh routing protocol using cluster label in the ZigBee network. *Proceedings of IEEE International Conference on Mobile Adhoc and Sensor Systems*, Washington, USA, 23–25 October 2006, pp. 801–806.

Yen LH and Tsai WT (2008) Flexible address configurations for tree-based ZigBee/IEEE 802.15.4 wireless networks. *Proceedings of 22nd International Conference on Advanced Information Networking and Applications*, Okinawa, Japan, 25–28 March 2008, pp. 395–402.

Websites

http://research.microsoft.com/mesh/ (Self-organizing neighborhood wireless mesh networks)
http://ti.com
http://iar.com
http://focus.ti.com/docs/toolsw/ folders/print/timac.html.
http://www.pages.cs.wise.edu, Ergen SC, ZigBee/IEEE 802.15.4 summary, 2004
http://www.ti-mac.com

Chapter 11
Tracking and Monitoring Software

11.1 Introduction

An application software has been developed to receive, process, analyze, and store data received from various end devices through routers and coordinator. The developed application software is named as "Tracking and Monitoring Software" (TMS). The software is installed in the surface computer placed in a control room and physically connected with the coordinator of a wireless network using RS232 cable. The software provides graphical presentation of miners and equipment locations and path. It is also helpful for generating miner's attendance report as daily, weekly, monthly, yearly, and actual duty hour basis. Safety of miner is another aspect, which is covered by this software, which is done by sending warning signals for each risky situation. The software also helps in establishing text message communication, online monitoring of environmental parameters, identifying trapped miners in case of disaster, and various other aspects.

11.2 Application Software

TMS software is developed in Visual Basic (VB) under windows as front-end tool and SQL Server as back-end support. VB is an object-oriented software package; therefore, various objects available in VB are used. Moreover, few functions and classes are designed in VB to integrate software. For reporting very powerful and extensive software, Crystal Report is used. The software is especially designed for different main purposes as described below.

(i) *Tracking of miners and vehicles*: This module allows continuously tracking miners' and mining equipment inside an underground mine in real time. Locations of miners are detected for each router and a complete miner path chart is drawn. Software keeps record of time when respective miner is going inside mine and coming back. It also helps mine management to keep record of attendance. Ultimately, this module helps in preparing computerized reporting on actual duty hours of each miner.

L.K. Bandyopadhyay et al., *Wireless Communication in Underground Mines*,
DOI 10.1007/978-0-387-98165-9_11, © Springer Science+Business Media, LLC 2010

(ii) *Route tracking in opencast mines*: This module allows tracking a dumper movement in opencast mine. Locations of dumper are detected for each router and a complete dumpers' path chart is drawn. Software keeps record of time when respective vehicle is going inside mine and coming back. It also helps mine management to keep record of vehicle operation duration. Ultimately, this module helps in preparing computerized reporting on actual duty hours of each dumper. This module can also be used for optimizing shovel-dumper performance and deploying optimum number of dumpers for a particular opencast mine.

(iii) *Preventing fatal accidents*: This module allows preventing fatal accident in underground mine by transmitting warning signal to miners and loco drivers. Software detects position and path of miner and moving loco, and automatically warns miner and loco driver when miner and loco are approaching nearer in same path.

(iv) *Preventing vehicle collisions*: This module allows monitoring locos' position inside an underground mine and transmitting warning signal to vehicle drivers approaching nearer. Software detects position and path of moving vehicle to and fro and then automatically warns driver after sensing vehicle reader tag, if both vehicles are in same path and approaching nearer.

(v) *Environmental monitoring*: This module allows wirelessly monitoring mine's environmental conditions using appropriate sensors. Software automatically collects data through coordinator of a router to which a sensor is coupled. Online environmental data and graph can be seen on monitor. System also gives warning when concentration of a particular gas reaches its permissible limit.

(vi) *Monitoring miner's unsafe practice*: This module monitors miners' unsafe practices and generates warning. System automatically detects router, which is placed in unsafe area. When any miner approaches near unsafe area router, then system warns the particular miner by audiovisual alarm and system automatically starts beeping at surface computer. This can help in ensuring miners' safety and generates a log file for keeping record with a particular date and time of the unsafe event.

(vii) *Sending alert message*: This module generates alarm messages on user-defined conditions. Different coded messages can also be sent to respective miner as and when required.

(viii) *Text message communication*: This module is utilized for text message communication among miners as well as with surface computer. Any number of text messages can be sent using keypad on requirement.

11.3 Operating Mode

Software operates online and off-line. In online (real-time) mode, central processing unit (CPU) of a computer is connected to coordinator by a serial port, and reads periodically tag's data through various routers by multi-hoping technique.

Coordinator receives signal from different routers, which are placed in strategic locations. Active RFID devices are attached to underground miners. Each device can transmit/receive signal to/from neighboring devices. Devices with ZigBee-complaint network interfaces can autonomously form network among themselves and with other static devices. The locations of tagged personnel are displayed numerically or graphically on the system and data are automatically saved in a database file. In this mode of operation, location of miner is detected for each router and the software draws a complete miner path chart as per request. It also helps mine management to keep record of attendance. Ultimately, this software helps in preparing computerized reporting on actual duty hours of each miner.

In off-line mode, one can display stored data by simply selecting file of required date. Graphical location of particular miner can also be displayed on screen. In both operating modes, software supports hard copy printing of data (numerical or graphical, as desired). Administrator can change software configurations, like working area of mine plan, unit of scale, router locations, router status, and different essential parameters. Enough security has been maintained using password.

11.4 Flowchart and Algorithm

The layout of TMS software is depicted in Fig. 11.1. Once the software is opened, different module of the software appears on the screen as shown in Fig. 11.1. Particular module is selected based on the requirement. Under each module, various functions are available for different uses. Figure 11.2 illustrates the data flow diagram of the software. It indicates flow of data, conditions applied at particular point, and various operations applied in the database, like retrieve, insert, delete, modify. Algorithm for different modules of the software is given in Appendix B.

11.5 Installation and Operation Procedures

TMS software operates in windows environment. Minimum requirements for running TMS software are (i) hardware requirements include CPU – Pentium IV, HDD – 40 GB (free), RAM – 512 MB, color monitor resolution – 1024 × 768, 32-bit color quality, one free serial communication port, and serial data cable; and (ii) software requirements include operating system – Windows NT/XP/2000/2003, SQL server 2000, and Crystal Report 8.5. The various steps for installation and operation of the software are described below.

11.5.1 Installation Procedures

Step 1: Set the whole system as shown in Fig. 11.3. The system includes server, coordinator, routers, and end devices. All the data are saved in server

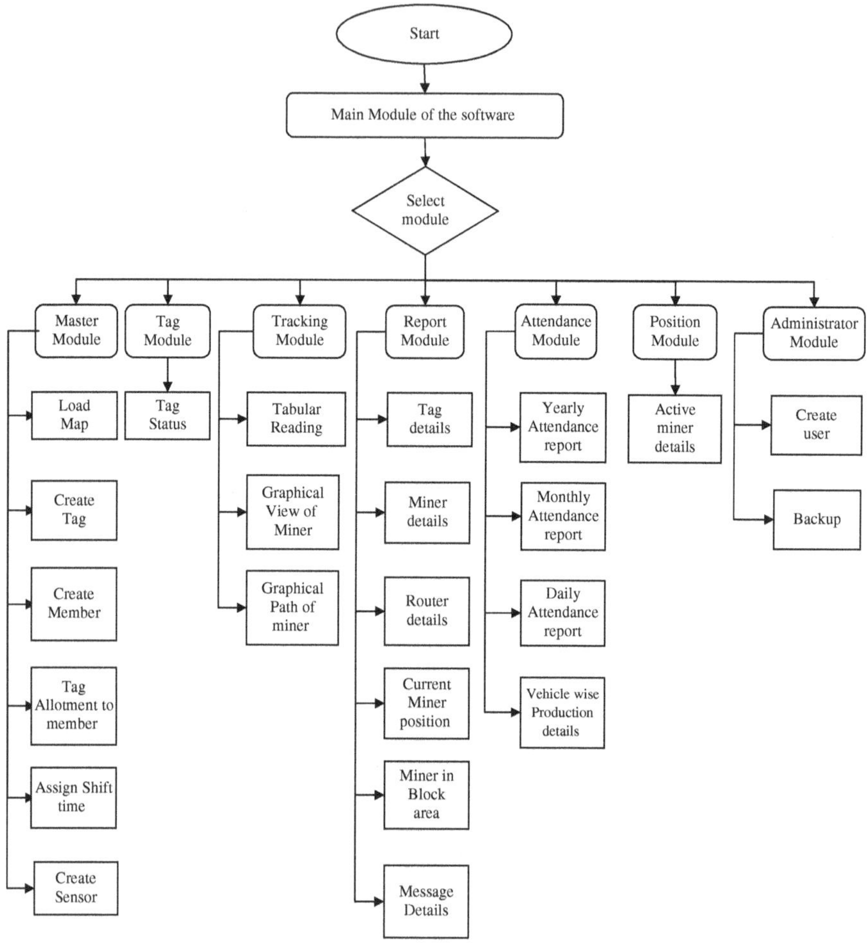

Fig. 11.1 Layout of TMS software

and is retrieved as and when required. Coordinator is directly connected to server with RS-232 port. Active routers are placed in static locations in underground mine and end devices are attached with the miners and moving vehicles. All routers and end devices are wirelessly connected and send data to coordinator. Coordinator transfers data through RS-232 port to the TMS software.

Step 2: Install the required software (Crystal Report and SQL Server), and create database for storing and retrieving information by giving a name such as MTWSN.

Step 3: Connect coordinator through RS-232 port with TMS software via cable as shown in Fig. 11.3. Routers are placed at different locations in under-

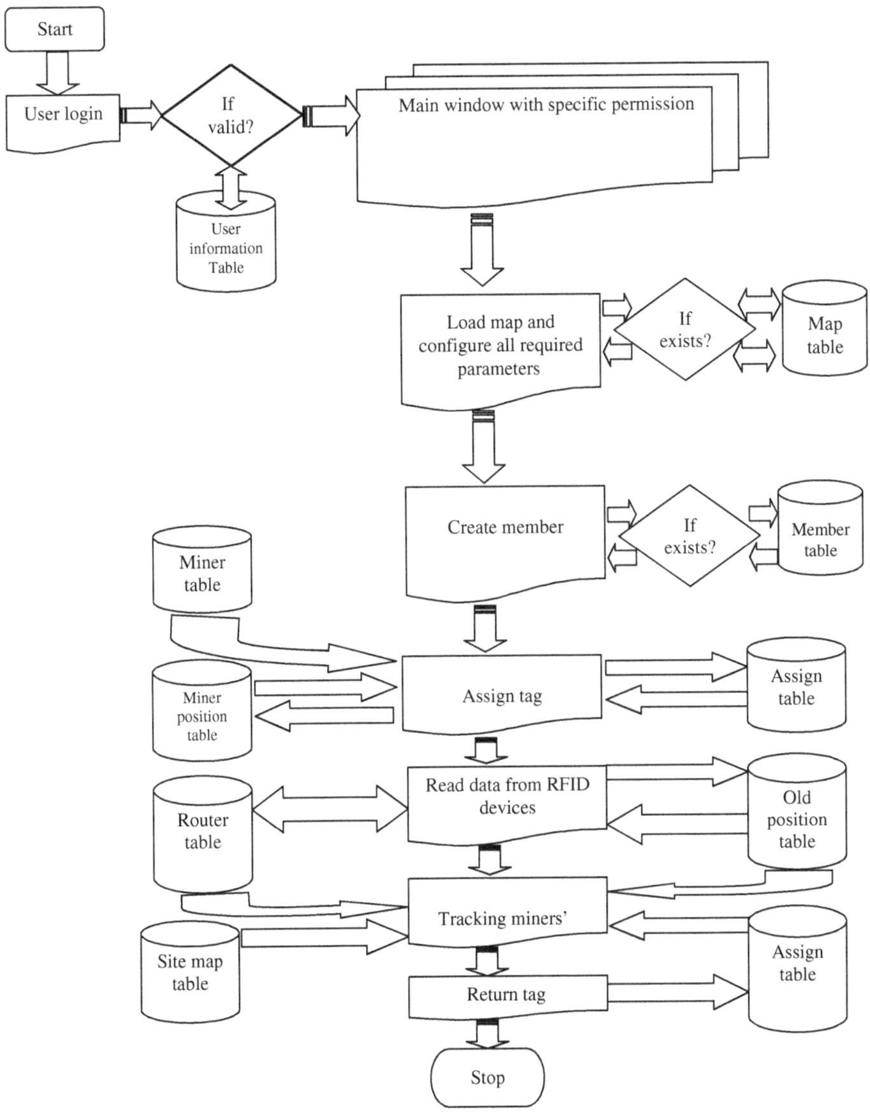

Fig. 11.2 Data flow diagram of TMS software

ground mines at an interval of 60–80 m distance depending on line-of-sight condition and size of open area. Switch on coordinator and routers after placing them in required locations.

Step 4: On starting the software, the first window, which comes into display, is shown in Fig.11.4. Next, user login page is displayed as shown in Fig. 11.5. Then the main window of the software comes into view.

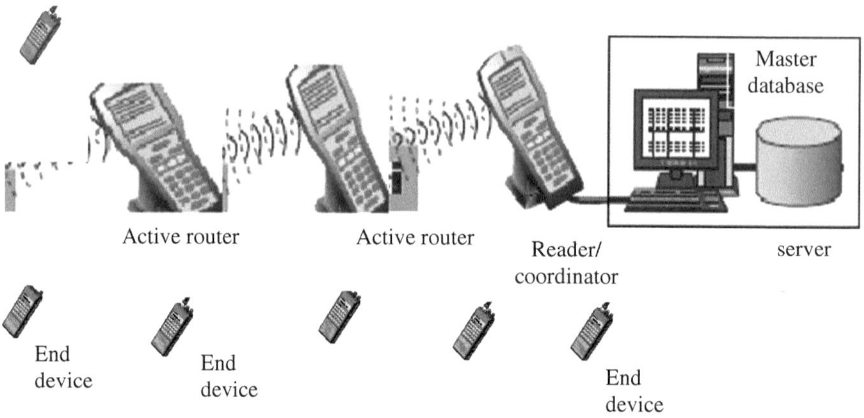

Fig. 11.3 Required setup of the whole system

Fig. 11.4 Database connectivity

Fig. 11.5 Login page with user type

Fig. 11.6 Window for loaded site map

Step 5: Load the mine plan (mine map) which is stored in computer with the extension of (∗.JPG/ ∗.BMP/ ∗.JPEG) and size 800×700 pixels. For this click Master menu bar and select Load Map option as shown in Fig. 11.6.

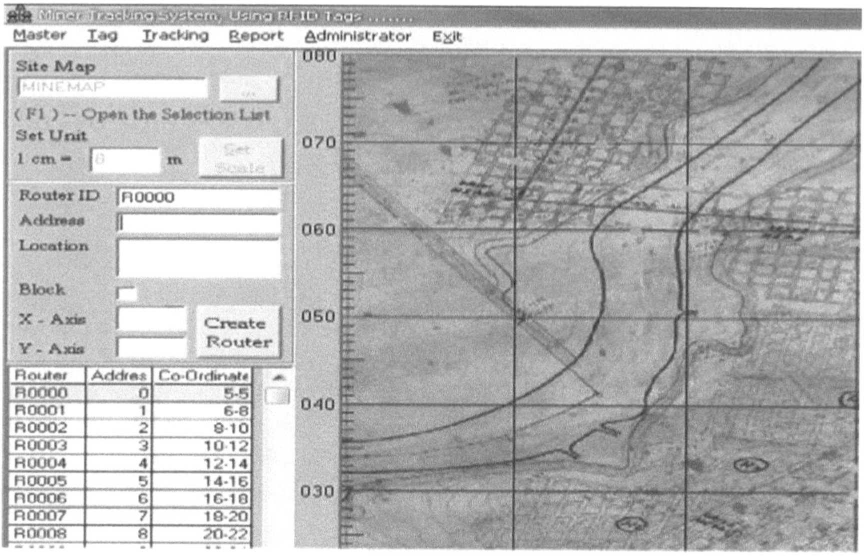

Fig. 11.7 Graphical view of site along with routers

Step 6: Enter site name/mine name in the text box and click brows button to select the map which is stored on software folder "PIC" and open it as shown in Fig. 11.7.

Step 7: Enter the Site Map scale and click on Set scale, then display grid scale on the map as shown in Fig. 11.7.

Step 8: Enter the router logical address, which is used for internal processing of computer. Then, enter location name of working area and *X*, *Y* coordinates, which are displayed on screen. Next click on Create Router button. Display router on the map in green-colored circle for router in normal locations and in red-colored circle for router in blocked areas.

Step 9: Select Create-Member module and add member details as shown in Fig. 11.8.

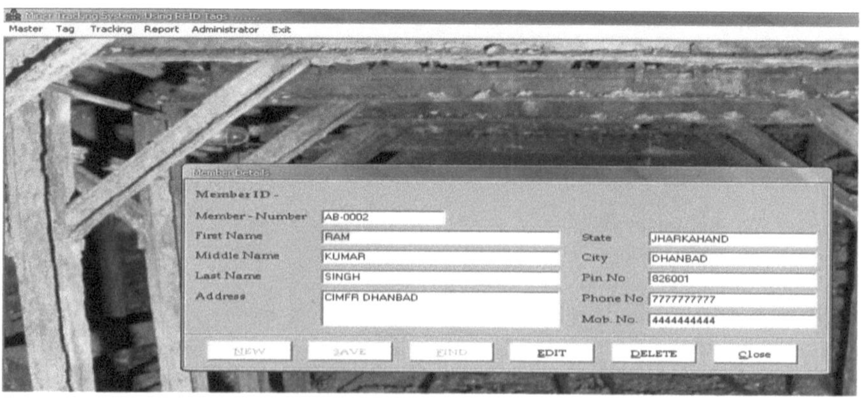

Fig. 11.8 Miners registration form

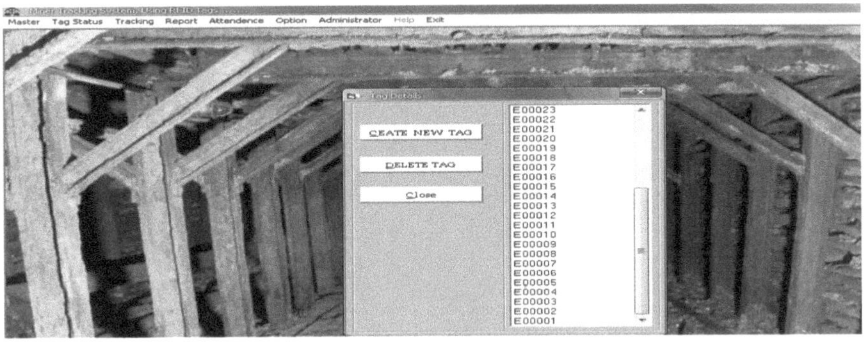

Fig. 11.9 Tag assignment form

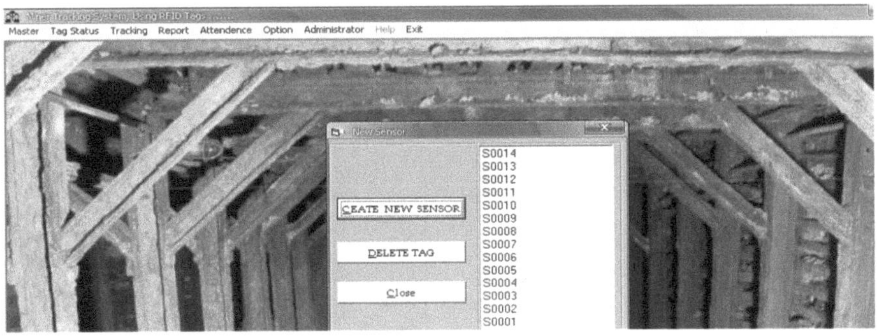

Fig. 11.10 Sensor assignment module

Step 10: Select Create New Tag module and add new tag as indicated in Fig. 11.9.
Step 11: Select Create New Sensor module and add new sensor as illustrated in Fig. 11.10.
Step 12: Allot tag to different members as depicted in Fig. 11.11.
Step 13: Set shift time of a mine according to Fig. 11.12.

11.5.2 Operation

After installation and functioning of the devices, all the data are received and saved in the database. Now the system is ready to retrieve the data through TMS software. Step-wise operation procedure is described below.

Step 1: Select Tracking menu and click on start-reading. Data received from device are arranged in tabular format as shown in Fig. 11.13.
Step 2: Select Tracking module for graphical presentation of site map with routers and end devices according to Fig. 11.14.

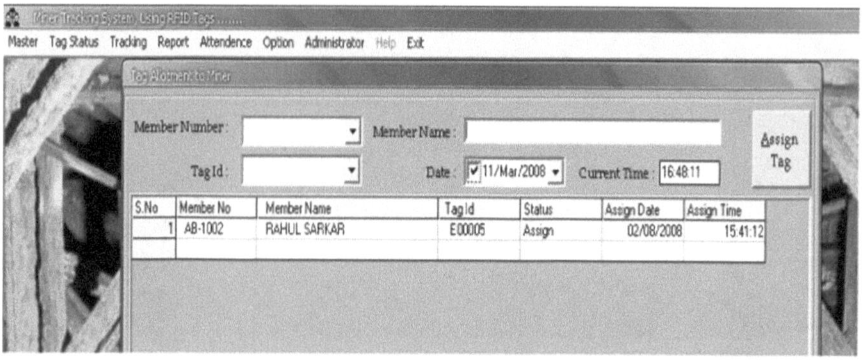

Fig. 11.11 Form for tag allotment to miners

Fig. 11.12 Shift time setting form

Step 3: Current location of underground miners can also be depicted as shown in Fig. 11.14.

Step 4: Path of miners during particular duration is graphically presented as shown in Fig. 11.15 with online data from active RFID device.

Step 5: Automatic alert messages with sound warning come from devices when the particular miner is in unsafe zone as shown in Fig. 11.16.

Step 6: Report of current position of the miners is generated as illustrated in Fig. 11.17.

Step 7: Attendance report through attendance menu can be obtained by choosing option for report as shown in Fig. 11.18.

Step 8: Different reports (e.g., Member details, Router details, Tag details, Current miner positions, Alert message, Blocked area, and different types attendance) can also be generated as shown in Figs. 11.19, 11.20, 11.21, 11.22, 11.23, 11.24, and 11.25.

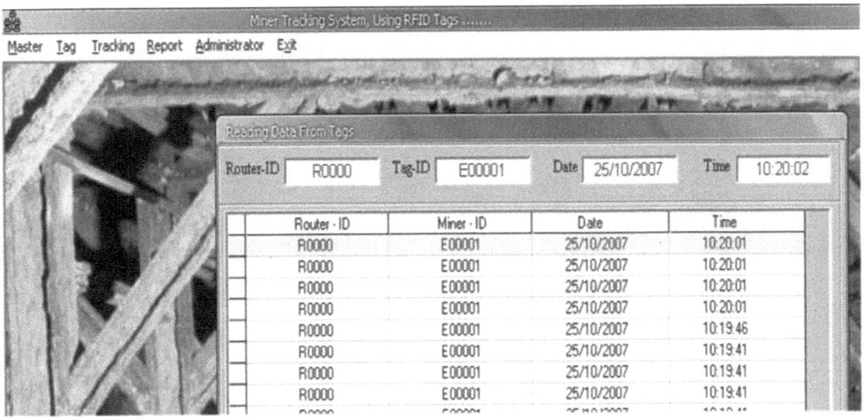

Fig. 11.13 Tabular format of received data from devices

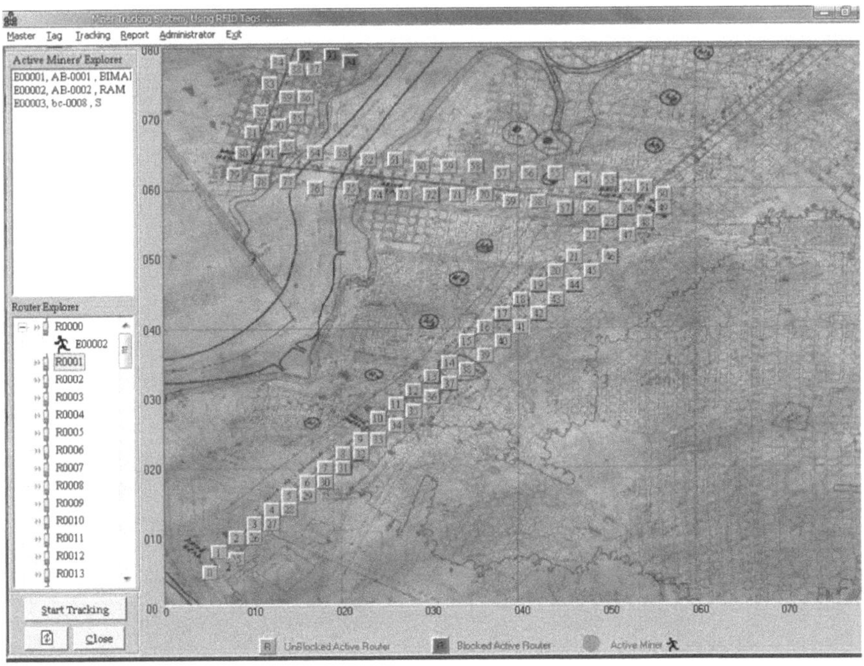

Fig. 11.14 Graphical presentation of site map with routers and end devices

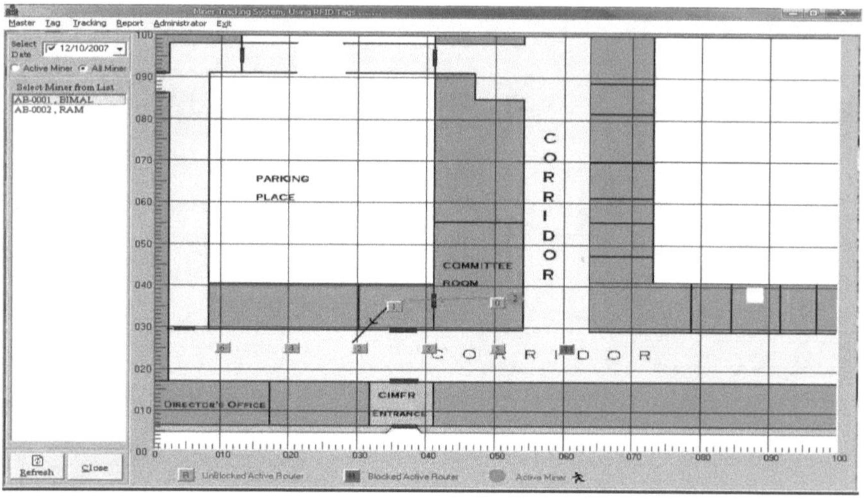

Fig. 11.15 Path of moving miners

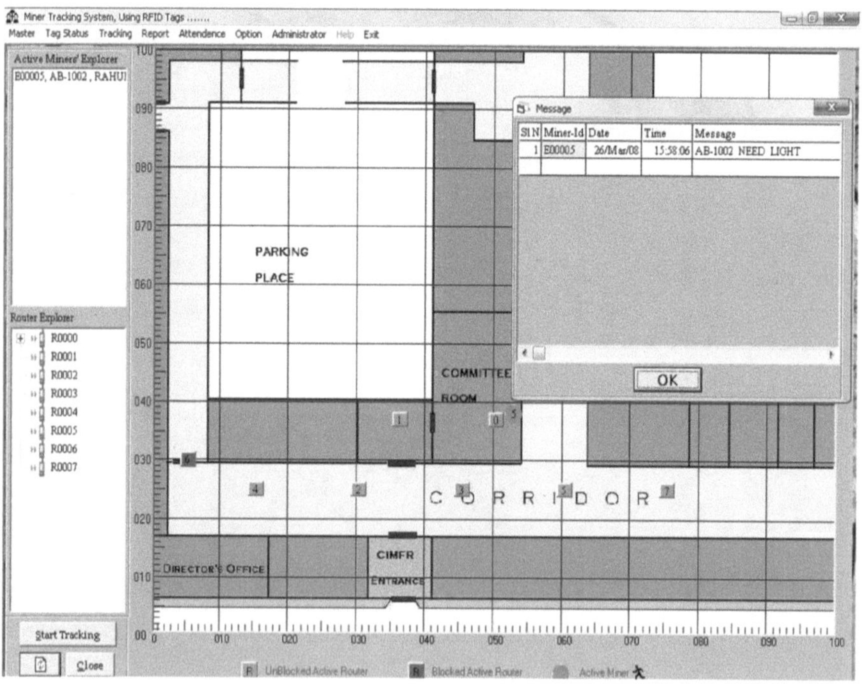

Fig. 11.16 Alert message with warning

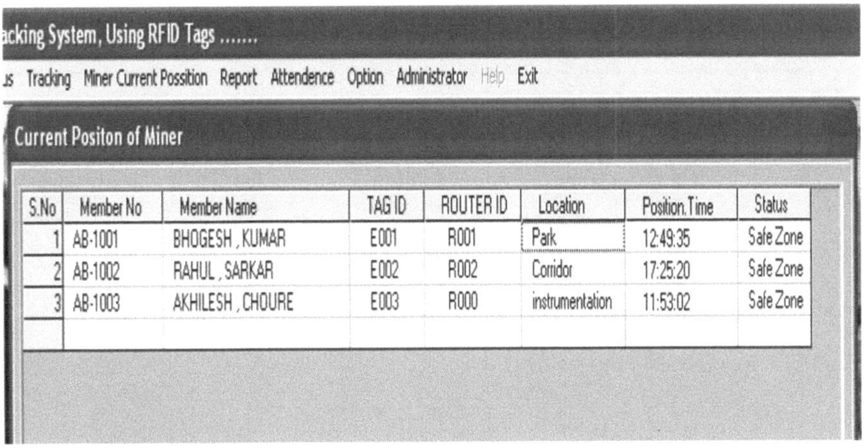

S.No	Member No	Member Name	TAG ID	ROUTER ID	Location	Position.Time	Status
1	AB-1001	BHOGESH , KUMAR	E001	R001	Park	12:49:35	Safe Zone
2	AB-1002	RAHUL , SARKAR	E002	R002	Corridor	17:25:20	Safe Zone
3	AB-1003	AKHILESH , CHOURE	E003	R000	instrumentation	11:53:02	Safe Zone

Fig. 11.17 Current position of miners

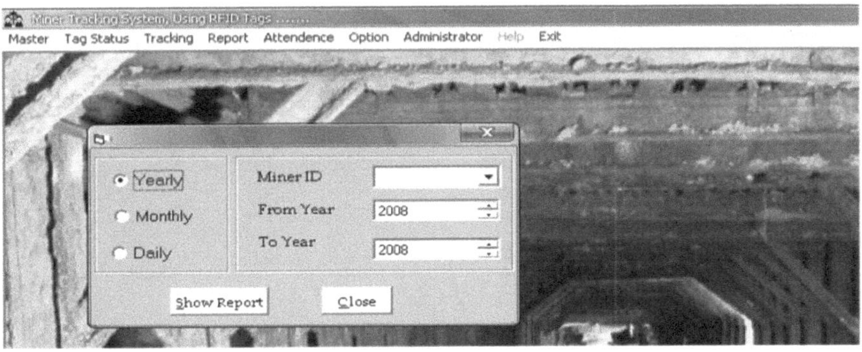

Fig. 11.18 Option for generating attendance report

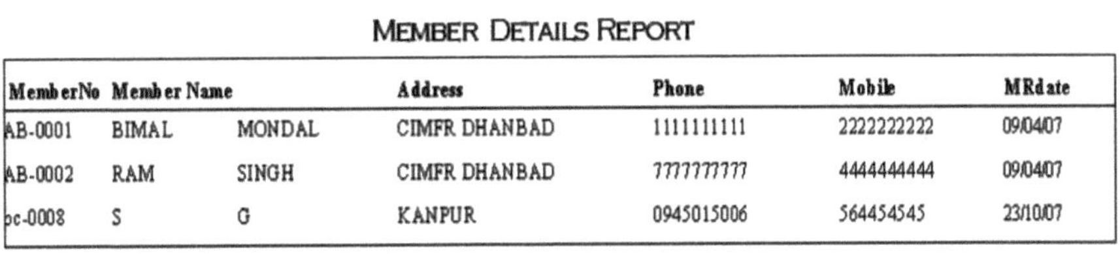

MEMBER DETAILS REPORT

MemberNo	Member Name		Address	Phone	Mobile	MRdate
AB-0001	BIMAL	MONDAL	CIMFR DHANBAD	1111111111	2222222222	09/04/07
AB-0002	RAM	SINGH	CIMFR DHANBAD	7777777777	4444444444	09/04/07
bc-0008	S	G	KANPUR	0945015006	564454545	23/10/07

Fig. 11.19 Member details report

ROUTER DETAILS REPORT

	Router ID	Router Address	Router Location	X-Y Coordinate	Block Status
Site Name CIMFR LAYOUT					
	R0000	0	Committe Room	50-36	
	R0001	5	Front Room	35-35	
	R0002	A	Left Corridor	30-25	
	R0003	3E8	Right Corridor	40-25	
	R0004	F	Director's cell	20-25	
	R0005	3ED	Up Stairs	50-25	
	R0006	14	Meeting Room	10-25	
	R0007	3F2	Ac Plant	60-25	Blocked

Fig. 11.20 Router details report

TIME WISE MINER DETAILS REPORT

25/10/07 - 25/10/07

Tag ID E00001 Site Name CIMFR LAYOUT

Member No. AB-0001 Assign Time 10:15:00

Member Name BIMAL , MONDAL Return Time 10:15:41AM

Router ID	Router Location	Time	Status
R0000	Committe Room	10:19:36	Safe Zone
R0000	Committe Room	10:19:41	Safe Zone
R0000	Committe Room	10:19:46	Safe Zone
R0000	Committe Room	10:20:01	Safe Zone
R0000	Committe Room	10:20:06	Safe Zone
R0000	Committe Room	10:20:11	Safe Zone

Fig. 11.21 Registered miners/members report

TAG DETAILS REPORT

S. No.	Tag ID	MemberNo	Miner Name
1	E00005	AB-1002	RAHUL KUMAR SARKAR
2	E00001	AB-1003	AKHILESH KUMAR CHOURE
3	E00005	AB-1003	AKHILESH KUMAR CHOURE
4	E00002	AB-1004	BIMAL KUMAR MANDAL

Fig. 11.22 Tag details report

MESSAGE DETAILS FROM END-DEVICE'S

26/03/2008 To 26/03/2008

TagId	Message	Message Date	Message Time
E00005			
	LIGHT	08 - Feb - 2008	16:37:03
	WATER	08 - Feb - 2008	16:37:19
	DOCTOR	08 - Feb - 2008	16:37:24
	DOCTOR	08 - Feb - 2008	16:37:34
	WATER	08 - Feb - 2008	16:37:49
	DOCTOR	08 - Feb - 2008	16:37:54
	LIGHT	08 - Feb - 2008	16:44:14

Fig. 11.23 Alert message report

MONTH WISE MINER ATTENDENCE

Mar - 2008

Miner Number: AB-1002   Miner Name: RAHUL SARKAR

Date	Assign Time	Return Time	Per day working hours of Underground
07 - Mar	09:42:43	09:45:01	- :3
24 - Mar	13:53:18	14:04:01	- :11
07 - Mar	09:45:12	09:47:02	- :2
12 - Mar	10:42:12	11:18:58	- :36
12 - Mar	11:26:28	11:48:03	- :22
24 - Mar	14:25:45	14:39:04	- :14
24 - Mar	14:41:07	15:01:00	- :20
07 - Mar	09:35:01	09:42:02	- :7
26 - Mar	15:54:18	16:11:31	- :17
		Total Working Hours in a month	2 : 12

Fig. 11.24 Month-wise miner report

First Shift **DAY WISE MINER ATTENDENCE** 03/Oct/08 Between 03/Oct/08

MemberNO	Member Name	Adate	Assigned Time	Return Time	Total working hours
AB-1002	RAHUL KUMAR SARKAR	30 / 09 / 2008	09:38:14	11:13:12	
AB-1003	AKHILESH KUMAR CHOURE	30 / 09 / 2008	09:38:02	11:13:12	
AB-1003		30 / 09 / 2008	11:13:31	14:26:57	
AB-1003		03 / 10 / 2008	11:08:36	11:11:02	
AB-1003		03 / 10 / 2008	11:12:31	11:22:03	
AB-1002	RAHUL KUMAR SARKAR	03 / 10 / 2008	11:08:31	11:11:02	
AB-1002		03 / 10 / 2008	11:12:31	11:22:03	
AB-1002		03 / 10 / 2008	11:24:41	16:17:16	
AB-1003	AKHILESH KUMAR CHOURE	03 / 10 / 2008	11:24:41	16:17:16	

Fig. 11.25 Day-wise miner report

11.6 Conclusions

TMS is an important application software for online tracking and monitoring of underground miners and moveable equipment. Use of the software would help in identifying the miners entering in underground mine to keep the track of the miners and maintain computerized attendance. This would help in monitoring equipment locations and their operation to improve productivity and reduce fatal collision accidents. This would help in locating and tracking the miners in case of disaster for speedy rescue operation. This would help in monitoring miners' unsafe practice and providing warning to the respective miner. This would further help in real-time monitoring of the environmental parameters in underground mine. This would also help in sending coded message to the concerned person in underground mine. This would further help in monitoring movement of dumpers in opencast mine, which will ultimately help in optimum shovel-dumper performance and improving productivity.

Chapter 12
Intrinsic Safety for Hazardous Area

12.1 Introduction

It is vital need to apply safety measures to the electrical apparatus used in explosive atmosphere as this impedes any damage to the property as well as reduces the risk of life of the workers on duty (Adams, 1990, 1991). There are various protection concepts deployed in sensitive and potentially explosive atmospheres. Among these, intrinsic safety (IS) offers the highest level of safety of entire explosion protection technique for electrical apparatus since it uses low voltage, low current, and low power. Intrinsic safety is a protection technique based on the restriction of electrical energy to a level below that which cannot cause ignition by either sparking or heating effects. It is achieved by limiting the amount of power available to the electrical device in the hazardous area. Thus, IS relies on the equipment being ignitior of a flammable gas, vapor, or dust that may occur in normal use, or under any conditions of fault likely to occur in practice. IS electrical circuits are designed with special consideration of proper printed circuit board (PCB) layout, minimum surface temperature, appropriate protection of electrical components, and suitable power supply to a device.

The basic principle on which IS works is to limit, under normal condition, the amount of electrical energy of a circuit operating in hazardous area circuits such that any sparks or arcs or high surface temperature cannot ignite the surrounding explosive atmosphere present in underground mines or any hazardous areas (Arnhold and Steilen, 2007). An IS circuit is designed in such a manner that its physical and electrical characteristics do not allow sufficient electrical energy (milli-joules) in an arc or spark to ignite or produce sufficient thermal energy from an overload condition to exceed the ignition temperature of a specific gas or vapor under normal operating conditions.

It may be stated that if something is intrinsically safe, it is safe by its own nature without any help from outside. This protection technique is not a matter of creating intrinsic safety, instead it is a matter of preserving the intrinsic safety, which the devices or apparatus already possesses. Preservation of intrinsic safety depends entirely on the strict control of energy in a circuit so that incendive spark or hot surfaces cannot possibly arise. This includes design so that it is unable to release

L.K. Bandyopadhyay et al., *Wireless Communication in Underground Mines*,
DOI 10.1007/978-0-387-98165-9_12,

sufficient energy, by either thermal or electrical means, to cause an restricting voltage and current to low values, and ensuring that no parts of the circuits can store or generate excessive level of energy.

The IS is a protection technique for safe operation of electronic equipment in explosive atmospheres (Gaunt, 1988; Towle, 1994). Electrical equipment in hazardous area must be designed to reduce the open circuit voltage and short circuit current to values that cannot cause ignition by opening or closing or earthing the circuit or by heating of any parts belonging to the circuit (Fry, 1995). The theory behind IS is to ensure that the available electrical and thermal energy in the system is always low enough such that ignition of the hazardous atmosphere cannot occur.

A device termed intrinsically safe is designed in such a way that it does not contain any components that produce sparks or which can hold enough energy to produce a spark of sufficient energy to cause an ignition (Bicchi et al., 2001). Another aspect of intrinsic safety is controlling abnormal small component temperatures. Under certain faulty conditions (such as an internal short circuit inside a semiconductor device), the temperature of a component case can rise to a much higher level than in normal use. Safeguards, such as current limiting by resistors and fuses, must be employed to ensure that in no case a component can reach a temperature that could cause autoignition of a combustible atmosphere. No single field device or wiring is intrinsically safe by itself (except for battery-operated, self-contained devices), but is intrinsically safe only when employed in a properly designed IS system (Bianco et al., 1994; Dutta, 2001; www.epsilonltd.com).

12.2 Concept of Intrinsic Safety

The gassy underground atmosphere is potentially explosive or in other words we can say that these hazardous areas in underground mines may come across various factors due to which fire or explosion may take place. These factors may be: (i) the ignition of flammable gases, (ii) vapors or dust in presence of oxygen, which may lead to explosion, and (iii) any rapid physical or chemical reaction accompanied by an increase of temperature and pressure. The basic requirements for explosion to take place are (i) flammable substance, (ii) oxygen, and (iii) source of ignition, which is shown by an ignition triangle (Fig. 12.1). The first two quantities may be insufficient to cause ignition, which are normally present in gassy coal mines (www.omega.com). Thus, in order to prevent explosions, we need to control the source of ignition, which is a critical parameter for underground coal mines. Therefore, current-carrying equipment (commonly known as Ex-equipment), which are placed in underground mines should be intrinsically safe to prevent explosion (Achari, 1991, 1997; Das, 1997; Pathak, 1997; Ramachandran, 1997; Das, 2001).

Ignition characteristics curve can be drawn for the fuel present in a hazardous area (Fig. 12.2). A minimum ignition energy (MIE) exists for every fuel that represents the ideal ratio of fuel to air. At this ratio, the mixture is

Fig. 12.1 Ignition triangle

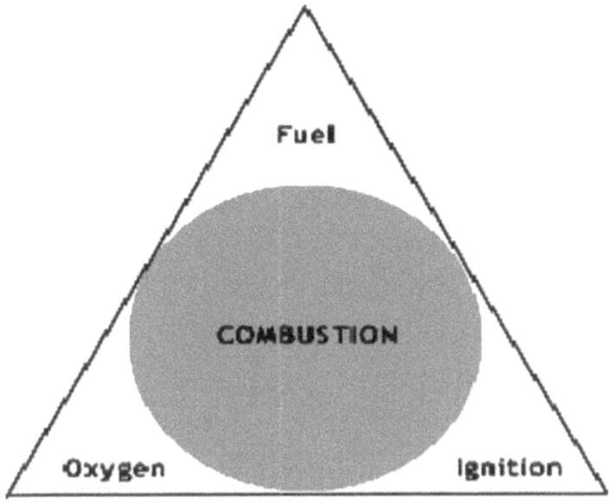

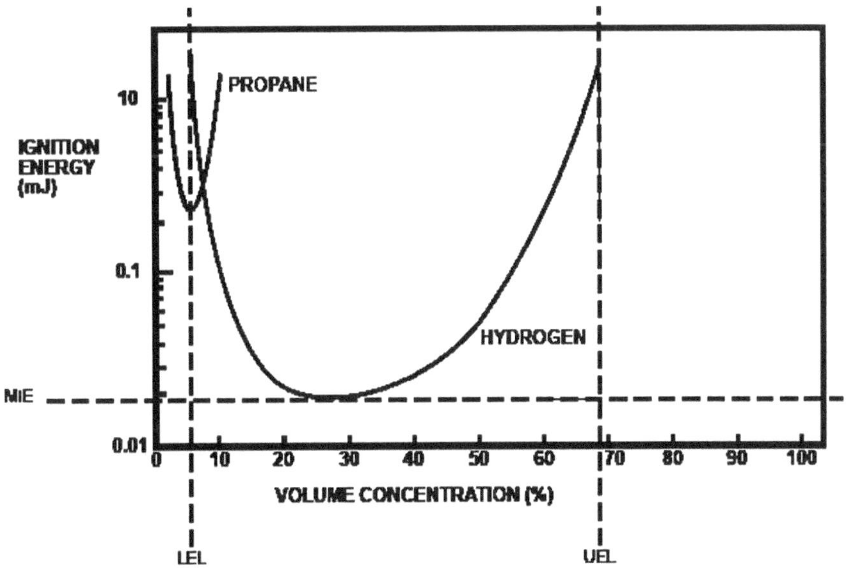

Fig. 12.2 Minimum ignition curve

most easily ignited. Below the MIE, ignition is impossible for any concentration (www.gminternationalsrl.com). For a concentration lower than the one corresponding to the MIE, the energy required to ignite the mixture increases until a concentration value is reached below which the mixture cannot be ignited due to the low quantity of fuel. This value is called the lower explosive limit (LEL). In the same way, when increasing the concentration, the energy requirement increases, and a concentration value is identified above which ignition cannot occur due to the low quantity of an oxidizer. This value is called the upper explosive limit (UEL), e.g., for methane LEL is 5% and for UEL it is 14%.

The basic concept of IS is to design the system in such a manner that electrical energy available from the circuit is less than the minimum ignition energy of hazardous gas mixture. A protection technique aims at preventing any ignition within the apparatus and in the interconnecting wiring by ensuring that any sparking that may occur in instrument or device exposed to a potentially explosive atmosphere is insensitive (Oliver, 1994). In normal working, the protection must be maintained during short circuit and breaks as can be foreseen. In designing an intrinsically safe circuit many components are interconnected and due to this, circuit can be complex or fairly simple according to the type of circuit and its electronic components, such as thyrister, logic gates, resistor, transistor microchips, inductor, capacitor, etc. Safety is achieved partly by limiting current and voltage, and partly by diverting or absorbing some of the energy released on collapse of magnetic field when an inductive circuit is broken. A resistor is used to limit the current. At the time of designing, designer must give thought to the effect of short circuit of component or failure of component in the circuit. For example, if a maximum of 60 V DC is applied to a system then the track separation in the PCB of the system should be 3 mm minimum without applying insulation and 1 mm minimum with insulation. Noncompliance to this means that the tracks will be considered to be under short circuit. One fault leading to a series of faults is considered to be single fault. It must also be remembered that current or voltage regulation by solid-state component, for example, by transistor or three terminal voltage regulators are always to be considered to be under fault (Ahirwal, 2006a; Ahirwal et al., 2007).

12.3 Classifications

Various safety factors are considered in installing an Ex-equipment in a hazardous area because of presence of flammable materials like various gases or vapors, dust particles, or fibers which cause ignition, thus hampering the safety (Magison, 1978; Casso, 1994; Clark et al., 1994; Fraczek, 1994; Schaefer et al., 2003; York et al., 2003; Qinghai and Longhua, 2001). All around the world, keeping this in mind, hazardous area is divided into various groups, classes, divisions, and zones. Classes define the general form of flammable materials in the atmosphere (Adamcik, 1990). Divisions define probability of the presence of flammable materials. Groups classify the exact flammable nature of material. Temperature identifications convey the maximum surface temperature of an apparatus based on 40°C room temperature. These temperature codes are selected carefully not to exceed the ignition temperature.

12.3.1 Hazardous Area

In United States, the National Electrical Code (NEC) defines a hazardous area by Zones and Divisions in Article 505, and in Article 500 by Division only (Toney et al., 2000), whereas in Canada, the Canadian Electrical Code, Section 18 formulated by Canadian Standard Association (CSA) defines a hazardous area by applying a three-part classification. In Europe, the classification is based on type of industry and degree of hazard by following the European Committee for Electrotechnical Standardization (CENELEC) and/or International Electrotechnical Commission (IEC) standards. In India, IEC standard is followed in harmonization with Bureau of Indian Standard (BIS) (Singh, 2006b). Table 12.1 summarizes the various classifications of hazardous area based on different standards (Kumar, 2001; Malhotra, 2001; Singh, 2006b).

Table 12.1 Classification of hazardous area

Parameter	Classification of hazardous area based on different standards	
	IEC, CENELEC, NEC 505 Codes, BIS	NEC 500, CSA Codes
Type of industry	Group I – Underground mining Group II – Surface industries	No classification, but mining and surface industries dealt with by different authorities
Type of fuel	Explosive mixtures of air with flammable gases or vapors and dust	Explosive mixtures of air with: Class I – Gases or vapors Class II – Dusts Class III – Fibers or filings
Degree of hazardous material present:		
• Continuously present	Zone 0	Division 1
• Intermittently present	Zone 1	Division 1
• Abnormally present	Zone 2	Division 2
Types of gas and vapors:		
• Acetylene	Group IIC	Class I/Group A
• Hydrogen	Group IIC	Class I/Group B
• Ethylene	Group IIB	Class I/Group C
• Propane	Group IIA	Class I/Group D
• Methane	Group I	–
Types of dust:		
• Metal	–	Class II/Group E
• Coal	–	Class II/Group F
• Grain	–	Class II/Group G
Fibers (All)	–	Class III

Dust is one of the dangerous fuels which enhances the ignition in hazardous area (Lloyd, 1994; Wenzel, 1994). It comprises small solid particles that settle out under their own weight but they may remain suspended for some time. This combustible dust when mixed with air and its concentration is within the lower explosive limit (LEL) and upper explosive limit (UEL) then it may cause major destruction to the processing industry. Dust cloud as well as dust layer may be a major reason of fire when it encounters an ignition source or reaches to its ignition temperature (Pidoll, 2001). Ignition temperature of the dust layer cloud is the lowest temperature of the dust at which ignition occurs under the defined condition of combustible dust layer/cloud. The atmosphere explosive (ATEX) directive (refer section "Standards Related to Intrinsic Safety") takes dust as important source of hazard from which the equipment must be safe. Hazardous area where combustible dust is taken as a risk factor is divided into zones. The ATEX directive takes zone 20, zone 21, and zone 22 for dust. In order to protect the equipment from dust, the ingress protection (IP) rating (refer section 12.7.2.1) of IP5X or IP6X is provided. The zone 21 area requires IP6X and zone 22 areas require minimum IP5X. The dust gets frequently ignited when it comes in contact with the hot surface. The dust ignition temperature is kept in mind before designing enclosures of electrical equipment so that it has maximum permitted surface temperature. It should be constructed as such to minimize the accumulation of dust layer on the surface, to minimize electrostatic charge, and to facilitate cleaning (Vishwakarma, 2005).

12.3.2 Temperature of Ex-Equipment

As mentioned earlier, the main cause of explosion in hazardous area is ignition by a hot surface. When a gas is heated above its ignition temperature it may spontaneously ignite. The maximum surface temperature is the highest attainable temperature by any Ex-equipment placed in the hazardous area. When the temperature crosses highest attainable temperature, it will cause an ignition to the surrounding atmosphere. Thus, based on the highest attainable temperature, Ex-equipment is classified into different temperature ("T") classes. Different surface temperature classes for electrical equipment adopted by the various countries are summarized in Table 12.2.

Table 12.2 Classification of Ex-equipment based on surface temperature

Maximum surface Temperature (°C)	Classification of equipment based on different standards	
	IEC, CENELEC, NEC 505 Codes, BIS	NEC 500, CSA Codes
450	T1	T1
300	T2	T2
280	–	T2A
260	–	T2B
230	–	T2C
215	–	T2D
200	T3	T3
180	–	T3A
165	–	T3B
160	–	T3C
135	T4	T4
120	–	T4A
100	T5	T5
85	T6	T6

According to the Bureau of Indian Standard, the surface temperature of the equipment in the underground mines shall be considered on the following features (Singh, 2006a):

- It totally depends upon the glow temperature of coal dust available in the mine.
- According to Indian Standard (IS: 8239), the maximum surface temperature should be restricted to class T4.
- The portable handheld equipment, like coal drills, and lamps, should not have surface temperature exceeding class T6.
- It is assumed that the glow temperature of coal dust is between 150 and 170°C.
- The critical temperature for Indian coal is around 80–120°C in most cases.

12.4 Standards Related to Intrinsic Safety

There are various standard-formulating agencies in different countries and these agencies have formulated different standards for the respective countries.

12.4.1 ATEX Directive

ATEX directive became applicable in European countries since July 2003. The ATEX stands for atmosphere explosives. ATEX directive deals with any control equipment used to ensure safe operation in a hazardous area. Under this directive both electrical and nonelectrical equipment to be used in hazardous area are "CE" marked and are covered under a single documentation. The positive aspect of this directive is that it also includes dust as a source of hazard for which the equipment must be safe. There are two ATEX (European) directives, concerning equipment and protective system for use in potentially explosive atmosphere and concerning the safety and health of workers potentially at risk from exposure to explosive atmospheres. It considers potentially explosive concentrations of gas, vapor, or mist in the air and concentrations of dust. It accounts for both electrical and mechanical sources of ignition. The two ATEX directives are (i) the ATEX directive "94/9/EC" concerning equipment and protective systems intended for use in potentially explosive atmospheres, and (ii) the ATEX directive "1999/92/EC" concerning the minimum requirements for improving safety and health protection of workers potentially at risk from exposure to explosive atmospheres.

Classification of equipment and hazardous area according to ATEX directive are as follows:

- Equipment is divided into two groups, namely (a) I for mining and (b) II for surface industries; and into categories (categories M1 and M2 for mining; and categories 1, 2, and 3 for all other industries).
- It classifies hazardous area into zones as summarized in Table 12.3.

Table 12.3 Categories and application of ATEX Group II

ATEX Group II categories and application				
Category	Design of safety	Design requirements	Application	Zone of use
1	Very high level of safety	Two independent means of protection or safe with two separate faults	Where explosive atmospheres are present continuously or for lengthy periods	Zone 0 (gas) Zone 20 (dust)
2	High level of safety	Safe with frequently occurring disturbances or with an operating fault	Where explosive atmospheres are likely to occur	Zone 1 (gas) Zone 21 (dust)
3	Normal level of safety	Safe in normal operation	Where explosive atmospheres are likely to occur infrequently for short duration	Zone 2 (gas) Zone 22 (dust)

12.4.1.1 ATEX Directive 94/9/EC

ATEX directive 94/9/EC (also known as ATEX 95 or ATEX 100A) gives details of design, manufacture, or sale of any equipment or protective system which is to be used in potentially explosive atmospheres and market those products in the European Union (EU). This directive covers equipment-1 and protective systems-2, which may be used in hazardous areas created due to the existence of flammable gases, vapors, mist, or dust.

The directive 94/9/EC is a "New Approach" directive laying down certain specifications for essential health and safety requirements and after that, primarily European standards give technical expression of the relevant requirements contained in the directive. Essential health and safety requirements are specific with respect to:

- Potential ignition sources of equipment intended for use in potentially explosive atmospheres;
- Autonomous protective systems intended to come into operation following an explosion with the prime objective to halt explosion immediately and/or limit the effects of explosion flames and pressures;
- Safety devices intended to contribute to safe functioning of such equipment with respect to ignition source and to the safe functioning of autonomous protective systems; and
- Components with no autonomous function essential to safe functioning of such equipment or autonomous protective system.

12.4.1.2 Directive 99/92/EC (ATEX 137)

ATEX directive 99/92/EC requires employers to protect workers from the risk of explosive atmospheres. An explosive atmosphere is defined as a mixture with air, under atmospheric conditions, of dangerous substances in the form of gases, vapor, mist, or dust in which after ignition has occurred, combustion spreads to the entire unburned mixture. With a clear focus on worker safety and employer responsibility, ATEX 137 demands careful consideration. ATEX 137 directive is based on three straightforward principles (Achari, 2006b; www.ul.com; www.ec.europe.eu):

- To prevent the formation of explosive atmospheres wherever possible;
- To prevent ignition where such atmospheres are unavoidable; and
- To ensure health and safety of workers by mitigating the effects of any explosions that can occur.

12.4.1.3 Authorities Approving Intrinsic Safety Standard

Canadian Standards Association (CSA): CSA sets standards that govern various aspects of product design and production, including eligibility for certification in hazardous locations within Canada. The CSA also evaluates, tests, and inspects products in order to ensure that they meet its intrinsic safety standards. Through interlaboratory agreements, they also assess products for conformity to American and European standards. The standard CSA C22.2, No. 157, (www.csa.ca) gives details of intrinsically safe and nonincendive equipment.

Factory Mutual (FM): FM Research Corporation determines whether equipment and materials used in hazardous locations meet National Electrical Code (NEC) standards for safety and reliability: NEC Standard 500 (Division classification) and NEC Standard 505 (Zone classification). Its services are also utilized in hazardous locations in the United States and elsewhere which attempts to harmonize with American and European classifications. Factory Mutual considers not only product performance but also ensures quality-control standards in place during production. Like CSA, Factory Mutual has interlaboratory agreements, which certify to Canadian and European as well as U.S. Standards. The standard FM Class 3610 Class I, II, III Division 2 (www.fmglobal.com) gives details of intrinsically safe apparatus and associated apparatus.

European Committee for Electrotechnical Standardization (CENELEC): The CENELEC has been officially recognized as the European Standards Organization by the European Commission Directives. These are some of the recognized European approval agencies that certify to CENELEC and IEC (International Electrotechnical Commission) standards for hazardous locations. EU countries have adopted CENELEC codes, which are identical to IEC, as CENELEC and IEC have cooperation agreement for harmonization of standards. Generally, IEC standards are used. However, in certain instances where IEC standards are considered too vague, CENELEC defines more precise requirements. The standard EN 50020 (www.cenelec.org) gives details of Electrical Apparatus for Potentially Explosive Atmosphere Intrinsic Safety "I" and IEC 60079-11 for Intrinsic Safety "i."

Bureau of Indian Standards (BIS): BIS is the national code-laying body in India. The role of BIS code in Indian hazardous location system is to cover all the aspects relating to area classification, equipment construction, testing, selection, installation, and maintenance of Ex-equipment. IEC is the international body for code laying. India accepts these international codes for Ex-equipment. This Indian Standard which is identical with IEC 60079-11(1999) Electrical apparatus for explosive gas atmospheres – Part 11: Intrinsic safety “I”, issued by the IEC was adopted by BIS on the recommendation of Electrical Apparatus for Explosive Atmospheres Sectional Committee and approval of Electrotechnical Division Council. Indian standard which deals with specifications of intrinsic safety is IS: 5780-2002 (Gupta, 2008).

As mentioned above, there are various standards applicable in different countries. Among them, the intrinsic safe standard (IEC 60079-11) formulated by the IEC is widely followed by European and many other countries for designing and testing intrinsically safe apparatus (Singh, 2006c; Towle, 2007). Different international IS testing laboratories (Table 12.4) commonly test Ex-equipment following IEC 60079-11 standard. Some of the laboratories also test Ex-equipment following other standards as per the requirement. Therefore, IEC 60079-11 standard is elaborated in detail in this chapter for common use.

Table 12.4 List of intrinsic safety testing laboratories in different countries

Country	Name of the laboratory
USA	• Mine Safety and Health Administration (MSHA)
	• Factory Mutual Research Corporation (FM)
	• Underwriters' Laboratories, Inc. (UL)
Australia	• Standard Association (SA)
Canada	• Canadian Standards Association (CSA)
South Africa	• South African Bureau of Standard (SABS)
Switzerland	• Schweizerischer Elektrotechnischer Verein (SEV)
India	• Central Institute of Mining and Fuel Research (CIMFR)
	• Electronic Regional Testing Laboratory (ERTL)
China	• Electosuisse
EUROPEAN COUNTRIES	
Austria	• TÜV-Österreich TÜV-A
Belgium	• Institut Scientifique Des Services Publics – Siège De Colfontaine (ISSEP)
	• INEX
Denmark	• Demko A/S
Finland	• VTT Automaatio (VTT Automation) VTT MDT
France	• Institut National de l'Environnement Industriel et des Risques (INERIS)
	• Laboratoire Central des Industries Électriques (lcie)
Germany	• Bundesanstalt für Materialforschung und Prüfung (BAM)
	• Deutsche Gesellschaft zur Zertifizierung von Management-Systemen mbH
	• DMT-Zertifizierungsstelle der DMT-Gesellschaft für Forschung und Prüfung mbH
	• Forschungsgesellschaft für angewandte Systemsicherheit und Arbeitsmedizin mbH (FSA)
	• IBExU – Institut für Sicherheits-technik GmbH

Table 12.4 (continued)

Country	Name of the laboratory
	• Physikalisch-Technische Bundesanstalt Braunschweig (PTB) • TÜV Hannover/Sachsen-Anhalt e.V. • TÜV CERT-Zertifizierungsstelle für Maschinen, Aufzugs- und Fördertechnik • TÜV Product Service GmbH • ZELM EX Prüf- und Zertifizierungsstelle
Italy	• Centro Elettrotecnico Sperimentale Italiano (CESI)
Luxembourg	• Service de l'Energie de l'Etat BP 10 • SNCH – Société Nationale de Certification et d'Homologation
Nederland	• Kema Registered Quality Bv
Norway	• Det Norske Veritas Classification AS
Spain	• Laboratorio Oficial José María de Madariaga (LOM)
Sweden	• Sveriges Provnings-Och Forskingsinstitut (SP)
United Kingdom	• Electrical Equipment Certification Service Health and Safety Executive • ITS Testing and Certification Ltd ITS House • Sira Certification Service Sira Test and Certification Limited • Baseefa

12.5 Important Definitions

Definitions of various terms used in intrinsic safety analysis as per the IEC60079-11, 2007 standard are given below.

- *Intrinsically safe circuit*: Circuit in which any spark or any thermal effect produced in the conditions, which include normal operation and specified fault condition, is not capable of causing ignition of a given explosive gas atmosphere.
- *Electrical apparatus*: Assembly of electrical components, electrical circuits, or parts of electrical circuits normally contained in a single enclosure.
- *Intrinsically safe apparatus*: Electrical apparatus in which all the circuits are intrinsically safe circuits.
- *Associated apparatus*: Associated apparatus, which contain both intrinsically and nonintrinsically safe circuits, and cannot adversely, affect the intrinsically safe circuits.
- *Normal operation*: Operation of intrinsically safe apparatus or associated apparatus such that it conforms electrically and mechanically the design specification produced by its manufacturer.
- *Infallible component or infallible assembly of components*: Component or assembly of components that is considered as not subject to certain fault modes. The probability of such fault modes occurring in service or storage is considered to be so low that they are not to be taken into account.
- *Infallible separation or insulation*: Separation or insulation between electrically conductive parts that are considered as not subject to shorts circuits. The probability of such fault modes occurring in service or storage is considered to be so low that they are not taken into account.

- *Fault*: Any defect of any component, separation, insulation, or connection between components, not defined as infallible, upon which the intrinsic safety of a circuit depends.
- *Countable fault*: Fault which occur in parts of electrical apparatus conforming to the constructional requirements.
- *Noncountable fault*: Fault which occurs in parts of electrical apparatus not conforming to the constructional features.
- *Simple apparatus*: Electrical component or combination of components of simple construction with well-defined electrical parameters, which is compatible with the intrinsic safety of the circuit in which it is used.
- *Internal wiring*: Wiring and electrical connections that are made within the apparatus by the manufacturer.
- *Minimum igniting current (MIC)*: Minimum current in resistive or inductive circuits that causes ignition of explosive test mixture in the spark-test apparatus.
- *Minimum igniting voltage*: Minimum voltage of capacitive and inductive circuits that causes ignition of explosive test mixture in the spark-test apparatus.
- *Maximum r.m.s. a.c. or d.c. voltage (Um)*: Maximum voltage that can be applied to the non intrinsically safe connection facilities of associated apparatus without invalidating intrinsic safety.
- *Maximum input voltage (Ui)*: Maximum voltage (peak a.c. or d.c.) that can be applied to the connection facilities for intrinsically safe circuits without invalidating intrinsic safety.
- *Maximum output voltage (Uo)*: Maximum output voltage (peak a.c. or d.c.) in an intrinsically safe circuit that can appear under open circuit condition at the connection facilities of the apparatus at any applied voltage up to the maximum voltage, including Um and Ui.
- *Maximum input current (Ii)*: Maximum current (peak a.c. or d.c.) that can be applied to the connection facilities for intrinsically safe circuits without invalidating intrinsic safety.
- *Maximum output current (Io)*: Maximum current (peak a.c. or d.c.) in an intrinsically safe circuit that can be taken from the connection facilities of the apparatus.
- *Maximum input power (Pi)*: Maximum input power in an intrinsically safe circuit that can be dissipated within the apparatus when it is connected to an external source without invalidating intrinsic safety.
- *Maximum output power (Po)*: Maximum electrical power in an intrinsically safe circuit that can be taken from the apparatus.
- *Maximum external capacitance (Co)*: Maximum external capacitance in an intrinsically safe circuit that can be connected to the connection facilities of the apparatus without invalidating intrinsic safety.
- *Maximum internal capacitance (Ci)*: Total equivalent internal capacitance of the apparatus which is considered as appearing across the connection facilities of the apparatus.
- *Maximum external inductance (Lo)*: Maximum value of inductance in an intrinsically safe circuit that can be connected to the connection facilities of the apparatus without invalidating intrinsic safety.

- *Maximum internal inductance (Li)*: Total equivalent internal inductance of the apparatus which is considered as appearing at the connection facilities of the apparatus.
- *Maximum external inductance to resistance ratio (Lo/Ro)*: Maximum value of the inductance (Lo) to resistance (Ro) of any external circuit that can be connected to the connection facilities of the electrical apparatus without invalidating intrinsic safety.
- *Maximum internal inductance to resistance ratio (Li/Ri)*: Maximum value of ratio inductance (Li) to resistance (Ri), which is considered as appearing at the external connection facilities of the electrical apparatus.
- *Clearance*: Shortest distance in air between two conductive parts that are exposed to the air and not to insulated parts within the casting compound.
- *Distance through casting compound*: Shortest distance through a casting compound between two conductive parts.
- *Distance through solid insulation*: Shortest distance through solid between two conductive parts.
- *Creepage distance in air*: Shortest distance along the surface of the insulating medium in contact with air between two conductive parts.
- *Creepage distance under coating*: Shortest distance between conductive parts along the surface of the insulating medium covered with insulating coating.
- *Fuse rating (In)*: Current rating of the fuse.
- *Sealed gas tight cell or battery:* Cell or battery remains closed and does not release either gas or liquid when operated within the limits of charge or temperature specified by the manufacturer. These types of cell or battery are equipped with a safety device to prevent dangerously high internal pressure. These cell or battery does not require addition to the electrolyte and is designed to operate during its life in its original sealed state.
- *Diode safety barrier*: These assemblies incorporating shunt diodes or diode chains (including Zener diodes) is protected by fuses or resistors or a combination of these, and are manufactured as an individual apparatus rather than as part of a large apparatus.
- *Thermal initiation time*: It is the time during which energy deposited by the spark accumulates in a small volume of gas around it without significant thermal dissipation.

12.6 Grouping and Classification of IS and Associated Apparatus

12.6.1 Categories of Electrical Apparatus

In general, there are two categories of intrinsically safe apparatus and associated apparatus. They are "ia" and "ib." The categories differ in two main aspects – first is the number of specified fault and second in the values of safety factor.

(i) Category "ia"
When Um and Ui are applied, the intrinsically safe circuit in electrical apparatus of category "ia" shall not be capable of causing ignition in each of the following circumstances:
(a) in normal operation and with noncountable fault;
(b) in normal operation and with one countable fault; and
(c) in normal operation and with two countable faults.
With the following safety factors applicable:
– for both (a) and (b) 1.5
– for (c) 1.0

(ii) Category "ib"
With Um and Ui applied, the intrinsically safe circuits in electrical apparatus of category "ib" shall not be capable of causing ignition in each of the following circumstances:
a) in normal operation and with noncountable faults; and
b) in normal operation and with one countable fault plus noncountable faults.

12.6.2 Simple Apparatus

There are some field devices that can be installed and considered as intrinsically safe or nonincendive even though they do not carry entity parameters. Such a device is known as simple apparatus. The entity parameters of any intrinsically safe apparatus are the most important factors. If any intrinsically safe equipment is to be connected with any other intrinsically safe equipment, it must have a matched entity parameter.

A simple apparatus does not generate or store more than the following values: >1.2 V, >100 mA, >20 μJ, and >25 mW. Intrinsic safety barrier is also applied to devices considered as simple apparatus.

Examples of simple apparatus (IEC 60079-11, 2007):

(a) Passive components, e.g., switches, junction boxes, resistors, and simple semiconductor devices.
(b) Sources of stored energy with well-defined parameters, e.g., capacitors or inductors whose value shall be considered as and when determining the overall safety of the system.
(c) Sources of generated energy, e.g., thermocouples and photocells, which do not generate more than 1.5 V, 100 mA, and 25 mW. Any inductance or capacitance present in these sources of energy is considered with well-defined parameters.

12.7 Designing of Intrinsically Safe Apparatus

12.7.1 Construction Features

Ex-equipment is primarily a low-energy device. To eliminate risk of explosion these devices operate at low voltage and amperage to ensure that during fault condition sufficient energy is not available for ignition. All the components, PCB track and wire cross section are chosen with high safety margins to prevent temperature rise under fault condition thus preventing ignition. In addition to these, many safety barriers are used to isolate field devices under fault condition. The other features considered are given in Table 12.5.

Table 12.5 Construction features of Ex-equipment

Features	Intrinsic safety
Cost	Cheap, but depends on the application
Maintenance	Live maintenance
Use	Easy
Special enclosures	No
Temperature rating	Yes
Ingress protection	Generally IP-54 for Group I, but depends on the application
Entity parameters	Very important
Cable gland	Only waterproof cable gland required
Cable length	Limited length
Weight	Light
Size	Optimum

The design of the intrinsically safe apparatus should also follow these criteria (Chandra, 1991; Ansari, 1997):

(i) Circuits shall be adequately separated from other circuits;
(ii) Temperature classification of intrinsically safe apparatus shall be carried out so as to ensure that ignition is not caused by hot surface; and
(iii) No spark ignition shall result when the circuit is tested or assessed.

12.7.2 Description of the Various Assemblies

12.7.2.1 Enclosures

Any special type of enclosure is not required for intrinsically safe circuit because the protection is employed within the circuit itself. However, where intrinsic safety can be impaired by access to conducting part, e.g., if the circuits contain infallible creepage distance in air, an enclosure with ingress protection (IP) of at least IP20 shall be provided as part of the apparatus under test. The IP65 and IP55 are also used in underground mines, whose information of the number is given in Table 12.6. For group I a degree of protection of IP54 is provided (Ritolia, 2001).

IP Rating: The IP Code defined in international standard IEC60529, 2004, classifies the degrees of protection provided against the intrusion of solid objects (including body parts like hands and fingers), dust, accidental contact, and water in electrical enclosures. It consists of the letters IP for "international protection rating" or "ingress protection rating" followed by two digits or three digits, where the third digit is the optional digit. Table 12.6 summarizes the data of the IP rating. An "X"

Table 12.6 Data of the IP rating

IP 1st number – protection against solid objects	IP 2nd number – protection against liquids	IP 3rd number (optional) – protection against mechanical impacts (not included in IEC60529)
0 No special protection	**0** No protection	**0** No protection
1 Protected against solid objects up to 5 mm, e.g., accidental touch by hands	**1** Protection against vertically falling drops of water, e.g., condensation	**1** Protects against impact of 0.225 joule (e.g., 150 g weight falling from 15 cm height)
2 Protected against solid objects up to 12 mm, e.g., fingers	**2** Protection against direct sprays of water up to 15° from the vertical	**2** Protected against impact of 0.375 joule (e.g., 250 g weight falling from 15 cm height)
3 Protected against solid objects over 2.5 mm tools and wires	**3** Protected against direct sprays of water up to 60° from the vertical	**3** Protected against impact of 0.5 joule (e.g., 250 g weight falling from 20 cm height)
4 Protection against solid objects over 1 mm (tools wires and small wires)	**4** Protection against water sprayed from all directions, limited ingress permitted	**4** Protected against impact of 2.0 joule (e.g., 500 g weight falling from 40 cm height)
5 Protected against dust limited ingress (no harmful deposit)	**5** Protected against low-pressure jets of water from all directions limited ingress	**5** Protected against impact of 6.0 joule (e.g., 1.5 kg weight falling from 40 cm height)
6 Totally protected against dust	**6** Protected against low-pressure jets of water, e.g., for use on ship decks – limited ingress permitted	**6** Protected against impact of 20.0 joule (e.g., 5 kg weight falling from 40 cm height)
	7 Protected against the effect of immersion between 15 cm and 1 m	
	8 Protects against long periods of immersion under pressure	

can be used for one of the digits if there is only one class of protection, i.e., IPX1, which addresses protection against vertically falling drops of water, e.g., condensation. The first digit indicates the protection against solid objects and the second digit indicates the protection against the liquids. For example, in the IP rating IP54, 5 indicates the level of protection from solid objects and 4 designates the level of protection from liquids.

Antistatic: Many of the fires and explosion in explosive industry and some other hazardous area have taken place due to electrostatic discharge (Pidoll, 2004; Pidoll et al., 2004). When this static electricity charge comes in contact with the live parts, it causes explosion (Pilkington, 1994). So, enclosures to be used in hazardous area should have the antistatic property. This property reduces the risk of accumulation of static electricity charges on the equipment, and thus, prevents fires and explosion. For selecting proper antistatic enclosure for different gas groups, following points should be taken into consideration:

(i) Electrical apparatus for gas group I

Any antistatic equipment to be designed for gas group I should comply with the following:-

- Enclosures with surface area more than 100 cm^2 projected in any direction should be designed in such a way that under normal conditions of use, maintenance, and cleaning, danger of ignition due to electrostatic charges is avoided.
- Proper selection of materials so that insulation resistance is not more than 1 GΩ at a temperature of (23 ± 2)°C and relative humidity of $(50 \pm 5)\%$.
- Shape, size and layout, or other protective method considered, such that dangerous electrostatic charges are not likely to occur.

(ii) Electrical apparatus for gas group II

Any antistatic equipment to be designed for gas group II should comply with the following:-

- Suitable selection of materials should be done so that insulation resistance of the materials is not more than 1 GΩ at a temperature of (23 ± 2)°C and relative humidity of $(50 \pm 5)\%$.
- Limitations of the surface area of the enclosures are
  - for gas group IIA and IIB, a maximum of 100 cm^2 surface area, it may be extended up to 400 cm^2 if exposed area is surround by conductive earthed frame.
  - for gas group IIC, surface area can be taken up to the maximum of 20 cm^2, it may be extended up to 100 cm^2 if extra protection is given against electrostatic charges.
- Shape, size and layout, or other protective method considered, such that dangerous electrostatic charges are not likely to occur.

12.7.2.2 Wiring and Small Component Temperature

Dust layers on group I equipment: The deposition of coal dust layer on the component shall not take place in group I equipment.

Table 12.7 Temperature classification of copper wiring (in a maximum ambient temperature of 40°C)

Diameter (mm)	Cross-sectional area (mm^2)	Maximum permissible current for temperature classification (A)		
		T1 to T4 and group I	T5	T6
0.035	0.000962	0.53	0.48	0.43
0.05	0.00196	1.04	0.93	0.84
0.1	0.00785	2.1	1.9	1.7
0.2	0.0314	3.7	3.3	3.0
0.35	0.0962	6.4	5.6	5.0
0.5	0.196	7.7	6.9	6.7

Note:
(i) The values given for maximum permissible current, in amperes, is the r.m.s. a.c. or d.c. value.
(ii) For stranded conductors, the cross-sectional area is taken as the total area of all strands of the conductor.
(iii) The table also applies to flexible flat conductors, such as in ribbon cable but not to printed circuit conductors.
(iv) Diameter and cross-sectional area are the nominal dimensions specified by the wire manufacturer.
(v) Where the maximum input power Pi does not exceed 1.3 W, the wiring can be awarded a temperature classification of T4 and is acceptable for group I.
Source: www.gminternationalsrl.com

Wiring within apparatus: The maximum permissible current corresponding to the maximum wire temperature due to self-heating can be calculated from the equation given below for metals in general or can also be taken from Table 12.7.

$$I = I_f[t(1 + aT)/T(1 + aT)]^{1/2}$$

where, a is the temperature coefficient of resistance of wire material (0.004265/K for copper), I the maximum permissible r.m.s. current in amperes, I_f the current at which the wire melts in an ambient temperature of 40°C in amperes, T the melting temperature of wire material in degree Celsius (1083°C for copper), wire temperature due to self-heating and ambient temperature in degree Celsius. The maximum current in insulated wire shall not exceed the rating specified by the manufacturer of the wire.

12.7.2.3 Printed Circuit Wiring

In designing a printed circuit board (PCB) the following parameters shall be taken care of:

Minimum thickness of the board – 0.5 mm
Minimum conducting track – 35 μm
Temperature and group classification – T4 and group I
Minimum input power – 1.3 W

Track width (mm)	Continuous current (A)
0.3	0.518
0.4	0.814
1	1.388
2.0	2.222

The permissible maximum current corresponding to the track thickness of PCB is given in Table 12.8 for different temperature classification of Ex-equipment.

Table 12.8 Temperature classification of copper wiring on PCB (in a maximum ambient temperature of 40°C)

	Maximum permissible current for temperature classification (A)		
Minimum track width (mm)	T1 to T4 and group	T5	T6
0.15	1.2	1.0	0.9
0.2	1.8	1.45	1.3
0.3	2.8	2.25	1.95
0.4	3.6	2.9	2.5
0.5	4.4	3.5	3.0
0.7	5.7	4.6	4.1
1.0	7.5	6.05	5.4
1.5	9.8	8.1	6.9
2.0	12.0	9.7	8.4
2.5	13.5	11.5	9.6
3.0	16.1	13.1	11.5
4.0	19.5	16.1	14.3
5.0	22.7	18.9	16.6
6.0	25.8	21.8	18.9

Note:

(i) The values given for maximum permissible current, in amperes, is the r.m.s. a.c. or d.c. value.

(ii) For stranded conductors, the cross-sectional area is taken as the total area of all strands of the conductor.

(iii) The table also applies to flexible flat conductor, such as in ribbon cable, but not to printed circuit conductors.

(iv) Diameter and cross-sectional area are the nominal dimensions specified by the wire manufacturer.

(v) Where the maximum input power Pi does not exceed 1.3 W, the wiring can be awarded a temperature classification of T4 and is acceptable for group I.

Source: www.gminternationalsrl.com

12.7.2.4 Small Components

The use of small components, e.g., transistors or resistors, in designing the circuits, whose temperature exceeds the permissible limit for the temperature classification, shall be acceptable only if they conform to one of the following:

a) it shall not cause ignition of flammable mixture and any deformation or deterioration caused by higher temperature shall not impair the type of protection.

Table 12.9 Assessment for T4 classification according to component size and ambient temperature

Total surfaces area excluding wires	Requirement for T4 and group I classification
$<20\ mm^2$	Surface temperature $\leq 275°C$
$>20\ mm^2$	Power dissipation ≤ 1.3 W*
$\leq 20\ mm^2 < 10\ cm^2$	Surface temperature $\leq 200°C$

*Reduced to 1.2 W with ambient temperature or 1.0 W with 80°C ambient temperature.
Source: www.isa.org

b) for T4 and group I classification, small components shall conform to Table 12.9.
c) for T5 classification, the surface temperature of a component with a surface area smaller than 10 cm^2 (excluding lead wires) shall not exceed 150°C.

12.7.2.5 Ways and Facilities for Connecting External Circuits

Differentiation of terminals of intrinsic safe circuits and nonintrinsic safe circuits shall be considered by any of the following methods:

i) when separation is accomplished by distance, then the clearance between terminals shall be at least 50 mm.
ii) when considering the insulating partition or an earthed metal partition between terminals with a common cover, the following factors are taken into account:
    a) Partitions used to separate terminals shall extend to within 1.5 mm of the enclosures walls, or have a minimum distance of 50 mm.
    b) Metal partitions shall be earthed and shall have sufficient strength and rigidity to ensure that they are not likely to be damaged during field wiring and shall be at least 0.45 mm thick, and if it has lesser thickness then it should have the capability to withstand a minimum force of 30 N applied by a 6 mm diameter solid test rod at the center of the partition for 10 s. Metal partitions shall have sufficient current-carrying capacity to prevent burning through or loss of earth connection under fault conditions.
    c) Nonmetallic insulating partitions shall be at least 0.9 mm thick, and if it is of lesser thickness then it should also, like metal partitions, withstand a minimum force of 30 N applied by a 6 mm diameter solid test rod.

12.7.2.6 Creepage and Clearance Distance

Creepage and clearance requirements are used to reduce the risk of interconnection leading to short and open circuit. The protection afforded to interconnection from short circuit connection is insulation. An interconnection is protected from an open circuit by its physical strength. The minimum of 6 mm clearance and 10 mm creepage distance is maintained when required. The minimum clearance between the bare conducting parts of external conductors connected to terminals and earthed metal or other conduction parts shall be 3 mm. The value of recommended clearance between bare conducting parts of terminals of separate intrinsically safe circuits is given in Table 12.10.

Table 12.10 Recommended values of clearance, creepage distance, and separation

	Voltage, peak value (V)														
Parameter	10	30	60	90	190	375	550	750	1,000	1,300	1,575	3.3 K	4.7 K	9.5 K	15.6 K
Clearance (mm)	1.5	2.0	3.0	4.0	5.0	6.0	7.0	8.0	10.0	14.0	16.0				
Separation distance through casting compound (mm)	0.5	0.7	1.0	1.3	1.7	2.0	2.4	2.7	3.3	4.6	5.3	9.0	12.0	20.0	33.0
Separation distance through solid insulation (mm)	0.5	0.5	0.5	0.7	0.8	1.0	1.2	1.4	1.7	2.3	2.7	4.5	6.0	10.0	16.5
Creepage distance in air (mm)	1.5	2.0	3.0	4.0	8.0	10.0	15.0	18.0	25.0	36.0	49.0	–	–	–	–
Creepage distance under coating (mm)	0.5	0.7	1.0	1.3	2.6	3.3	5.0	6.0	8.3	12.0	16.3	–	–	–	–
Comparative tracking index, CTI:															
– ia	100	100	100	175	175	275	275	275	275	275	–	–	–	–	–
– ib	100	100	100	175	175	175	175	175	175	175	–	–	–	–	–

Source: www.gminternationalsrl.com

12.7.2.7 Plugs and Sockets

Plugs and sockets, which shall be used for connection of external intrinsically safe circuits, shall be separate from and noninterchangeable with those for nonintrinsically safe circuits. If intrinsically safe apparatus or associated apparatus is fitted with more than one plug and socket for external connections and interchange, it will adversely affect the type of protection. Such plugs and sockets shall be arranged:

i) By keying, so that interchange is not possible.
ii) By mating in which plugs and sockets shall be identified by marking or color coding, to make interchange identification obvious.

iii) If a connector is used and it carries earthed circuits then the safety depends on the earth connection for :

a) "ia" category of apparatus, the connectors shall contain three independent connecting elements.
b) "ib" category it shall have two independent connecting elements.

These elements shall be connected in parallel as shown in Fig. 12.3.

Fig. 12.3 Parallel connection

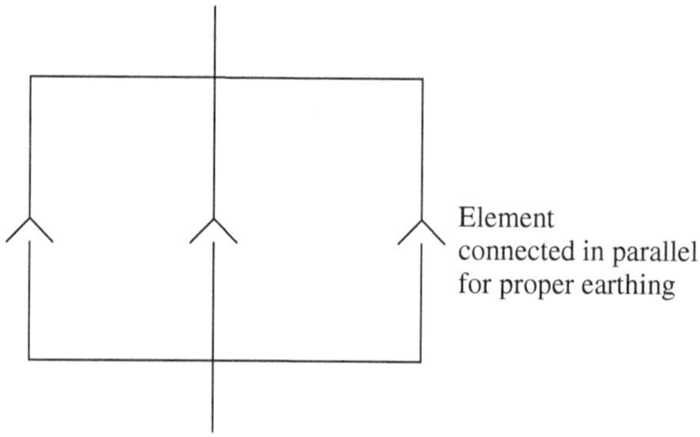

12.7.2.8 Separation Distance

Separation of Conductive Parts

If the protection depends on separation of conductive part between:

i) intrinsically safe and nonintrinsically safe circuits, or
ii) different intrinsically safe circuit, or
iii) a circuited and earthed or isolated metal parts,

then, it shall be taken as described below:-

- Separation distance shall be measured or assessed taking into account any possible movement of the conductors or conductive parts. Manufacturing tolerance shall not reduce the distance by more than 10% or 1 mm, whichever is smaller.
- A partition shall be at least 0.45 mm thick and attached to a rigid earthed metal portion of the devices.
- Smaller separation distances shall be considered as countable short circuit fault if its value exceeds one-third of the value specified in Table 12.10. If its value is less than one-third of the values, then it is considered as noncountable short circuit fault if it impairs intrinsic safety.

- If the separation is according to the values given in Table 12.10, then it shall be considered as not subject to failure to lower insulation resistance.
- Separation distance is not applicable to earthed metal which separates an intrinsically safe circuit to nonintrinsically safe circuit provided that when any breakdown takes place it does not hampers the safety and the earthed conductive parts can carry the maximum current that would flow under the fault condition.

Voltage Between Conductive Parts

The voltage between any conductive parts for which the separation has an effect on the type of protection of the circuit under consideration, e.g.,

The voltage between an intrinsically safe circuit and

- parts of the same circuit which is not intrinsically safe or
- nonintrinsically safe circuits or
- other intrinsically safe circuits.

The value of the voltage is taken as applicable:

i) For galvanically separated circuits within the apparatus, the highest voltage is considered and is derived from:

- the rated voltages of the circuits, or
- the maximum voltages specified by the manufacturer which may be supplied to the circuits, or
- any voltage generated within the same circuits.

For such sinusoidal voltages, peak voltage shall be considered as:

$\sqrt{2} \times$ r.m.s. value of the rated voltage.

ii) Between parts of the circuits: The total sum of the voltage from different sources gives the maximum value of the voltage that occurs in either part of the circuit.

12.7.2.9 Requirement for Assembled Printed Circuit Board

When intrinsic safety of the apparatus is affected by creepage and clearance distance, the printed circuit board is covered by conformal coating and thus clearance and creepage distance shall be applied to any conductive parts which lie outside coating, e.g.,

- tracks which emerge from coating;
- the free surface of a printed circuit which is coated on one side only; and
- bare part of components able to protrude through the coating.

12.7.2.10 Electrical Strengths Test

The insulation between an intrinsically safe circuit and frame of the electrical apparatus or parts which may be earthed shall normally be capable of withstanding an r.m.s. a.c. test voltage or twice the voltage of the intrinsically safe circuit or 500 V, whichever is greater.

The insulation between an intrinsically safe circuits and nonintrinsically safe circuits shall be capable of withstanding an r.m.s. a.c. test voltage of $2U + 1{,}000$ V with a minimum of 1,500 V r.m.s.; where U is the sum of the r.m.s. value of voltage of the intrinsically safe and nonintrinsically safe circuits.

12.7.2.11 Relays

In normal operation when the coil of the relay is connected to an intrinsically safe circuit, it should have maximum current of 5 A r.m.s., voltage of 250 V, power of 100 VA, and will follow the separation distance according to Table 12.10. If the current rating exceeds this value but is less than 10 A or 500 VA, than the creepage and clearance distance for the relevant voltage in Table 12.10 shall be doubled.

For higher values, intrinsically safe circuits and nonintrinsically safe circuits shall be connected to the same relay only if it is separated by an earthed metal barrier. The relay shall be designed such that broken or damaged contacts arrangements cannot dislodge and impair the integrity of separation between intrinsically safe and nonintrinsically safe circuits. The relay coil should be shunted with duplicate or triplicate diodes to nullify the incendive charge of inductance.

12.7.2.12 Earth Conductors, Connection, and Terminals

When earthing is required to maintain protection of some of the equipment like enclosures, conductors, metal screen, printed wiring board conductors, segregation contacts of plug-in-connectors, and diode safety barriers then conductors, connectors, and terminals used for this purpose shall have cross-sectional area such that they are rated to carry the maximum possible current.

When a connector carriers earthed circuits and the type of protection depends on the earthed circuits, the connectors shall comprise at least three independent connecting terminals for "ia" circuits and at least two for "ib" circuits. These elements shall be connected in parallel.

Terminals shall be fixed in their mountings without possibility of self-loosing and shall be constructed so that the conductors cannot slip out from their intended location. Proper contact shall be assured without deterioration of the conductors, even if multi-stranded cores are used in terminals, which are intended for direct clamping of the cores. The contact made by the terminals shall not be appreciably impaired by temperature change in normal service. Terminals for conductors having a smaller cross section up to 4 mm^2 shall also be suitable for the effective connection of conductors having a smaller cross section. The following type of terminals should not be used (Laxminarayanan, 2001):

i) Terminals with sharp edge, which can damage the conductors;
ii) Terminals, which may turn, be twisted, or permanently deformed by normal tightening; and
iii) Insulating materials, which transmit contact pressure in terminals.

12.7.2.13 Encapsulation Used for the Exclusion of a Potentially Explosive Atmosphere

When a casting compound is used to exclude a potentially explosive atmosphere from components and intrinsic safe circuits, e.g., fuses, piezoelectric devices, and energy storage devices with their suppression components, separation distance, and requirement of casting compound shall be required.

In addition, where a casting compound is used to reduce the ignition capability of hot components, e.g., diodes and resistors, the volume and thickness of the casting compound shall reduce the maximum surface temperature to the desired value.

12.7.2.14 Piezoelectric Devices

Capacitance and voltage appearing across the piezoelectric device should be measured first. The maximum energy stored by the capacitance of the crystal at the maximum measured voltage shall not exceed the following values:

- for group I apparatus: 1,500 μJ
- for group IIA apparatus: 950 μJ
- for group IIB apparatus: 250 μJ
- for group IIC apparatus: 50 μJ

When the electrical output of the piezoelectric device is limited by protective components (including guards), these components shall not be damaged by the impact in such a way as to allow the type of protection to be invalidated. Protection shall be given to the apparatus in order to prevent it from external physical impact so that there is no further increase in the impact energy.

12.7.2.15 Determination of Lo/Ro for Resistance Limited Power Source

The maximum external inductance to resistance ratio (Lo/Ro), which may be connected to a resistance limited power source shall be calculated using the given formula:

$$\frac{\mathrm{Lo}}{\mathrm{Ro}} = \frac{8e\mathrm{Ri} + (64e^2\mathrm{R}^2\mathrm{i} - 72\mathrm{U}^2\mathrm{o}e\mathrm{Li})^{1/2}}{4.5\mathrm{U}^2\mathrm{o}} \mathrm{H}/\Omega$$

Where, e is the minimum spark-test apparatus ignition energy in joules, and its value should be as follows:

– Group I apparatus: 525 μJ
– Group IIA apparatus: 320 μJ
– Group IIB apparatus: 160 μJ
– Group IIC apparatus: 40 μJ

Ri is the minimum output resistance of the power source, in ohms;
Uo is the maximum open circuit voltage, in volts;
Li is the maximum inductance present at the power source terminals, in henries.

if Li = 0, then

$$\frac{\text{Lo}}{\text{Ro}} = \frac{32e\text{Ri}}{9\text{Uo}^2}\text{H}/\Omega$$

12.7.2.16 Effect of Reactive Components on Intrinsic Safety

Reactive components constitute inductors and capacitors, and special care should be taken when IS circuits involve inductors and capacitors. According to Faraday's law "whenever an inductive circuit is interrupted, a back emf is generated which is proportional to the frequency of make and break." So this back emf is added up to the source voltage to produce a discharge at much higher voltage at the make and break contacts.

The inductance energy $\frac{1}{2} LI^2$ may be increased during the frequent operation of inductive intrinsically safe apparatus. For nullifying the effect of inductance, reverse diodes in duplicate or triplicate are provided which creates a short circuit path to the back emf so that this cannot be added up to the source voltage.

Protective current-limiting resistors shall be used in series with the inductive circuits to reduce current to a safe level. But if a specific minimum voltage is required for the inductive devices such as relays and solenoids, two shunt diodes or Zener diodes shall be placed in parallel with the inductive device to serve as spark-suppression devices in case the power lead to the inductive circuits is broken.

In case of a circuit containing a capacitor, the capacitor will charge at the source voltage and may discharge under a short circuit condition. Since the energy released during a capacitive discharge is $\frac{1}{2} CV^2$ joules, where V is in volts and C is in Farads, it will largely depend on the capacitance and voltage. If C is high, the discharge energy may become incendive. Similarly, if V is high, the discharge energy can also be high in case of the capacitor. Therefore, a resistor or a voltage-limiting diode should be connected in series with a capacitor in order to reduce the charging voltage.

12.7.3 Designing of Simple Circuits

12.7.3.1 Simple Inductive Circuits

Let us consider for group IIC consisting of a power supply comprising a 22 V battery with suitably mounted infallible 300 Ω current-limiting resistors feeding into an 1,100 Ω, 100 mH inductor as shown in Fig. 12.4.

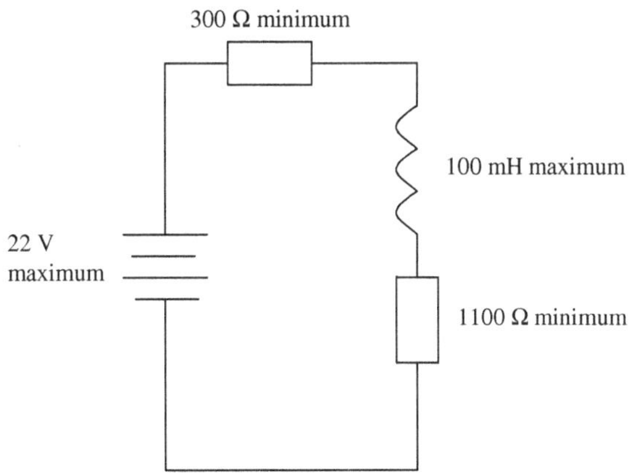

Fig. 12.4 A simple inductive circuit

The 300 Ω and 1,100 Ω values are the minimum and 100 mH is the maximum value. Two assessments shall be taken care of:

i) to ensure that the power supply is intrinsically safe, and
ii) to take into account of the effect of the connected load.

(i) Power supply

a) The minimum value of the current-limiting resistor connected to the power supply shall be taken as 300 Ω.
b) The maximum short circuit current is (22 / 300) mA = 73.3 mA.
c) For group IIC, the minimum igniting current for a resistive circuit at 22 V is 337 mA.
   Therefore, 73.3 mA of current, which is less than 337 mA of current at 22 V and power supply, is intrinsically safe in regard to spark ignition.

(ii) Connection of load

a) Since the maximum battery voltage is 22 V and the minimum resistances are 300 Ω and 1,100 Ω, the maximum possible current is 22/(300 + 1,100) mA = 15.7 mA.
b) Including the safety factor of 1.5, the current in the circuit is increased to 1.5×15.7 mA = 23.6 mA.
c) For group IIC, a 100 mH inductor has a minimum igniting current of 28 mA for a source of 24 V as given in Fig. 12.5.

The calculated value of current is 23.6 mA, which is less than the minimum igniting current of 28 mA, so the circuit is intrinsically safe in regard to spark ignition for group II ignition.

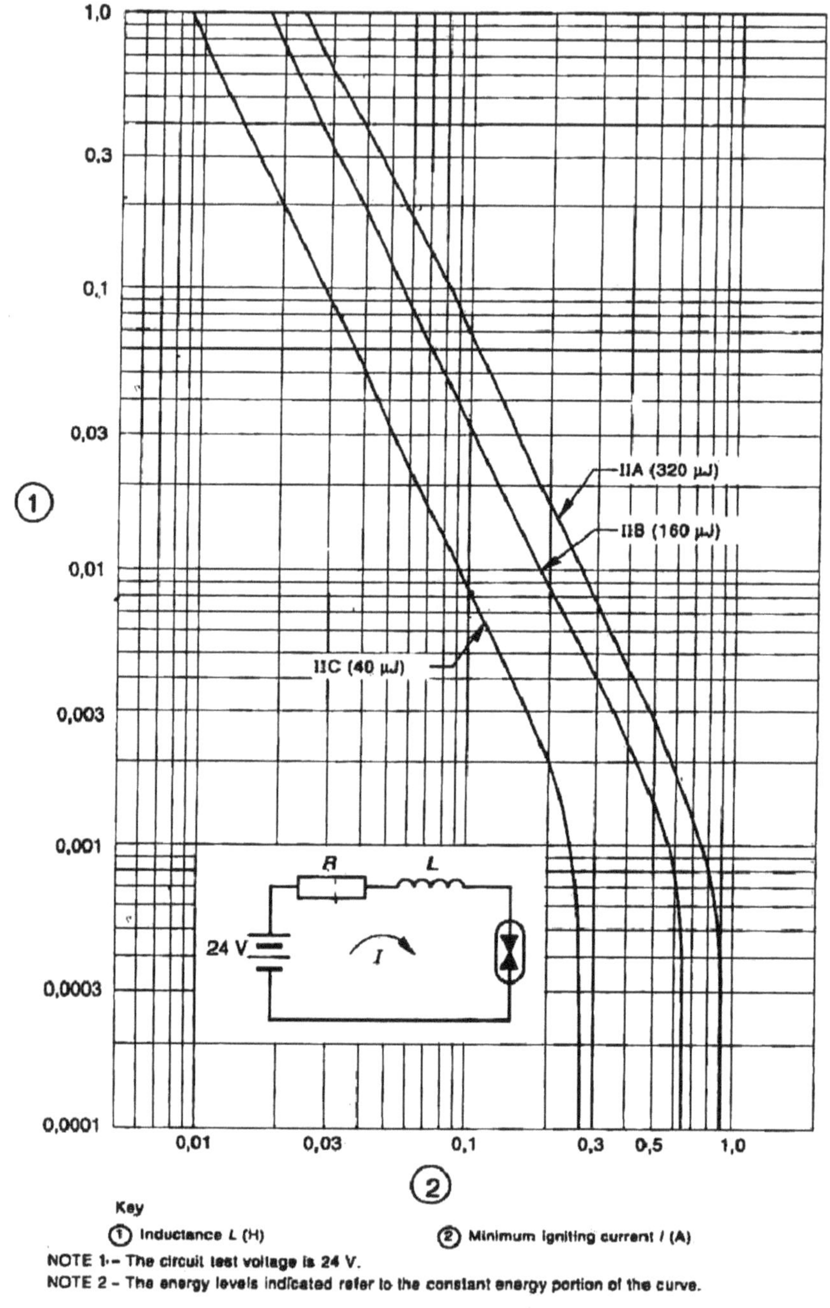

Fig. 12.5 Group II inductive circuit (www.gminternationalsrl.com)

12.7.3.2 Simple Capacitive Circuits

Let us consider a group I application as shown in Fig. 12.6. It consist of a 30 V battery connected to a 10 μF capacitor through a suitably mounted infallible 10 kΩ resistor. The value of battery and capacitor is taken as the maximum value and 10 kΩ resistor is taken as the minimum value. Two separate assessments should be done to ensure that:

i) the power supply is intrinsically safe, and
ii) proper capacitor is used.

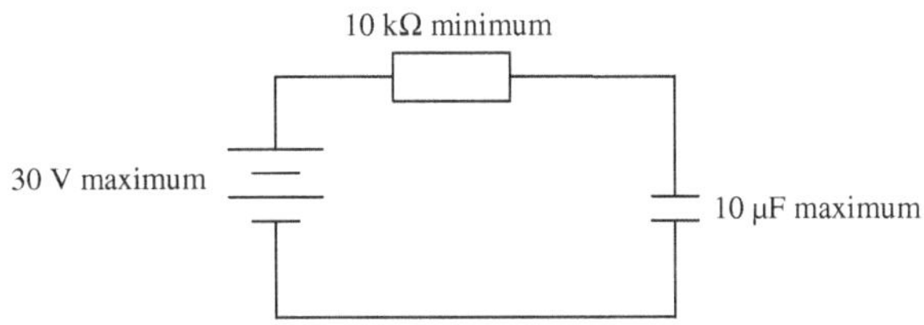

Fig. 12.6 A simple capacitive circuit

(i) Power supply

a) The minimum value of the current-limiting resistor connected to the power supply is taken as 10 kΩ.
b) The maximum short circuit current is 30 V/10 kΩ = 3 mA.
On comparing with the available data of the minimum short circuit current for group I at 30 V, 3 mA is less, thus the power supply is intrinsically safe.

(ii) Capacitor

a) The maximum battery voltage is 30 V and the maximum capacitance value is 10 μF. Since 10 kΩ resistor is infallible, no fault condition shall arise.
b) Considering the safety factor the voltage is raised to 1.5 × 30 V = 45 V.
c) According to Fig. 12.7 at 45 V, the minimum value of capacitance to give ignition is 3 μF and at 30 V it is 8 μF.

So, 10 μF is greater than the minimum capacitance value 3 μF at 45 V, therefore, the circuit cannot be assessed as intrinsically safe.

Similarly, in designing the resistive circuit and group I inductive circuit, the values of minimum igniting current can be taken from Figs. 12.8 and 12.9, respectively (Widginton, 1988). Figure 12.8 gives the value of the minimum igniting current to the corresponding voltage for resistive circuit. Figure 12.9 gives the maximum inductance value corresponding to the minimum igniting current for group I inductive circuit. Figure 12.10 gives the capacitance value corresponding to the minimum igniting voltage for group II capacitive circuits. For group II capacitive circuit, the minimum igniting voltage can be taken from Fig. 12.10.

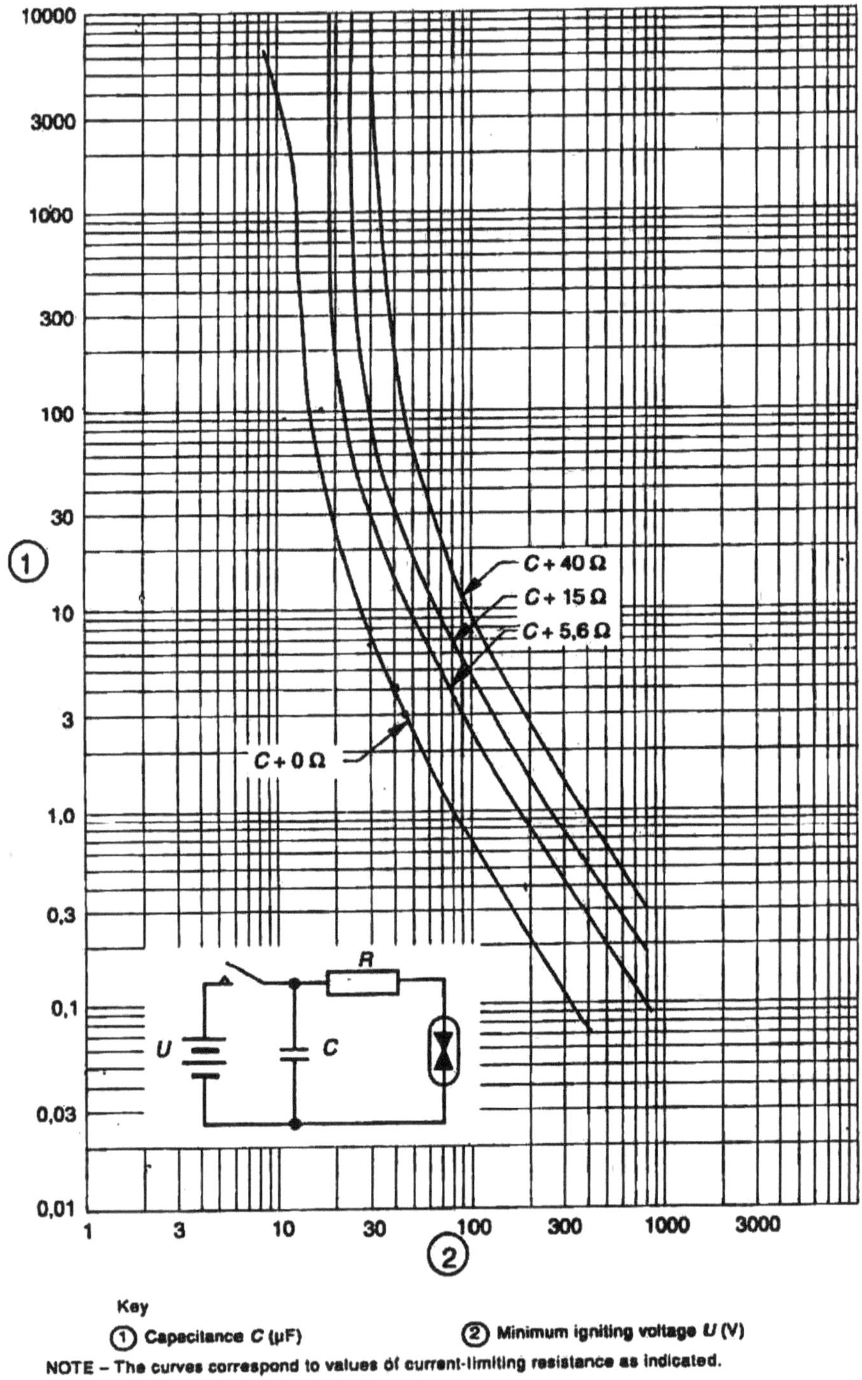

Fig. 12.7 Group I capacitive circuits (www.gminternationalsrl.com)

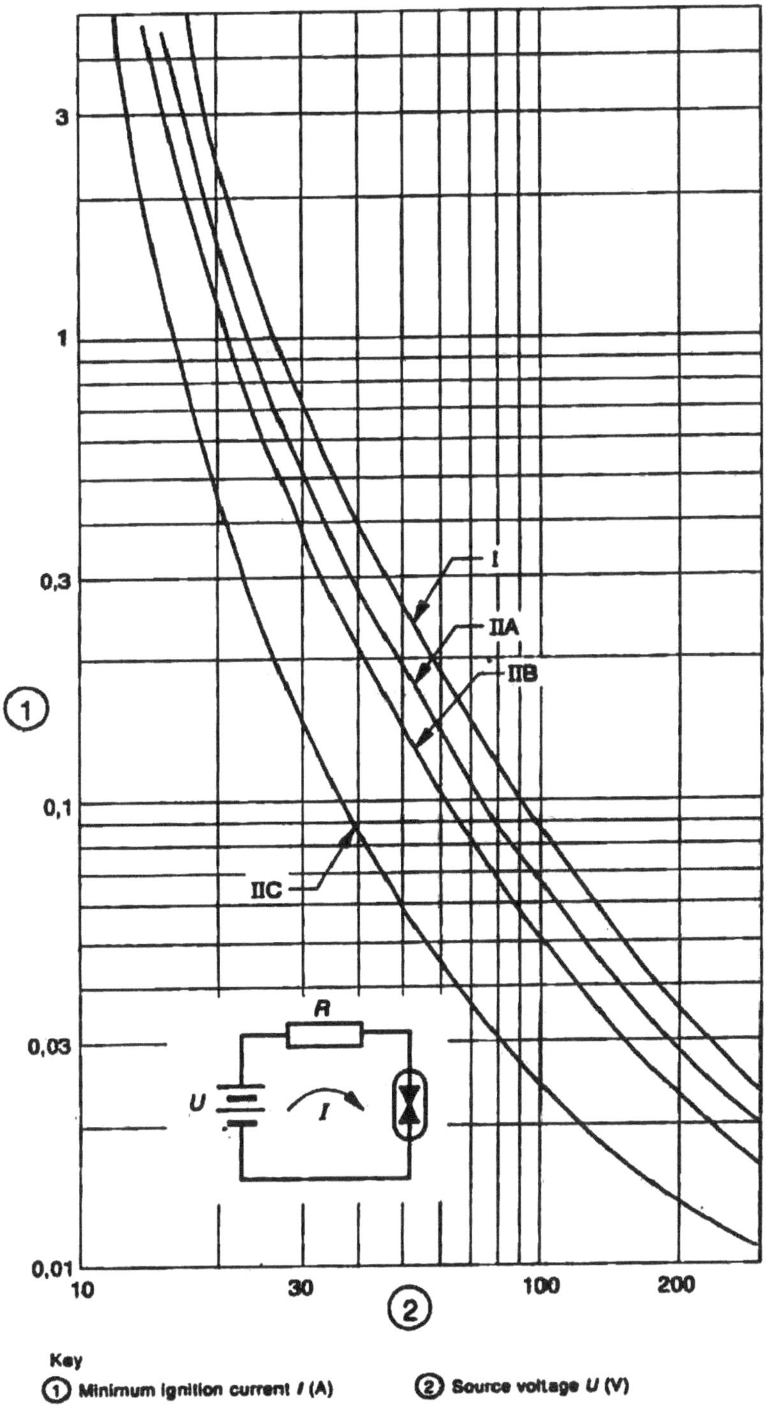

Fig. 12.8 Resistive circuit (www.gminternationalsrl.com)

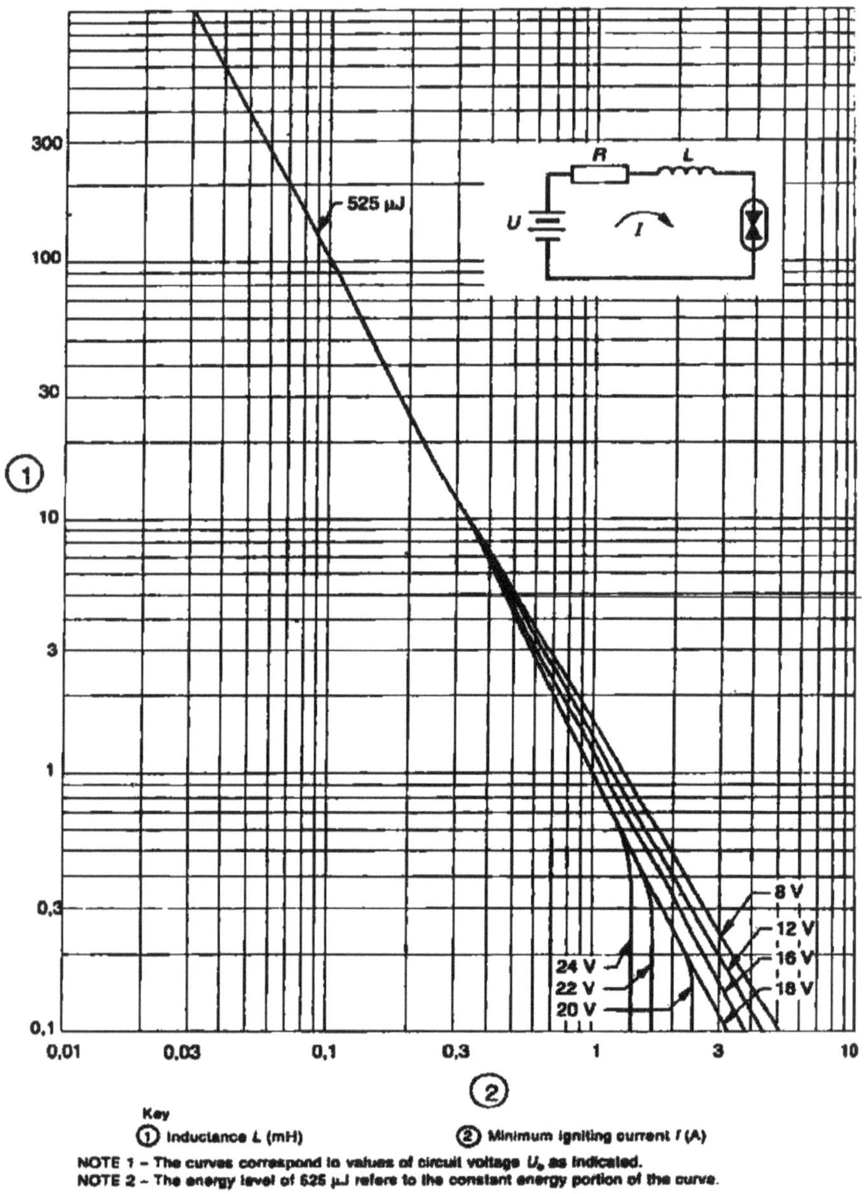

Fig. 12.9 Group I inductive circuits (www.gminternationalsrl.com)

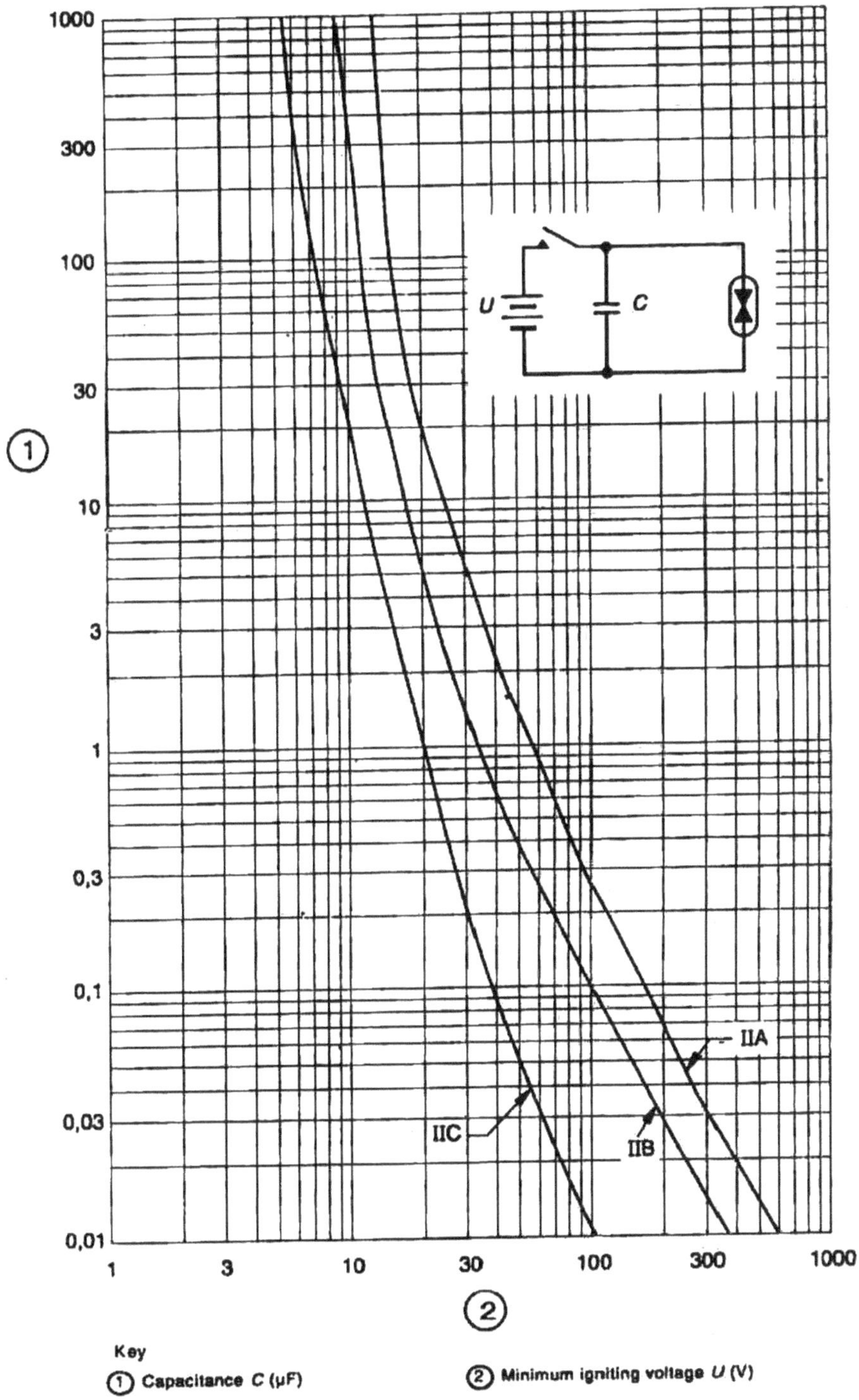

Fig. 12.10 Group II capacitive circuits (www.gminternationalsrl.com)

12.7.4 Analysis by the Parameters

(i) For the analysis of any intrinsically safe designed circuit for gas group II, the values of short circuit current corresponding to the voltage can be taken from Tables 12.11 to 12.12 with an application of the safety factor, unity (1.0) or 1.5 as per the requirement.
(ii) For designing an intrinsically safe circuit for gas group II, the values of minimum permitted capacitance corresponding to voltage can be taken from Tables 12.13 to 12.18 with an application of safety factor, 1 or 1.5 as per the requirement.

Table 12.11 Values of short circuit current corresponding to voltage for group II

	Permitted short circuit current (A)					
	For group IIC apparatus with a safety factor		For group IIB apparatus with a safety factor		For group IIA apparatus with a safety factor	
Voltage	× 1.0	× 1.5	× 1.0	× 1.5	× 1.0	× 1.5
12.0						
12.1	5.00	3.33				
12.2	4.72	3.15				
12.3	4.46	2.97				
12.4	4.21	2.81				
12.5	3.98	2.65				
12.6	3.77	2.51				
12.7	3.56	2.37				
12.8	3.37	2.25				
12.9	3.19	2.13				
13.0	3.02	2.02				
13.1	2.87	1.91				
13.2	2.72	1.81				
13.3	2.58	1.72				
13.4	2.45	1.63				
13.5	2.32	1.55	5.00	3.33		
13.6	2.21	1.47	4.86	3.24		
13.7	2.09	1.40	4.72	3.14		
13.8	1.99	1.33	4.58	3.05		
13.9	1.89	1.26	4.45	2.97		
14.0	1.80	1.20	4.33	2.88		
14.1	1.75	1.16	4.21	2.80		
14.2	1.70	1.13	4.09	2.73		
14.3	1.65	1.10	3.98	2.65		
14.4	1.60	1.07	3.87	2.58		
14.5	1.55	1.04	3.76	2.51		
14.6	1.51	1.01	3.66	2.44		
14.7	1.47	0.98	3.56	2.38		
14.8	1.43	0.95	3.47	2.31	5.00	3.33
14.9	1.39	0.93	3.38	2.25	4.86	3.24

Table 12.11 (continued)

	Permitted short circuit current (A)					
	For group IIC apparatus with a safety factor		For group IIB apparatus with a safety factor		For group IIA apparatus with a safety factor	
Voltage	× 1.0	× 1.5	× 1.0	× 1.5	× 1.0	× 1.5
15.0	1.35	0.900	3.29	2.19	4.73	3.15
15.1	1.31	0.875	3.20	2.14	4.60	3.07
15.2	1.28	0.851	3.12	2.08	4.48	2.99
15.3	1.24	0.828	3.04	2.03	4.36	2.91
15.4	1.21	0.806	2.96	1.98	4.25	2.83
15.5	1.18	0.784	2.89	1.92	4.14	2.76
15.6	1.15	0.769	2.81	1.88	4.03	2.69
15.7	1.12	0.744	2.74	1.83	3.92	2.62
15.8	1.09	0.724	2.68	1.78	3.82	2.55
15.9	1.06	0.705	2.61	1.74	3.72	2.48
16.0	1.03	0.687	2.55	1.70	3.63	2.42
16.1	1.00	0.669	2.48	1.66	3.54	2.36
16.2	0.98	0.652	2.42	1.61	3.45	2.30
16.3	0.95	0.636	2.36	1.57	3.36	2.24
16.4	0.93	0.620	2.31	1.54	3.28	2.19
16.5	0.91	0.604	2.25	1.50	3.20	2.13
16.6	0.88	0.589	2.20	1.47	3.12	2.08
16.7	0.86	0.575	2.15	1.43	3.04	2.03
16.8	0.84	0.560	2.10	1.40	2.97	1.98
16.9	0.82	0.547	2.05	1.37	2.90	1.93
17.0	0.80	0.533	2.00	1.34	2.83	1.89
17.1	0.78	0.523	1.96	1.31	2.76	1.84
17.2	0.77	0.513	1.93	1.28	2.70	1.80
17.3	0.75	0.503	1.89	1.26	2.63	1.76
17.4	0.74	0.493	1.85	1.24	2.57	1.72
17.5	0.73	0.484	1.82	1.21	2.51	1.68
17.6	0.71	0.475	1.79	1.19	2.45	1.64
17.7	0.70	0.466	1.75	1.17	2.40	1.60
17.8	0.69	0.457	1.72	1.15	2.34	1.56
17.9	0.67	0.448	1.69	1.13	2.29	1.53
18.0	0.66	0.440	1.66	1.11	2.24	1.49

Source: www.gminternationalsrl.com

Table 12.12 Values of short circuit current corresponding to voltage for group II

	Permitted short circuit current (mA)					
	For group IIC apparatus with a safety factor		For group IIB apparatus with a safety factor		For group IIA apparatus with a safety factor	
Voltage	× 1.0	× 1.5	× 1.0	× 1.5	× 1.0	× 1.5
18.1	648	432	1,630	1,087	2,188	1,459
18.2	636	424	1,601	1,068	2,139	1,426
18.3	625	417	1,573	1,049	2,091	1,394
18.4	613	409	1,545	1,030	2,045	1,363

Table 12.12 (continued)

	Permitted short circuit current (mA)					
	For group IIC apparatus with a safety factor		For group IIB apparatus with a safety factor		For group IIA apparatus with a safety factor	
Voltage	× 1.0	× 1.5	× 1.0	× 1.5	× 1.0	× 1.5
18.5	602	402	1,518	1,012	2,000	1,333
18.6	592	394	1,491	995	1,967	1,311
18.7	581	387	1,466	977	1,935	1,290
18.8	571	380	1,441	960	1,903	1,269
18.9	561	374	1,416	944	1,872	1,248
19.0	551	367	1,392	928	1,842	1,228
19.1	541	361	1,368	912	1,812	1,208
19.2	532	355	1,345	897	1,784	1,189
19.3	523	348	1,323	882	1,755	1,170
19.4	514	342	1,301	867	1,727	1,152
19.5	505	337	1,279	853	1,700	1,134
19.6	496	331	1,258	839	1,673	1,116
19.7	448	325	1,237	825	1,648	1,098
19.8	480	320	1,217	811	1,622	1,081
19.9	472	314	1,197	798	1,597	1,065
20.0	464	309	1,177	785	1,572	1,048
20.1	456	304	1,158	772	1,549	1,032
20.2	448	299	1,140	760	1,525	1,016
20.3	441	294	1,122	748	1,502	1,001
20.4	434	289	1,104	736	1,479	986
20.5	427	285	1,087	724	1,457	971
20.6	420	280	1,069	713	1,435	957
20.7	413	275	1,053	702	1,414	943
20.8	406	271	1,036	691	1,393	929
20.9	400	267	1,020	680	1,373	915
21.0	394	262	1,004	670	1,353	902
21.1	387	258	989	659	1,333	889
21.2	381	254	974	649	1,314	876
21.3	375	250	959	639	1,295	863
21.4	369	246	945	630	1,276	851
21.5	364	243	930	620	1,258	839
21.6	358	239	916	611	1,240	827
21.7	353	235	903	602	1,222	815
21.8	347	231	889	593	1,205	804
21.9	342	228	876	584	1,189	792
22.0	337	224	863	575	1,172	781
22.1	332	221	851	567	1,156	770
22.2	327	218	838	559	1,140	760
22.3	322	215	826	551	1,124	749
22.4	317	211	814	543	1,109	739
22.5	312	208	802	535	1,093	729
22.6	308	205	791	527	1,078	719
22.7	303	202	779	520	1,064	709
22.8	299	199	768	512	1,050	700
22.9	294	196	757	505	1,036	690

Table 12.12 (continued)

	Permitted short circuit current (mA)					
	For group IIC apparatus with a safety factor		For group IIB apparatus with a safety factor		For group IIA apparatus with a safety factor	
Voltage	× 1.0	× 1.5	× 1.0	× 1.5	× 1.0	× 1.5
23.0	290	193	747	498	1,022	681
23.1	287	191	736	491	1,008	672
23.2	284	189	726	484	995	663
23.3	281	187	716	477	982	655
23.4	278	185	706	471	969	646
23.5	275	183	696	464	956	638
23.6	272	182	687	458	944	629
23.7	270	180	677	452	932	621
23.8	267	178	668	445	920	613
23.9	264	176	659	439	908	605
24.0	261	174	650	433	896	597
24.1	259	173	644	429	885	590
24.2	256	171	637	425	873	582
24.3	253	169	631	421	862	575
24.4	251	167	625	416	852	568
24.5	248	166	618	412	841	561
24.6	246	164	612	408	830	554
24.7	244	163	606	404	820	547
24.8	241	161	601	400	810	540
24.9	239	159	595	396	800	533
25.0	237	158	589	393	790	527
25.1	234	156	583	389	780	520
25.2	232	155	578	385	771	514
25.3	230	153	572	381	762	508
25.4	228	152	567	378	752	502
25.5	226	150	561	374	743	496
25.6	223	149	556	371	734	490
25.7	221	148	551	367	726	484
25.8	219	146	546	364	717	478
25.9	217	145	541	360	708	472
26.0	215	143	536	357	700	467
26.1	213	142	531	354	694	463
26.2	211	141	526	350	688	459
26.3	209	139	521	347	683	455
26.4	207	138	516	344	677	451
26.5	205	137	512	341	671	447
26.6	203	136	507	338	666	444
26.7	202	134	502	335	660	440
26.8	200	133	498	332	655	437
26.9	198	132	493	329	649	433
27.0	196	131	489	326	644	429
27.1	194	130	485	323	639	426
27.2	193	128	480	320	364	422
27.3	191	127	476	317	629	419
27.4	189	126	472	315	624	416
27.5	188	125	468	312	619	412

Table 12.12 (continued)

	Permitted short circuit current (mA)					
	For group IIC apparatus with a safety factor		For group IIB apparatus with a safety factor		For group IIA apparatus with a safety factor	
Voltage	× 1.0	× 1.5	× 1.0	× 1.5	× 1.0	× 1.5
27.6	186	124	464	309	614	409
27.7	184	123	460	306	609	406
27.8	183	122	456	304	604	403
27.9	181	121	452	301	599	399
28.0	180	120	448	299	594	396
28.1	178	119	444	296	590	393
28.2	176	118	440	293	585	390
28.3	175	117	436	291	581	387
28.4	173	116	433	288	576	384
28.5	172	115	429	286	572	381
28.6	170	114	425	284	567	378
28.7	169	113	422	281	563	375
28.8	168	112	418	279	559	372
28.9	166	111	415	277	554	370
29.0	165	110	411	274	550	367
29.1	163	109	408	272	546	364
29.2	162	108	405	270	542	361
29.3	161	107	401	268	538	358
29.4	159	106	398	265	534	356
29.5	158	105	395	263	530	353
29.6	157	105	392	261	526	351
29.7	155	104	388	259	522	348
29.8	154	103	385	257	518	345
29.9	153	102	382	255	514	343
30.0	152	101	379	253	510	340
30.2	149	99.5	373	249	503	335
30.4	147	97.9	367	245	496	330
30.6	145	96.3	362	241	489	326
30.8	142	94.8	356	237	482	321
31.0	140	93.3	350	233	475	317
31.2	138	92.2	345	230	468	312
31.4	137	91.0	339	226	462	308
31.6	135	89.9	334	223	455	303
31.8	133	88.8	329	219	449	299
32.0	132	87.8	324	216	442	295
32.2	130	86.7	319	213	436	291
32.4	129	85.7	315	210	431	287
32.6	127	84.7	310	207	425	283
32.8	126	83.7	305	204	419	279
33.0	124	82.7	301	201	414	276
33.2	123	81.7	297	198	408	272
33.4	121	80.8	292	195	403	268
33.6	120	79.8	288	192	398	265
33.8	118	78.9	284	189	393	262
34.0	117	78.0	280	187	389	259

Table 12.12 (continued)

Voltage	Permitted short circuit current (mA)					
	For group IIC apparatus with a safety factor		For group IIB apparatus with a safety factor		For group IIA apparatus with a safety factor	
	× 1.0	× 1.5	× 1.0	× 1.5	× 1.0	× 1.5
34.2	116	77.2	277	185	384	256
34.4	114	76.3	274	183	380	253
34.6	113	75.4	271	181	376	251
34.8	112	74.6	269	179	372	248
35.0	111	73.8	266	177	368	245
35.2	109	73.0	263	175	364	242
35.4	108	72.2	260	174	360	240
35.6	107	71.4	258	172	356	237
35.8	106	70.6	255	170	352	235
36.0	105	69.9	253	166	348	232
36.2	104	69.1	250	167	345	230
36.4	103	68.4	248	165	341	227
36.6	102	67.7	245	164	337	225
36.8	100	66.9	243	162	334	223
37.0	99.4	66.2	241	160	330	220
37.2	98.3	65.6	238	159	327	218
37.4	97.3	64.9	236	157	324	216
37.6	96.3	64.2	234	156	320	214
37.8	95.3	63.6	231	154	317	211
38.0	94.4	62.9	229	153	314	209
38.2	93.4	62.3	227	151	311	207
38.4	92.5	61.6	225	150	308	205
38.6	91.5	61.0	223	149	304	203
38.8	90.6	60.4	221	147	301	201
39.0	89.7	59.8	219	146	298	199
39.2	88.8	59.2	217	145	296	197
39.4	88.0	58.6	215	143	293	195
39.6	87.1	58.1	213	142	290	193
39.8	86.3	57.5	211	141	287	191
40.0	85.4	57.0	209	139	284	190
40.5	83.4	55.6	205	136	278	185
41.0	81.4	54.3	200	133	271	181
41.5	79.6	53.0	196	131	265	177
42.0	77.7	51.8	192	128	259	173
42.5	76.0	50.6	188	125	253	169
43.0	74.3	49.5	184	122	247	165
43.5	72.6	48.4	180	120	242	161
44.0	71.0	47.4	176	117	237	158
44.5	69.5	46.3	173	115	231	154
45.0	68.0	45.3	169	113	227	151

Source: www.gminternationalsrl.com

Table 12.13 Values of permitted capacitance corresponding to voltage for group II

	Permitted capacitance (μF)					
	For group IIC apparatus with a safety factor		For group IIB apparatus with a safety factor		For group IIA apparatus with a safety factor	
Voltage	× 1.0	× 1.5	× 1.0	× 1.5	× 1.0	× 1.5
5.0		100				
5.1		88				
5.2		79				
5.3		71				
5.4		65				
5.5		58				
5.6	1,000	54				
5.7	860	50				
5.8	750	46				
5.9	670	43				
6.0	600	40		1,000		
6.1	535	37		880		
6.2	475	34		790		
6.3	420	31		720		
6.4	370	28		650		
6.5	325	25		570		
6.6	285	22		500		
6.7	250	19.6		430		
6.8	220	17.9		380		
6.9	200	16.8		335		
7.0	175	15.7		300		
7.1	155	14.6		268		
7.2	136	13.5		240		
7.3	120	12.7		216		
7.4	110	11.9		195		
7.5	100	11.1		174		
7.6	92	10.4		160		
7.7	85	9.8		145		
7.8	79	9.3		130		
7.9	74	8.8		115		
8.0	69	8.4		100		
8.1	65	8.0		90		
8.2	61	7.6		81		
8.3	56	7.2		73		
8.4	54	6.8		66		
8.5	51	6.5		60		
8.6	49	6.2		55		
8.7	47	5.9		50		1,000
8.8	45	5.5		46		730
8.9	42	5.2		43		590
9.0	40	4.9	1,000	40		500
9.1	38	4.6	920	37		446
9.2	36	4.3	850	34		390
9.3	34	4.1	790	31		345
9.4	32	3.9	750	29		300
9.5	30	3.7	700	27		255

Table 12.13 (continued)

	Permitted capacitance (μF)					
	For group IIC apparatus with a safety factor		For group IIB apparatus with a safety factor		For group IIA apparatus with a safety factor	
Voltage	× 1.0	× 1.5	× 1.0	× 1.5	× 1.0	× 1.5
9.6	28	3.6	650	26		210
9.7	26	3.5	600	24		170
9.8	24	3.3	550	23		135
9.9	22	3.2	500	22		115
10.0	20.0	3.0	450	20.0		100
10.1	18.7	2.87	410	19.4		93
10.2	17.8	2.75	380	18.7		88
10.3	17.1	2.63	350	18.0		83
10.4	16.4	2.52	325	17.4		79
10.5	15.7	2.41	300	16.8		75
10.6	15.0	2.32	280	16.2		72
10.7	14.2	2.23	260	15.6		69
10.8	13.5	2.14	240	15.0		66
10.9	13.0	2.05	225	14.4		63
11.0	12.5	1.97	210	13.8		60
11.1	11.9	1.90	195	13.2		57.0
11.2	11.4	1.84	180	12.6		54.0
11.3	10.9	1.79	170	12.1		51.0
11.4	10.4	1.71	160	11.7		48.0
11.5	10.0	1.64	150	11.2		46.0
11.6	9.6	1.59	140	10.8		43.0
11.7	9.3	1.54	130	10.3		41.0
11.8	9.0	1.50	120	9.9		39.0
11.9	8.7	1.45	110	9.4		37.0
12.0	8.4	1.41	100	9.0		36.0
12.1	8.1	1.37	93	8.7		34.0
12.2	7.9	1.32	87	8.4		33.0
12.3	7.6	1.28	81	8.1		31.0
12.4	7.2	1.24	75	7.9		30.0
12.5	7.0	1.20	70	7.7		28.0
12.6	6.8	1.15	66	7.4		27.0
12.7	6.6	1.10	62	7.1		25.4
12.8	6.4	1.06	58	6.8		24.2
12.9	6.2	1.03	55	6.5		23.2
13.0	6.0	1.00	52	6.2	1,000	22.5
13.1	5.7	0.97	49	6.0	850	21.7
13.2	5.4	0.94	46	5.8	730	21.0
13.3	5.3	0.91	44	5.6	630	20.2
13.4	5.1	0.88	42	5.5	560	19.5
13.5	4.9	0.85	40	5.3	500	19.0
13.6	4.6	0.82	38	5.2	450	18.6
13.7	4.4	0.79	36	5.0	420	18.1
13.8	4.2	0.76	34	4.9	390	17.7
13.9	4.1	0.74	32	4.7	360	17.3
14.0	4.0	0.73	30	4.60	330	17.0
14.1	3.9	0.71	29	4.49	300	16.7

Table 12.13 (continued)

	Permitted capacitance (μF)					
	For group IIC apparatus with a safety factor		For group IIB apparatus with a safety factor		For group IIA apparatus with a safety factor	
Voltage	× 1.0	× 1.5	× 1.0	× 1.5	× 1.0	× 1.5
14.2	3.8	0.70	28	4.39	270	16.4
14.3	3.7	0.68	27	4.28	240	16.1
14.4	3.6	0.67	26	4.18	210	15.8
14.5	3.5	0.65	25	4.07	185	15.5
14.6	3.4	0.64	24	3.97	160	15.2
14.7	3.3	0.62	23	3.86	135	14.9
14.8	3.2	0.61	22	3.76	120	14.6
14.9	3.1	0.59	21	3.65	110	14.3
15.0	3.0	0.58	20.2	3.55	100	14.0
15.1	2.90	0.57	19.7	3.46	95	13.7
15.2	2.82	0.55	19.2	3.37	91	13.4
15.3	2.76	0.53	18.7	3.28	88	13.1
15.4	2.68	0.521	18.2	3.19	85	12.8
15.5	2.60	0.508	17.8	3.11	82	12.5
15.6	2.52	0.497	17.4	3.03	79	12.2
15.7	2.45	0.487	17.0	2.95	77	11.9
15.8	2.38	0.478	16.6	2.88	74	11.6
15.9	2.32	0.469	16.2	2.81	72	11.3
16.0	2.26	0.460	15.8	2.75	70	11.0
16.1	2.20	0.451	15.4	2.69	68	10.7

Source: www.gminternationalsrl.com

Table 12.14 Values of permitted capacitance corresponding to voltage for group II

	Permitted capacitance					
	For group IIC apparatus with a safety factor		For group IIB apparatus with a safety factor		For group IIA apparatus with a safety factor	
	× 1.0	× 1.5	× 1.0	× 1.5	× 1.0	× 1.5
Voltage	μF	nF	μF	μF	μF	μF
16.2	2.14	442	15.0	2.63	66	10.5
16.3	2.08	433	14.6	2.57	64	10.2
16.4	2.02	424	14.2	2.51	62	10.0
16.5	1.97	415	13.8	2.45	60	9.8
16.6	1.92	406	13.4	2.40	58	9.6
16.7	1.88	398	13.0	2.34	56	9.4
16.8	1.84	390	12.6	2.29	54	9.3
16.9	1.80	382	12.3	2.24	52	9.1
17.0	1.76	375	12.0	2.20	50	90
17.1	1.71	367	11.7	2.15	48	8.8
17.2	1.66	360	11.4	2.11	47	8.7
17.3	1.62	353	11.1	2.06	45	8.5
17.4	1.59	346	10.8	2.02	44	8.4

Table 12.14 (continued)

	Permitted capacitance					
	For group IIC apparatus with a safety factor		For group IIB apparatus with a safety factor		For group IIA apparatus with a safety factor	
	× 1.0	× 1.5	× 1.0	× 1.5	× 1.0	× 1.5
Voltage	μF	nF	μF	μF	μF	μF
17.5	1.56	339	10.5	1.97	42	8.2
17.6	1.53	333	10.2	1.93	40	8.1
17.7	1.50	327	9.9	1.88	39	8.0
17.8	1.47	321	9.6	1.84	38	7.9
17.9	1.44	315	9.3	1.80	37	7.7
18.0	1.41	309	9.0	1.78	36	7.6
18.1	1.38	303	8.8	1.75	35	7.45
18.2	1.35	297	8.6	1.72	34	7.31
18.3	1.32	291	8.4	1.70	33	7.15
18.4	1.29	285	8.2	1.69	32	7.00
18.5	1.27	280	8.0	1.67	31	6.85
18.6	1.24	275	7.9	1.66	30	6.70
18.7	1.21	270	7.8	1.64	29	6.59
18.8	1.18	266	7.6	1.62	28	6.48
18.9	1.15	262	7.4	1.60	27	6.39
19.0	1.12	258	7.2	1.58	26	6.30
19.1	1.09	252	7.0	1.56	25.0	6.21
19.2	1.06	251	6.8	1.55	24.2	6.12
19.3	1.04	248	6.6	1.52	23.6	6.03
19.4	1.02	244	6.4	1.51	23.0	5.95
19.5	1.00	240	6.2	1.49	22.5	5.87
19.6	0.98	235	6.0	1.47	22.0	5.80
19.7	0.96	231	5.9	1.45	21.5	5.72
19.8	0.94	227	5.8	1.44	21.0	5.65
19.9	0.92	223	5.7	1.42	20.5	5.57
20.0	0.90	220	5.6	1.41	20.0	5.50
20.1	0.88	217	5.5	1.39	19.5	5.42
20.2	0.86	213	5.4	1.38	19.2	5.35
20.3	0.84	209	5.3	1.36	18.9	5.27
20.4	0.82	206	5.2	1.35	18.6	5.20
20.5	0.80	203	5.1	1.33	18.3	5.12
20.6	0.78	200	5.0	1.32	18.0	5.05
20.7	0.76	197	4.9	1.31	17.7	4.97
20.8	0.75	194	4.8	1.30	17.4	4.90
20.9	0.74	191	4.7	1.28	17.2	4.84
21.0	0.73	188	4.6	1.27	17.0	4.78
21.1	0.72	185	4.52	1.25	16.8	4.73
21.2	0.71	183	4.45	1.24	16.6	4.68
21.3	0.70	181	4.39	1.23	16.4	4.62
21.4	0.69	179	4.32	1.22	16.2	4.56
21.5	0.68	176	4.25	1.20	16.0	4.50
21.6	0.67	174	4.18	1.19	15.8	4.44
21.7	0.66	172	4.11	1.17	15.6	4.38
21.8	0.65	169	4.04	1.16	15.4	4.32

Source: www.gminternationalsrl.com

Table 12.15 Values of permitted capacitance corresponding to voltage for group II

	Permitted capacitance					
	For group IIC apparatus with a safety factor		For group IIB apparatus with a safety factor		For group IIA apparatus with a safety factor	
	× 1.0	× 1.5	× 1.0	× 1.5	× 1.0	× 1.5
Voltage	nF	nF	μF	μF	μF	μF
21.9	640	167	3.97	1.15	15.2	4.26
22.0	630	165	3.90	1.14	15.0	4.20
22.1	620	163	3.83	1.12	14.8	4.14
22.2	610	160	3.76	1.11	14.6	4.08
22.3	600	158	3.69	1.10	14.4	4.03
22.4	590	156	3.62	1.09	14.2	3.98
22.5	580	154	3.55	1.08	14.0	3.93
22.6	570	152	3.49	1.07	13.8	3.88
22.7	560	149	3.43	1.06	13.6	3.83
22.8	550	147	3.37	1.05	13.4	3.79
22.9	540	145	3.31	1.04	13.2	3.75
23.0	530	143	3.25	1.03	13.0	3.71
23.1	521	140	3.19	1.02	12.8	3.67
23.2	513	138	3.13	1.01	12.6	3.64
23.3	505	136	3.08	1.00	12.4	3.60
23.4	497	134	3.03	0.99	12.2	3.57
23.5	490	132	2.98	0.98	12.0	3.53
23.6	484	130	2.93	0.97	11.8	3.50
23.7	478	128	2.88	0.96	11.6	3.46
23.8	472	127	2.83	0.95	11.4	3.42
23.9	466	126	2.78	0.94	11.2	3.38
24.0	460	125	2.75	0.93	11.0	3.35
24.1	454	124	2.71	0.92	10.8	3.31
24.2	448	122	2.67	0.91	10.7	3.27
24.3	442	120	2.63	0.90	10.5	3.23
24.4	436	119	2.59	0.89	10.3	3.20
24.5	430	118	2.55	0.88	10.2	3.16
24.6	424	116	2.51	0.87	10.0	3.12
24.7	418	115	2.49	0.87	9.9	3.08
24.8	412	113	2.44	0.86	9.8	3.05
24.9	406	112	2.40	0.85	9.6	3.01
25.0	400	110	2.36	0.84	9.5	2.97
25.1	395	108	2.32	0.83	9.4	2.93
25.2	390	107	2.29	0.82	9.3	2.90
25.3	385	106	2.26	0.82	9.2	2.86
25.4	380	105	2.23	0.81	9.1	2.82
25.5	375	104	2.20	0.80	9.0	2.78
25.6	370	103	2.17	0.80	8.9	2.75
25.7	365	102	2.14	0.79	8.8	2.71
25.8	360	101	2.11	0.78	8.7	2.67
25.9	355	100	2.08	0.77	8.6	2.63
26.0	350	99	2.05	0.77	8.5	2.60
26.1	345	98	2.02	0.76	8.4	2.57
26.2	341	97	1.99	0.75	8.3	2.54

Table 12.15 (continued)

	Permitted capacitance					
	For group IIC apparatus with a safety factor		For group IIB apparatus with a safety factor		For group IIA apparatus with a safety factor	
	× 1.0	× 1.5	× 1.0	× 1.5	× 1.0	× 1.5
Voltage	nF	nF	μF	μF	μF	μF
26.3	337	97	1.96	0.74	8.2	2.51
26.4	333	96	1.93	0.74	8.1	2.48
26.5	329	95	1.90	0.73	8.0	2.45
26.6	325	94	1.87	0.73	8.0	2.42
26.7	321	93	1.84	0.72	7.9	2.39
26.8	317	92	1.82	0.72	7.8	2.37
26.9	313	91	1.80	0.71	7.7	2.35
27.0	309	90	1.78	0.705	7.60	2.33
27.1	305	89	1.76	0.697	7.50	2.31
27.2	301	89	1.74	0.690	7.42	2.30
27.3	297	88	1.72	0.683	7.31	2.28
27.4	293	87	1.71	0.677	7.21	2.26
27.5	289	86	1.70	0.672	7.10	2.24

Source: www.gminternationalsrl.com

Table 12.16 Values of permitted capacitance corresponding to voltage for group II

	Permitted capacitance					
	For group IIC apparatus with a safety factor		For group IIB apparatus with a safety factor		For group IIA apparatus with a safety factor	
	× 1.0	× 1.5	× 1.0	× 1.5	× 1.0	× 1.5
Voltage	nF	nF	μF	nF	μF	μF
27.6	285	86	1.69	668	7.00	2.22
27.7	281	85	1.68	663	6.90	2.20
27.8	278	84	1.67	659	6.80	2.18
27.9	275	84	1.66	654	6.70	2.16
28.0	272	83	1.65	650	6.60	2.15
28.1	269	82	1.63	645	6.54	2.13
28.2	266	81	1.62	641	6.48	2.11
28.3	263	80	1.60	636	6.42	2.09
28.4	260	79	1.59	632	6.36	2.07
28.5	257	78	1.58	627	6.30	2.05
28.6	255	77	1.57	623	6.24	2.03
28.7	253	77	1.56	618	6.18	2.01
28.8	251	76	1.55	614	6.12	2.00
28.9	249	75	1.54	609	6.06	1.98
29.0	247	74	1.53	605	6.00	1.97
29.1	244	74	1.51	600	5.95	1.95

Table 12.16 (continued)

	Permitted capacitance					
	For group IIC apparatus with a safety factor		For group IIB apparatus with a safety factor		For group IIA apparatus with a safety factor	
	× 1.0	× 1.5	× 1.0	× 1.5	× 1.0	× 1.5
Voltage	nF	nF	μF	nF	μF	μF
29.2	241	73	1.49	596	5.90	1.94
29.3	238	72	1.48	591	5.85	1.92
29.4	235	71	1.47	587	5.80	1.91
29.5	232	71	1.46	582	5.75	1.89
29.6	229	70	1.45	578	5.70	1.88
29.7	226	69	1.44	573	5.65	1.86
29.8	224	68	1.43	569	5.60	1.85
29.9	222	67	1.42	564	5.55	1.83
30.0	220	66	1.41	560	5.50	1.82
30.2	215	65	1.39	551	5.40	1.79
30.4	210	64	1.37	542	5.30	1.76
30.6	206	62.6	1.35	533	5.20	1.73
30.8	202	61.6	1.33	524	5.10	1.70
31.0	198	60.5	1.32	515	5.00	1.67
31.2	194	59.6	1.30	506	4.90	1.65
31.4	190	58.7	1.28	497	4.82	1.62
31.6	186	57.8	1.26	489	4.74	1.60
31.8	183	56.9	1.24	482	4.68	1.58
32.0	180	56.0	1.23	475	4.60	1.56
32.2	177	55.1	1.21	467	4.52	1.54
32.4	174	54.2	1.19	460	4.44	1.52
32.6	171	53.3	1.17	452	4.36	1.50
32.8	168	52.4	1.15	444	4.28	1.48
33.0	165	51.5	1.14	437	4.20	1.46
33.2	162	50.6	1.12	430	4.12	1.44
33.4	159	49.8	1.10	424	4.05	1.42
33.6	156	49.2	1.09	418	3.98	1.41
33.8	153	48.6	1.08	412	3.91	1.39
34.0	150	48.0	1.07	406	3.85	1.37
34.2	147	47.4	1.05	401	3.79	1.35
34.4	144	46.8	1.04	397	3.74	1.33
34.6	141	46.2	1.02	393	3.69	1.31
34.8	138	45.6	1.01	390	3.64	1.30
35.0	135	45.0	1.00	387	3.60	1.28
35.2	133	44.4	0.99	383	3.55	1.26
35.4	131	43.8	0.97	380	3.50	1.24
35.6	129	43.2	0.95	376	3.45	1.23
35.8	127	42.6	0.94	373	3.40	1.21
36.0	125	42.0	0.93	370	3.35	1.20
36.2	123	41.4	0.91	366	3.30	1.18
36.4	121	40.8	0.90	363	3.25	1.17

Source: www.gminternationalsrl.com

Table 12.17 Values of permitted capacitance corresponding to voltage for group II

	Permitted capacitance					
	For group IIC apparatus with a safety factor		For group IIB apparatus with a safety factor		For group IIA apparatus with a safety factor	
	× 1.0	× 1.5	× 1.0	× 1.5	× 1.0	× 1.5
Voltage	nF	nF	nF	nF	μF	nF
36.6	119	40.2	890	359	3.20	1,150
36.8	117	39.6	880	356	3.15	1,130
37.0	115	39.0	870	353	3.10	1,120
37.2	113	38.4	860	347	3.05	1,100
37.4	111	37.9	850	344	3.00	1,090
37.6	109	37.4	840	340	2.95	1,080
37.8	107	36.9	830	339	2.90	1,070
38.0	105	36.4	820	336	2.85	1,060
38.2	103	35.9	810	332	2.80	1,040
38.4	102	35.4	800	329	2.75	1,030
38.6	101	35.0	790	326	2.70	1,020
38.8	100	34.6	780	323	2.65	1,010
39.0	99	34.2	770	320	2.60	1,000
39.2	98	33.8	760	317	2.56	980
39.4	97	33.4	750	314	2.52	970
39.6	96	33.1	750	311	2.48	960
39.8	95	32.8	740	308	2.44	950
40.0	94	32.5	730	305	2.40	940
40.2	92	32.2	720	302	2.37	930
40.4	91	31.9	710	299	2.35	920
40.6	90	31.6	700	296	2.32	910
40.8	89	31.3	690	293	2.30	900
41.0	88	31.0	680	290	2.27	890
41.2	87	30.7	674	287	2.25	882
41.4	86	30.4	668	284	2.22	874
41.6	85	30.1	662	281	2.20	866
41.8	84	29.9	656	278	2.17	858
42.0	83	29.7	650	275	2.15	850
42.2	82	29.4	644	272	2.12	842
42.4	81	29.2	638	269	2.10	834
42.6	79	28.9	632	266	2.07	826
42.8	78	28.6	626	264	2.05	818
43.0	77	28.4	620	262	2.02	810
43.2	76	28.1	614	259	2.00	802
43.4	75	27.9	608	257	1.98	794
43.6	74	27.6	602	254	1.96	786
43.8	73	27.3	596	252	1.94	778
44.0	72	27.1	590	250	1.92	770
44.2	71	26.8	584	248	1.90	762
44.4	70	26.6	578	246	1.88	754
44.6	69	26.3	572	244	1.86	746

Table 12.17 (continued)

	Permitted capacitance					
	For group IIC apparatus with a safety factor		For group IIB apparatus with a safety factor		For group IIA apparatus with a safety factor	
	× 1.0	× 1.5	× 1.0	× 1.5	× 1.0	× 1.5
Voltage	nF	nF	nF	nF	μF	nF
44.8	68	26.1	566	242	1.84	738
45.0	67	25.9	560	240	1.82	730
45.2	66	25.7	554	238	1.80	722
45.4	65	25.4	548	236	1.78	714
45.6	64	25.1	542	234	1.76	706
45.8	63	24.9	536	232	1.74	698
46.0	62.3	24.7	530	230	1.72	690

Source: www.gminternationalsrl.com

Table 12.18 Values of permitted capacitance corresponding to voltage for group II

	Permitted capacitance					
	For group IIC apparatus with a safety factor		For group IIB apparatus with a safety factor		For group IIA apparatus with a safety factor	
	× 1.0	× 1.5	× 1.0	× 1.0	× 1.5	× 1.0
Voltage	nF	nF	nF	nF	μF	nF
46.2	61.6	24.4	524	228	1.70	682
46.4	60.9	24.2	518	226	1.68	674
46.6	60.2	23.9	512	224	1.67	666
46.8	59.6	23.7	506	222	1.65	658
47.0	59.0	23.5	500	220	1.63	650
47.2	58.4	23.2	495	218	1.61	644
47.4	57.8	22.9	490	216	1.60	638
47.6	57.2	22.7	485	214	1.59	632
47.8	56.6	22.5	480	212	1.57	626
48.0	56.0	22.3	475	210	1.56	620
48.2	55.4	22.0	470	208	1.54	614
48.4	54.8	21.8	465	206	1.53	609
48.6	54.2	21.5	460	205	1.52	604
48.8	53.6	21.3	455	203	1.50	599
49.0	53.0	21.1	450	201	1.49	594
49.2	52.4	20.8	445	198	1.48	589
49.4	51.8	20.6	440	197	1.46	584
49.6	51.2	20.4	435	196	1.45	579
49.8	50.6	20.2	430	194	1.44	574
50.0	50.0	20.0	425	193	1.43	570
50.5	49.0	19.4	420	190	1.40	558
51.0	48.0	19.0	415	187	1.37	547
51.5	47.0	18.6	407	184	1.34	535
52.0	46.0	18.3	400	181	1.31	524
52.5	45.0	17.8	392	178	1.28	512

Table 12.18 (continued)

	Permitted capacitance					
	For group IIC apparatus with a safety factor		For group IIB apparatus with a safety factor		For group IIA apparatus with a safety factor	
	× 1.0	× 1.5	× 1.0	× 1.0	× 1.5	× 1.0
Voltage	nF	nF	nF	nF	μF	nF
53.0	44.0	17.4	385	175	1.25	501
53.5	43.0	17.0	380	172	1.22	490
54.0	42.0	16.8	375	170	1.20	479
54.5	41.0	16.6	367	168	1.18	468
55.0	40.0	16.5	360	166	1.16	457

Source: www.gminternationalsrl.com

12.8 Parameters Considered for Intrinsic Safety

12.8.1 Ratings of Components

After considering the normal operation and the fault condition, the other remaining components on which the type of protection depends shall not operate at more than two-third of their maximum current, voltage, and power related to the rating of the device, with the specification of mounting conditions and temperature range. However, some devices like transformers, fuses, thermal trips, relays, and switches are allowed to operate at their normal ratings in order to function correctly.

12.8.2 Connectors

Connectors for internal connections, plug-in cards, and components shall be designed in such a manner that an incorrect connection or interchangeability with other connectors in the same electrical apparatus is not possible unless it does not result in an unsafe condition. If the type of protection depends on a connection, the failure to a high resistance or open circuit of the connection shall be a countable fault. If the connector carries earthed circuit and the type of protection depends on the earth connection then the connector shall comprise of:

- three independent connecting elements – for the "ia" category of circuits;
- two independent connecting elements – for "ib" category of circuits.

These elements shall be connected in parallel.

12.8.3 Fuses

Fuses are used to protect other component. The current value of 1.7 In and a voltage rating of at least Um (or Ui in intrinsically safe apparatus and circuits) shall be used here. Its time—current characteristics shall ensure that, the transient rating of protected components are not exceeded. Encapsulation is used for the protection of fuse located in explosive atmosphere and a current-limiting device is used to limit the current to a value which is less than the rated breaking capacity of the fuse. The rated values shall be at least:

– current rating $1.5 \times 1.7 \times In$
– voltage rating Um or Ui
– power rating $1.5 \times (1.7 \times In)^2 \times$ resistance of current limiting device.

12.8.4 Cells and Batteries

12.8.4.1 Primary and Secondary Cells and Batteries

Cell and battery are confirmed before putting in use for the safety of intrinsically safe apparatus and associated apparatus. Thus, some cell or batteries are of type that there can be no spillage of electrolyte or they are enclosed to prevent damage by the electrolyte to the component on which intrinsic safety depends. Gas tight or valve-regulated cell or batteries fulfill this requirement.

12.8.4.2 Considering Features

- Cell/batteries maximum voltage shall be considered as the open circuit voltage just after a full charge.
- By taking the onerous value of the short circuit current and the maximum open circuit voltage, the internal resistance of the cell or batteries shall be determined.
- Use of series current-limiting resistors at the battery output to ensure that output of the battery is intrinsically safe.
- Use of current-limiting device in order to ensure the safety of the battery, which can be done by using:
  - two diodes for – category "ib"
  - three diodes for – category "ia"

  and the whole circuit shall be protected by a rated fuse.
- Maximum of Um input voltage shall be applied to the connection facilities in order to charge the battery.
- Cells' and batteries' external charging contacts are taken care of such that any type of short circuit is prevented and is also prohibited in delivering any type of ignition-capable energy.
- If any reverse polarity power sources are present in the design which can produce battery aiding current, two shunt diodes are connected across the output lines of the battery assembly to shunt aiding currents around the battery assembly.
- Table 12.19 indicates the cell voltage for different types of cells and batteries.

Table 12.19 Cell voltage

IEC type	Cell type	Peak open circuit voltage for spark ignition hazard	Nominal voltage for component surface temperature assessment
K	Nickel–cadmium	1.5	1.3
	Lead--acid (dry)	2.35	2.2
	Lead–acid (wet)	2.67	2.2
L	Alkaline–manganese	1.65	1.5
M	Mercury–zinc	1.37	1.35
N	Mercury–manganese dioxide–zinc	1.6	1.4
	Silver–zinc	1.63	1.55
S	Zinc–air	1.55	1.4
A	Lithium–manganese dioxide	3.7	3.0
C	Zinc–manganese dioxide (Zinc–carbon: Leclanche)	1.725	1.5
	Nickel–hydride	1.6	1.3

Source: www.isa.org

12.8.4.3 Battery Construction

The cell and battery should be constructed in any of the following types as per the requirement:

a) Sealed (gas tight) cell or batteries;
b) Sealed (valve regulated) cell or batteries; and
c) Cells or batteries sealed but without any pressure-relief devices. In this type, there is no requirement of addition of electrolyte lifelong and shall have a sealed metallic and plastic enclosures.

12.8.5 Semiconductors

In associated apparatus, semiconductor devices should be capable of withstanding the maximum AC and DC current. Any type of transient effect generated due to the power source shall be ignored. Semiconductors can be used as shunt voltage limiters and series current limiters.

12.8.5.1 Shunt Voltage Limiters

Semiconductors should be capable of carrying without open circuiting 1.5 times the current which would flow at their place of installation, if they failed in the short circuit mode. On considering the following case, this shall be confirmed from their manufacturer's data by:

a) diodes, diode connected transistors, thyristors, and equivalent semiconductor devices having a forward current rating of at least 1.5 times the maximum possible short circuit current.
b) Zener diode being rated as follows:

   i. in the Zener direction (reverse direction) at 1.5 times power that would be dissipated in the Zener mode, and
   ii. in the forward direction at 1.5 times the maximum current when short circuited.

12.8.5.2 Series Current Limiters

The use of three series-blocking diodes in circuits of category "ia" is permitted. However, other semiconductors and controllable semiconductors devices shall be used as series current-limiter devices only in "ib" apparatus.

Semiconductors and controllable semiconductor as current-limiting device shall not be used for "ia" category of apparatus, but for "ib" category of apparatus.

Semiconductors and controllable semiconductor as shunt-voltage-limiting device, used for "ia" category of apparatus.

12.8.6 Failure

Components and connections may fail if following points are not considered carefully:

i) In both normal operation and after application of the fault condition, any remaining components on which the type of protection depends, are not operated at more than two-third of their maximum current, voltage, and power related to the rating of the devices, the mounting conditions and the temperature range specified. But some devices like transformers fuses, thermal trips, relays, and switches, need to operate with normal rating in order to function correctly. If the above protections are not considered, the failure shall be a countable fault.
ii) Where a fault can lead to a subsequent fault, then the two are considered as a single fault.
iii) The failure of resistors to any value of resistance between open circuit and short circuit shall be a countable fault.
iv) Semiconductor devices shall be considered to fail to short circuit or to open circuit and also to the state due to the failure of other component. Integrated circuit can fail so that any combination of short and open circuits can exist between their external connections.

Although any combination can be assumed, once that fault is applied it cannot be changed and if any capacitance or inductance is connected to the device under the fault condition, then it is onerous connection.

v) Connection shall be considered to fail to open circuit and if free to move, may connect to any part of the circuit within the range of movement. The initial break is one countable fault and the reconnection is a second countable fault.
vi) Clearance, creepage, and separation distance shall be considered.
vii) Failures of capacitors to open circuit, short circuit, and any value less than the maximum value shall be considered.
viii) Failure of inductors to open circuit and any value between nominal resistance and short circuit but only to inductance to resistance ratios lower than that derived from the inductance specification shall be considered.
ix) Open circuit failure of any wire or printed circuit track, including its connection, is single countable fault.

12.9 Infallible Components

Infallible components, infallible assemblies of components, and infallible connections imply that they are safe to be used in an intrinsic safe circuit. No fault condition arises due to these components. They are used in interconnecting the intrinsic safe circuits. These components are used to limit the voltage, current, and power, which are transferred between the circuits. These component needs to be properly rated.

12.9.1 Transformers

Transformers have primary and secondary windings. If any short circuit between the winding takes place, then the current is limited and finally the transformer is no longer serviceable and must be discarded or rewound. However, the main transformer which is infallible does not fall to a short circuit between any winding supplying an intrinsic safe circuit and any other windings. Short circuits within and open circuits of windings shall be considered to occur, but this combination of fault, which may result in an increased output voltage, shall not be taken into account.

12.9.1.1 Transformers Construction

All windings for supplying intrinsically safe circuits shall be separated from all other windings by any one of the following types of construction:

(i) For type I construction, the winding shall be placed, either (a) on one leg of the core, side by side, or (b) on different legs of the core, maintaining the separation parameters in the windings according to the Table 12.10.

(ii) For type II construction, the winding shall be wounded one over another with, either (a) solid insulation in accordance with the separation parameters between the windings, or (b) an earthed screen (made of copper foil) between the windings or equivalent wire windings (wire screen). The thickness of copper foil or the wire screen is given in Table 12.20 with respect to rated current of fuse.

Table 12.20 Minimum foil thickness or minimum wire diameter of the screen

Parameter	Rating of the fuse (A)					
	0.1	0.5	1	2	3	5
Minimum thickness of foil screen (mm)	0.05	0.05	0.075	0.15	0.25	0.3
Minimum diameter of wire of screen (mm)	0.2	0.45	0.63	0.9	1.12	1.4

Source: www.gminternationalsrl.com

12.9.1.2 Transformer Protective Measures

Safety is a major consideration for any transformer. Electrical contact between primary and secondary windings must not be allowed under any realistic fault condition. The input circuit of mains transformers intended for supplying intrinsically safe circuits shall be protected either by a fuse or a rated circuit breaker. Overheating of transformers is protected by an embedded thermal fuse or thermal device. An earthed metal screen is used to protect input and output windings and nonearthed input line shall be protected by a fuse or circuit breaker.

The transformers along with its associated devices like fuses, circuit breakers, thermal devices, and resistors connected to winding terminals shall maintain a safety electrical isolation between power supply and intrinsically safe circuits even if any one of the output windings is short circuited and all other output windings are subjected to the maximum rated electrical load. The safe electrical isolation voltage for transformers shall be according to Table 12.21. The Un is the highest rated voltage of any winding. The input current shall be adjusted to 1.7 In (www.msha.gov, www.sound.westhost.com/xfmr2.htm).

Table 12.21 AC test voltage of the transformer

Test voltage (a.c. r.m.s.)	Where applied
4 Un or 2,500 V, whichever is greater	Between input and output windings
2 Un or 1,000 V, whichever is greater	Between all output windings and the core and shield
2 Un + 1,000 or 1,500 V, whichever is greater	Between each winding supplying intrinsic safe circuits and every other output windings

12.9.1.3 Damping Windings

Damping windings are used as short circuit turns to minimize the effect of inductance. They shall be considered not subjected to open circuit fault, if their mechanical construction is reliable, e.g., seamless metal tubes or winding of bare wire continuously short circuits by soldering.

12.9.2 Current-Limiting Resistors

An infallible current-limiting resistor shall be constructed in such a manner that no fault mode (usually burnout) can cause it to become short circuit or become more conductive. An infallible resistor must also have a continuous power rating of at least 1.5× the power that it will dissipate, and thus it is considered as one countable fault as it shall fail to open circuit condition. It shall be any of the following types:

a) Metal film type.
b) Wire wound type with protection to prevent unwinding of wire in the event of breakage.
c) Printed resistors as used in hybrid and similar circuits covered by coating.

Carbon resistor shall not be used.

12.9.3 Blocking Capacitors

The blocking capacitor is used between intrinsically safe and nonintrinsically safe circuits, including its transients for blocking d.c. voltage. Series connection is applied by taking the most onerous value and applying the safety factor of 1.5. Blocking capacitors shall be of high-reliability solid dielectric type. Electrolytic or tantalum capacitors shall not be used. Its insulation is tested by electrical strength test. Either of the two series capacitors in an infallible arrangement of blocking capacitors shall be considered of failing to short or open circuit. A high-reliability solid dielectric-type blocking capacitors shall be taken.

12.10 Use of Shunt Safety Assemblies

An assembly of components shall be considered as a shunt safety assembly, when it ensures the intrinsic safety of a circuit by utilization of shunt components, e.g., diodes or Zener diodes. These are used as shunt components in an infallible shunt safety assembly; they shall form at least two parallel paths of diodes. It shall be connected as such that if either of the shunt path becomes disconnected, the circuit or component being protected becomes disconnected at the same time. Features to be considered for utilization of a shunt safety assembly:

- Either of the shunt paths in the assembly shall be considered as failing to an open circuit condition.
- The voltage of the assembly shall be that of the highest voltage shunt path.
- The failure of either shunt path to short circuit shall be considered as one fault.
- A safety factor shall be used in the application of all the fault counts.

12.10.1 Safety Shunts

A shunt safety assembly shall be considered as a safety shunt when it ensures that the electrical parameters of a specified component or parts of an intrinsically safe apparatus are controlled to values, which do not invalidate intrinsic safety. An assembly of suitably rated bridge-connected diodes shall be considered as an infallible safety shunt. It shall be subjected to transient analysis when connected to power supplies defined by Um, except when used as follows:

- for the limitation of discharge from energy-storing devices, e.g., inductors or piezoelectric devices; and
- for the limitation of voltage to energy-storing devices, e.g., capacitors.

12.10.2 Shunt Voltage Limiters

A shunt safety assembly shall be considered as a shunt voltage limiter when it ensures that defined voltage level is applied to an intrinsically safe circuit. Shunt voltage limiters shall be subjected to the required analysis of transients when they are connected to power supplies defined by Um. But it shall not be accepted when applied from any of the following devices:

i) an infallible transformer,
ii) a diode safety barrier,
iii) a battery, and
iv) an infallible shunt safety assembly.

12.10.3 Wirings and Connection

Wiring including its connection which forms parts of the apparatus shall be considered as infallible against open circuit failure in the following cases:

a) for wires:-

    i) where two wires are parallel.

ii) where a single wire has a diameter of 0.5 mm and has an unsupported length of 50 mm or is mechanically secured at the point of connection.
iii) where a single wire is of stranded or flexible ribbon-type construction, has a cross-sectional area of at least 0.125 mm^2 (0.4 mm diameter), is not flexed in service, and is either less than 50 mm long or is secured adjacent to its point of connection.

b) for printed board tracks:

i) where two tracks of minimum width of 1 mm are in parallel.
ii) where a single track is at least 2 mm wide or has a width of 1% of its length whichever is greater.
iii) where each track is formed from copper cladding having a normal thickness of minimum 35 μm.

c) for connections (excluding plugs, sockets, and terminals):

i) where there are two terminals in parallel.
ii) where there is a single connection which is screwed or bolted.

12.10.4 Galvanically Separating Components

Optocouplers or relays are isolating components which can provide infallible separation to separate intrinsically safe circuits if these conditions are considered:

i) it should have proper rating of the component.
ii) it should undergo electric strength test before application of Um and Ui.

For proper separation between the intrinsically safe and nonintrinsically safe circuits, the separation distance is also applicable to the isolating components along with the above points.

The protection of nonintrinsically safe terminals shall be done by inclusion of a single-shunt Zener diode protected by a suitably rated fuse so that rating of the component is not exceeded. The Zener diode power rating shall be at least 1.7 In times the diode maximum Zener voltage. For the above, separation distance is not applicable.

Galvanically separating relay, when used, shall have one contact with intrinsically safe circuits and other contact with nonintrinsically safe circuits. The intrinsically safe and nonintrinsically safe circuits shall be separated by insulating or earth metal barrier along with the application of separation distance as given in Table 12.10. The relay shall be designed such that broken or damaged contact arrangements cannot dislodge and impair the integrity of the separation between intrinsically safe and nonintrinsically safe circuits.

12.10.5 Safety Barriers

Safety barriers are protection devices placed between hazardous and nonhazardous area interconnected apparatus with the purpose of limiting the voltage applied to the intrinsically safe circuits. It is followed by an infallible current-limiting resistor that limits the current which can flow into the circuits when there shall be a fault condition and this can be done by (i) diverting the fault energy to earth or (ii) by blocking the fault energy with isolating elements. During fault conditions, voltage and current levels, which can appear in hazardous area, are limited to safe values.

12.10.5.1 Zener Barriers

Principle of Zener

Zener diodes are widely used to regulate the voltage across a circuit. A Zener diode like the normal diode permits current to flow in the forward direction as well as in the reverse direction, when the voltage is larger than the breakdown voltage known as "Zener knee voltage" or "Zener voltage" (Fig. 12.11). A Zener diode contains a heavily doped p-n junction. When sufficient reverse voltage is applied, a p-n junction will experience a rapid avalanche breakdown, electrons tunnels from the valence band of the p-type material to the conduction band of the n-type material and will conduct current in the reverse direction. A reverse-biased Zener diode will exhibit a controlled breakdown and let the current flow to keep the voltage across the Zener

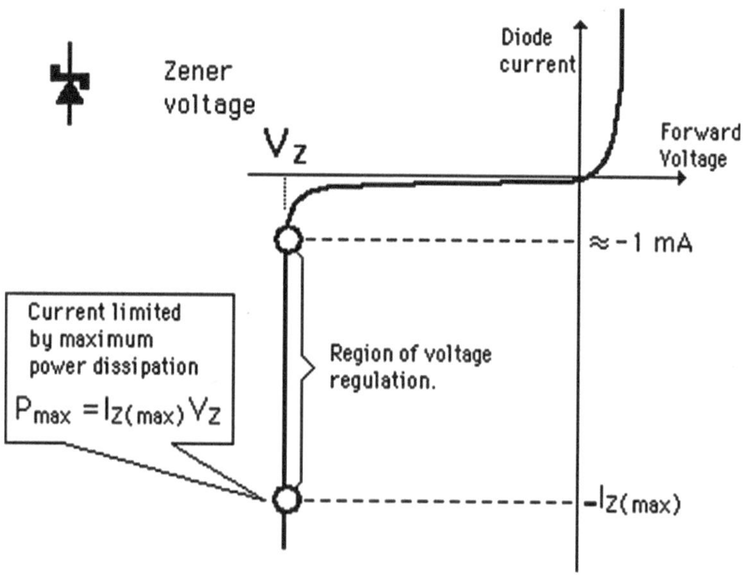

Fig. 12.11 V–I characteristic of Zener

diode at the Zener voltage. For example, a diode with a Zener breakdown voltage of 3.2 V will exhibit a voltage drop of 3.2 V if reverse bias voltage applied across it is more than its Zener voltage.

When this process is taking place, very small changes in voltage can cause very large changes in current. The breakdown process depends upon the applied electric field, so by changing the thickness of the layer to which the voltage is applied, Zener diode can be formed which breaks down at voltages from about 4 V to several hundred volts.

Design Parameter of Zener Barrier

Intrinsic safety barriers are devices that limit the current, voltage, and total energy delivered to the device in a hazardous area or flammable environment in order to prevent an explosion. It uses energy-diversion concept, consisting of very simple network of components. In its normal operating conditions, the barrier passes electrical signals, in both directions, without shunting them. The Zener barrier is designed on the basis of operating voltage, operating current, type of load, type of gas group, and temperature class as described below.

Voltage: The voltage of the nonhazardous and hazardous locations at the barrier should be as follows: (i) on the nonhazardous location of the barrier, the maximum operating voltage should be limited to 250 V a.c. r.m.s. and (ii) on the hazardous location of barrier, the voltage should be limited to a maximum value of +50 V to –50 V depending on the use of positive or negative barrier, subject to condition that energy transfer is below the ignition level of hazardous gases or vapors or dust. The recommended voltages for standardization are +10 V, +20 V, +30 V, –10 V, –20 V, –30 V, etc.

Current: The maximum permissible short circuit current shall be limited to 150 mA subject to condition that maximum short circuit power transfer is below the ignition level of the hazardous gases under reference (short circuit power generally less than 1.230 mW). The current flowing to hazardous areas shall be resistive only.

Temperature: The temperature within barrier enclosure shall not exceed 55°C.

Load: Field equipment used shall be resistive only. If otherwise, these shall be made resistive by adding a redundant diode in parallel with inductance components or preventing capacitance discharge to field wiring by use of resistor and series diode.

Casing: Barriers shall be encapsulated in epoxy resin or into plastic polyester or other suitable resin or in the FLP enclosure by maintaining creepage distance and clearance (Ahirwal, 2005).

During the fault voltage (250 Vrms max), the resulting high current flows to ground through the fuse and Zener diodes. The fuse rating prevents the failure of Zener diodes and isolates it during fault condition.

During fault transient, the open circuit voltage (Voc) at the hazardous area terminals of the barrier is clamped to Zener voltage, while the short circuit current

(Isc), in hazardous area, is limited by the output resistor (Rlim). These values, Voc and Isc, are relevant to assess maximum allowable capacitance and inductance, for the gas groups, which are safe by these values. The efficiency of a barrier depends on a good ground connection, which must provide a return path for the fault current thus preventing any substantial increase in the voltage and current (www.gminternationalsrl.com). This can be done by using a good conductor, which must be run separately from any other structural ground, to the reference ground point (Fig. 12.12). The resistance from the farthest barrier to the ground point must be maintained at less than 1 Ω and standard requirements are for a minimum size of 4 mm^2. "Zener Barriers" are simple, reliable, and low-cost devices. Proper care should be taken in using Zener barriers like:

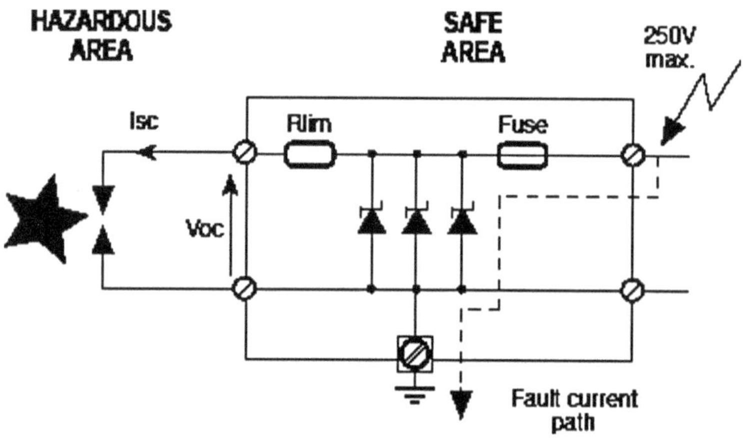

Fig. 12.12 Concept of Zener barrier

- A good ground connection must be provided and maintained.
- Field devices must be isolated from ground (and maintained as such).
- Voltage drop across the barriers can make some applications difficult.
- Improper connection or voltage surges could blow the fuse.
- Very poor common mode rejection (common mode rejection is the immunity of a device to interfering voltages applied at both input terminals with respect to ground).

12.10.5.2 Galvanically Isolated Barriers

"Galvanically isolated barrier" is based on concept of isolation (www.gminternationalsrl.com). The basic difference lies in providing isolation between hazardous and safe area circuits by using components such as transformers, relays, and optocouplers that must comply with requirements of safety standards to guarantee safety (Fig. 12.13). When properly designed, "galvanic isolator barriers" shall not permit the fault voltage (250 Vrms max*)* to reach

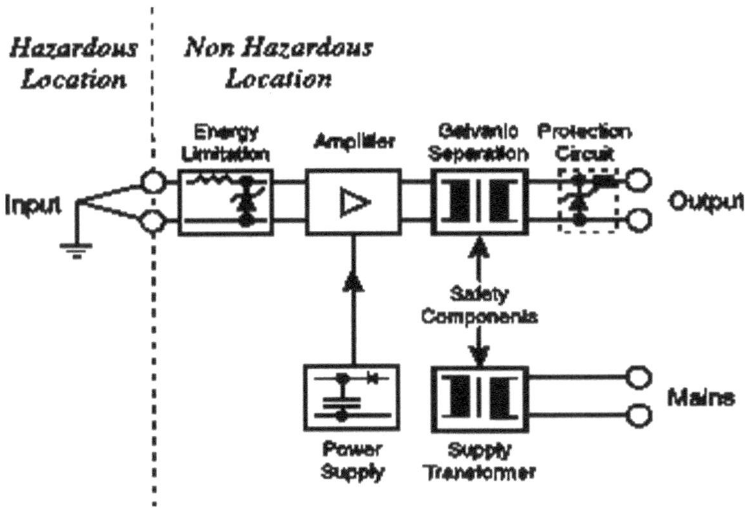

Fig. 12.13 Concept of galvanically isolated barriers

the energy limitation circuit that must be able to withstand only the maximum voltage at the secondary side. Galvanic isolation allows the energy limitation circuit to be floating from ground; thus a ground connection, as well as a protective fuse, for this circuit is not needed. Safety parameters, Voc and Isc, are determined in a similar way to that used for "Zener barriers." The main features of "galvanic isolator barriers" are (Mitra, 1997; Newton, 2007)

- A dedicated ground connection is not required and field devices can be connected to ground.
- Full voltage is available to field devices.
- Signal conditioning and circuit protection are combined in a single unit.
- Simple installation and commissioning with elimination of ground loops.
- High common mode voltage can be tolerated.

12.11 Electromagnetic and Ultrasonic Energy Radiating Equipment

In case of electromagnetic and ultrasonic energy radiating equipment (Stolarczyk, 1989; Hind, 1994; Schultz, 2007), the energy levels shall not exceed the values as given below.

12.11.1 Radio Frequency Sources

The threshold power of radio frequency (10 kHz–300 GHz) for continuous transmission and for pulsed transmission whose pulse durations exceed the thermal initiation time shall not exceed the values shown in Table 12.22. Programmable or software control intended for setting by the user shall not be permitted. Thermal initiation time is the time during which energy deposited by the spark accumulates in a small volume of gas around it without significant thermal dissipation.

Table 12.22 Radio frequency power thresholds

Apparatus for	Threshold power (W)	Thermal initiation time, Averaging period (μs)
Group I	6	200
Group IIA	6	100
Group IIB	3.5	80
Group IIC	2	20
Group III	6	200

Source: www.isa.org

For pulsed radar and other transmissions where the pulses are short compared with the thermal initiation time, the threshold energy values (Z_{th}) shall not exceed those given in Table 12.23.

Table 12.23 Radio frequency energy thresholds

Apparatus for	Threshold energy, Z_{th} (μJ)
Group I	1,500
Group IIA	950
Group IIB	250
Group IIC	50
Group III	1,500

Source: www.isa.org

12.12 Advice to the First-Time Designer

Intrinsic safety is the protection method, which is more favorable than the other existing safety concepts and is based on the principle of limitation of energy stored in an electrical circuit. Many points are to be kept in mind before designing an intrinsically safe circuit for installation in a hazardous area. The basic of these are discussed in the next section.

12.12.1 Basic Design Concepts

If apparatus is intended to be sold and certified for use in hazardous areas, then there are a number of factors to be taken into account to avoid long delays and escalating costs. The following are the possible procedures (Newton, 2007):

i. At an early stage, consult with the certification body. In particular, make sure that your existing quality-control system is acceptable to that organization. The problems created in trying to get a finalized design certified can be insurmountable, and should only be attempted by the brave or desperate. The certification body has to avoid offering a design consultancy service (because of the potential conflict of interests) but they can help to avoid basic mistakes.
ii. Keep things simple. Use little power and low voltage as much as possible. Things are easier at less than 10 V and 500 mW.
iii. From a system viewpoint, ensure that the residual effective capacitance and inductance at the apparatus terminals is less than 1 nF and 10 μH, respectively. Wherever possible, use only resistive limited sources of power.
iv. Create a large flat space for the label.
v. Use of infallible power (mains) transformers to insure that a safe limit voltage appears in the low-voltage circuitry of the devices.
vi. Inclusion of a fused redundant Zener diode barrier to insure that an excessive voltage cannot appear in the low-voltage circuitry of the device.
vii. Where a relay output is installed, installation of a relay that provides adequate coil to contract spacing and isolation.
viii. Installation of an infallible current-limiting resistor in series.
ix. Maintain adequate creepage and clearance distances on circuit boards.
x. An infallible resistor must be constructed, such that no failure mode (usually burnout) can cause it to short circuit or become more conductive. An infallible resistor must also have a continuous power rating of at least 1.5× the power that it would dissipate.
xi. Generally, a minimum of 6 mm clearance and 10 mm creepage distances are maintained where required. However, in some cases, 10 mm creepage requirement becomes impossible to maintain on the circuit board. To resolve this problem, a grounded (earthed) and noninsulated guard conductor shall be interposed (Achari, 2006a).

12.12.1.1 Example to Design the Power Supply

Let the value of current-limiting resistor "R" and Zener diode "ZD" in Fig. 12.14 be assumed to design the intrinsically safe power supply for safe use in group IIB, where F1 and F2 are fuses, R3 is resistor (16 Ω/30 W), D6, D7, and D8 are Zener diodes (6 W), and I/P is input power supply (12 V DC).

(i) On calculating the value of current-limiting resistor (CLR):

$$\text{DC voltage} = \text{AC} \times \sqrt{2} = 12 \times \sqrt{2}\ \text{V DC} = 16.9\ \text{V DC}$$

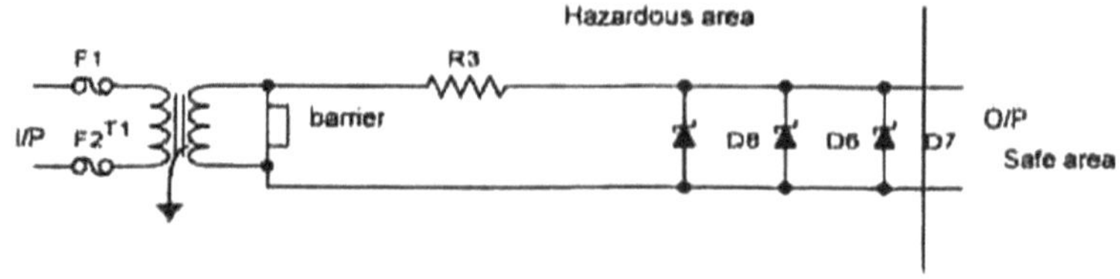

Fig. 12.14 Concept of power supply

Short circuit current = V/R = 16.9 V/16 Ω = 1.06 A
Permitted short current = 1.37 Amp for 16.9 V for group IIB as per Tables 12.10 and 12.11.
Wattage of resistor = $I^2 R$ = 1.06 ×1.06 ×16 W = 17.9 W
With a safety factor = 17.5 × 1.5 W = 26.9 W resistor is required.

(ii) Similarly on calculation for the value of ZD:

Power rating (wattage) of Zener diode = VI_Z , where I_Z is current passing through Zener diode.
I_Z = (V input – VZ)/R = V/R = (16.0 – 12)/16 A = 4.9/16 A = 0.306 A
Power rating of ZD = 12 × 0.306 W = 3.675 W
With 1.5 safety factor = 3.675 × 15 W = 5.5 W Zener diode is required.

The value of resistor and Zener diode is assessed and found adequate with 1.5 safety factor for intrinsically safe supply as shown in Fig. 12.14 for group IIB (Ahirwal, 2006a).

12.12.2 Safety Analysis

The main focus in designing an intrinsically safe equipment is to ensure that it is functional and safe, cost effective, and fulfills the market demands. In designing, manufacturing, and supplying an intrinsically safe equipment, its certification process should always be kept in mind. After designing, any circuit safety analysis should be carried out by considering the following points step by step, so that the designed circuit is intrinsically safe:

i) The temperature classification of the intrinsically safe apparatus shall be done to ensure that ignition is not caused by the hot surface, which shall be done by estimating the maximum surface temperatures of components from knowledge of their thermal behavior and the maximum power to which they may be subjected under the appropriate fault conditions.
ii) The principle of intrinsic safety is energy limitation. Therefore, proper care should be taken in choosing the appropriate power source which is suitable for

the circuit such that it has controlled output voltage and current, and it does not hampers the intrinsic safety, reducing the risk of ignition in hazardous area. Reverse polarity protection by fuses should be checked.

iii) For the safety analysis of a printed circuit board, its parameters should be considered.

iv) Identify the component on which intrinsic safety depends (current and voltage limitations) and calculate the power dissipation across the resistance, thus giving the minimum wattage rating.

v) Analysis for current and power ratings shall be done.

vi) Creepage and clearance distance around the component (and through insulation) on which intrinsic safety depends should be verified. This is to ensure that these components cannot be bypassed effectively disabling the protection.

vii) Component on which segregation depends should also be assessed for suitability. These may include transformers, capacitors, relays, optocouplers, etc.

viii) For determining the rating of the fuse, calculate the maximum current and power as follows:

$$I\,\mathrm{max} = 1.5 \times 1.7 \times \mathrm{In};$$
$$P\,\mathrm{max} = 1.5 \times (1.7 \times \mathrm{In})^2 \times R \text{ (current limit resistors)}$$

ix) Rating of the diodes shall be 2/3 (of the manufactures rating) × 1.5 (of the safety factors).

x) Piezoelectric devices used as audible alarm contains a crystal which when impacted can generate a peak voltage. Thus, proper care shall be taken in order to isolate piezo. The safety of the device also depends on the impact testing. Piezo capacitance is also taken into account.

xi) In general, marking and its connection facilities are important for the identification of the component in order for the safety analysis.

xii) The standard IEC 60079-0: 1998 specifies the minimum marking for the intrinsically safe apparatus and associated apparatus.

- Generally, the marking shall include

  For associated apparatus "Ex ia" or "Ex ib" shall be enclosed in square brackets.
  All relevant parameters shall be marked, e.g., Um, Li, Ci, Lo, Co, etc.
  Italic characters or subscript shall be restricted in presentation and simplified way shall be used. Like Uo rather than *Uo*.

- Marking of connection facilities:

  Clear identification and marking shall be provided to the connection facilities, terminals boxes, plugs, and sockets. The color used shall be light blue.

Some examples are given subsequently.

(a) Self-contained intrinsically safe apparatus

C TOME LTD.
PAGING RECEIVER TYPE 3
Ex ia IIC T4
-25°C ≤ Ta ≤ +50°C
ACB No. Ex95****
Serial No. xxxx

(b) Intrinsically safe apparatus designed to be connected to other apparatus

M HULOT
TRANSDUCTEUR TYPE 12
Ex ib IIB T4
ACB No. Ex98****

Li: 10 μH	Ci: 1200 pF
Ui: 28 V	Ii: 250 mA
Pi: 1.3 W	

(c) Associated apparatus

J SCHMIDT A. G.
STROMVERSORGUNG TYPE 4
[Ex ib]
ACB No. Ex98****

Um: 250 V	Po: 0.9 W
Io: 150 mA	Uo: 24 V
Lo: 20 mH	Co: 5.5 μF

(d) Associated apparatus protected by a flameproof enclosure

PIZZA ELECT. SpA
Ex ia IIB T6
ACB No: Ex98****

Um: 250 V	Po: 0.9 W
Uo : 36 V	Io: 100 mA
Co: 0.31 μF	Lo: 15 mH

Serial No. XXXX

Where, ABC represents the initials of the certifying body.

12.12.3 Factors Affecting Intrinsic Safety

While designing an intrinsically safe circuit following safety factors should be taken into account:

i) Proper group, zone, and temperature classification shall be accounted.
ii) Proper care should be taken in the construction of the cell and batteries so that any type of spillage of electrolytes shall not take place, which ensures the safety of the component on which intrinsic safety depends.
iii) Appropriate temperature classification is given to the PCB board depending on the track width.
iv) Maintain adequate creepage and clearance distance on circuit boards.
v) Small components like transistors and resistors shall be used with proper temperature classification.
vi) In installing the external apparatus, proper care should be taken for the connection facilities so that intrinsic safety is not invalidated.
vii) In making the circuit intrinsically safe, the reactive component plays an important part due to the stored charges. Thus, we shall take a value of inductance and capacitance as a maximum allowable connection value to the intrinsically safe circuits.
viii) Precautions must be taken in installation and maintenance as per the standards.
ix) During the repairing of any electrical and electronics part of equipment – if proper rating/matching components are not available, the whole equipment should be replaced by a new one – and it should be repaired by an authorized firm or manufacturer.
x) Replacement of faulty component by an identical part obtained from the manufacturer for that purpose – such as coils, resistors, ICs, capacitors, diodes, Zener diode – may be permissible.

12.12.4 Maintenance and Handling of the Ex-Equipment

Proper selection, installation, and maintenance of electrical apparatus are necessary part of any industry (Nutter, 1983). By good knowledge of maintenance, fault can be detected easily as well as some factor can be minimized like breakdown, accident, failure of equipment, etc. (Adjaye, 1994; Fox et al., 1994; Hillebrand and Schwarz, 2006). In maintaining the electrical equipment following points should be taken care of:

- Always try to replace same rating of spare parts or equipment.
- Wherever it is required to modify the control circuit of any equipment/device, information should be given to all concerned staff.

- At the time of maintenance of substation, shut down should be taken from the power distribution station authority.
- Danger tags board should be provided where work is going on.
- Machine and equipment are always in motion with electrical power and they require maintenance to avoid breakdowns.

Selection, installation, and maintenance for hazardous electrical equipment requires more commensurate care (Kumar, 1997; Achari, 2001a, b; Ahirwal, 2005, 2006b; Laik, 2006; Vishwakarma, 2006). Some of the important points are given below:

- Electrical apparatus should not be opened in energized condition in hazardous area.
- Rotating machines should not be opened in an isolated running condition because it involves the back emf.
- Prior to taking the apparatus into operation in the said atmosphere, attention must be paid that any unused entry is plugged with certified flameproof plug.
- Aluminum paint must not be applied on any external surface of the equipment to prevent incendive frictional sparking.
- Cable conductor must not be subjected to any tensile stress during use.
- The equipment must not be tampered with during its normal operating condition.
- The equipment shall be kept free as far as possible from accumulation of dust and dirt to protect excessive rise of temperature.
- The damaged or missed bolts/screws should be replaced with their proper size.
- No component should be added without prior intimation to the manufacture.
- Insulation to provide basic protection against electric shock. The standard defines levels of insulation required in terms of constructional requirements (creepage distance and clearance) and electrical requirement (compliance with electric strength tests). Basic insulation is considered to be shorted under single fault condition. The actual value required depends on the working voltage to which the insulation is subjected, as well as other factors.
- Cable to be used in hazardous area must have inner sheath of extruded PVC.
- Less than 16 mm^2 aluminum cable is not permitted to be used in hazardous area.
- Earth electrode must be isolated from the grid which measures the earth resistance.
- Any discrepancies or defects of electrical or mechanical nature observed in the equipment during their functioning must be brought to the notice of the manufacturer.
- The permission to work certificate should be issued by the engineer in-charge of the operation to the supervisor in-charge of the plant/plant.
- Any kind of alternation or modification to the tested product must not be done during its service in the user premises without the knowledge of testing authority.
- Documentation is very important for effective and efficient maintenance.
- Only certified and approved equipment for hazardous area should be used.

12.13 Guidelines for Designing Intrinsic Safety Circuit

To design an intrinsically safe circuit of an electrical system, first decide the purpose of the system and type of gas group in which the system will be used. Decide the source of power supply, i.e., whether the power supply should be assembled inside the system or there shall be a separate intrinsically safe power supply to the system. If power supply is assembled inside the system, then the system enclosure should be flameproof (FLP) and should follow the required IP standard (Saha, 1997; Singh, 2006a). If the IS power supply is fed externally to the system then FLP enclosure is not required, but enclosure should be provided as per the IP standard. Then design the circuit keeping in mind the related IS standard for the particular gas group where the system will be used. Select the value of components as per the voltage, current, and power requirement at the respective points and incorporating recommended safety factor as per the IS standard for the particular gas group.

While designing the circuit, following points should be considered to make the circuit intrinsically safe:

i) Always visualize the worst conditions that may arise due to different faults in the system and incorporate the safety measures to make the system intrinsically safe, e.g., maximum current that can flow in the system due to short circuit, maximum circuit voltage, upper and lower limits of frequency, upper and lower ratings of variable components, etc.
ii) Voltage and current of power supply should not exceed rated value, for which a Zener barrier/diode, fuse, and current-limiting resistor (CLR) should be incorporated as per the requirement between the power supply and circuit of the system as well as for other components which generate and store energy, such as piezoelectric buzzer, capacitor, and inductor. An example of Zener barrier is shown in Fig. 12.15 for a circuit having both positive and negative supplies.

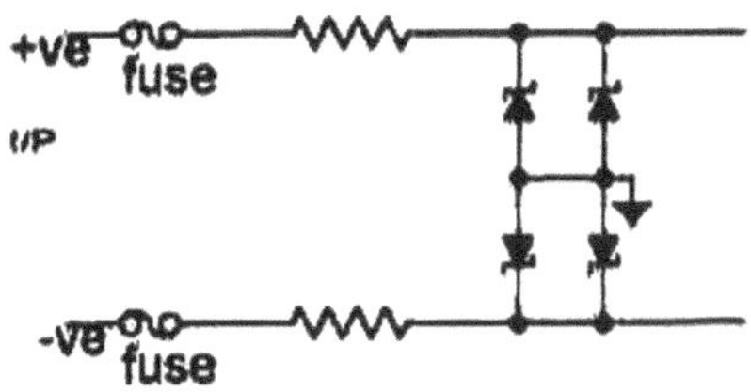

Fig. 12.15 Zener barrier for a circuit

iii) Zener diode voltage rating should be just above the supply voltage, e.g., in case of 3 V supply Zener voltage should be 3.3 V. Zener diode wattage rating should be taken after considering the safety factor 1.5 with Imax rating of power supply, i.e., in case of 3 V/150 mA power supply, the Zener diode of 3.3 V and wattage of at least $3.3\times(1.5\times150\times10^{-3})$ W should be used.
iv) Incorporate a fuse as per the maximum current requirement in the circuit, i.e., if power supply is of 3 V/150 mA rating, but circuit needs only 50 mA, then

incorporate a fuse of 50 mA in the circuit. The fuse should be antistatic and antisurge (normally HBC fuses are used).

v) If there is an intrinsically safe power supply, which provides limited current and a Zener diode with a safety factor of 1.5 is provided in the circuit, then CLR is not required. But, CLR is necessary if any voltage regulator IC is used with IS power supply output (Fig. 12.16), e.g., let IS power supply is providing 9 V/150 mA and a voltage regulator IC say 7805 is used to provide a voltage of 5 V to the circuit, then a CLR between IS supply and voltage regulator should be incorporated with the following rating: $R = (V$ of IS power supply $- V$ of regulator$) / I$ of IS power supply $= (9\text{–}5) / (150\times10^{-3})\ \Omega = 27\ \Omega$.

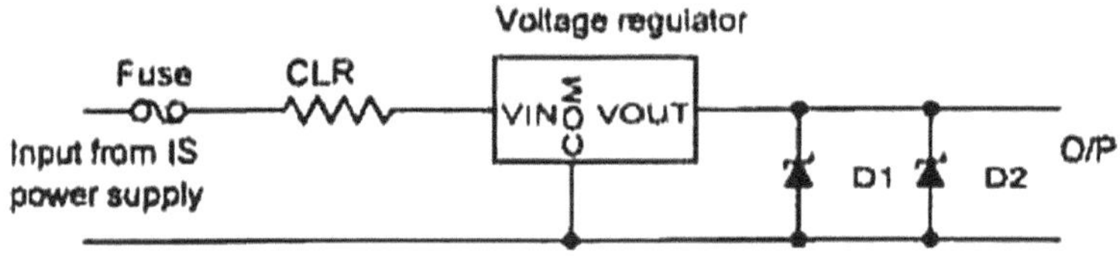

Fig. 12.16 Schematic diagram of IS protection

vi) A diode is inserted in the circuit to maintain safety from opposite voltage polarity connection. Figure 12.17 illustrates a circuit provided with diodes to protect it from opposite voltage polarity connection.

Fig. 12.17 Use of diode to protect a circuit from opposite voltage polarity connection

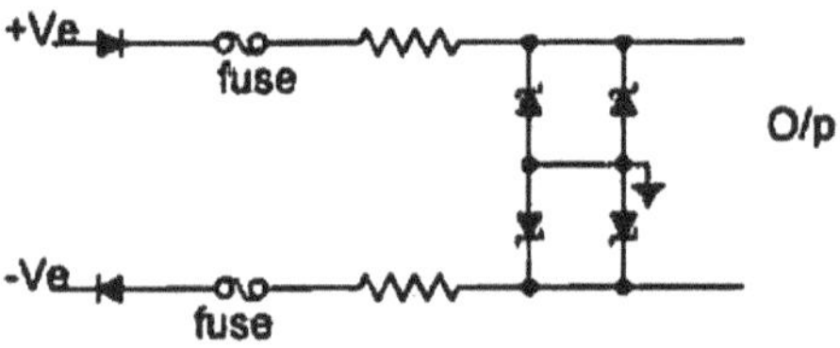

vii) The total capacitance and inductance should be of limited value; either with or without resistor and at a given voltage, should not exceed the value as per the Figs. 12.7–12.10. The capacitor value should be as low as possible. Use of ceramic capacitor should be preferred.

viii) Blocking capacitors should be of dielectric type and voltage rating should be selected taking 1.5 safety factor, i.e., voltage rating of capacitor $= 1.5 \times$ maximum input voltage to the circuit.

ix) Every resistor should have wattage rating with safety factor of 1.5, i.e., wattage rating should be $> 1.5 \times V \times$ Imax.

x) In case of AC or not complete DC, calculation of all components should be done at peak rating, i.e., in case of AC if r.m.s. value is 12 V, then peak value of voltage is $= \sqrt{2} \times 12$ V.

xi) Piezoelectric elements such as piezo-buzzer should be in a cabinet to protect it from impact. Two diodes with piezo-buzzer or piezoelectric element are used to drop or block the voltage that may develop across the piezo by impact. In case of piezo element, its capacitance should be measured and its energy storage (i.e., $\frac{1}{2}CV^2$) should be limited as per the IS specification, i.e., 1,500 μJ for group I apparatus, 950 μJ for group IIA apparatus, 250 μJ for group IIB apparatus, and 50 μJ for group IIC apparatus.

xii) All resistors should be either metal film type or wire wound type.

xiii) For getting two separate voltages from a transformer, both windings should be separately wound with proper insulation and ground between the windings.

xiv) Creepage distance and clearance should be maintained as per Table 12.10.

xv) In case of relay, one or two diodes of high voltage rating are connected in parallel with input supply terminal of relay (Fig. 12.18) to block reverse voltage due to relay inductance, e.g., in case of 12 V/200 Ω relay, use IN 4007, IN 4001, BY 127, or similar rating diode. The relay contact current should not exceed the 5 A r.m.s. value or 250 V r.m.s. value or 100 VA value. When the value exceeds the above rated values, but not exceeds 10 A or 500 VA, the values of creepage distance and clearance for the relevant voltage as given in Table 12.10 should be doubled. For further higher rating refer to relay section.

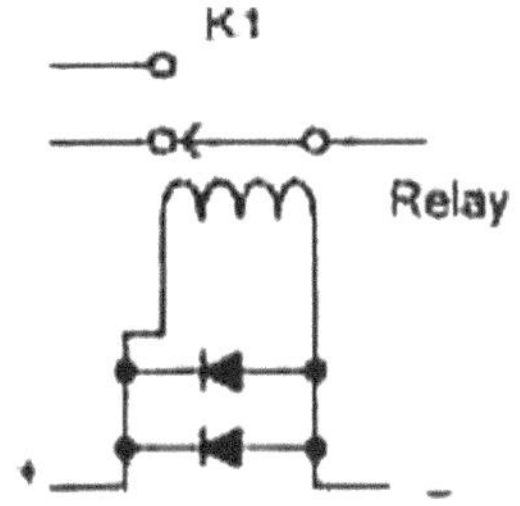

Fig. 12.18 Use of relay with diodes

xvi) When only fuse is used to limit the current in the circuit, then the current, which can flow through fuse is calculated as (1.5 × 1.7 × fuse current rating) A. But only fuse should not be used to limit the current from the battery or any other type non-IS power supply.

xvii) In case of frequency-based circuit, calculation of current is done using L impedance (Z), i.e., $I = V/Z$, not using the equation $I = V/R$. Take $Xc = 1/\omega C$ and $X_L = 1/\omega L$, and $Z = \sqrt{R^2} + (XL\text{–}XC)^2$, where X_C is the capacitive reactance, X_L the inductive reactance, ω the cyclic frequency, C the capacitance value, and L the inductance value. Thus, for calculating Imax, first consider maximum frequency and calculate Imax, and then consider minimum frequency and calculate Imax. Maximum of these two values are considered as Imax value for IS calculation.

xviii) A resistor is placed in series with capacitor or inductor so as to limit the unbounded current that can flow due to storage of charge in capacitor or electric field in inductor. The resistor value should be such that it may limit the current at a given voltage as per Fig. 12.8 of resistive circuits, e.g., as per the Fig. 12.8, 30 V/0.12 A is enough for ignition, so if a capacitor is operated at 30 V then there must be a resistor in series with it and the value of resistor should be >30 V/0.12 A = 250 Ω.

xix) The whole circuit inductance (Li) and capacitance (Ci), and the power supply inductance (Lo) and capacitance (Co) should be measured and should be clearly mentioned in the safety aspect document for further verification.

xx) The cables used in the circuit for any type of connection should follow the specification as mentioned in Table 12.7 for current rating.

xxi) The resistance, capacitance, and inductance of power connection cable should always be kept in mind, as these affect the maximum allowable distance the system can be kept from an IS power supply. These also affect the cost and weight of the total system.

xxii) Selection of battery, if used any, should be done as per Table 12.19.

xxiii) Power connection cable's complete specification, i.e., length, diameter, current-carrying capacity, capacitance, inductance, resistance per meter, etc., should be mentioned for further verification.

xxiv) For small size components, having surface area <20 mm^2 to <10 cm^2, Table 12.9 should be followed for the limitation of temperature range, so that components should work properly.

xxv) On printed circuit board of at least 0.5 mm thickness having a conducting track of 35 μm on one or both sides, a temperature classification T4 for group I shall be given to the printed tracks subject to fulfilling the following conditions: For maximum current rating that can flow on PCB wiring of various types in different gas group should follow Table 12.8.

Minimum track width (mm)	Continuous current (A)
0.3	0.518
0.4	0.814
1.0	1.388
2.0	2.222

xxvi) See all the components' data sheet used in the circuit so that no one avoids Imax or Vmax rating in any operating condition.

12.14 Basic Protection Concepts of Intrinsic Safety

(i) Why resistor is essential with capacitor?

A capacitor is an energy storage element, when connected to a voltage source it is charged up to the maximum voltage of power source and stores charge as much as rated capacity of the capacitor. So, in case of fault, there appears a low-resistance path in between the terminals of capacitor, then the current in this low-resistance path is given as $I = V/R$ and if $R \geq 0$, then the capacitor operates as voltage source and in that case the current is very large for small time (depending upon the storage capacity of the capacitor). This large current may cause a spark in its surrounding, which may cause ignition in a gaseous environment. Thus, from the above discussion it is clear that there should be a current-limiting resistance (CLR) in series with capacitor so that it may not violate the ignition current at a particular voltage specified by the intrinsic safety standard. The value of resistor with capacitor should be selected as follows:

$$R = \frac{\text{Voltage that is applied across the capacitor}}{\text{Minimum current required for ignition at that particular voltage}}$$

(ii) Why resistor is essential with inductors?

Similar to capacitor, inductor is also an energy storage element, which can store maximum of $\frac{1}{2}LI^2$ energy. So, it can behave as an energy source when power is removed from circuit. So, according to Figs. 12.5 and 12.9 for inductance value versus ignition current, the additional resistance required with inductor is as follows:

$$R = \frac{\text{Voltage applied to inductor}}{\text{Ignition current at that particular value of inductance}} - R \text{ of inductance coil.}$$

Another aspect is also considered while using inductors in the circuit. When power source is connected to the circuit and if somehow the coil of inductor is not providing any resistive path inside, then the current through inductor may be very large. So, there must be a CLR with inductor so that it limits current from going beyond the ignition current level at a particular voltage. So, resistance with inductor should be selected as follows:

$$R = \frac{\text{Voltage applied across the inductor}}{\text{Minimum ignition current required for ignition at that particular voltage}}$$

From the above two calculations, the resistance should be inserted of that value, which is larger.

(iii) Why diodes with relays (not solid-state relays)?

Relays are switching elements that provide and break connection between terminals, which rapidly apply and remove voltage from relay's inductive coil. Also, normally relays are operated at the same voltage as of supply. So, considering these situations, diodes should be placed in parallel with relay to avoid large back emf, which is produced when the voltage is removed from the relay's coil. Thus, the main purpose of providing diodes with relays is to prevent production of back emf in the coil of relays due to switching action with inductive coil, i.e., $V = L(\mathrm{d}i/\mathrm{d}t)$. So, if rate of current change ($\mathrm{d}i/\mathrm{d}t$) is large, then the voltage is also large. Keeping this point in mind the diode should be selected of high voltage rating.

(vi) Why diodes with piezo elements?

Piezo elements have piezoelectric property according to which a voltage is produced when there is mechanical deformation across it and vice versa. Thus, voltage across the piezo elements may occur when subject to mechanical impact/deformation. Keeping this in mind, piezo elements or components containing piezo elements should be properly kept or covered, to save it from mechanical impact. Also there should be electrical protection in the circuit to prevent any voltage developed across it, as this voltage may sum up or may provide back emf to the circuit. So, two diodes should be provided with piezo elements to stop summing of voltage and to avoid back emf.

(v) Why insulation in between windings of transformer?

Transformers are subjected to coil binding, in which the coils are wound on its primary and secondary windings. Also from the basic physics we know that when the electric field strength in the air exceeds beyond 300,000 V/m, then, there is breakdown of the air dielectric and current flows through this high-field region. To avoid this type of failures between windings of transformer coil, there should be a minimum clearance between windings of a transformer. So, Table 12.10 is provided with the intrinsic safety standard, which is prepared taking 5–6 times safety factor with respect to the electric field, corresponding current and voltage of a transformer.

(vi) Why external Zener diodes are provided after the voltage regulators?

Voltage regulators are generally IC based, which may fail in case of worst condition. Therefore, there must be some arrangement to avoid flowing of large current due to high voltage applied to a circuit in absence of regulator. Considering this type of failure Zener diodes are provided after regulator. The rating of Zener diodes should be just higher than the rating of the regulator, so that in case of regulator failure, these diodes provide short circuit through them, which will stop flowing current into remaining portion of the circuit. The number of diodes depends upon the use of the circuit in a particular hazardous zone.

(vii) Why current-limiting resistor in series to circuit?

Current-limiting resistor (CLR) is provided to stop any unbounded current to the circuit in case of any short circuit path in the circuit. In worst conditions, CLR protects the Zener from burnout by limiting current across it, when Zener is used to protect voltage regulator.

12.15 Conclusions

The intrinsic safety of an electrical circuit is essentially dependent on the safe limitation of current and voltage, and consequently of the power supplied, so that in neither normal operation nor under specified fault conditions ignition-capable sparks can be produced by making or breaking circuits, or when there is a short circuit to earth.

To avoid spark ignition, the energy stored in the circuit of course has to be limited. Even small amounts of additional energy can be sufficient to impair intrinsic safety. Besides spark ignition, also thermal ignition due to hot surfaces must be avoided. Therefore, it has to be ensured that the maximum current, voltage, and power available within the intrinsically safe circuit will not lead to unacceptably high surface temperatures at apparatus, components, and cables located in the hazardous area, in normal operation and under fault conditions.

For compliance with these criteria not only the individual device but also the complete interconnection and interoperation of all apparatus in the intrinsically safe circuit, including the connecting cables, must be considered. The standard for electrical installations consequently requires a verification of intrinsic safety for intrinsically safe circuits, which should ideally be carried out during planning and design, and includes the selection of apparatus, suitable for interconnection used.

For simple intrinsically safe circuits with only one source supplying current, voltage, and power to the circuit, verification of intrinsic safety can easily be made by comparison of the safety values as shown in Fig. 12.19.

Intrinsic safety is dependent on a circuit operating with low power and acceptable temperatures so that it does not have the required energy to ignite a flammable atmosphere. For most of the circuit this can be achieved at a relatively low cost

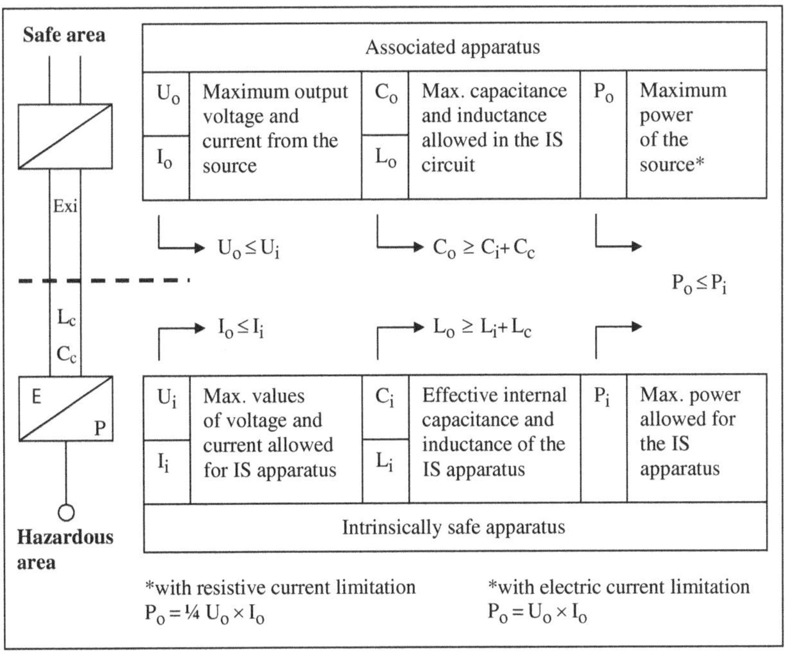

Fig. 12.19 Criteria for the verification of a simple intrinsically safe circuit

if certain design techniques are employed. Relative to maintenance costs, intrinsic safety is most advantageous because this method allows live maintenance with no need for plant shutdown. Thus, it is the most economical technique for installation and maintenance. Intrinsic safety is also more reliable due to the use of infallible components. When designing or certifying an intrinsically safe circuit, all possible scenarios for connections of components are considered and many safety factors shall be included when calculating the safe circuit parameters. Certain electricity rules are also followed while designing and installing the intrinsic safe system in the hazardous area (Ghosh, 1997; Prasad, 2001; Mishra, 2001; Sharma and Karmakar, 2005; www.bently.com; www.epsilonltd.com; www.furzy.com; www.futek.com; www.rtkinstruments.com). It is this combination of factors that allows intrinsically safe electrical equipment to be successfully sited in the most hazardous areas.

References

Achari J (1991) Intrinsic safety and its application for the hazardous area instrumentations. *Proceedings of Safety Course on Oil Mining*, Human Resource Development, Central Mining Research Institute, India, 15–19 July 1991, pp. 1–10.

Achari J (1997) Protected equipment for hazardous area applications with reference to Indian standard norms. *Proceedings of Mines Acts, Statutory Requirements and Safe Working Practices*, Human Resource Development, Central Mining Research Institute, India, 8–12 December 1997, pp. 8–22.

Achari J (2001a) Explosion proof electrical equipment – preventive maintenance. *Proceedings of Electrical Maintenance Procedures for Explosion Proof and Flame Proof Equipment Statutory Requirements Governing the Above and Prevention of Oil Fires*, Human Resource Development, Central Mining Research Institute, India, 8–12 October 2001, pp. 125–137.

Achari J (2001b) Testing and certification of explosion proof electrical equipment at CMRI. *Proceedings of Electrical Maintenance Procedures for Explosion Proof and Flame Proof Equipment Statutory Requirements Governing the Above and Prevention of Oil Fires*, Human Resource Development, Central Mining Research Institute, India, 8–12 October 2001, pp. 51–68.

Achari J (2006a) An introduction to flame proofing with special reference to Ex-equipment. *Proceedings of Executive Development Program on FLP Equipment – Basic Design and Safety Analysis*, Human Resource Development, Central Mining Research Institute, India, 20–24 February 2006, pp. 10–14.

Achari J (2006b) ATEX directive. *Proceedings of Executive Development Program on FLP Equipment – Basic Design and Safety Analysis*, Human Resource Development, Central Mining Research Institute, India, 20–24 February 2006, pp. 81–85.

Adamcik LJ (1990) Electrical area classification definitions, applications and alternatives. *Proceedings of IEEE 37th Annual Petroleum and Chemical Industry Conference of Industry Applications Society*, Houston, USA, 10–12 September 1990, pp. 145–150.

Adams JM (1990) Electrical apparatus for flammable atmospheres. *Power Engineering Journal*, **4**(1): 57–60.

Adams JM (1991) Electrical apparatus for flammable atmospheres: intrinsic safety. *Power Engineering Journal*, **5**(6): 278–282.

Adjaye RE (1994) Design and optimized operation with reliability centered maintenance. *Proceedings of Fifth International Conference on Electrical Safety in Hazardous Environments*, London, UK, 19–21 April 1994, pp. 165–171.

Ahirwal B (2005) General requirement of maintenance, selection and installation of Ex-equipment. *Proceedings of Executive Training Development Program on FLP Equipment*,

Human Resource Development, Central Mining Research Institute, India, 3–7 October 2005, pp. 53–65.

Ahirwal B (2006a) Basic concept of intrinsically safe circuit design. *Proceedings of Executive Development Program on FLP Equipment – Basic Design and Safety Analysis*, Human Resource Development, Central Mining Research Institute, India, 20–24 February 2006, pp. 15–20.

Ahirwal B (2006b) General requirements of maintenance, selection and installation of Ex-equipment. *Proceedings of Executive Development Program on FLP Equipment – Basic Design and Safety Analysis*, Human Resource Development, Central Mining Research Institute, India, 20–24 February 2006, pp. 53–65.

Ahirwal B, Singh AK, Vishwakarma RK and Sinha A (2007) Intrinsically safe equipment in oil mines. *Electrical India*, December, pp. 88–96.

Ansari AS (1997) Explosion hazards due to electrical equipment and their prevention. *Proceedings of Mines Acts, Statutory Requirements and Safe Working Practices*, Human Resource Development, Central Mining Research Institute, India, 8–12 December 1997, pp. 36–44.

Arnhold T and Steilen A (2007) Energy from deep underground protection against explosions in modern drilling equipment. *Ex-Magazine*, **33**: 8–19.

Bianco L, Feudale F, Razza F, Sica M, Contin A and Rabach G (1994) High voltage induction motors for use in hazardous areas: evolution of legal aspects and technical investigations related to safety problems. *Proceedings of Fifth International Conference on Electrical Safety in Hazardous Environments*, London, UK, 19–21 April 1994, pp. 84–92.

Bicchi A, Rizzini SL and Tonietti G (2001) Compliant design for intrinsic safety: general issues and preliminary design. *Proceedings of IEEE International Conference on Intelligent Robots and Systems*, Osaka, Japan, Vol. 4, 29 October–3 November 2001, pp. 1864–1869.

Casso CG (1994) Current application of oil well logging equipment in hazardous areas. *Proceedings of Fifth International Conference on Electrical Safety in Hazardous Environments*, London, UK, 19–21 April 1994, pp. 172–179.

Chandra D (1991) Instrumentation for hazardous industry. *Proceedings. of Safety Course on Oil Mining*, Human Resource Development, Central Mining Research Institute, India, 15–19 July 1991, pp. 47–55.

Clark PE, Glew CN and Regan RH (1994) Air-gap sparking-cause and cure. *Proceedings of Fifth International Conference on Electrical Safety in Hazardous Environments*, London, UK, 19–21 April 1994, pp. 98–103.

Das SK (1997) Frictional ignition risks with light metals and their alloys. *Proceedings of Mines Acts, Statutory Requirements and Safe Working Practices*, Human Resource Development, Central Mining Research Institute, India, 8–12 December 1997, pp. 60–68.

Das SK (2001) Ignition hazard due to frictional incendivity of light alloy. *Proceedings of Electrical Maintenance Procedures for Explosion Proof and Flame Proof Equipment Statutory Requirements Governing the Above and Prevention of Oil Fires*, Human Resource Development, Central Mining Research Institute, India, 8–12 October 2001, pp. 71–78.

Dutta MK (2001) Application of intrinsic safety hazardous area instrumentation. *Proceedings of Electrical Maintenance Procedures for Explosion Proof and Flame Proof Equipment Statutory Requirements Governing the Above and Prevention of Oil Fires*, Human Resource Development, Central Mining Research Institute, India, 8–12 October 2001, pp. 147–164.

Fox J, Hammond P, Elsdon D, Khabaza T, Montgomery A, Khabaza I, Mantil R, Susskind R, Tapper C, Swaffield G, Burnott R, Das S and Bell J (1994) Expert systems for safety-critical applications: theory, technology and applications. *IEE Colloquium on Knowledge-Based Systems for Safety Critical Applications*, London, UK, 2–3 February 1994, Vol. 5, pp. 1–5.

Fraczek JK (1994) Assessment of hazard caused by intrinsically safe systems. *Proceedings of Fifth International Conference on Electrical Safety in Hazardous Environments*, London, UK, 19–21 April 1994, pp. 191–197.

Fry D (1995) Intrinsic safety: explosion proof circuitry. *Proceedings of IEEE Technical Applications Conference and Workshops*, Portland, USA, 10–12 October 1995, pp. 363–366.

Gaunt DR (1988) Intrinsic safety-simplicity itself. *Fourth International Conference on Electrical Safety in Hazardous Areas*, London, UK, 22–24 November 1988, pp. 103–105.

Ghosh KM (1997) Electricity rules and statutory requirements. *Proceedings of Mines Acts, Statutory Requirements and Safe Working Practices*, Human Resource Development, Central Mining Research Institute, India, 8–12 December 1997, pp. 45–59.

Gupta BK (2008) *Overview of Indian Hazardous Location System*. www.exprotecta.com/Overview%20of%20Indian%20Hazardous%20Location%20System.pdf.

Hillebrand R and Schwarz G (2006) Saving on installations in hazardous areas. *Proceedings of 53rd Annual Conference on Petroleum and Chemical Industry of Industry Applications Society*, Baden, Germany, 11 September – 15 October 2006, pp. 1–6.

Hind DJ (1994) Radio frequency identification and tracking systems in hazardous areas. *Proceedings of Fifth International Conference on Electrical Safety in Hazardous Environments*, London, UK, 19–21 April 1994, pp. 215–227.

IEC 60529 (2004) *Degrees of Protection Provided by Enclosures (IP Code) (Identical National Adoption)*. International Electrotechnical Commission, Geneva, Switzerland.

IEC60079-11 (2007) *Standard on Electrical Apparatus for Explosive Gas Atmospheres – Intrinsic Safety "i" – Specification*. International Electrotechnical Commission, Geneva, Switzerland.

Kumar D (1997) Electrical equipment in hazardous atmosphere: selection, installation, inspection and maintenance. *Proceedings of Mines Acts, Statutory Requirements and Safe Working Practices*, Human Resource Development, Central Mining Research Institute, India, 8–12 December 1997, pp. 80–98.

Kumar T (2001) Hazardous zones in petroleum operations causes of hazards and precautionary measures. *Proceedings of Electrical Maintenance Procedures for Explosion Proof and Flame Proof Equipment Statutory Requirements Governing the Above and Prevention of Oil Fires*, Human Resource Development, Central Mining Research Institute, India, 8–12 October 2001, pp. 79–86.

Laik S (2006) Safety in oil-field production and installation. *Proceedings of Executive Development Program on FLP Equipment – Basic Design and Safety Analysis*, Human Resource Development, Central Mining Research Institute, India, 20–24 February 2006, pp. 21–23.

Laxminarayanan G (2001) Design and selection for suitability of electrical cables for use in underground mines/hazardous areas. *Proceedings of Electrical Maintenance Procedures for Explosion Proof and Flame Proof Equipment Statutory Requirements Governing the Above and Prevention of Oil Fires*, Human Resource Development, Central Mining Research Institute, India, 8–12 October 2001, pp. 139–145.

Lloyd FC (1994) Area classification for combustible dusts-principles, standardization and practice. *Proceedings of Fifth International Conference on Electrical Safety in Hazardous Environments*, London, UK, 19–21 April 1994, pp. 117–122.

Magison EC (1978) Intrinsically safe systems: trends in design and installation standards. *IEEE Transactions on Industry Applications*, **14**(1): 87–89.

Malhotra RL (2001) Safety measures on on-shore and off-shore oil and gas installation. *Proceedings of Electrical Maintenance Procedures for Explosion Proof and Flame Proof Equipment Statutory Requirements Governing the Above and Prevention of Oil Fires*, Human Resource Development, Central Mining Research Institute, India, 8–12 October 2001, pp. 15–44.

Mishra BN (2001) Provision of Indian electricity rules to cover oil mines and neighbouring areas. *Proceedings of Electrical Maintenance Procedures for Explosion Proof and Flame Proof Equipment Statutory Requirements Governing the Above and Prevention of Oil Fires*, Human Resource Development, Central Mining Research Institute, India, 8–12 October 2001, pp. 1–6.

Mitra DK (1997) Intrinsically safe apparatus and circuits. *Proceedings of Mines Acts, Statutory Requirements and Safe Working Practices*, Human Resource Development, Central Mining Research Institute, Dhanbad, India, 8–12 December 1997, pp. 104–109.

Newton B (2007) *Intrinsic Safety – Designing of Intrinsically Safe Circuits*. www.auburnsys.com.

Nutter RS (1983) Hazard evaluation methodology for computer-controlled mine monitoring/ control systems. *IEEE Transactions on Industry Applications*, **19**(3): 445–449.

Oliver GR (1994) Cost effective explosion safety. *Proceedings of Fifth International Conference on Electrical Safety in Hazardous Environments*, London, UK, 19–21 April 1994, pp. 33–39.

Pathak AK (1997) Oil mine's fire: causes, case histories and safety rules. *Proceedings of Mines Acts, Statutory Requirements and Safe Working Practices*, Human Resource Development, Central Mining Research Institute, India, 8–12 December 1997, pp. 69–79.

Pidoll UV (2001) The ignition of clouds of sprays, powders and fibers by flames and electric sparks. *Journal of Loss Prevention in the Process Industries*, **14**: 103–109.

Pidoll UV (2004) Testing products and processes with regard to electrostatic hazards. *Proceedings of Joint Symposium on Electrostatics*, Tokyo, 7–10 November 2004, pp. 245–254.

Pidoll UV, Brzostek E and Fröchtenigt HR (2004) Determining the incendivity of electrostatic discharges without explosive gas mixtures. *IEEE Transactions on Industry Applications*, **40**: 1467–1475.

Pilkington GA (1994) Static electricity-friend or foe. *Proceedings of Fifth International Conference on Electrical Safety in Hazardous Environments*, London, UK, 19–21 April 1994, pp. 63–68.

Prasad R (2001) Statutory provisions for use of electricity in mines and oil mines. *Proceedings of Electrical Maintenance Procedures for Explosion Proof and Flame Proof Equipment Statutory Requirements Governing the Above and Prevention of Oil Fires*, Human Resource Development, Central Mining Research Institute, India, 8–12 October 2001, pp. 93–119.

Qinghai M and Longhua M (2001) Study on characteristics of low energy arc discharge. *Proceedings of International Symposium on Electrical Insulating Materials*, Himeji, Japan, 19–22 November 2001, pp. 773–776.

Ramachandran R (1997) Electrical fires, causes and various case studies. *Proceedings of Mines Acts, Statutory Requirements and Safe Working Practices*, Human Resource Development, Central Mining Research Institute, India, 8–12 December 1997, pp. 30–35.

Ritolia SK (2001) Weather protection/environmental protection for electrical apparatus. *Proceedings of Electrical Maintenance Procedures for Explosion Proof and Flame Proof Equipment Statutory Requirements Governing the Above and Prevention of Oil Fires*, Human Resource Development, Central Mining Research Institute, India, 8–12 October 2001, pp. 15–44.

Saha SP (1997) Flameproof lighting fitting and their use in hazardous area. *Proceedings of Mines Acts, Statutory Requirements and Safe Working Practices*, Human Resource Development, Central Mining Research Institute, India, 8–12 December 1997, pp. 99–103.

Schaefer B, Borges GH and Malohn DP (2003) Global installation of intrinsically safe and non-incendive systems. *Proceedings of 50th Annual Petroleum and Chemical Industry Conference, IEEE Industry Applications Society*, Houston, USA, 15–17 September 2003, pp. 349–351.

Schultz S (2007) The use of wireless technology in hazardous areas. *Ex-Magazine*, **33**: 20–25.

Sharma VP and Karmakar JP (2005) Oil mines regulations. *Proceedings of Executive Training Development Program on FLP Equipment*, Human Resource Development, Central Mining Research Institute, India, 3–7 October 2005, pp. 112–164.

Singh AK (2006a) Essential aspects of design for flameproof equipment. *Proceedings of Executive Development Program on FLP Equipment – Basic Design and Safety Analysis*, Human Resource Development, Central Mining Research Institute, India, 20–24 February 2006, pp. 29–43.

Singh AK (2006b) Area classification and type of protection for hazardous area. *Proceedings of Executive Development Program on FLP Equipment – Basic Design and Safety Analysis*, Human Resource Development, Central Mining Research Institute, India, 20–24 February 2006, pp. 1–9.

Singh AK (2006c) Testing and approval requirements of flameproof equipment. *Proceedings of Executive Development Program on FLP Equipment – Basic Design and Safety Analysis*, Human Resource Development, Central Mining Research Institute, India, 20–24 February 2006, pp. 66–80.

Stolarczyk LG (1989) Emergency and quick operational radio communications in Utah Power and Light Company's Deer Creek Coal Mine. *Proceedings of IEEE 39th Vehicular Technology Conference*, San Francisco, USA, Vol. 1, 1–3 May 1989, pp. 383–397.

Toney MK, Griffith T, Hyde W, Schram PJ and Soffrin DE (2000) A history of electrical area classification in the United States. *Proceedings of IEEE 47th Annual Petroleum and Chemical Industry Conference of Industry Applications Society*, Texas, USA, 11–13 September, pp. 273–280.

Towle LC (1994) Intrinsic safety-the way forward. *Proceedings of Fifth International Conference on Electrical Safety in Hazardous Environments*, London, UK, 19–21 April 1994, pp. 198–203.

Towle C (2007) The current state of the IEC intrinsically safe standards. *Proceedings of Hazardous Area Conference on IDC Technology*, www.iceweb.com.au

Vishwakarma RK (2005) Dust ignition protection in hazardous locations. *Proceedings of Executive Training Development Program on FLP Equipment*, Human Resource Development, Central Mining Research Institute, India, 3–7 October 2005, pp. 86–90.

Vishwakarma RK (2006) Operation, selection, installation and maintenance of flame arresters in hazardous area. *Proceedings of Executive Development Program on FLP Equipment – Basic Design and Safety Analysis*, Human Resource Development, Central Mining Research Institute, India, 20–24 February 2006, pp. 91–97.

Wenzel H (1994) Electrical equipment for use in the presence of explosive dust atmosphere-experiences in Germany since 1980. *Proceedings of Fifth International Conference on Electrical Safety in Hazardous Environments*, London, UK, 19–21 April 1994, pp. 111–116.

Widginton DW (1988) Intrinsic safety reference curves: some recent considerations. *Proceedings of Fourth International Conference on Electrical Safety in Hazardous Areas*, London, UK, 22–24 November 1988, pp. 50–54.

York TA, Smit Q, Davidson JL and Grieve BD (2003) An intrinsically safe electrical tomography system. *Proceedings of IEEE International Symposium on Industrial Electronics*, Rio de Janerio, Brazil, Vol. 2, 9–11 June 2003, pp. 946–951.

Websites

www.bently.com (Hazardous area and standard installation of electrical equipment)

www.cenelec.org

www.csa.ca

www.ec.europa.eu/enterprise/atex/guide/index.htm

www.epsilon-ltd.com/Web%20Documents/New%20WEB%20Articles/An%20introduction%20to%20Intrinsic%20Safety.pdf

www.epsilonltd.com/Web%20Documents/New%20WEB%20Articles/An%20introduction%20to%20Intrinsic%20Safety.pdf, (Clarke S, *Basic Techniques of Intrinsically Safe Circuit Assessment*)

www.fmglobal.com

www.furzy.com

www.futek.com (Applying intrinsic safety – about standard, hazardous area)

www.gminternationalsrl.com/?p=intrinsic_safety_basic_principles

www.isa.org/Content/Microsites147/SP12_0,_Gen__Requirements_for_Elec__Equip.../IEC_Pub_60079-00e-ed5_dr1.pdf

www.msha.gov/TECHSUPP/ACC/StandardTestProcs/ASTP2220.pdf

www.omega.com

www.rtkinstruments.com (The Basic of Intrinsic Safety)

www.sound.westhost.com/xfmr2.htm (Transformer)

www.ul.com/hazloc/potential.html

Appendix A
Program for RFID Devices

1 Programming for Coordinator

The programming procedures for CC2430 RFID chips having 48 and 22 pins are discussed and subsequently program codes are listed. CC2430 RFID chip having 48 pins is manufactured by Texas Instruments, USA (www.ti.com) and CC2430 RFID chip having 22 pins is manufactured by Shenzhen Technology Co. Ltd., China (www.soczigbee.com).

1.1 Programming with CC2430 RFID Chip (48 Pins)

For making of coordinator, download IEEE 802.15.4 medium access control (MAC) CC2430-Swrc-069b zip file from the Texas Instruments website (http://focus.ti.com/docs/toolsw/ folders/print/timac.html). Then extract Swrc-069b zip file. After extracting Swrc-069b file run the Swrc-069b setup. Now all the required TI-MAC libraries are installed in the system, namely HAL, OSAL, MSA, MAC, etc. These program codings require an IAR Embedded Workbench (8051 EW) compiler for debugging code. Therefore, install IAR Embedded Workbench (8051 EW) and run the setup file. Now, open IAR Embedded Workbench and create New Project and subsequently select Empty Project. Save this file in a particular computer drive by entering a new file name. Then, select the project file displayed in the left margin of window and right click on the file to open the options and configure the language setting, compiler setting, file setting, and input–output setting. Subsequently, import the supporting libraries/header files, namely Application, HAL, OSAL, MAC layers, etc., by using add grouping in the options. In Application layer, attach some header files, like msa.c, msa.h, msa_Main.c, and msa_Osal.c, available in Swrc-069b files. After following the above procedure, attach the supporting files for other header files, like HAL (Common, Drivers, and Target), OSAL (OSAL.c, OSAL_Memor, OSAL_Tasks.c, and OSAL_Timers.c), and MAC (High Level, Include, and Low Level).

The msa.c file is the main file to be modified for programming the different devices. Therefore, the program codes for different devices should be saved in

L.K. Bandyopadhyay et al., *Wireless Communication in Underground Mines*,
DOI 10.1007/978-0-387-98165-9,

different names and pasted in the msa.c file while programming the respective device. There is no need for changing other files. The program code is debugged to the respective device using USB port.

The algorithm for programming a coordinator is as follows:

i. TI-MAC library is installed to design the MAC application.
ii. IAR Embedded Workbench (8051EW) is installed.
iii. IAR Workbench is opened and language setting, compiler setting, file setting, and input–output setting are configured.
iv. New workspace is created and all the header files for supporting HAL, OSAL, MAC layers are imported.
v. In MAC application layer a new main program is created, and it automatically initializes all the drivers and layers, devices, data packet, LED, and timer function.
vi. All the associate parameters of the devices are assigned, as total number of routers, channel number 11–28, beacon order, packet length, maximum results, and ADC (analog to digital converter) resolution.
vii. Add device (coordinator) information as PAN_ID and PAN address.
viii. Timer for reboot transmission is set and its radio transmitting power range is adjusted.
ix. Coordinator device is connected to a PC through USB port, and the program is compiled and debugged.
x. The USB port is disconnected, and the coordinator is ready.

To make a program for coordinator paste the following program codes in the msa.c file.

```
********************Include Driver******************************
Include necessary drivers as described above.
/*** Include HAL (Hardware Abstraction Layer)
#include "hal_key.h"
#include "hal_timer.h"
#include "hal_drivers.h"
#include "hal_led.h"
#include "hal_adc.h"
/***Include OSAL (Operating System Abstraction Layer)***/
#include "OSAL.h"
#include "OSAL_Tasks.h"
#include "OSAL_PwrMgr.h"
/*** Application Includes ***/
#include "OnBoard.h"
/*** MAC Application Interface ***/
#include "mac_api.h"
/*** Application ***/
#include "msa.h"
```

```
**************************Define constant*********************
#define MSA_DEV_SHORT_ADDR 0x0000 /* Device initial short address - This will change after association */
#define MSA_HEADER_LENGTH 4 /* Header includes Data Length + Device ShortAddr + Sequence */
#define MSA_ECHO_LENGTH 8 /* Echo packet */
#define MSA_MAC_MAX_RESULTS 5 /* Maximum number of scan result that will be accepted */
#define MSA_MAX_DEVICE_NUM 7 /* Maximum number of devices can associate with the coordinator*/
#if defined (HAL_BOARD_CC2430DB)
#define MSA_HAL_ADC_CHANNEL HAL_ADC_CHANNEL_0 /* AVR - Channel 0 and Resolution 10 */
#define MSA_HAL_ADC_RESOLUTION HAL_ADC_RESOLUTION_10
#elif defined (HAL_BOARD_DZ1611)
|| defined (HAL_BOARD_DZ1612)
|| defined (HAL_BOARD_DRFG4618)
#define MSA_HAL_ADC_CHANNEL HAL_ADC_CHANNEL_0 /* DZ1611 and DZ1612 - Channel 0 and Resolution 12 */
#define MSA_HAL_ADC_RESOLUTION HAL_ADC_RESOLUTION_12
#else
#define MSA_HAL_ADC_CHANNEL HAL_ADC_CHANNEL_7 /* CC2430 EB & DB - Channel 7 and Resolution 14 */
#define MSA_HAL_ADC_RESOLUTION HAL_ADC_RESOLUTION_14
#endif
#define PIN_HIGH () st(P1DIR |= BV(1); /* set P1.1 output */ \
P1_1 = 1;)
#define PIN_LOW () st(P1DIR |= BV(1); /* set P1.1 output */ \
P1_1 = 0;)
/* Size table for MAC structures */
const CODE uint8 msa_cbackSizeTable [] =
{0, /* unused */
sizeof(macMlmeAssociateInd_t), /* MAC_MLME_ASSOCIATE_IND */
sizeof(macMlmeAssociateCnf_t), /* MAC_MLME_ASSOCIATE_CNF */
sizeof(macMlmeDisassociateInd_t), /* MAC_MLME_DISASSOCIATE_IND */
sizeof(macMlmeDisassociateCnf_t), /* MAC_MLME_DISASSOCIATE_CNF */
sizeof(macMlmeBeaconNotifyInd_t), /* MAC_MLME_BEACON_NOTIFY_IND */
```

```
41. sizeof(macMlmeOrphanInd_t),
    /* MAC_MLME_ORPHAN_IND */
42. sizeof(macMlmeScanCnf_t),
    /* MAC_MLME_SCAN_CNF */
43. sizeof(macMlmeStartCnf_t),
    /* MAC_MLME_START_CNF */
44. sizeof(macMlmeSyncLossInd_t),
    /* MAC_MLME_SYNC_LOSS_IND */
45. sizeof(macMlmePollCnf_t),
    /* MAC_MLME_POLL_CNF */
46. sizeof(macMlmeCommStatusInd_t),
    /* MAC_MLME_COMM_STATUS_IND */
47. sizeof(macMcpsDataCnf_t),
    /* MAC_MCPS_DATA_CNF */
48. sizeof(macMcpsDataInd_t),
    /* MAC_MCPS_DATA_IND */
49. sizeof(macMcpsPurgeCnf_t),
    /* MAC_MCPS_PURGE_CNF */
50. sizeof(macEventHdr_t)
    /* MAC_PWR_ON_CNF */
51. };
*************************Local Variables**********************
/*Extended address of the device, for coordinator, it will
be msa_ExtAddr1
For any device, it will be msa_ExtAddr2 with
the last 2 bytes from a ADC read */
52. sAddrExt_t  msa_ExtAddr;
53. sAddrExt_t  msa_ExtAddr1 = {0x11, 0x22, 0x33, 0x44,
    0x55, 0x66, 0x77, 0x88};
54. sAddrExt_t  msa_ExtAddr2 = {0xA0, 0xB0, 0xC0, 0xD0,
    0xE0, 0xF0, 0x00, 0x00};
/* Coordinator and Device information */
55. uint16 msa_PanId = MSA_PAN_ID:
56. ;
/*
/* Predefined packet that will be sent out by the
coordinator*/
57. uint8 msa_Data1[MSA_PACKET_LENGTH];
/* /* TRUE and FALSE value */
58. bool msa_MACTrue = TRUE;
59. bool msa_MACFalse = FALSE;
/* Beacon payload, this is used to determine if the device
is not ZigBee device */
60. uint8 msa_BeaconPayloadLen = 3;
```

```
/* Contains pan descriptor results from scan */
macPanDesc_t msa_PanDesc[MSA_MAC_MAX_RESULTS];
/* flags used in the application */
bool msa_IsCoordinator = FALSE; /* True if the device
  is started as a Pan  Coordinate */
bool msa_IsStarted = FALSE; /* True if the device
  started, either as Pan Coordinator or device */
bool msa_IsSampleBeacon = FALSE; /* True if the beacon
  payload match with the predefined */
bool msa_IsDirectMsg = FALSE; /* True if the messages
  will be sent as direct messages */
uint8 msa_State = MSA_IDLE_STATE; /* Either IDLE state or
  SEND state */
/* Task ID */
uint8 MSA_TaskId;
uint8 MSA_UART_Write_TaskID;
//uint16 p_address,address;
uint8 MSA_UART_Write_TaskID
#define MAXUARTBUF 4
#define MAXUARTBUFLENGTH 31
uint8 currentBufLevel=0;
#define MAXCHILDREN 5
#define MAXBUFFERLENGTH 20
#define MAXPACKETLENGTH 20
#define MAXDATASTORE 5 // imp need to change code if this
  value is changed
#define MAXMSGHOLDTIME 50000
#define MAXTIMEEXPECTPACKET 60000 // 5 minutes
uint16 myChildrenAddress[MAXCHILDREN];
int numChildren=-1;
int sendToChild=-1;
bool receivedPacket = FALSE;
#define MAXNAMELENGTH 5
uint16 myAddress=0;
uint8 myName[] = "C00";
uint16 myParentAddress;
#define MAXUARTINPUTBUFFER 20
uint8 UARTInputBuffer[MAXUARTINPUTBUFFER];
uint8 UARTInputBufferLevel = 0;
uint8 UARTInputMessageCount = 0;
bool UARTInputRead=TRUE;
bool UARTInputReadComplete = FALSE;
void processUARTInput(uint8 *b, uint16 l);
```

```
******************Local Function Prototype*******************
/* Setup routines */
void MSA_CoordinatorStartup(void);
void MSA_DeviceStartup(void);
/* MAC related routines */
void MSA_AssociateReq(void);
void MSA_AssociateRsp(macCbackEvent_t* pMsg);
void MSA_McpsDataReq(uint8* data, uint8 dataLength,
    bool directMsg, uint16 dstShortAddr);
void MSA_McpsPollReq(void);
void MSA_ScanReq(uint8 scanType, uint8 scanDuration);
void MSA_SyncReq(void);
/* Support */
bool MSA_BeaconPayLoadCheck(uint8* pSdu);
bool MSA_DataCheck(uint8* data, uint8 dataLength);
***********************************************************
@fn MSA_Init
@brief Initialize the application
@param taskId - taskId of the task after it was
    added in the OSAL task queue
@return none
-------------------------------------------------------------
void MSA_Init(uint8 taskId)
{
MSA_TaskId = taskId;
/* initialize MAC features */
MAC_InitDevice();
MAC_InitCoord();
/* Initialize MAC beacon */
MAC_InitBeaconDevice();
MAC_InitBeaconCoord();
/* Reset the MAC */
MAC_MlmeResetReq(TRUE);
// AD
initCOM1Port();
// ~AD
HalLedBlink (HAL_LED_4, 0, 40, 200);
PIN_LOW();
// AD
writeMsg("INIT Complete.",16);
k=0;
// ~AD
}
/*
-------------------------------------------------------------
```

```
* @fn MSA_ProcessEvent
* @brief This routine handles events
    a. * @param taskId - ID of the application task when
         it registered with the OSAL
    b. events - Events for this task
* @return 16bit - Unprocessed events
---------------------------------------------------------------
 */
 uint16 MSA_ProcessEvent(uint8 taskId, uint16 events)
 {
 uint8* pMsg;
 macCbackEvent_t* pData;
  if (events & SYS_EVENT_MSG)
   {
    while ((pMsg = osal_msg_receive(MSA_TaskId)) != NULL)
    {
     switch ( *pMsg )
     {
       case MAC_MLME_ASSOCIATE_IND:
        writeMsg("got associate indicate.",25);
 //      MSA_AssociateRsp((macCbackEvent_t*)pMsg);
        break;
       case MAC_MLME_ASSOCIATE_CNF:
        /* Retrieve the message */
        pData = (macCbackEvent_t *) pMsg;
        break;
       case MAC_MLME_COMM_STATUS_IND:
        break;
       case MAC_MLME_BEACON_NOTIFY_IND:
        /* Retrieve the message */
        pData = (macCbackEvent_t *) pMsg;
  break;
          case MAC_MLME_COMM_STATUS_IND:
            break;
          case MAC_MLME_BEACON_NOTIFY_IND:
  break;
       case MAC_MLME_START_CNF:
  pData = (macCbackEvent_t *) pMsg;
  /* Set some indicator for the Coordinator */
       if ((!msa_IsStarted) &&
             (pData->startCnf.hdr.status == MAC_SUCCESS))
       {
       msa_IsStarted = TRUE;
       msa_IsCoordinator = TRUE;
       writeMsg("PAN coordinator!",16);
```

```
        writeMsg("My short add is:",18);
        writeVal(myAddress);
        osal_start_timer (MSA_POLL_EVENT, 400);
        osal_start_timer (MSA_SEND_EVENT,
            MAXTIMEEXPECTPACKET);
        HalLedSet ( HAL_LED_4, HAL_LED_MODE_ON);
        }
        break;
        case MAC_MLME_SCAN_CNF:
        /* Check if there is any Coordinator out there */
        pData = (macCbackEvent_t *) pMsg;
         MSA_CoordinatorStartup();
        break;
       case MAC_MCPS_DATA_CNF:
        pData = (macCbackEvent_t *) pMsg;
     /* If last transmission completed, ready to send the
        next one */
             if ((pData->dataCnf.hdr.status ==
                  MAC_SUCCESS) ||
              (pData->dataCnf.hdr.status ==
              MAC_CHANNEL_ACCESS_FAILURE) ||
          (pData->dataCnf.hdr.status == MAC_NO_ACK))
             {
               if ( sendToChild > -1 &&
                    sendToChild <= numChildren )
               {
MSA_McpsDataReq((uint8*)msa_Data1,MSA_PACKET_LENGTH,TRUE,
             myChildrenAddress[sendToChild++]);
               }
               else
                 sendToChild=-1;
   }
   osal_msg_deallocate((uint8 *) pData->dataCnf.pDataReq);
             break;
    case MAC_MCPS_DATA_IND:
             pData = (macCbackEvent_t*)pMsg;
   receivedPacket = TRUE;
             if (MSA_DataCheck ( pData->dataInd.msdu.p,
                                 pData->dataInd.msdu.len ))
             {
   if ( BUILD_UINT16(pData->dataInd.msdu.p[2],
        pData->dataInd.msdu.p[1]) == myAddress )
               {
                 uint8 printBuffer[MAXBUFFERLENGTH];
                 //uint8 receivedName[MAXNAMELENGTH];
```

```
                 uint8 i=0,j=0;
                 printBuffer[0]='*';
                 printBuffer[1]='*';
                 while ( pData->dataInd.msdu.p[i+6] != '#'
                        && (i+6) < pData->dataInd.msdu.len
                        && (i+4) < MAXBUFFERLENGTH)
                 {
                   printBuffer[i+2] =
                     pData->dataInd.msdu.p[i+6];
     i++;
                 }
                 printBuffer[i+2] = '*';
                 printBuffer[i+3] = '*';
                 if ( pData->dataInd.msdu.p[5] == 5 )
                 {
                   for ( j=0; j< MAXNAMELENGTH; j++ )
                   {
                     if ( myName[j] == '\0' || (i+4+j) >
                          MAXBUFFERLENGTH - 2 )
                       break;
                     printBuffer[i+4+j] = myName[j];
                   }
                   printBuffer[i+4+j]='*';
                   printBuffer[i+4+j+1]='*';
                   j+=2;
                 }
                 writeMsg(printBuffer,i+4+j);
                 for ( i =0 ;i <= numChildren ; i++ )
                 {
                   if ( myChildrenAddress[i] ==
                        pData->dataInd.mac.srcAddr.
                               addr.shortAddr )
                     break;
                 }
                 if ( i > numChildren )
                 {
                   // unique address
                   for ( i=MAXCHILDREN -1; i>0; i- )
                   {
                     myChildrenAddress[i]=
                       myChildrenAddress[i-1];
                   }
                   myChildrenAddress[0]=
                     pData->dataInd.mac.srcAddr.
                           addr.shortAddr;
```

```

                    if ( numChildren < MAXCHILDREN - 1)
                      numChildren++;
                  }
}
}
            break;
          }
osal_msg_deallocate((uint8 *) pMsg);
    }
    return events ^ SYS_EVENT_MSG;
  }
  /* This event handles polling for in-direct message device */
  if (events & MSA_POLL_EVENT)
  {
clearWDTimer();
    osal_start_timer (MSA_POLL_EVENT, 400);
return events ^ MSA_POLL_EVENT;
  }
if (events & MSA_SEND_EVENT)
  {
if ( receivedPacket == FALSE )
    {
      writeMsg("No PKT received-reboot",22);
      systemReset();
    }
    else
      receivedPacket = FALSE;
    osal_start_timer (MSA_SEND_EVENT, MAXTIMEEXPECTPACKET);
    return events ^ MSA_SEND_EVENT;
  }
     if (events & MSA_MSG1_EVENT)
{
return events ^ MSA_MSG1_EVENT;
  }
  if (events & MSA_MSG2_EVENT)
  {
return events ^ MSA_MSG2_EVENT;
  }
  if (events & MSA_MSG3_EVENT)
  {
return events ^ MSA_MSG3_EVENT;
  }
```

```
  if (events & MSA_MSG4_EVENT)
  {
return events ^ MSA_MSG4_EVENT;
}
  if (events & MSA_MSG5_EVENT)
  {
return events ^ MSA_MSG5_EVENT;
  }
return 0;
}
*****************************************************

* @fn MAC_CbackEvent
* @brief This callback function sends MAC events
    to the application.
a. The application must implement this function.
    A typical
b. implementation of this function would allocate an
    OSAL
c. message, copy the event parameters to the message, and
d. send the message to the application's OSAL event
    handler.
e. This function may be executed from task or interrupt
f. context and therefore must be reentrant.
* @param pData - Pointer to parameters structure.
* @return None.
--------------------------------------------------------------
*/
void MAC_CbackEvent(macCbackEvent_t *pData)
{
macCbackEvent_t *pMsg = NULL;
uint8 len = msa_cbackSizeTable[pData->hdr.event];
switch (pData->hdr.event)
{
case MAC_MLME_BEACON_NOTIFY_IND:
len += sizeof(macPanDesc_t) +  pData->beaconNotifyInd.
        sduLength +
MAC_PEND_FIELDS_LEN(pData->beaconNotifyInd.pendAddrSpec);
if ((pMsg = (macCbackEvent_t *)
osal_msg_allocate(len))
   != NULL)
{
/* Copy data over and pass them up */
osal_memcpy(pMsg, pData, sizeof
           (macMlmeBeaconNotifyInd_t));
```

```
pMsg->beaconNotifyInd.pPanDesc = (macPanDesc_t *)
   ((uint8 *) pMsg + sizeof(macMlmeBeaconNotifyInd_t));
osal_memcpy(pMsg->beaconNotifyInd.pPanDesc,
   pData->beaconNotifyInd.pPanDesc, sizeof(macPanDesc_t));
pMsg->beaconNotifyInd.pSdu = (uint8 *)
   (pMsg->beaconNotifyInd.pPanDesc + 1);
osal_memcpy(pMsg->beaconNotifyInd.pSdu,
   pData->beaconNotifyInd.pSdu,
   pData->beaconNotifyInd.sduLength);
}
break;
case MAC_MCPS_DATA_IND:
pMsg = pData;
break;
default:
if ((pMsg = (macCbackEvent_t *)
osal_msg_allocate(len))
     != NULL)
{
osal_memcpy(pMsg, pData, len);
}
break;
    }
if (pMsg != NULL)
{
osal_msg_send(MSA_TaskId, (byte *) pMsg);
}
}
/*
--------------------------------------------------------------
* @fn MAC_CbackCheckPending
* @brief Returns the number of indirect messages
             pending in the application
* @param None
@return  Number of indirect messages in the application
--------------------------------------------------------------
*/
uint8 MAC_CbackCheckPending(void)
{
return (0);
}
/*
--------------------------------------------------------------
* @fn MSA_CoordinatorStartup()
* @brief Update the timer per tick
```

```
* @param beaconEnable: TRUE/FALSE
*/
void MSA_CoordinatorStartup()
{
macMlmeStartReq_t   startReq;
MAC_MlmeSetReq(MAC_EXTENDED_ADDRESS, &msa_ExtAddr);
/* Setup MAC_SHORT_ADDRESS */
MAC_MlmeSetReq(MAC_SHORT_ADDRESS, &msa_CoordShortAddr);
/* Setup MAC_BEACON_PAYLOAD_LENGTH */
/* Setup MAC_BEACON_PAYLOAD */
MAC_MlmeSetReq(MAC_BEACON_PAYLOAD, &msa_BeaconPayload);
/* Enable RX */
MAC_MlmeSetReq(MAC_RX_ON_WHEN_IDLE, &msa_MACTrue);
/* Setup MAC_ASSOCIATION_PERMIT */
MAC_MlmeSetReq(MAC_ASSOCIATION_PERMIT, &msa_MACTrue);
MAC_MlmeSetReq(MAC_PAN_ID, &msa_PanId);
/* Fill in the information for the start request
   structure */
startReq.startTime = 0;
startReq.panId = msa_PanId;
startReq.logicalChannel = MSA_MAC_CHANNEL;
startReq.beaconOrder = msa_Beacon_Order;
startReq.superframeOrder = msa_SuperFrame_Order;
startReq.panCoordinator = TRUE;
startReq.batteryLifeExt = FALSE;
startReq.coordRealignment = FALSE;
startReq.realignSec.securityLevel = FALSE;
startReq.beaconSec.securityLevel = FALSE;
/* Call start request to start the device as
   a coordinator */
MAC_MlmeStartReq(&startReq);
/*
------------------------------------------------------------
* @fn MSA_DeviceStartup()
* @brief Update the timer per tick
* @param beaconEnable: TRUE/FALSE
a. @return  None
------------------------------------------------------------
*/
void MSA_DeviceStartup()
{
uint16 AtoD = 0;
AtoD = HalAdcRead (MSA_HAL_ADC_CHANNEL,
                   MSA_HAL_ADC_RESOLUTION);
/* Setup MAC_EXTENDED_ADDRES */
```

```
#if defined (HAL_BOARD_CC2420DB) ||
  defined (HAL_BOARD_DZ1611) ||
  defined (HAL_BOARD_DZ1612) ||
  defined (HAL_BOARD_DRFG4618)
/* Use HI and LO of AtoD on CC2420 and MSP430 */
msa_ExtAddr2[6] = HI_UINT16( AtoD );
msa_ExtAddr2[7] = LO_UINT16( AtoD );
#else
/* On 2430 Use the MSB to ensure the stability */
msa_ExtAddr2[6] = HI_UINT16( AtoD ) » 4;
msa_ExtAddr2[7] = HI_UINT16( AtoD ) » 3;
#endif
sAddrExtCpy(msa_ExtAddr, msa_ExtAddr2);
MAC_MlmeSetReq(MAC_EXTENDED_ADDRESS, &msa_ExtAddr);
/* Setup MAC_BEACON_PAYLOAD_LENGTH */
MAC_MlmeSetReq(MAC_BEACON_PAYLOAD_LENGTH,
               &msa_BeaconPayloadLen);
/* Setup MAC_BEACON_PAYLOAD */
MAC_MlmeSetReq(MAC_BEACON_PAYLOAD, &msa_BeaconPayload);
/* Setup PAN ID */
MAC_MlmeSetReq(MAC_PAN_ID, &msa_PanId);
/* This device is setup for Direct Message */
if (msa_IsDirectMsg)
MAC_MlmeSetReq(MAC_RX_ON_WHEN_IDLE, &msa_MACTrue);
else
MAC_MlmeSetReq(MAC_RX_ON_WHEN_IDLE, &msa_MACTrue);
/* Setup Coordinator short address */
MAC_MlmeSetReq(MAC_COORD_SHORT_ADDRESS,
               &msa_AssociateReq.coordAddress.
               addr.shortAddr);
/* this is a router, allow association. */
MAC_MlmeSetReq(MAC_ASSOCIATION_PERMIT, &msa_MACTrue);
   // comment this line for an end-device.
/* Power saving */
MSA_PowerMgr (MSA_PWR_MGMT_ENABLED);
}
/*
--------------------------------------------------------------
* @fn MSA_AssociateReq()
a.   @brief
* @param None
a.   @return  None
--------------------------------------------------------------
*/
void MSA_AssociateReq(void)
```

```
{
}
/*
--------------------------------------------------------------
* @fn MSA_AssociateRsp()
* @brief This routine is called by Associate_Ind
            in order to return the response to the device
* @param pMsg - pointer to the structure received
            by MAC_MLME_ASSOCIATE_IND
a.   @return  None
--------------------------------------------------------------
*/
void MSA_AssociateRsp(macCbackEvent_t* pMsg)
{
}
/**********************************************
* @fn MSA_McpsDataReq()
* @brief This routine calls the Data Request
* @param data - contains the data that
                   would be sent
a. dataLength - length of the data that will be sent
b. @return  None
--------------------------------------------------------------
*/
void MSA_McpsDataReq(uint8* data, uint8 dataLength,
bool directMsg, uint16 dstShortAddr)
{
macMcpsDataReq_t  *pData;
static uint8 handle = 0;
if ((pData = MAC_McpsDataAlloc(dataLength,
    MAC_SEC_LEVEL_NONE, MAC_KEY_ID_MODE_NONE)) != NULL)
{
pData->mac.srcAddrMode = SADDR_MODE_SHORT;
pData->mac.dstAddr.addrMode = SADDR_MODE_SHORT;
pData->mac.dstAddr.addr.shortAddr = dstShortAddr;
pData->mac.dstPanId = msa_PanId;
pData->mac.msduHandle = handle++;
pData->mac.txOptions = MAC_TXOPTION_ACK;
/* If it's the coordinator and the device is
   in-direct message */
if (msa_IsCoordinator)
{
if (!directMsg)
{
pData->mac.txOptions |= MAC_TXOPTION_INDIRECT;
```

```
}
}
/* Copy data */
osal_memcpy (pData->msdu.p, data, dataLength);
/* Send out data request */
MAC_McpsDataReq(pData);
}   }
void MSA_UART_Write_Init(uint8 taskID)
{
MSA_UART_Write_TaskID=taskID;
}
*/
-----------------------------------------------------------------
* @fn MSA_McpsPollReq()
* @brief Performs a poll request on the coordinator
* @param None
a.   @return  None
-----------------------------------------------------------------
*/
void MSA_McpsPollReq(void)
{
}
*/
-----------------------------------------------------------------
* @fn MacSampelApp_ScanReq()
* @brief Performs active scan on specified channel
* @param None
a. @return  None
*/
void MSA_ScanReq (uint8 scanType, uint8 scanDuration)
{  macMlmeScanReq_t scanReq;
/* Fill in information for scan request structure */
scanReq.scanChannels = (uint32) 1 << MSA_MAC_CHANNEL;
scanReq.scanType = scanType;
scanReq.scanDuration = scanDuration;
scanReq.maxResults = MSA_MAC_MAX_RESULTS;
scanReq.result.pPanDescriptor = msa_PanDesc;
/* Call scan request */
MAC_MlmeScanReq(&scanReq);
}
/*
-----------------------------------------------------------------
@fn MSA_SyncReq()
* @brief Sync Request
* @param None
```

```
    a. @return  None
----------------------------------------------------------------
*/
void MSA_SyncReq(void)
{
macMlmeSyncReq_t syncReq;
/* Fill in information for sync request structure */
syncReq.logicalChannel = MSA_MAC_CHANNEL;
syncReq.channelPage    = MAC_CHANNEL_PAGE_0;
syncReq.trackBeacon    = TRUE;
/* Call sync request */
MAC_MlmeSyncReq(&syncReq);
}
/*
----------------------------------------------------------------
* @fn MSA_BeaconPayLoadCheck()
* @brief Check if the beacon comes from MSA but
     not zigbee
* @param pSdu - pointer to the buffer that contains
           the data
    a. @return  TRUE or FALSE
----------------------------------------------------------------
*/
bool MSA_BeaconPayLoadCheck(uint8* pSdu)
{
}
return TRUE;
}
/*
----------------------------------------------------------------
* @fn MSA_DataCheck()
* @brief Check if the data match with the predefined
        data
* @param data - pointer to the buffer where the data
        will be checked against the predefined data
    a. dataLength - length of the data
    b. @return  TRUE if the data matched else it's the
       response
----------------------------------------------------------------
          / echo packet
*/
bool MSA_DataCheck(uint8* data, uint8 dataLength)
{
return TRUE;
}
```

```
/*
-------------------------------------------------------------
@fn      MSA_HandleKeys
* @brief Callback service for keys
* @param keys - keys that were pressed
a.   state - shifted
b.   @return  void
-------------------------------------------------------------
*/
void MSA_HandleKeys(uint8 keys, uint8 shift)
{
if ( keys & HAL_KEY_SW_1 )
{
// AD
/* disabled since on delhi board we have no buttons.
   the   device will come up upon init now.
// Start the device as a direct message device and
   beacon disabled
if (!msa_IsStarted)
{
// Decide if direct or indirect messaging is used
msa_IsDirectMsg = MSA_DIRECT_MSG_ENABLED;
if (msa_IsDirectMsg)
{
// Start the device as an DIRECT messaging device
if (msa_BeaconOrder != 15)
MSA_ScanReq(MAC_SCAN_PASSIVE, MSA_MAC_BEACON_ORDER + 1);
else
MSA_ScanReq(MAC_SCAN_ACTIVE, 3);
}
else
{
// Start the device as an INDIRECT messaging device and
   beacon disabled
// Beacon network doesn't work with polling
if (!msa_IsStarted)
{
msa_IsDirectMsg = FALSE;
MSA_ScanReq(MAC_SCAN_ACTIVE, 3);
}
}
}
*/
//~AD
}
```

```
if ( keys & HAL_KEY_SW_2 )
/* Start sending message */
//if (msa_IsStarted)
//{
/* Set App's state to SEND or IDLE */
//if (msa_State == MSA_IDLE_STATE)
//msa_State = MSA_SEND_STATE;
//else
//msa_State = MSA_IDLE_STATE;
/* Send Event to the App to carry out the changes */
//osal_start_timerEx ( MSA_TaskId, MSA_SEND_EVENT, 100);
// }
}
if ( keys & HAL_KEY_SW_3 )
{
}
if ( keys & HAL_KEY_SW_4 )
{
}
if ( keys & HAL_KEY_SW_5 )
{
}
}
//AD
/*
---------------------------------------------------------------
* @fn keylessStart
 a. @brief
 b. This function starts the device. It mimics the
    functionality
 c. of what was done when SW1 button was pressed.
    In delhi board
 d. we have no SW1 button, so the device starts upon INIT.
* @param void
 a. @return  none
---------------------------------------------------------------
void keylessStart()
{
   msa_IsDirectMsg = MSA_DIRECT_MSG_ENABLED;
   MSA_ScanReq(MAC_SCAN_ACTIVE, 7);
   writeMsg("Keyless start initiated.",26);
   HalLedBlink(HAL_LED_4, 0, 50, 300);
}
uint16 MSA_UART_Write_ProcessEvent(uint8 taskID,
                                   uint16 events)
```

```
{
  if ( events != 0 && taskID == MSA_UART_Write_TaskID)
  {
    if (UARTInputReadComplete == TRUE)
      {
        bool UARTInputError = TRUE;
        uint8 i;
        UARTInputReadComplete = FALSE;
        //writeMsg(UARTInputBuffer,6);
        if (UARTInputBuffer[0] == '*')
        {
          i=1;
          while ( UARTInputBuffer[i] != '*'
                  && i < MAXUARTINPUTBUFFER - 1
                  && i+6 < MAXPACKETLENGTH)
          {
            msa_Data1[i+5] = UARTInputBuffer[i];
            i++;
          }
          //writeMsg(&msa_Data1[6],2);
         if ( UARTInputBuffer[i] == '*' )
          {
            msa_Data1[i+5] = '*'; // end of name
            msa_Data1[i+6] = UARTInputBuffer[i+1];
                              // One byte message id/
           // writeMsg(&msa_Data1[6],4);
            if ( UARTInputBuffer[i+2] == '*' )
          {
            msa_Data1[i+7] = '*'; // end of message
            msa_Data1[i+8] = UARTInputBuffer[i+3];
                              // One byte message/
            if (UARTInputBuffer[i+4] == '#')
            {
              msa_Data1[0] = MSA_PACKET_LENGTH;
              msa_Data1[1] = HI_UINT16(0xFFFF);
              msa_Data1[2] = LO_UINT16(0xFFFF);
              msa_Data1[3] = HI_UINT16(myAddress);
              msa_Data1[4] = LO_UINT16(myAddress);
              msa_Data1[5] = 3;
              //msa_Data1[i+9] = UARTInputMessageCount++;
              // to prevent end device from processing
              same msg twice.
              msa_Data1[i+9] = '#';
              UARTInputError = FALSE;
            }
```

```
        }
        }
      }
      if ( UARTInputError == FALSE )
      {
        if (numChildren >= 0)
        {
        sendToChild=1;
MSA_McpsDataReq((uint8*)msa_Data1,MSA_PACKET_LENGTH,
                  TRUE,myChildrenAddress[0]);
        }
            uint8 printDataOK[MAXBUFFERLENGTH],i;
              printDataOK[0]='*';
               printDataOK[1]='*';
                printDataOK[2]='P';
              for (i=0;i<MAXNAMELENGTH;i++)
              {
                 if ( msa_Data1[i+6] == '*' )
                   break;
                 printDataOK[i+3]=msa_Data1[i+6];
              }
              printDataOK[i+3] = '*';
              printDataOK[i+4] = '*';
              printDataOK[i+5] = msa_Data1[i+7];
              printDataOK[i+6] = '*';
              printDataOK[i+7] = '*';
              writeMsg(printDataOK,i+8);
      }
      else
      {
        if ( strncmp(UARTInputBuffer,
             "REBOOT",6) == 0 )
          systemReset();
        else
        {
          writeMsg("I!",2);
          writeMsg(UARTInputBuffer,10);
        }
      }
      //UARTInputRead = TRUE;
    }
    else
      return 0;
  }
  return 0;
```

```
}
void writeMsg(uint8 *buf,uint8 len)
{
  //flag=0;
  if ( len < MAXUARTBUFLENGTH - 1 )
  {
    if ( write2COM1Port(buf,len) != len )
    {
      currentBufLevel++;
      if ( currentBufLevel > MAXUARTBUF / 2 )
      {
        currentBufLevel=0;
         HalUARTClose(HAL_UART_PORT_0);
          HalUARTInit();
          initCOM1Port();
          write2COM1Port("Recover",7);
          HalLedBlink(HAL_LED_4, 0, 50, 300);
      }
    }
    else
    {
      write2COM1Port("\n",1);
      currentBufLevel=0;
    }
  }
  void writeVal(uint16 val)
{
  uint8 ascAdd[5],i;
  for(i=0;i<5;i++)
  {
    ascAdd[4-i]=val%10 + 48;
    val/=10;
  }
  writeMsg(ascAdd,5);
}
//Reading UART inputs characterwise and sending to specified address as a string
void processUARTInput(uint8 *b, uint16 len)
{
//  writeMsg(b,1);
//  osal_start_timer(MSA_SEND_EVENT,100);
//  if ( UARTInputRead == TRUE )
//  {
  uint16 i;
```

```
    for (i=0;i<len;i++)
    {
      UARTInputBuffer[UARTInputBufferLevel] = b[i];
      //writeVal(len);
      UARTInputBufferLevel++;
      if (b[i] == '#' || UARTInputBufferLevel >=
          MAXUARTINPUTBUFFER )
      {
//        UARTInputRead = FALSE;
        UARTInputReadComplete = TRUE;
        UARTInputBufferLevel = 0;
     osalTaskRec_t *taskPtr=osalFindTask
       (MSA_UART_Write_TaskID);taskPtr->events=1;

      }
    }
//  }
}
*****************************************************************
```

1.2 *Programming with CC2430 Chip (22 Pins)*

For making of coordinator using CC2430 RFID chip (22 pins) follow the same process as CC2430 RFID chip (48 pins) as discussed above. There is one difference in Mac_NoDebug library. Open the IAR, go to Project option, choose C/C++ complier, and then select Preprocessor option and include Mac_NoDebug library in the defined symbol. The few differences in the algorithm steps are given below.

The algorithm for programming a coordinator is as follows:

i. The algorithm is same as above up to 9 steps. Step 10 is different from above.
ii. Coordinator device is connected to 10 pin wires ribbon cable and this ribbon cable is connected to a simulator.
iii. The same simulator is connected to a PC through USB port.
iv. The program is compiled and debugged.
v. The USB port and simulator are disconnected, and the coordinator is ready.

The coding for coordinator in CC2430 RFID chip (22 pins) is the same as that for CC2430 RFID chip (48 pins).

2 Programming for Router

2.1 Programming with CC2430 Chip (48 Pins)

The algorithm for programming a router is as follows:

i. All the associated parameters are assigned accordingly as mentioned in the algorithm steps iii – v for coordinator.
ii. Device (router) information (ROUTER_ID, router address, maximum number of associated routers) to be fixed, maximum number of associated end devices to be fixed, and maximum buffer size to be fixed.
iii. Timer for reboot transmission is set and radio transmitting power range is adjusted.
iv. The choke point for the block area network is set.
v. Router device is connected to a PC through USB port, the program is compiled and debugged.
vi. USB port is disconnected, and router is ready.

For programming the router, copy the above coordinator programming file and rename the file. All library and drivers files are same as used in coordinator. Coding process for router is same as coordinator up to line 54 given above. After line 54, following changes are required in the respective intermittent lines.

```
***********************************************************
**************** Change in lines 55 to 61****************
/*Coordinator and Device information */
uint16 msa_PanId = MSA_PAN_ID:
uint16 msa_CoordShortAddr = MSA_COORD_SHORT_ADDR;
uint16 msa_DevShortAddr   = MSA_DEV_SHORT_ADDR;
/*
  List of short addresses that the coordinator will use
  to assign to each the device that associates to it
*/
uint16 msa_DevShortAddrList [] =
{0x0001, 0x0002, 0x0003, 0x0004, 0x0005, 0x0006, 0x0007};
/* Current number of devices associated to the coordinator
*/
uint8 msa_NumOfDevices = 0;
/* Predefined packet that will be sent out
by the coordinator */
uint8 msa_Data1[MSA_PACKET_LENGTH];
uint8 msa_sendToChild[MSA_PACKET_LENGTH];
/* Predefined packet that will be sent back to the
   coordinator by the device */
uint8  msa_Data2[MSA_ECHO_LENGTH];
```

```
/* Beacon payload, this is used to determine if the
   device is
not zigbee device */
uint8  msa_BeaconPayload[] = {0x22, 0x33, 0x44};
/* Contains pan descriptor results from scan */
macPanDesc_t  msa_PanDesc[MSA_MAC_MAX_RESULTS+1];
/********************Add code after line 71***************/
uint16 p_address,address;
/**************** Change in line 74***********************/
uint8 UART_Buffer[30][31];
/************* Change in lines 76 to 96*******************/
#define MAXDATASTORE 5
// imp need to  change code if this value is changed
#define MAXMSGHOLDTIME 50000
#define MAXPACKETLENGTH 20
#define MAXNAMELENGTH 5
#define ROUTERHEALTHCHKTIME 3000
uint8 dataStore[MAXDATASTORE][MAXPACKETLENGTH];
#define SENDTOPARENT 1
#define SENDTOOTHER 0
uint16 myAddress=10;
uint8 myName[]="R001";
uint16 myParentAddress=0xFFFF;
uint16 myPanID;

uint16 possibleParent[MAXPARENTSCAN];
uint8 possibleParentLinkQ[MAXPARENTSCAN];
uint8 curLevel = 0;
uint8  healthChk[MSA_PACKET_LENGTH];
uint16 myAddress=10;
uint8 myName[]="R001";
uint16 myParentAddress=0xFFFF;
uint16 myPanID;
uint16 possibleParent[MAXPARENTSCAN];
uint8 possibleParentLinkQ[MAXPARENTSCAN];
uint8 curLevel = 0;
uint8  healthChk[MSA_PACKET_LENGTH];
uint16 myChildrenAddress[MAXCHILDREN];
int numChildren=-1;
int sendToChild = -1;
uint8 contPacketMiss = 0;
uint8 msgHandlerTaskID=0;
uint8 msgBuffer[30][31];
uint8 j,msg[30],l=0;
uint8 *temp;
```

```
/*****************Add codes after line 109*****************/
void msgHandler_Init(uint8 taskId)
{
  msgHandlerTaskID=taskId;
  uint8 i;
  for ( i=0;i<30;i++)
  {
     msgBuffer[i][0]='\0';
     msgBuffer[i][30]=0;
  }
}
/************** Change in lines 124 to 135*****************/
  // to make our health chk data.
  healthChk [0] = MSA_PACKET_LENGTH;
  healthChk [1] = 0;
  healthChk [2] = 0;
  healthChk [3] = HI_UINT16(myAddress);
  healthChk [4] = LO_UINT16(myAddress);
  healthChk [5] = 1; // msg type
  healthChk [6] = 'T';
  for (i=0;i<MAXNAMELENGTH;i++)
  {
   if ( myName[i] == '\0' || (i + 9) >= MAXPACKETLENGTH - 1 )
     break;
   healthChk[i+7] = myName[i];
  }
  healthChk[i+7] = '#';
  healthChk[i+8] = HI_UINT16(myParentAddress);
  healthChk[i+9] = LO_UINT16(myParentAddress);
  //healthChk[i+9] = '#';
  msa_BeaconOrder = MSA_MAC_BEACON_ORDER;
  msa_SuperFrameOrder = MSA_MAC_SUPERFRAME_ORDER;
  // AD
  initCOM1Port();
  //~AD
  for (i=0; i < MAXDATASTORE;i++)
    dataStore[i][0]='-';
  //HalLedBlink (HAL_LED_4, 0, 40, 200);
  // AD
    writeMsg("INIT Complete.",16);
  //~AD
}
 There are some changes in  lines 174 & 175
  msa_IsCoordinator = FALSE;
      writeMsg("Router!",7);
```

```
******************** Change in line 185*********************
pData = (macCbackEvent_t *) pMsg;
if (curLevel == 0)//&& (pData->scanCnf.hdr.status ==
    MAC_NO_BEACON)) ||
(!msa_IsSampleBeacon))
          {
            keylessStart();
          }
          else
          {
            uint8 i,threshold=0;
            for ( i=0; i < curLevel ; i++ )
            {
              if ( possibleParentLinkQ[i] > threshold
                   && possibleParent[i] < myAddress )
              {
                myParentAddress =  possibleParent[i];
                threshold = possibleParentLinkQ[i];
              }
              writeMsg(&possibleParentLinkQ[i],1);
              writeVal(possibleParent[i]);
            }
            if ( myParentAddress != 0xFFFF
                 && myParentAddress != myAddress)
            {
              curLevel=0;
              MSA_CoordinatorStartup();
              osal_start_timer(MSA_SEND_EVENT,1000);
            }
            else
              keylessStart();
          }
          break;
***************** Change in line 192*********************
{
            if ( pData->dataCnf.msduHandle == SENDTOPARENT )
            {
              if (pData->dataCnf.hdr.status != MAC_SUCCESS)
              {
               if ( contPacketMiss > 2 )
                keylessStart();
              else
                contPacketMiss++;
              }
              else if (pData->dataCnf.hdr.status
```

```
                        == MAC_SUCCESS)
                  {
                    contPacketMiss=0;
                  }
                }
******************* Change in line 204*********************
writeMsg("Data Received!",14);
**************** Change in lines 208 to 239*****************
if (MSA_DataCheck ( pData->dataInd.msdu.p,
                   pData->dataInd.msdu.len ))
         {
           uint16 destAdd,sourceAdd;
           int i;
           uint8 receivedName[MAXNAMELENGTH];
           destAdd = BUILD_UINT16(pData->dataInd.msdu.p[2],
           pData->dataInd.msdu.p[1]);
           sourceAdd=BUILD_UINT16(pData->dataInd.msdu.p[4],
           pData->dataInd.msdu.p[3]);
           bool receivedPktForMe = TRUE;
           if (pData->dataInd.msdu.p[5] == 3)
           {
             for (i=0; i<MAXNAMELENGTH;i++)
             {
               if (myName[i] == '\0')
                 break;
             //receivedName[i] = pData->dataInd.msdu.p[i+6];
               if ( pData->dataInd.msdu.p[i+6] != myName[i] )
                 receivedPktForMe = FALSE;
             }
             if ( pData->dataInd.msdu.p[i+6] != '*' )
               receivedPktForMe = FALSE;
           }
           else
             receivedPktForMe = FALSE;
           if ( destAdd == myAddress
               || receivedPktForMe == TRUE)
           {
             writeMsg("from:",5);
             writeVal(sourceAdd);
             writeMsg("contents",8);
             writeMsg(pData->dataInd.msdu.p+7+i,1);
             writeMsg("-----------",17);
           }
           else if ( destAdd < myAddress )
           {
```

```
            //writeMsg("dest<myadd",10);
            for ( i =0 ;i <= numChildren ; i++ )
            {
              if ( myChildrenAddress[i] == pData->dataInd.
                   mac.srcAddr.addr.shortAddr )
                break;
            }
            if ( i > numChildren )
            {
              // unique address
              for ( i=MAXCHILDREN - 1; i>0; i- )
              {
                myChildrenAddress[i]=myChildrenAddress[i-1];
              }
              myChildrenAddress[0]=
                pData->dataInd.mac.srcAddr.addr.shortAddr;

              if ( numChildren < MAXCHILDREN - 1)
                numChildren++;
            }
            if ( pData->dataInd.msdu.p[5] == 5 )
            {
}
****************** Change in lines 250 to 300***************
if ( pData->dataInd.msdu.p[5] == 5 )
       {
         // packet is from an end device.
         for (i=0; i < 5; i++)
           msa_Data1[i] = pData->dataInd.msdu.p[i];
         msa_Data1[5] = 2;
         for (i=0; i < MAXPACKETLENGTH; i++)
         {
           if ( pData->dataInd.msdu.p[i+6] == '#' )
             break;
           msa_Data1[i+6] = pData->dataInd.msdu.p[i+6];
           if ( i < MAXNAMELENGTH )
             receivedName[i] = pData->dataInd.msdu.p[i+6];
         }
  msa_Data1[i+6] = '*';
        msa_Data1[i+7] = '*';
        for ( j=0; j < MAXNAMELENGTH; j++ )
        {
                 if ( myName[j] == '\0'|| (i + 9 + j)
                  > = MAXPACKETLENGTH )
```

```
        break;
      msa_Data1[i+8+j] = myName[j];
    }
    msa_Data1[i+8+j] = '#';
     // we now check if any messages are meant
        for this guy.
    uint16 destAddrDS =
       pData->dataInd.mac.srcAddr.addr.shortAddr;
    bool sendPacket;
    for ( i=0; i< MAXDATASTORE;i++ )
    {
      if (dataStore[i][0] != '-')
      {
        //destAddrDS =
          BUILD_UINT16(dataStore[i][2],
                       dataStore[i][1]);
        sendPacket=TRUE;
        for (j=0; j< MAXNAMELENGTH;j++)
        {
          if ( dataStore[i][j+6] == '*' )
            break;
          if ( dataStore[i][j+6]
              != receivedName[j+1] )
          {
            sendPacket = FALSE;
              break;
          }
        }
        if ( sendPacket == TRUE && j > 0 )
        {
  MSA_McpsDataReq(dataStore[i],MSA_PACKET_LENGTH,
                  TRUE,destAddrDS);
          dataStore[i][0]='-';
          switch ( i)
          {
            case 0:
              osal_stop_timerEx(MSA_TaskId,
                  MSA_MSG1_EVENT);
              break;
            case 1:
              osal_stop_timerEx(MSA_TaskId,
                  MSA_MSG2_EVENT);
              break;
```

```
                    case 2:
                      osal_stop_timerEx(MSA_TaskId,
                          MSA_MSG3_EVENT);
                      break;
                    case 3:
                      osal_stop_timerEx(MSA_TaskId,
                          MSA_MSG4_EVENT);
                      break;
                    case 4:
                      osal_stop_timerEx(MSA_TaskId,
                          MSA_MSG5_EVENT);
                      break;
                  }
                }
              }
            }
          }
          else
          {
            for (i=0;i<MAXPACKETLENGTH;i++)
              msa_Data1[i]=pData->dataInd.msdu.p[i];
          }
           //writeMsg("sending",7);
MSA_McpsDataReq((uint8*)msa_Data1,MSA_PACKET_LENGTH,TRUE,
   myParentAddress);
        }
        else
        {
          if ( pData->dataInd.msdu.p[6] == 'E' )
          {
            // we need to store the data
            int i,j;
            for (i=0;i<MAXDATASTORE;i++)
            {
              if (dataStore[i][0] == '-')
                break;
            }
            if ( i < MAXDATASTORE )
            {
              for ( j=0;j<MAXPACKETLENGTH;j++)
              {
                dataStore[i][j]=pData->dataInd.msdu.p[j];
              }
              switch ( i)
              {
```

```
                    case 0:
    osal_start_timerEx(MSA_TaskId,MSA_MSG1_EVENT,
        MAXMSGHOLDTIME);
                      break;
                    case 1:
   osal_start_timerEx(MSA_TaskId,MSA_MSG2_EVENT,
       MAXMSGHOLDTIME);
                      break;
                    case 2:
  osal_start_timerEx(MSA_TaskId,MSA_MSG3_EVENT,
      MAXMSGHOLDTIME);
                      break;
                    case 3:
  osal_start_timerEx(MSA_TaskId,MSA_MSG4_EVENT,
      MAXMSGHOLDTIME);
                      break;
                    case 4:
  osal_start_timerEx(MSA_TaskId,MSA_MSG5_EVENT,
      MAXMSGHOLDTIME);
                      break;
                  }
                }
              }
              // we now send to all our children.
              for (i=0; i<MSA_PACKET_LENGTH;i++)
              {
                msa_sendToChild[i]=pData->dataInd.msdu.p[i];
              }
              sendToChild=1;
              MSA_McpsDataReq((uint8 *)msa_sendToChild,
                 MSA_PACKET_LENGTH,TRUE, myChildrenAddress[0]);
            }
          }
          break;
      }
      /* Deallocate */
      osal_msg_deallocate((uint8 *) pMsg);
    }
    return events ^ SYS_EVENT_MSG;
  }
 /* This event handles polling for in-direct message device
*/
  if (events & MSA_POLL_EVENT)
  {
    MSA_McpsPollReq();
```

```
//    osal_start_timer (MSA_POLL_EVENT, MSA_WAIT_PERIOD);
    return events ^ MSA_POLL_EVENT;
  }
  /* This event will blink a LED then send out a DataReq and
     continue to do so every WAIT_PERIOD */
  if (events & MSA_SEND_EVENT)
  {
    //writeMsg("sending first pkt",17);
MSA_McpsDataReq((uint8*)healthChk,MSA_PACKET_LENGTH,TRUE,
   myParentAddress);
    osal_start_timer(MSA_SEND_EVENT,ROUTERHEALTHCHKTIME);
    return events ^ MSA_SEND_EVENT;
  }
   if (events & MSA_MSG1_EVENT)
  {
    dataStore[0][0] = '-';
    return events ^ MSA_MSG1_EVENT;
  }
  if (events & MSA_MSG2_EVENT)
  {
    dataStore[1][0] = '-';
    return events ^ MSA_MSG2_EVENT;
  }
  if (events & MSA_MSG3_EVENT)
  {
    dataStore[2][0] = '-';
    return events ^ MSA_MSG3_EVENT;
  }
  if (events & MSA_MSG4_EVENT)
  {
    dataStore[3][0] = '-';
    return events ^ MSA_MSG4_EVENT;
  }
  if (events & MSA_MSG5_EVENT)
  {
    dataStore[4][0] = '-';
    return events ^ MSA_MSG5_EVENT;
  }
  return 0;
}
****************** Change in lines 358 to 359**************
uint16 AtoD = 0;
  AtoD = HalAdcRead (MSA_HAL_ADC_CHANNEL,
    MSA_HAL_ADC_RESOLUTION);
```

```
  /*
    Setup MAC_EXTENDED_ADDRESS
  */
#if defined (HAL_BOARD_CC2420DB) || defined (HAL_BOARD_DZ1611)
    || defined (HAL_BOARD_DZ1612)
    || defined (HAL_BOARD_DRFG4618)
  /* Use HI and LO of AtoD on CC2420 and MSP430 */
  msa_ExtAddr2[6] = HI_UINT16( AtoD );
  msa_ExtAddr2[7] = LO_UINT16( AtoD );
#else
  /* On 2430 Use the MSB to ensure the stability */
  msa_ExtAddr2[6] = HI_UINT16( AtoD ) » 4;
  msa_ExtAddr2[7] = HI_UINT16( AtoD ) » 3;
#endif
/* Setup MAC_EXTENDED_ADDRESS */
  sAddrExtCpy(msa_ExtAddr, msa_ExtAddr1);
******************** Change in lines 426 to 429**************
void MSA_AssociateReq(void)
{
  MAC_MlmeAssociateReq (&msa_AssociateReq);
}
**************** Change in lines 476 to 479******************
void MSA_McpsPollReq(void)
{
  macMlmePollReq_t  pollReq;
  /* Fill in information for poll request */
  pollReq.coordAddress.addrMode = SADDR_MODE_SHORT;
  pollReq.coordAddress.addr.shortAddr = msa_CoordShortAddr;
  pollReq.coordPanId = msa_PanId;
  pollReq.sec.securityLevel = MAC_SEC_LEVEL_NONE;
  /* Call poll reuqest */
  MAC_MlmePollReq(&pollReq);
}
*************** Change in lines 515 to 520*******************
bool MSA_BeaconPayLoadCheck(uint8* pSdu)
{
  uint8 i = 0;
  for (i=0; i<msa_BeaconPayloadLen; i++)
  {
    if (pSdu[i] != msa_BeaconPayload[i])
    {
      return FALSE;
    }
  }
  return TRUE;
```

```
}
***********Comment lines 536 to 566 ************************
***********Comment lines 568 to 578 ************************
*************** Change in lines 605 to 741****************
if ( transmitInProgress() == FALSE)
      {
        write2COM1Port(UART_Buffer[0],UART_Buffer[0][30]);
        for ( uint8 i=0; i < currentBufLevel-1; i++)
        {
          osal_memcpy(UART_Buffer[i],UART_Buffer[i+1],
            UART_Buffer[i+1][30]);
          UART_Buffer[i][30]=UART_Buffer[i+1][30];
        }
        if ( currentBufLevel > 0 )
        currentBufLevel-;
        if ( currentBufLevel == 0 )
        {
         //osalTaskRec_t *taskPtr=osalFindTask
            (MSA_UART_Write_TaskID);
         return 0;
         //taskPtr->events=0;
        }
      }
      return 1;
    }
    return 0
}
void writeMsg(uint8 *buf,uint8 len)
{
  //flag=0;
  if ( currentBufLevel != 30 && len < 31 )
  {
     osal_memcpy(UART_Buffer[currentBufLevel],buf,len+2);
     UART_Buffer[currentBufLevel][len]=10;
     UART_Buffer[currentBufLevel][len+1]=13;
     UART_Buffer[currentBufLevel][30]=len+2;
     currentBufLevel++;
     osalTaskRec_t
        *taskPtr=osalFindTask(MSA_UART_Write_TaskID);
     taskPtr->events=1;
  }
}
void writeStr(uint8 *buf,uint8 len)
{
  //flag=1;
```

```
  if ( currentBufLevel != 30 && len < 31 )
  {
   osal_memcpy(UART_Buffer[currentBufLevel],buf,len);
   //      UART_Buffer[len]=10;
   //      UART_Buffer[len+1]=13;
   UART_Buffer[currentBufLevel][30]=len;
   currentBufLevel++;
   osalTaskRec_t
      *taskPtr=osalFindTask(MSA_UART_Write_TaskID);
   taskPtr->events=1;
  }
}
void writeVal(uint16 val)
{
  uint8 ascAdd[5],i;
  for(i=0;i<5;i++)
  {
    ascAdd[4-i]=val%10 + 48;
    val/=10;
  }
  writeMsg(ascAdd,5);
}
```

After replacing the above lines in the copied coordinator file, the total number of lines in the router file may vary depending upon the purpose of the router. After incorporating the above changes, there is no need for any other change. Then paste the modified program code in the msa.c file and debug the program into a router using USB port. Now the router "R001" is created. For programming the other routers, simply change the router ID (e.g., 2 for "R002", 3 for "R003", etc.) in the above msa.c file and debug the program into the respective router.

2.2 Programming with CC2430 Chip (22 Pins)

The algorithm for programming a router is as follows:

i. Follow the same procedure up to step 4. Changes are required after step 4.
ii. Router device is connected to 10 pin wires ribbon cable and this ribbon cable is connected to a simulator.
iii. The same simulator is connected to a PC through USB port.
iv. The program is compiled and debugged.
v. USB port and simulator are disconnected, and router is ready.

The coding for router in CC2430 RFID chip (22 pins) is the same as that for CC2430 RFID chip (48 pins).

3 Programming for End Device

3.1 Programming with CC2430 Chip (48 Pins)

The algorithm for programming an end device is as follows:

i. All the associated parameters are assigned accordingly as mentioned in the algorithm steps iii–v for coordinator.
ii. Device (end device) information (end device_ID, end device address associated routers address, and maximum buffer size) are to be fixed, and number of current associate devices are declared.
iii. The proper message at end device is set, joystick is used for sending messages to the coordinator (maximum five user-defined messages are stored inside the end device, and whenever necessary it can be used).
iv. Timer for reboot transmission is set, and radio transmitting power range is adjusted.
v. The end device is connected to a PC through USB port, the program is compiled and debugged.
vi. The USB port is disconnected, and the end device is ready.

For programming the end device, copy the above coordinator programming file and rename the file. All library and driver files are same as used in coordinator. Coding process for end device is same as that of coordinator up to line 54. After line 54, the following changes are required in the respective intermittent lines.

```
/* Coordinator and Device information */
*************Add codes after line 55 *************************
uint16 msa_CoordShortAddr = MSA_COORD_SHORT_ADDR;
uint16 msa_DevShortAddr = MSA_DEV_SHORT_ADDR;
/*
List of short addresses that the coordinator will use to
   assign to each the device that associates to it
*/
uint16 msa_DevShortAddrList []
   = {0x0001, 0x0002, 0x0003, 0x0004, 0x0005, 0x0006, 0x0007};
/* Current number of devices associated to the coordinator */
uint8 msa_NumOfDevices = 0;
****************Add codes after line 57*********************
/* Predefined packet that will be sent back to the
coordinator
   by the device */
uint8    msa_Data2 [MSA_ECHO_LENGTH];
******************** Change in line 60**********************
* Beacon payload, this is used to determine if the device is
```

```
  not zigbee device */
uint8    msa_BeaconPayload[] = {0x22, 0x33, 0x44};
uint8    msa_BeaconPayloadLen = 3;
structure that used for association request */
macMlmeAssociateReq_t msa_AssociateReq;
******************Add code after line 67*******************
/* Structure that used for association request */
macMlmeAssociateReq_t msa_AssociateReq;
/* Structure that used for association response */
macMlmeAssociateRsp_t msa_AssociateRsp;
/* Structure that contains information about the device that
   associates with the coordinator */
typedef struct
{
 Uint16 devShortAddr;
 Uint8 isDirectMsg;
}
msa_DeviceInfo_t;
/* Array contains the information of the devices */
//static msa_DeviceInfo_t msa_DeviceRecord[MSA_MAX_DEVICE_NUM];
uint8 msa_SuperFrameOrder;
uint8 msa_BeaconOrder;
***************** Change in lines 70 to 96 *****************
uint8 UART_Buffer[30][31]; // we will have a queue up to 30
uint8 currentBufLevel=0;
uint16 myAddress=1000;
uint8 myName[] = "E001";
uint8 lastMsgId=0;
#define MAXPARENTSCAN 5
#define MAXNAMELENGTH 5
#define MAXPACKETLENGTH 20
#define MAXDEVICEONTIME 10
uint16 possibleParent[MAXPARENTSCAN];
uint8 possibleParentLinkQ[MAXPARENTSCAN];
uint8 curLevel = 0;
bool processComplete = FALSE;
uint16 myParentAddress=0xFFFF;
uint16 myPanID;
uint8 msgHandlerTaskID=0;
uint8 msgBuffer[30][31];
uint16 msaCoordShortAddress;
extern bool sleepTime;
*******************Add codes after line 109*****************
void sendPacket();
void msgHandler_Init(uint8 taskId)
```

```
{
  msgHandlerTaskID=taskId;
  uint8 i;
  for ( i=0;i<30;i++)
  {
    msgBuffer[i][0]='\0';
    msgBuffer[i][30]=0;
  }
}
extern void powerm(void);
****************** Change in lines 124 to 127****************
void MSA_Init(uint8 taskId)
{
  uint8 i;
  /* Initialize the task id */
  MSA_TaskId = taskId;
  /* initialize MAC features */
  MAC_InitDevice();
  MAC_InitCoord();
  /* Initialize MAC beacon */
  MAC_InitBeaconDevice();
  MAC_InitBeaconCoord();
  /* Reset the MAC */
  MAC_MlmeResetReq(TRUE);
  /* Initialize the data packet */
  for (i=MSA_HEADER_LENGTH; i<MSA_PACKET_LENGTH; i++)
  {
   msa_Data1[i] = i-MSA_HEADER_LENGTH;
  }
  /* Initialize the echo packet */
  for (i=0; i<MSA_ECHO_LENGTH; i++)
  {
   msa_Data2[i] = 0xEE;
  }
  msa_BeaconOrder = MSA_MAC_BEACON_ORDER;
  msa_SuperFrameOrder = MSA_MAC_SUPERFRAME_ORDER;
  // AD
  initCOM1Port();
********************Comment line 128*************************
******************** Change in lines 162 to 167************
writeMsg("Got beacon not ind.",21);
msa_IsSampleBeacon = MSA_BeaconPayLoadCheck
      (pData->beaconNotifyInd.pSdu);
          /* If it's the correct beacon payload, retrieve the
             data for association req */
```

```
        /* // EndDevice is not Associate with
                Coordinator*/ //
 if(((!msa_IsStarted) && curLevel < MAXPARENTSCAN))//
     && ( pData->beaconNotifyInd.pPanDesc->coordAddress.
    addr.shortAddr!=0)
        {
 possibleParent[curLevel]=pData->beaconNotifyInd.pPanDesc->
    coordAddress.addr.shortAddr; possibleParentLinkQ[curLevel]
    =pData->beaconNotifyInd.pPanDesc->linkQuality;curLevel++;
        }
        break;
**************** Change in lines 177 to 179****************
writeVal (msa_CoordShortAddr);
***************** Change in lines 185 to 187****************
 if (curLevel == 0)//&& (pData->scanCnf.hdr.status
    == MAC_NO_BEACON)) || (!msa_IsSampleBeacon))
        {
          sleepTime = TRUE;
        }
        else
        {
          uint8 i,threshold=0;
          for ( i=0; i < curLevel ; i++ )
          {
            if ( possibleParentLinkQ[i] > threshold &&
                 possibleParent[i] < myAddress )
            {
              myParentAddress =  possibleParent[i];
              threshold = possibleParentLinkQ[i];
            }
            //writeMsg(&possibleParentLinkQ[i],1);
            //writeVal(possibleParent[i]);
          }
          if ( myParentAddress != 0xFFFF )
          {
            curLevel=0;
            processComplete = FALSE;
            MSA_DeviceStartup();
            HalLedSet ( HAL_LED_4, HAL_LED_MODE_ON);
            sendPacket();
            osal_start_timer(MSA_SEND_EVENT,
                             MAXDEVICEONTIME);
            msa_IsStarted = TRUE;
```

```
        }
        else
          sleepTime=TRUE;
      }
      break;
****************** Change in lines 204 to 299***************
writeMsg("Data Received!",14);
      if (MSA_DataCheck ( pData->dataInd.msdu.p,
          pData->dataInd.msdu.len ))
      {
        uint16 destAdd,sourceAdd;
        int i;
        bool receivedPktForMe = TRUE;
        destAdd = BUILD_UINT16(pData->dataInd.msdu.p[2],
        pData->dataInd.msdu.p[1]);
        sourceAdd = BUILD_UINT16(pData->dataInd.msdu.p[4],
        pData->dataInd.msdu.p[3]);
        destAdd++;
        if (pData->dataInd.msdu.p[5] == 3)
        {
          for (i=0; i<MAXNAMELENGTH;i++)
          {
            if (myName[i] == '\0')
              break;
            //receivedName[i] = pData->dataInd.msdu.p[i+6];
            if ( pData->dataInd.msdu.p[i+6] != myName[i] )
              receivedPktForMe = FALSE;
          }
          if ( pData->dataInd.msdu.p[i+6] != '*' )
            receivedPktForMe = FALSE;
        }
        else
          receivedPktForMe = FALSE;
        if ( destAdd == myAddress ||
             receivedPktForMe == TRUE)
        {
          if ( ( *(pData->dataInd.msdu.p+7+i)
              != lastMsgId ))
          {
          processComplete = FALSE;
          lastMsgId = *(pData->dataInd.msdu.p+7+i);
        writeMsg("from:",5);
        writeVal(sourceAdd);
        writeMsg("through",7);
```

```
            writeVal(pData->dataInd.mac.srcAddr.
              addr.shortAddr);
            writeMsg("contents",8);
            writeMsg(pData->dataInd.msdu.p+9+i,1);
            writeMsg("of message id:",14);
            writeMsg(pData->dataInd.msdu.p+7+i,1);
            writeMsg("-----------",17);
            if(*(pData->dataInd.msdu.p+9+i)=='5'
                  ||*(pData->dataInd.msdu.p+9+i)=='0')
            {
               PIN_HIGH();
               osal_stop_timer(MSA_SEND_EVENT);
               osal_start_timer(MSA_SEND_EVENT,3000);
               //osal_start_timer (MSA_GO_EVENT, 1000);
            }
            // we now send the acknowledgement back to
               the coord.
              uint8 j;
              msa_Data1[0] = MSA_PACKET_LENGTH;
              msa_Data1[1] = 0;
              msa_Data1[2] = 0;
              msa_Data1[3] = HI_UINT16(myAddress);
              msa_Data1[4] = LO_UINT16(myAddress);
              msa_Data1[5] = 4;
              msa_Data1[6]='A';
              for ( j=0; j<MAXNAMELENGTH;j++)
              {
                if (myName[j] == '\0'
                    || j+10 > MAXPACKETLENGTH)
                  break;
                msa_Data1[j+7] = myName[j];
              }
              msa_Data1[j+7]='*';
              msa_Data1[j+8]='*';
              msa_Data1[j+9]=*(pData->dataInd.msdu.p+7+i);
              msa_Data1[j+10]='#';
MSA_McpsDataReq((uint8*)msa_Data1,MSA_PACKET_LENGTH,TRUE,
   myParentAddress);
            }
            }
//            osal_stop_timer(MSA_SEND_EVENT);
//            sleepTime=TRUE;
          }
          break;
      }
```

```
      /* Deallocate */
      osal_msg_deallocate((uint8 *) pMsg);
    }
    return events ^ SYS_EVENT_MSG;
  }
if(events & MSA_SLEEP_EVENT)
  //{ if(i==6)
  {
    PIN_LOW();
    powerm();
    MAC_MlmeResetReq(TRUE);
    MAC_MlmeSetReq(MAC_EXTENDED_ADDRESS, &msa_ExtAddr);
    MAC_MlmeSetReq(MAC_BEACON_PAYLOAD_LENGTH,
        &msa_BeaconPayloadLen);
    MAC_MlmeSetReq(MAC_BEACON_PAYLOAD, &msa_BeaconPayload);
    MAC_MlmeSetReq(MAC_PAN_ID, &msa_PanId);
    MAC_MlmeSetReq(MAC_RX_ON_WHEN_IDLE, &msa_MACTrue);
    MAC_MlmeSetReq(MAC_COORD_SHORT_ADDRESS,
        &msaCoordShortAddress);
    MAC_MlmeSetReq(MAC_SHORT_ADDRESS, &msa_DevShortAddr);
    osal_start_timerEx ( MSA_TaskId, MSA_SEND_EVENT, 100);
    // }
    //  i++;
   // osal_start_timer (MSA_SLEEP_EVENT, 3000);
  return events ^ MSA_SLEEP_EVENT;
  }
if (events & MSA_GO_EVENT)
  {
    PIN_LOW();
    return events ^ MSA_GO_EVENT;
  }
  /* This event handles polling for in-direct message device
*/
  if (events & MSA_POLL_EVENT)
  {
    MSA_McpsPollReq();
//    osal_start_timer (MSA_POLL_EVENT, MSA_WAIT_PERIOD);
    return events ^ MSA_POLL_EVENT;
  }
  /* This event will blink a LED then send out a DataReq and
     continue to do so every WAIT_PERIOD */
  if (events & MSA_SEND_EVENT)
  {
    //if ( processComplete == TRUE )
      PIN_LOW();
```

```
      sleepTime=TRUE;
    //else
    //  osal_start_timer (MSA_SEND_EVENT, 300);
    return events ^ MSA_SEND_EVENT;
  }
  return 0;
}
****************** Change in lines 426 to 428***************
void MSA_AssociateReq(void)
{
 MAC_MlmeAssociateReq(&msa_AssociateReq);
}
****************** Change in lines 476 to 478***************
void MSA_McpsPollReq(void)
{
  macMlmePollReq_t  pollReq;
  /* Fill in information for poll request */
  pollReq.coordAddress.addrMode = SADDR_MODE_SHORT;
  pollReq.coordAddress.addr.shortAddr = msa_CoordShortAddr;
  pollReq.coordPanId = msa_PanId;
  pollReq.sec.securityLevel = MAC_SEC_LEVEL_NONE;
  /* Call poll reuqest */
  MAC_MlmePollReq(&pollReq);
}
****************** Change in lines 515 to 520*************
bool MSA_BeaconPayLoadCheck(uint8* pSdu)
{
  uint8 i = 0;
  for (i=0; i<msa_BeaconPayloadLen; i++)
  {
    if (pSdu[i] != msa_BeaconPayload[i])
    {
      return FALSE;
    }
  }
  return TRUE;
}
****************** Change in lines 534 to 591***************
void MSA_HandleKeys(uint8 keys, uint8 shift)
{
  uint8 i;
  msa_Data1[0] = MSA_PACKET_LENGTH;
  msa_Data1[1] = 0;
  msa_Data1[2] = 0;
  msa_Data1[3] = HI_UINT16(myAddress);
```

```
  msa_Data1[4] = LO_UINT16(myAddress);
  msa_Data1[5] = 4;
  msa_Data1[6]='M';
  for ( i=0; i<MAXNAMELENGTH;i++)
  {
    if (myName[i] == '\0' || i+9 > MAXPACKETLENGTH)
      break;
    msa_Data1[i+7] = myName[i];
  }
  msa_Data1[i+7]='*';
  msa_Data1[i+8]='*';
  msa_Data1[i+10]='#';
  if ( keys & HAL_KEY_SW_1 )
  {
    //msa_Data1[i+9]='1';
  }
  if ( keys & HAL_KEY_SW_2 )
  {
   // msa_Data1[i+9]='2';
  }
  if ( keys & HAL_KEY_SW_3 )
  {
   // msa_Data1[i+9]='3';
  }
  if ( keys & HAL_KEY_SW_4 )
  {
   // msa_Data1[i+9]='4';
  }
  if ( keys & HAL_KEY_SW_5 )
  {
   // msa_Data1[i+9]='5';
  }
  if ( keys & HAL_KEY_SW_6 || keys & HAL_KEY_SW_5)
  {
    msa_Data1[i+9]='5';
    osal_start_timerEx(MSA_TaskId,MSA_POLL_EVENT,100);
  }
  MSA_McpsDataReq((uint8*)msa_Data1,MSA_PACKET_LENGTH,TRUE,
     myParentAddress);
}
**************** Change in lines 605 to 741*****************
      {
      if(transmitInProgress() ==FALSE)write2COM1Port
        (UART_Buffer[0], UART_Buffer[0][30]);
```

```
        for ( uint8 i=0; i < currentBufLevel-1; i++)
        {
          osal_memcpy(UART_Buffer[i],UART_Buffer[i+1],
                      UART_Buffer[i+1][30]);
          UART_Buffer[i][30]=UART_Buffer[i+1][30];
        }
        if ( currentBufLevel > 0 )
        currentBufLevel-;
        if ( currentBufLevel == 0 )
        {
         //osalTaskRec_t
            *taskPtr=osalFindTask(MSA_UART_Write_TaskID);
         return 0;
         //taskPtr->events=0;
        }
      }
      return 1;
    }
    return 0;
}
void writeMsg(uint8 *buf,uint8 len)
{
  //flag=0;
  if ( currentBufLevel != 30 && len < 31 )
  {
     osal_memcpy(UART_Buffer[currentBufLevel],buf,len+2);
     UART_Buffer[currentBufLevel][len]=10;
     UART_Buffer[currentBufLevel][len+1]=13;
     UART_Buffer[currentBufLevel][30]=len+2;
     currentBufLevel++;
     osalTaskRec_t *taskPtr=osalFindTask
       (MSA_UART_Write_TaskID); taskPtr->events=1;

  }
}
void writeStr(uint8 *buf,uint8 len)
{
  //flag=1;
  if ( currentBufLevel != 30 && len < 31 )
  {
     osal_memcpy(UART_Buffer[currentBufLevel],buf,len);
//     UART_Buffer[len]=10;
//     UART_Buffer[len+1]=13;
     UART_Buffer[currentBufLevel][30]=len;
     currentBufLevel++;
```

```
        osalTaskRec_t *taskPtr=osalFindTask(MSA_UART_Write_TaskID);
        taskPtr->events=1;
    }
}
void writeVal(uint16 val)
{
    uint8 ascAdd[5],i;
    for(i=0;i<5;i++)
    {
        ascAdd[4-i]=val%10 + 48;
        val/=10;
    }
    writeMsg(ascAdd,5);
}
void sendPacket()
{
    uint8 i;
    msa_Data1[0] = MSA_PACKET_LENGTH;
    msa_Data1[1] = 0;
    msa_Data1[2] = 0;
    msa_Data1[3] = HI_UINT16(myAddress);
    msa_Data1[4] = LO_UINT16(myAddress);
    msa_Data1[5] = 5;
    msa_Data1[6]='T';
    for ( i=0; i<MAXNAMELENGTH;i++)
    {
        if (myName[i] == '\0' || i+8 > MAXPACKETLENGTH)
            break;
        msa_Data1[i+7] = myName[i];
    }
    msa_Data1[i+7]='#';

    MSA_McpsDataReq((uint8*)msa_Data1,MSA_PACKET_LENGTH,TRUE,
                    myParentAddress);
}
```

After replacing the above lines of coordinator programming file, the total number of lines in the end device file may vary depending upon the purpose of the end device. After incorporating the above changes, there is no need for any other change. Then paste the modified program code in the msa.c file and debug the program into an end device using USB port. Now the end device "E001*" is created. For programming the other end devices, simply change the end device ID (e.g., 2 for E002*, 3 for E003*, etc.) in the above msa.c file and debug the program into the respective end device.

3.2 Programming with CC2430 Chip (22 Pins)

The algorithm for programming a proximity warning device is as follows:

i. All the associated parameters are assigned accordingly as mentioned in the algorithm steps iii–v for coordinator.
ii. Device (end device) information (end device_ID, end device address associated routers router address, and maximum buffer size) are to be fixed, and number of current associate devices are declared.
iii. The proper message at end device is set, joystick is used for sending messages to the coordinator (maximum five user-defined messages are stored inside the end device, and whenever necessary it can be used).
iv. Timer for reboot transmission is set, and radio transmitting power range is adjusted.
v. End device is connected to 10 pin wires ribbon cable and this ribbon cable is connected to a simulator.
vi. The same simulator is connected to a PC through USB port.
vii. The program is compiled and debugged.
viii. USB port and simulator are disconnected, and end device is ready.

The coding for end device in CC2430 RFID chip (22 pins) is same as that for CC2430 RFID chip (48 pins).

4 Programming for Proximity Warning Device

4.1 Programming with CC2430 Chip (48 Pins)

The algorithm for programming a proximity warning device is as follows:

i. All the associate parameters of the devices are assigned accordingly as mentioned in the algorithm steps iii–v for coordinator.
ii. Total number of devices, channel number 11–28, beacon order, packet length, maximum results, and ADC (analog to digital converter) resolution are assigned.
iii. Device information (device_ID, device address, maximum number of associated device, and maximum buffer size) are to be fixed, and number of current associate devices are declared.
iv. Power variable is assigned as Boolean.
v. Timer for reboot transmission is set, and radio transmitting power range is adjusted.
vi. Device is connected to PC through USB port, the program is compiled and debugged.
vii. USB port is disconnected, and the device is ready.
viii. Another device in the same channel for receiving and transmitting signal is created.

For programming the proximity warning device, copy the above coordinator programming file and rename the file. All library and driver files are same as used in coordinator. Coding process for proximity warning device is same as given above for coordinator up to line 319. After line 319, the following changes are required in the respective intermittent lines.

```
*************** Change in lines 320 to 345*****************
  case MAC_MCPS_DATA_IND:
  pData = (macCbackEvent_t*)pMsg;
  if (MSA_DataCheck ( pData->dataInd.msdu.p,
        pData->dataInd.msdu.len ))
     {
     if(pData->dataInd.msdu.p[5]==0)
     {
     uint8 i=0;
     while(*(pData->dataInd.msdu
     .p+6+i)!='#')
     {
     i++;
     }
     writeMsg(pData->dataInd.msdu.p+6,i);
     }
     else if(pData->dataInd.msdu.p[5]==1)
     {
     uint8 i=0;
     while(*(pData->dataInd.msdu.p+6+i)!='#')
     {
     i++;
     }
     writeMsg(pData->dataInd.msdu.p+6,i);
     }
     }
     break;
     }
*************** Change in lines 352 to 369*****************
{
   msa_Data1[0] = MSA_PACKET_LENGTH;
   msa_Data1[1] = HI_UINT16(msa_myParentAdd);
  msa_Data1[2] =HI_UINT16(msa_myParentAdd);
       // send to co-ord  HI_UINT16(msa_CoordShortAddr);
  msa_Data1[3] = HI_UINT16(msa_myAdd);
 msa_Data1[4]=LO_UINT16(msa_myAdd);
       //LO_UINT16(msa_CoordShortAddr);
```

```
msa_Data1[5] = 0;
        msa_Data1[6]='E';  ⎫
        msa_Data1[7]='V';  ⎪
        msa_Data1[8]='1';  ⎬ This block show the Data format
        msa_Data1[9]='#';  ⎪
        msa_Data1[10]='X'; ⎭
      MSA_McpsDataReq((uint8*)msa_Data1,
          MSA_PACKET_LENGTH,  TRUE,
     msa_myParentAdd);
```

After replacing the above lines of coordinator programming file, the total number of lines in the proximity warning device file may vary depending upon the purpose. After incorporating the above change, there is no need for any other change. Then paste the modified program code in the msa.c file and debug the program into a proximity warning device using USB port. Now the proximity warning device "EV1" is created. For programming the other proximity warning devices, simply change the proximity warning device ID (e.g., 2 for "EV2", 3 for "EV3", etc.) in the above msa.c file and debug the program into the respective proximity warning device.

4.2 Programming with CC2430 Chip (22 Pins)

The algorithm for programming a proximity warning device is as follows:

i. All the associate parameters of the devices are assigned accordingly as mentioned in the algorithm steps iii–v for coordinator.
ii. Total number of devices, channel number 11–28, beacon order, packet length, maximum results, and ADC (analog to digital converter) resolution are assigned.
iii. Device information (device_ID, device address, maximum number of associated device, and maximum buffer size) are to be fixed, and number of current associate devices are declared.
iv. Power variable is assigned as Boolean.
v. Timer for reboot transmission is set, and radio transmitting power range is adjusted.
vi. Device is connected to 10 pin wires ribbon cable and this ribbon cable is connected to a simulator.
vii. The same simulator is connected to a PC through USB port.
viii. The program is compiled and debugged.
ix. USB port and simulator are disconnected, and device is ready.
x. Another device in the same channel for receiving and transmitting signal is created.

The coding for proximity warning device in CC2430 RFID chip (22 pins) is the same as that for CC2430 RFID chip (48 pins).

5 Programming for Router with Sensor Device

5.1 Programming with CC2430 Chip (48 Pins)

The algorithm for programming a router with sensor device is as follows:

i. All the associate parameters of the devices are assigned accordingly as mentioned in the algorithm steps iii–v for coordinator.
ii. Total number of devices, channel number 11–28, beacon order, packet length, maximum results, and ADC (analog to digital converter) resolution are assigned.
iii. Device information (device_ID, device address, maximum number of associated device, and maximum buffer size) are to be fixed, and number of current associate devices are declared.
iv. Sensor power variable and data format are assigned.
v. Timer for reboot transmission is set, and radio transmitting power range is adjusted.
vi. Device is connected to a PC through USB port, the program is compiled and debugged.
vii. USB port is disconnected, and device is ready.

For programming the router with sensor device, copy the above coordinator programming file and rename the file. All library and driver files are same as used in coordinator. Coding process of router with sensor device is same as given above for coordinator up to line 16. After line 16, following changes are required in the respective intermittent lines.

```
**********Change in line 17 to define constant **************
#define MSA_MAC_MAX_RESULTS 0 /* Maximum number of scan
result
that will be accepted */
******************Comments line 29 to 31*********************
*****************Add codes after line 51 ********************
#define PIN_HIGH() st(P1DIR |= BV(1); /* set P1.1 output */ \
 P1_1 = 1 ;)
#define PIN_LOW() st(P1DIR |= BV(1); /* set P1.1 output */ \
 P1_1 = 0 ;)
/*Configuring ADC port here port P0_4 has been activated for
performing ADC operations*/
```

```
#define ADC_PORT_CONFIG()st(P0DIR |= BV(1);
/* set P0.1 output */ \ P0_0 = 0;
/* drive P0.1 inactive (high) */ \ halSleepWait(180);
/* delay 180us */ \ P0DIR &= ~BV(1);
/* set P0.1 input */ \ )
/* set P0.1 tri-state */
*****************Add code after line 55***********************
uint16 msa_CoordShortAddr = MSA_COORD_SHORT_ADDR;
uint16 msa_DevShortAddr   = MSA_DEV_SHORT_ADDR;
*************Add code after line 60***************************
/* Beacon payload, this is used to determine if the device is
   not zigbee device */
uint8 msa_BeaconPayload[] = {0x22, 0x33, 0x44};
uint8 msa_BeaconPayloadLen = 3;
******************** Change in lines 70 to 96 **************
uint8 UART_Buffer[30][31]; // we will have a queue up to 30
uint8 currentBufLevel=0;
uint16 myAddress=200;
uint8 myName[] = "E1";
uint8 temp[3];
uint8 lastMsgId=0;
uint16 AtoD;
#define MAXPARENTSCAN 5
#define MAXNAMELENGTH 5
#define MAXPACKETLENGTH 20
#define MAXDEVICEONTIME 100
#define MAXSENSOR 1
uint16 possibleParent[MAXPARENTSCAN];
uint8 possibleParentLinkQ[MAXPARENTSCAN];
uint8 curLevel = 0;
bool processComplete = FALSE;
uint16 myParentAddress=0xFFFF;
uint16 myPanID;
uint8 currentSensor;
uint8 msgHandlerTaskID=0;
uint8 msgBuffer[30][31];
uint16 msaCoordShortAddress;
extern bool sleepTime;
****************** Change in lines 109*********************
void msgHandler_Init(uint8 taskId)
{
  msgHandlerTaskID=taskId;
  uint8 i;
  for ( i=0;i<30;i++)
  {
```

```
      msgBuffer[i][0]='\0';
      msgBuffer[i][30]=0;
   }
}
extern void powerm(void);
void sendVal(uint16 val);
*********************** Change in line 124*****************
  /* Reset the MAC */
  MAC_MlmeResetReq(TRUE);
  /* Initialize the data packet */
  for (i=MSA_HEADER_LENGTH; i<MSA_PACKET_LENGTH; i++)
  {
    msa_Data1[i] = i-MSA_HEADER_LENGTH;
  }
  /* Initialize the echo packet */
  for (i=0; i<MSA_ECHO_LENGTH; i++)
  {
    msa_Data2[i] = 0xEE;
  }
  msa_BeaconOrder = MSA_MAC_BEACON_ORDER;
  msa_SuperFrameOrder = MSA_MAC_SUPERFRAME_ORDER;
*********************Comments line 128*********************
*******************Add code after line 166 ****************
 PIN_LOW();
    ADC_PORT_CONFIG();
  //~AD
}
  /* Retrieve the message */
          pData = (macCbackEvent_t *) pMsg;
          writeMsg("Got beacon not ind.",21);
          msa_IsSampleBeacon = MSA_BeaconPayLoadCheck
            (pData->beaconNotifyInd.pSdu);
          /* If it's the correct beacon payload, retrieve
             the data for association req */
          /* // EndDevice is not Associate with
                        Coordinator*/
          if ( (!msa_IsStarted) && curLevel < MAXPARENTSCAN )
          {
            possibleParent[curLevel]
              = pData->beaconNotifyInd.pPanDesc->coordAddress.
                  addr.shortAddr;
            possibleParentLinkQ[curLevel]
              = pData->beaconNotifyInd.pPanDesc->linkQuality;
            curLevel++;
          }
```

```
****************** Change in lines 177 to 181***************
  writeMsg("My short add is:",18);
            writeVal(msa_CoordShortAddr);
            HalLedSet ( HAL_LED_4, HAL_LED_MODE_ON);
          }
********************Add code after line 184 ***************
/* Check if there is any Coordinator out there */
          pData = (macCbackEvent_t *) pMsg;
          if (curLevel == 0)
            //&& (pData->scanCnf.hdr.status == MAC_NO_BEACON))
            || (!msa_IsSampleBeacon))
          {
            sleepTime = TRUE;
          }
          else
          {
            uint8 i,threshold=0;
            for ( i=0; i < curLevel ; i++ )
            {
              if ( possibleParentLinkQ[i] > threshold
                   && possibleParent[i] < myAddress )
              {
                myParentAddress =  possibleParent[i];
                threshold = possibleParentLinkQ[i];
              }
              //writeMsg(&possibleParentLinkQ[i],1);
              //writeVal(possibleParent[i]);
            }
            if ( myParentAddress != 0xFFFF )
            {
              curLevel=0;
              processComplete = FALSE;
              MSA_DeviceStartup();
              HalLedSet ( HAL_LED_4, HAL_LED_MODE_ON);
              // lets sense and send the packet.
              currentSensor=0;
              sendADCValue(currentSensor);
              //osal_start_timer (MSA_SEND_EVENT, 1000);
              osal_start_timer(MSA_SEND_EVENT,MAXDEVICEONTIME);
              msa_IsStarted = TRUE;
            }
            else
              sleepTime=TRUE;
          }
```

```
****************** Change in lines 188 to 298*****************
pData = (macCbackEvent_t *) pMsg;
          /* If last transmission completed, ready to send
                the next one */
          if ((pData->dataCnf.hdr.status == MAC_SUCCESS) ||
              (pData->dataCnf.hdr.status
                == MAC_CHANNEL_ACCESS_FAILURE) ||
              (pData->dataCnf.hdr.status == MAC_NO_ACK))
          {
            //osal_start_timer (MSA_SEND_EVENT,
              MSA_WAIT_PERIOD);
            if (currentSensor > 0
                  && currentSensor < MAXSENSOR)
              sendADCValue(currentSensor);
            else
            processComplete = TRUE;
          }
          osal_msg_deallocate((uint8 *)
            pData->dataCnf.pDataReq);
          break;
        case MAC_MCPS_DATA_IND:
          pData = (macCbackEvent_t*)pMsg;
          writeMsg("Data Received!",14);
          if (MSA_DataCheck ( pData->dataInd.msdu.p,
              pData->dataInd.msdu.len ))
          {
            uint16 destAdd,sourceAdd;
            int i;
            bool receivedPktForMe = TRUE;
            destAdd = BUILD_UINT16(pData->dataInd.msdu.p[2],
                      pData->dataInd.msdu.p[1]);
            sourceAdd = BUILD_UINT16(pData->dataInd.msdu.p[4],
                        pData->dataInd.msdu.p[3]);
            destAdd++;
            if (pData->dataInd.msdu.p[5] == 3)
            {
              for (i=0; i<MAXNAMELENGTH;i++)
              {
                if (myName[i] == '\0')
                  break;
               //receivedName[i] = pData->dataInd.msdu.p[i+6];
                if ( pData->dataInd.msdu.p[i+6] != myName[i] )
                  receivedPktForMe = FALSE;
              }
```

```
              if ( pData->dataInd.msdu.p[i+6] != '*' )
                receivedPktForMe = FALSE;
            }
            else
              receivedPktForMe = FALSE;
            if ( destAdd == myAddress
                || receivedPktForMe == TRUE)
            {
              if ( ( *(pData->dataInd.msdu.p+7+i)
                     != lastMsgId ))
              {
              processComplete = FALSE;
              lastMsgId = *(pData->dataInd.msdu.p+7+i);
            }
            }
//            osal_stop_timer(MSA_SEND_EVENT);
//            sleepTime=TRUE;
          }
          break;
      }
      /* Deallocate */
      osal_msg_deallocate((uint8 *) pMsg);
    }
    return events ^ SYS_EVENT_MSG;
  }
if(events & MSA_SLEEP_EVENT)
  //{ if(i==6)
  {
    powerm();
    MAC_MlmeResetReq(TRUE);
    MAC_MlmeSetReq(MAC_EXTENDED_ADDRESS, &msa_ExtAddr);
    MAC_MlmeSetReq(MAC_BEACON_PAYLOAD_LENGTH,
        &msa_BeaconPayloadLen);
    MAC_MlmeSetReq(MAC_BEACON_PAYLOAD, &msa_BeaconPayload);
    MAC_MlmeSetReq(MAC_PAN_ID, &msa_PanId);
    MAC_MlmeSetReq(MAC_RX_ON_WHEN_IDLE, &msa_MACTrue);
    MAC_MlmeSetReq(MAC_COORD_SHORT_ADDRESS,
        &msaCoordShortAddress);
    MAC_MlmeSetReq(MAC_SHORT_ADDRESS, &msa_DevShortAddr);
    osal_start_timerEx ( MSA_TaskId, MSA_SEND_EVENT, 100);
 // }
//  i++;
   // osal_start_timer (MSA_SLEEP_EVENT, 3000);
  return events ^ MSA_SLEEP_EVENT;
  }
```

```
if (events & MSA_GO_EVENT)
  {
    PIN_LOW();
    return events ^ MSA_GO_EVENT;
  }
  /* This event handles polling for in-direct message device
*/
  if (events & MSA_POLL_EVENT)
  {
   // MSA_McpsPollReq();
//     osal_start_timer (MSA_POLL_EVENT, MSA_WAIT_PERIOD);
      if ( currentSensor++ >= MAXSENSOR )
    currentSensor=0;
  if (currentSensor > 0)
    sendPacket(currentSensor+47,last3Digits(AtoD));
    return events ^ MSA_POLL_EVENT;
  }
  /* This event will blink a LED then send out a DataReq and
     continue to do so every WAIT_PERIOD */
  if (events & MSA_SEND_EVENT)
  {
    if ( processComplete == TRUE )
      sleepTime=TRUE;
    else
      osal_start_timer (MSA_SEND_EVENT, 100);
    return events ^ MSA_SEND_EVENT;
  }
  return 0;
}
*****************Add code after line 426********************
void MSA_AssociateReq(void)
{
  MAC_MlmeAssociateReq(&msa_AssociateReq);
}
*****************Add code after line 476 *******************
{
  macMlmePollReq_t  pollReq;
  /* Fill in information for poll request */
  pollReq.coordAddress.addrMode = SADDR_MODE_SHORT;
  pollReq.coordAddress.addr.shortAddr = msa_CoordShortAddr;
  pollReq.coordPanId = msa_PanId;
  pollReq.sec.securityLevel = MAC_SEC_LEVEL_NONE;
  /* Call poll reuqest */
  MAC_MlmePollReq(&pollReq);
}
```

```
Add line after 515
{
  uint8 i = 0;
  for (i=0; i<msa_BeaconPayloadLen; i++)
  {
    if (pSdu[i] != msa_BeaconPayload[i])
    {
      return FALSE;
    }
  }
  return TRUE;
}
**************** Change in lines 535 to 741*****************
void keylessStart()
{
   msa_IsDirectMsg = MSA_DIRECT_MSG_ENABLED;
   msa_IsStarted = FALSE;
   curLevel = 0;
   myParentAddress = 0xFFFF;
   MAC_MlmeResetReq(TRUE);
   MSA_ScanReq(MAC_SCAN_ACTIVE, 7);
   writeMsg("Keyless start initiated.",26);
   HalLedBlink(HAL_LED_4, 0, 50, 300);
}
uint16 MSA_UART_Write_ProcessEvent(uint8 taskID,uint16 events)
{
    if ( events != 0 && taskID == MSA_UART_Write_TaskID)
    {
      if ( transmitInProgress() == FALSE)
      {
        write2COM1Port(UART_Buffer[0],UART_Buffer[0][30]);
        for ( uint8 i=0; i < currentBufLevel-1; i++)
        {
          osal_memcpy(UART_Buffer[i],UART_Buffer[i+1],
                      UART_Buffer[i+1][30]);
          UART_Buffer[i][30]=UART_Buffer[i+1][30];
        }
        if ( currentBufLevel > 0 )
        currentBufLevel-;
        if ( currentBufLevel == 0 )
        {
         //osalTaskRec_t *taskPtr=osalFindTask
            (MSA_UART_Write_TaskID);
         return 0;
         //taskPtr->events=0;
```

```
            }
          }
          return 1;
        }
        return 0;
}
void writeMsg(uint8 *buf,uint8 len)
{
  //flag=0;
  if ( currentBufLevel != 30 && len < 31 )
  {
     osal_memcpy(UART_Buffer[currentBufLevel],buf,len+2);
     UART_Buffer[currentBufLevel][len]=10;
     UART_Buffer[currentBufLevel][len+1]=13;
     UART_Buffer[currentBufLevel][30]=len+2;
     currentBufLevel++;
     osalTaskRec_t
       *taskPtr=osalFindTask(MSA_UART_Write_TaskID);
     taskPtr->events=1;
  }
}
void writeStr(uint8 *buf,uint8 len)
{
  //flag=1;
  if ( currentBufLevel != 30 && len < 31 )
  {
     osal_memcpy(UART_Buffer[currentBufLevel],buf,len);
//     UART_Buffer[len]=10;
//     UART_Buffer[len+1]=13;
     UART_Buffer[currentBufLevel][30]=len;
     currentBufLevel++;
     osalTaskRec_t
       *taskPtr=osalFindTask(MSA_UART_Write_TaskID);
     taskPtr->events=1;
  }
}
void writeVal(uint16 val)
{
  uint8 ascAdd[5],i;
  for(i=0;i < 5;i++)
  {
    ascAdd[4-i]=val%10 + 48;
    val/=10;
  }
  writeMsg(ascAdd,5);
```

```
}
void sendPacket(uint8 sensorNum,uint8* sensorValue)
{
  uint8 i;
  msa_Data1[0] = MSA_PACKET_LENGTH;
  msa_Data1[1] = 0;
  msa_Data1[2] = 0;
  msa_Data1[3] = HI_UINT16(myAddress);
  msa_Data1[4] = LO_UINT16(myAddress);
  msa_Data1[5] = 5;
  msa_Data1[6]='S';
  msa_Data1[7]='0';
  msa_Data1[8]='0';
  msa_Data1[9]='1';
  msa_Data1[10]=sensorValue[0];
  msa_Data1[11]=sensorValue[1];
  msa_Data1[12]=sensorValue[2];
  msa_Data1[13]='#';
  MSA_McpsDataReq((uint8*)msa_Data1,MSA_PACKET_LENGTH,TRUE,
                  myParentAddress);
}
//Reading UART inputs characterwise and sending to specified
  address as a string
/**********************************************************/
uint8* last3Digits(uint16 val)
{
  uint8 i;
  for(i=0;i<3;i++)
  {
    temp[2-i]=val%10+48;
    val/=10;
  }
  return temp;
}
void sendADCValue(uint8 sensor)
{
  switch ( sensor )
  {
  case 0:
    AtoD = HalAdcRead (HAL_ADC_CHANNEL_0 ,
                       HAL_ADC_RESOLUTION_8);
    break;
  }
  osal_start_timer(MSA_POLL_EVENT,100);
  }
```

After replacing the above lines of coordinator programming file, the total number of lines in the router with sensor device file may vary depending upon the purpose. After incorporating the above change, there is no need for any other change. Then paste the modified program code in the msa.c file and debug the program into a router with sensor device using USB port. Now the router with sensor device "SE1**" is created. For programming the other router with sensor devices, simply change the router with sensor device ID (e.g., 2 for "S002**", 3 for "S003**", etc.) in the above msa.c file and debug the program into the respective router with sensor device.

5.2 Programming with CC2430 Chip (22 Pins)

The algorithm for programming a router with sensor device is as follows:

i. All the associate parameters of the devices are assigned accordingly as mentioned in the algorithm steps iii–v for coordinator.
ii. Total number of devices, channel number 11–28, beacon order, packet length, maximum results, and ADC (analog to digital converter) resolution are assigned.
iii. Device information (device_ID, device address, maximum number of associated device, and maximum buffer size) are to be fixed, and number of current associate devices are declared.
iv. Sensor power variable and data format are assigned.
v. Timer for reboot transmission is set, and radio transmitting power range is adjusted.
vi. Device is connected to 10 pin wires ribbon cable and this ribbon cable is connected to a simulator.
vii. The same simulator is connected to a PC through USB port.
viii. The program is compiled and debugged.
ix. USB port and simulator are disconnected, and device is ready.

The coding for router with sensor device in CC2430 RFID chip (22 pins) is the same as that for CC2430 RFID chip (48 pins).

6 Programming for End Device with Data Logger

6.1 Programming with CC2430 Chip (48 Pins)

The algorithm for programming an end device with data logger is as follows:

i. All the associate parameters of the devices are assigned accordingly as mentioned in steps iii–v for coordinator.

ii. Total number of devices, channel number 11–28, beacon order, packet length, maximum results, and ADC (analog to digital converter) resolution are assigned.
iii. Device information (device_ID, device address, maximum number of associated device, and maximum buffer size) are to be fixed, and number of current associate devices are declared.
iv. Data format is assigned.
v. Timer for reboot transmission is set, and radio transmitting power range is adjusted.
vi. Device is connected to a PC through USB port, the program is compiled and debugged.
vii. USB port is disconnected, and device is ready.

For programming an end device with data logger, copy the above coordinator programming file. All library and driver files are same as used in coordinator. Coding process of data logger is same as given above for coordinator up to line 319. After line 319, the following changes are required in the respective intermittent lines.

```
****************Change in lines 320 to 345*******************
            case MAC_MCPS_DATA_IND:
            pData = (macCbackEvent_t*)pMsg;
               if (MSA_DataCheck ( pData->dataInd.msdu.p,
                       pData->dataInd.msdu.len ))
           {
             if(pData->dataInd.msdu.p[5]==0)
             {
                 osal_memcpy(tempName, pData->
                            dataInd.msdu.p+6,3);
                             tempName =tempName+3;
             }
                           }
*****************Add in line 783****************************
if ( keys & HAL_KEY_SW_6 )
  {
    writeStr (p,29);
  }
****************Change in lines 971 to 982*******************
msa_Data2[0] = MSA_PACKET_LENGTH;
            msa_Data1[1] = HI_UINT16(msa_myParentAdd);
            msa_Data1[2] =HI_UINT16(msa_myParentAdd);
              // send to co-ord HI_UINT16(msa_CoordShortAddr);
            msa_Data1[3] = HI_UINT16(myAddress);
            msa_Data1[4] = LO_UINT16(myAddress);
```

```
        //LO_UINT16(msa_CoordShortAddr);
      msa_Data2[5] = 1;
      msa_Data2[6] = '*';
      for(uint8 d=0;d<1;d++)
{
       msa_Data2[d+7] = msg[d+1];
}
       msa_Data2[1+7]='#';
```

After replacing the above lines in coordinator programming file, the total number of lines in the router file may vary depending upon the purpose of the end device with data logger. After incorporating the above changes, there is no need for any other change. Then paste the modified program code in the msa.c file and debug the program into an end device with data logger using USB port. Now the data logger "1*" is created. For programming the other end device with data logger, simply change the end device with data logger ID (e.g., 2 for "2*", 3 for "3*", etc.) in the above msa.c file and debug the program into the respective end device with data logger.

6.2 Programming with CC2430 Chip (22 Pins)

The algorithm for programming an end device with data logger is as follows:

i. All the associate parameters of the devices are assigned accordingly as mentioned in steps iii–v for coordinator.
ii. Total number of devices, channel number 11–28, beacon order, packet length, maximum results, and ADC (analog to digital converter) resolution are assigned.
iii. Device information (device_ID, device address, maximum number of associated device, and maximum buffer size) are to be fixed, and number of current associate devices are declared.
iv. Data format is assigned.
v. Timer for reboot transmission is set, and radio transmitting power range is adjusted.
vi. Device is connected to 10 pin wires ribbon cable and this ribbon cable is connected to a simulator.
vii. The same simulator is connected to a PC through USB port.
viii. The program is compiled and debugged.
ix. USB port and simulator are disconnected, and device is ready.

The coding for end device with data logger in CC2430 RFID chip (22 pins) is the same as that for CC2430 RFID chip (48 pins).

7 Programming for Messaging Device

7.1 Use of Keil uVersion3 Software

Install the Keil uVersion3 software to a computer. Open Keil uVersion3. Write coded program on this platform. Follow the steps to write the code on Keil uVersion3 platform as given below:

- Open Project menu and click on Targets, Groups and Files.
- Type a new target name into the Add box and Add it.
- Click on the New Target name in the available Targets box and set as Current Target.
- Click on Target 1 name and remove Target. Select the Group/Add Files tab.
- Type a new group name in Group to Add box and Add it.
- Click on the Source Group 1 name in the available Group box and remove Group.
- Click on the New Group name and click on Add Files to Group button.
- Type in the name of the new source file in the File name box, using name.a51 format.
- Select Assembly Source files from the Files of type box. Click the Add button.
- Close all windows and note that all names have changed to the new ones selected.

Then again open the Keil uVersion, and open Created file.

- Write the code of the program on this created file.
- Compile that file and save that file as an extension of .hex.

7.2 Use of UniPro Software

For flashing the code to controller chip, open flashing software, UniPro (a universal Chip programmer Kit), in the following way:

- Open the UniPro software, open the saved HEX file, and connect UniPro kit.
- Insert the controller chip on it.
- Then press Program, to flash the program into chip.

7.3 Program Code

The program codes for messaging device are as follows:

```
;;;;r5 used to give output on the port p1 pins, i.e., to make
    one row low at a time.
```

```
;;;;r6 is used as a flag to move the cursor forward or not.
;;;;r2 is used to store the cursor position on LCD,i.e, where
    to place cursor on LCD.
;;;;;address 21h is used as an address to store the value of
     pulse count on p3.5 pin,i.e timer1 pin .
baudnum equ 0fdh
dlylsb equ 0f4h   ;;set delay LSB in register A.
dlymsb equ 01h    ;;set delay MSB in register B.
delay equu 0a6h    ;;delay constant for 'softime' routine.
bounce equ 32h     ;;bounce time delay is 50d milliseconds.
newkey equ 60h     ;;RAM variable to store valid new key.
newkey1 equ 61h  ;;RAM variable to store key value to find key
                   repetition.
next equ 64h
newkeyflag equ 00h ;;main program newkeyflag (bit addressable),
                     used to identify if any key is pressed.
flag equ 01h       ;;used in convert subroutine
                     (bit addressable),used to see not more
                     then one key is in pressed condition.
upflg equ 02h      ;;used to identify the next call of 'over'
                     for confirmation of same key remain
                     pressed,i.e to avoid-
;;;;    ;;-noise (bit addressable as above bits are).
maininterruptflag equ 03h  ;;interrupt flag of AND logic gate.
org 0000h
ajmp startpointx
startpoint1:
org 0003h     ;;external interrupt0 address.
mov ie,#00h   ;;disable all interrupts.
setb maininterruptflag
reti
;;;;;;;;;;;;;;;;;;;;;;;;;;;;;;;;;;;;;;;;;;;;;;;;;;;;;;;;;;;;;;;;;
org 000bh      ;;T0 interrupt routine address.
clr tr0
acall timer0interrupt
reti
;;;;;;;;;;;;;;;;;;;;;;;;;;;;;;;;;;;;;;;;;;;;;;;;;;;;;;;;;;;;;;;;;
;;;;;;;;;;;;;;;;;;;;;;;;;;;;;;;;;;;;;;;;;;;;;;;;;;;;;;;;;;;;;;;;;
;;;;;;;;;;;;;;
org 0040h
startpointx:
clr p0.2   ;;;;to 'off' the buffer, i.e 'off' the small
               microcontroller .
;;clr p0.7   ;;;; to 'off' the buzzer connected on pin p0.7  .
```

```
mov sp,#50h
mov r0,#24h
clearmemory:
cjne r0,#46h,clearmemory1
mov r0,#24h
ajmp clearmemory2
clearmemory1:
mov @r0,#' '
inc r0
ajmp clearmemory
clearmemory2:
mov newkey1,#'a'  ;;;store an odd value in newkey1 so that
                     there may be no any match with previous
                     keys.
mov a,#3ch
acall command
mov a,#0eh
acall command
mov a,#06h
acall command
mov a,#01h
acall command
mov a,#0fh   ;;;;to on the cursor blinking.
acall command
mov r2,#80h   ;;;;r2 is used to store the cursor position
                 on LCD,i.e where to place cursor is
                 decided by it's content.
mov a,r2
acall command
mov a,#'R'  ;;;it will be displayed on r2 i.e 80h.
inc r2
acall display
mov a,#'E' ;;;it will be displayed on r2 i.e 81h.
inc r2
acall display
mov a,#'A' ;;;;it will be displayed on r2 i.e 82h.
inc r2
acall display
mov a,#'D'    ;;;;it will be displayed on r2 i.e 83h
inc r2
acall display
mov a,#'Y' ;;; it will be displayed on r2 i.e 84h
inc r2   ;;;;;finally r2 becomes 85h by this.
acall display
mov a,#' '  ;;;it will be displayed on r2 i.e 80h.
```

```
inc r2
acall display
mov a,#'C' ;;;it will be displayed on r2 i.e 81h.
inc r2
acall display
mov a,#'I' ;;;;it will be displayed on r2 i.e 82h.
inc r2
acall display
mov a,#'M'    ;;;;it will be displayed on r2 i.e 83h
inc r2
acall display
mov a,#'F' ;;; it will be displayed on r2 i.e 84h
inc r2   ;;;;;finally r2 becomes 85h by this.
acall display
mov a,#'R'  ;;;it will be displayed on r2 i.e 80h.
inc r2
acall display
mov a,#'#' ;;;it will be displayed on r2 i.e 81h.
inc r2
acall display
mov a,#dlylsb
mov b,#dlymsb
acall softime
mov a,#dlylsb
mov b,#dlymsb
acall softime
mov a,#dlylsb
mov b,#dlymsb
acall softime
;acall softime
;;;;;;;;;;;;;;;;;;
mov a,#3ch
acall command
mov a,#0eh
acall command
mov a,#06h
acall command
mov a,#01h
acall command
mov a,#0fh   ;;;;to on the cursor blinking.
acall command
mov r2,#80h   ;;;;r2 is used to store the cursor position on
                  LCD,i.e where to place cursor is decided
                  by it's content.
mov a,r2
```

```
acall command
;;;;;;;;;;;;;;;;;;;;;;;
startpoint:
clr maininterruptflag
orl tcon,#01h ;;enable external interrupt 0 to be triggered
                by a falling edge signal on '/INIT0' .
orl ie,#81h        ;;;enable global and external
                      interrupt0 also .
;;;initially make p1.0 to p1.3 low so that any key switch may
   trigger an external interrupt (i.e /INIT0).
clr p1.0
clr p1.1
clr p1.2
clr p1.3
partofstartpoint:
jnb maininterruptflag,partofstartpoint
;;;;;make pins p1.0 to p1.3 high before entering to the main
     program loop for key detection.
setb p1.0
setb p1.1
setb p1.2
setb p1.3
;;
startpoint2:
clr upflg    ;;upflg is used for noise detection.
clr flag    ;;flag is used in convert subroutine.
;;;;;;;;;;;;;;;;;;;;;;;;;;;;;;;;;;;;;;;;;;;;;;;;;;;;;;;;;;;;;;;;
;;;;;;;;;;;;;;;;;;;;;;;;;;;;;;;;;;;;;;;;;;;;;;;;;;;;;;;;;;;;;;;;
over:
clr newkeyflag  ;;used to see if any key is pressed.
mov r5,#0feh     ;;initialize row 0 low.
mov tmod,#01h    ;;set T0 as a 16 bit counter.
mov tl0,#30h     ;;set for counting 2000 micro-seconds.
mov th0,#0ffh    ;; initially '0f8h'.
mov ie,#82h      ;;enable global and T0 interrupt.
setb Tr0         ;;START T0
;;;;;;;;;;;;;;;;;;;;;;;
;;;;;;;;;;;;;;;;;;;;;;
main:
jnb newkeyflag,main
clr newkeyflag
jb upflg,next_keydetermination
jmp keydetermination
;;ajmp startpoint2
;;;;;;;;;;;;;;;;;;;;;;;;;;;;;;;;;;;;;;;;;;;;;;;;;;;;;
```

```
keydetermination:
;;;;;
setb p0.3
jnb p0.3,vendit1xx        ;;;i.e increment the cursor when bit
                            p0.3 i.e p3.7 of small
                            microcontroller is cleared
                            otherwise not.
jmp vendit1xxx
vendit1xx:
mov newkey1,#'a'
;;;;;
vendit1xxx:
acall convert
jbc flag,over
mov newkey,a      ;;save key and wait for debounce time.
mov b,#00h
mov a,#bounce     ;;debounce key.
acall softime   ;;delay debounce time.
setb upflg    ;;'upflg' flag is set so that next time from
               'over'subroutine which is going to be called
               after this
;;;-line for the confirmation of same key remain pressed
    doesn't again call 'over' subroutine.
ajmp over   ;;see if key is still down,i.e do all the
              operations from staring and again find the
              pressed key  ;bhogesh
next_keydetermination:
clr upflg
acall convert              ;;again determine the key number
                             from the convert subroutine.
jbc flag,startpoint2         ;;jbc flag,over.
cjne a,newkey,startpoint2    ;;;if last key == stored key ,
                               then valid key.
mov newkey1,newkey ;;;store the valid key in newkey1 to find
                              key repetition.
clr tr0        ;;modified on date 23rd jan 2009 .
mov ie,#00h
mov r3,newkey1
cjne r3,#'b',vendittt ;;;; here 'b' is returned in newkey1 via
                            accumulator when clear key of keypad
                            is pressed .
acall clear
ajmp waitingloop
vendittt:
acall vendit              ;;;for valid key to be displayed.
```

```
;;;;;;;;;;;;;;;start of wait for all keys up;;;;;;;;;;;;;;;;;;
waitingloop:
clr p1.0  ;;make all rows low.
clr p1.1
clr p1.2
clr p1.3
setb p1.4       ;;make p1.4,p1.5 and p1.6 as input bits.
setb p1.5
setb p1.6
mov a,p1     ;;read p1
orl a,#8fh   ;;to only keep relation with p1.4,p1.5 and p1.6,
               so make one all other bits in accumulator.
;;mov r1,a     ;;save accumulator in r1 to use further in
                 convert subroutine.
cpl a        ;;complement accumulator
jnz waitingloop
mov b,#00h
mov a,#next
acall softime
mov b,#00h
mov a,#next
acall softime
mov a,p1
orl a,#8fh
cpl a
jnz waitingloop
;;;;;;;;;;;;;;;;;;;;;end of wait for all keys up.
ajmp startpoint
;;;;;;;;;;;;;;;;;;;;;;;;;;;;;;;;;;;;;;;;;;;;;;;;;;;;;;
;;;start of softime ;;;;;;;;;;;;;;;;;;;;;;;;;;;;;;;;;;;
softime:
push 07h
push acc
orl a,b
cjne a,#00h,ok
pop acc
sjmp done
ok:
 pop acc
timer:
mov r7,#delay
onemil:
nop
nop
nop
```

```
nop
djnz r7,onemil
nop
nop
dec a
cjne a,#0ffh,noroll
dec b
noroll:
cjne a,#00h,timer
cjne a,b,timer
done:
pop 07h
ret
;;;end of softimer ;;;;;;;;;;;;;;;
;;;;;;;;;start of convert routine;;;;;;;;;;;;;;;;;;;;;;;;;;;;;
convert:
;;;;;;;;;;;;;;;;;;;;;;;;;;
cjne r5,#0feh,convert1
cjne r1,#0bfh,one    ;;check for p1.6 to be low.
mov r4,newkey1
cjne r4,#'1',partzero1
mov a,#'A'
mov r6,#01h   ;;i.e not move the cursor.
ret
partzero1:
cjne r4,#'A',partzero2
mov a,#'B'
mov r6,#01h   ;;i.e not move the cursor.
ret
partzero2:
cjne r4,#'B',partzero3
mov a,#'C'
mov r6,#01h   ;;i.e not move the cursor.
ret
partzero3:
cjne r4,#'C',partzero4
mov a,#'1'
mov r6,#01h
ret
partzero4:
mov a,#'1'
mov r6,#00h   ;; move the cursor.
ret
one:
cjne r1,#0dfh,two
```

```
mov r4,newkey1
cjne r4,#'2',partone1
mov a,#'D'
mov r6,#01h   ;;i.e not move the cursor.
ret
partone1:
cjne r4,#'D',partone2
mov a,#'E'
mov r6,#01h   ;;i.e not move the  cursor.
ret
partone2:
cjne r4,#'E',partone3
mov a,#'F'
mov r6,#01h   ;;i.e not move the cursor.
ret
partone3:
cjne r4,#'F',partone4
mov a,#'2'
mov r6,#01h
ret
partone4:
mov a,#'2'
mov r6,#00h   ;;move the cursor.
ret
convert1:
ajmp convert1x
two:
cjne r1,#0efh,bad2   ;;check for p1.4 to be low.
mov r4,newkey1
cjne r4,#'3',parttwo1
mov a,#'G'
mov r6,#01h    ;;i.e not move the cursor.
ret
parttwo1:
cjne r4,#'G',parttwo2
mov a,#'H'
mov r6,#01h    ;;i.e not move the cursor.
ret
parttwo2:
cjne r4,#'H',parttwo3
mov a,#'I'
mov r6,#01h   ;;i.e not move the cursor.
ret
parttwo3:
cjne r4,#'I',parttwo4
```

```
mov a,#'3'
mov r6,#01h
ret
parttwo4:
mov a,#'3'
mov r6,#00h   ;;i.e move the cursor.
ret
;;;;;;;;;;;;;;;;;;;;;;;;;;;;;;;;
bad2:
setb flag
ret
convert1x:
cjne r5,#0fdh,convert2
cjne r1,#0bfh,one1      ;;check for p1.6 to be low.
mov r4,newkey1
cjne r4,#'4',partzero11
mov a,#'J'
mov r6,#01h    ;;i.e not move the cursor.
ret
partzero11:
cjne r4,#'J',partzero21
mov a,#'K'
mov r6,#01h     ;;i.e not move the cursor.
ret
partzero21:
cjne r4,#'K',partzero31
mov a,#'L'
mov r6,#01h   ;; i.e not move the cursor.
ret
partzero31:
cjne r4,#'L',partzero41
mov a,#'4'
mov r6,#01h
ret
partzero41:
mov a,#'4'
mov r6,#00h    ;;i.e move the cursor.
ret
one1:
cjne r1,#0dfh,two1   ;;check for p1.5 to be low.
mov r4,newkey1
cjne r4,#'5',partone11
mov a,#'M'
mov r6,#01h     ;;i.e not move the cursor.
ret
```

```
partone11:
cjne r4,#'M',partone21
mov a,#'N'
mov r6,#01h   ;;i.e not move the cursor.
ret
partone21:
cjne r4,#'N',partone31
mov a,#'O'
mov r6,#01h   ;;i.e not move the cursor.
ret
partone31:
cjne r4,#'O',partone41
mov a,#'5'
mov r6,#01h
ret
partone41:
mov a,#'5'
mov r6,#00h  ;;i.e move the cursor.
ret
convert2:
ajmp convert2x
two1:
cjne r1,#0efh,bad1         ;;check for p1.4 to be low.
mov r4,newkey1
cjne r4,#'6',parttwo11
mov a,#'P'
mov r6,#01h          ;;i.e not move the cursor.
ret
parttwo11:
cjne r4,#'P',parttwo21
mov a,#'Q'
mov r6,#01h      ;;i.e not move the cursor.
ret
parttwo21:
cjne r4,#'Q',parttwo31
mov a,#'R'
mov r6,#01h   ;;i.e not move the cursor.
ret
parttwo31:
cjne r4,#'R',parttwo41
mov a,#'6'
mov r6,#01h
ret
parttwo41:
mov a,#'6'
```

```
mov r6,#00h  ;;i.e move the cursor.
ret
bad1:
setb flag
ret
;;;;;;;;;;;;;;;;
convert2x:
cjne r5,#0fbh,convert3
cjne r1,#0bfh,one2    ;;;check for p1.6 to be low.
mov r4,newkey1
cjne r4,#'7',partzero12
mov a,#'S'
mov r6,#01h   ;;i.e not move the cursor.
ret
partzero12:
cjne r4,#'S',partzero22
mov a,#'T'
mov r6,#01h    ;;i.e not move the cursor.
ret
partzero22:
cjne r4,#'T',partzero32
mov a,#'U'
mov r6,#01h    ;;i.e not move the cursor.
ret
partzero32:
cjne r4,#'U',partzero42
mov a,#'7'
mov r6,#01h
ret
partzero42:
mov a,#'7'
mov r6,#00h  ;;i.e move the cursor.
ret
one2:
cjne r1,#0dfh,two2   ;;;check for p1.5 to be low.
mov r4,newkey1
cjne r4,#'8',partone12
mov a,#'V'
mov r6,#01h   ;;i.e not move the cursor.
ret
partone12:
cjne r4,#'V',partone22
mov a,#'W'
mov r6,#01h  ;;i.e not move the cursor.
ret
```

```
partone22:
cjne r4,#'W',partone32
mov a,#'X'
mov r6,#01h   ;;i.e not move the cursor.
ret
partone32:
cjne r4,#'X',partone42
mov a,#'8'
mov r6,#01h
ret
partone42:
mov a,#'8'
mov r6,#00h   ;;i.e move the cursor.
ret
two2:
cjne r1,#0efh,bad   ;;check for p1.4 to be low.
mov r4,newkey1
cjne r4,#'9',parttwo12
mov a,#'Y'
mov r6,#01h   ;;i.e not move the cursor.
ret
parttwo12:
cjne r4,#'Y',parttwo22
mov a,#'Z'
mov r6,#01h   ;;i.e not move the cursor.
ret
parttwo22:
cjne r4,#'Z',parttwo32
mov a,#','
mov r6,#01h   ;;i.e not move the cursor.
ret
parttwo32:
cjne r4,#',',parttwo42
mov a,#'9'
mov r6,#01h
ret
parttwo42:
mov a,#'9'
mov r6,#00h  ;;i.e move the cursor.
ret
;;;;;;;;;;;;;;;;
convert3:
cjne r5,#0f7h,bad
cjne r1,#0bfh,one3   ;;check for p1.6 to be low.
mov r4,newkey1
```

```
cjne r4,#'0',partzero13
mov a,#'.'
mov r6,#01h    ;;i.e not move the cursor.
ret
partzero13:
cjne r4,#'.',partzero23
mov a,#'/'
mov r6,#01h   ;;i.e not move the cursor.
ret
partzero23:
cjne r4,#'/',partzero33
mov a,#' '
mov r6,#01h   ;;i.e not move the cursor.
ret
partzero33:
cjne r4,#' ',partzero43
mov a,#'0'
mov r6,#01h
ret
partzero43:
mov a,#'0'
mov r6,#00h  ;;i.e move the cursor.
ret
one3:
cjne r1,#0dfh,two3  ;;check for p1.5 to be low.
mov a,#'b'
;mov r6,#01h     ;;i.e not move the cursor.
ret
;;;;;;;;;;;;;;;;;;;;;;;;;;;;;;;;*****************;;;;;;;
;;;;;;;;;;;;;;;;;;;;;;;;;;;;;;*********************;;;;;;
two3:
cjne r1,#0efh,bad   ;;check for p1.4 to be low.
ajmp transmit
ret
bad:
setb flag
ret
;;;end of convert routine ;;;;;;;;;;;;;;;;;;;;;;;;;;;;;;;;;;;;
vendit:
cjne r6,#01h,vendit1   ;;;r6 content of '01h' is to not move
                          the cursor.
vendit2:
dec r2
mov a,r2     ;;;;;to keep cursor on the same position,i.e not
                         move the cursor.
```

```
acall command
inc r2
mov a,newkey
inc r0
mov @r0,a
;;;;;;;;;
inc r0
mov @r0,#'#'
dec r0
;;;;;
acall display
setb tr0        ;;;modified on date 23rd jan 2009.
clr p0.2
acall wait1
setb p0.2
ret
vendit1:
cjne r2,#90h,vendit1zz
mov r2,#0c0h
mov a,r2
acall command
dec r2
vendit1zz:
inc r2
mov a,newkey
cjne r0,#46h,vendit1cc
setb p0.7                ;;;;sound the buzzer which is connected
                             on pin p0.7
mov a,#dlylsb
mov b,#dlymsb
acall softime
mov a,#dlylsb
mov b,#dlymsb
acall softime
mov a,#dlylsb
mov b,#dlymsb
acall softime
clr p0.7            ;;;off sound of buzzer
ret
vendit1cc:
inc r0
mov @r0,a
;;;;
inc r0
mov @r0,#'#'
```

```
dec r0
;;;;
acall display
clr p0.2
acall wait1
setb p0.2  ;;; i.e on the small microcontroller to start
               counting .
ret
clear:
dec r2
mov a,r2
acall command
inc r2
mov a,#' '
acall display
dec r2
mov a,r2
acall command
cjne r0,#24h,clear1
ret
clear1:
mov @r0,#' '
dec r0
ret
;LCD strobe subroutines
command:
acall ready
mov p2,a
clr p0.0
clr p0.1  ;;;;initially p3.3 .
setb p3.4
clr p3.4
ret
display:
acall ready
mov p2,a
setb p0.0
clr p0.1   ;;;;initially p3.3 .
setb p3.4
clr p3.4
ret
ready:
clr p3.4
mov p2,#0ffh
clr p0.0
```

```
setb p0.1  ;;;;initially p3.3 .
wait:
clr p3.4
setb p3.4
jb p2.7,wait
clr p3.4
ret
;;;;;;
timer0interrupt:
mov tl0,#30h    ;;set for counting 2000 microsecond .
mov th0,#0ffh   ;;initially '0f8h'.
mov p1,r5       ;;make first row low
setb p1.4       ;;make p1.4,p1.5 and p1.6 as input bits .
setb p1.5
setb p1.6
mov a,p1     ;;read p1
orl a,#8fh   ;;to only keep relation with p1.4,p1.5 and p1.6,
               so make one all other bits in accuumlator.
mov r1,a     ;;save accumulator in r1 to use further in convert
               subrountine.
cpl a        ;;complement accumulator
jz next1         ;;if not any key is pressed  then move to
                   next1
setb newkeyflag ;;if any key is pressed then set the 'newflg'
                  flag and without starting of timer return
                  back from interrupt-
ret     ;;-subroutine and do the keydetermination process .
next1:
cjne r5,#0f7h,next_next1 ;;to make only p1.0 to p1.3 low
                           successively one at a time and then
                           rotate again and again, so when
;;;;;;p1.3 becomes low load r5 with 7fh so that in next loop
                           p1.0 again becomes low.
mov r5,#7fh     ;;to resume the first row low condition by the
                  following lines of codes of this line.
next_next1:
mov a,r5
rl a
mov r5,a      ;;now r5 is ready to make next row low.
setb tr0
ret
;;;;;;;;;;;;;;;;;;
;;;;;;;;;;;;;;;;;;;;;;;;
wait1:
mov r3,#0ffh
```

```
wait1x:
djnz r3,wait1x
ret
;;;;;;;;;;;;;;;;;;;
transmit:
mov a,#3ch
acall command
mov a,#0eh
acall command
mov a,#06h
acall command
mov a,#01h
acall command
mov a,#0fh   ;;;;to on the cursor blinking.
acall command
;;;;;
mov a,#80h
acall command
mov a,#'T'
acall display
mov a,#'R'
acall display
mov a,#'A'
acall display
mov a,#'N'
acall display
mov a,#'S'
acall display
mov a,#'M'
acall display
mov a,#'I'
acall display
mov a,#'T'
acall display
mov a,#'T'
acall display
mov a,#'I'
acall display
mov a,#'N'
acall display
mov a,#'G'
acall display
mov a,#0c0h
acall command
mov a,#'P'
```

```
acall display
mov a,#'L'
acall display
mov a,#'E'
acall display
mov a,#'A'
acall display
mov a,#'S'
acall display
mov a,#'E'
acall display
mov a,#' '
acall display
mov a,#'W'
acall display
mov a,#'A'
acall display
mov a,#'I'
acall display
mov a,#'T'
acall display
mov a,#'.'
acall display
mov a,#'.'
acall display
mov a,#'.'
acall display
mov a,#'.'
acall display

 ;mov r2,#45h

 mov r2,#25h
 ljmp overxxx
 serial:
  org 0023h
  clr ti
  ;;;;; give delay between transmission of characters.
  mov r6,#05h
  loop3:
  mov r5,#0ffh
  loop2:
  mov r1,#0ffh
  loop1:
```

```
 djnz r1,loop1
 djnz r5,loop2
 djnz r6,loop3
 ;;;;;;;;;;;;;;transmission delay complete.;;;;;;;;;;;;;;;;;
 inc r2
 push acc
 mov a,r2
 mov r0,a
 pop acc
 mov sbuf,@r0
 reti
 org 0500h
overxxx:
 orl pcon,#80h      ;;;u can do orl pcon,#80h to double the
                       baud rate, suggestion
                       date 15 dec 2008...
 anl tmod,#30h
 orl tmod,#20h
 mov th1,#baudnum
 setb tr1
 mov scon,#40h
 orl ie,#90h
 mov 76h,a           ;address 76h is used to temporarily store
                      the value of a
 mov a,r2
 mov r0,a
 mov a,76h
 mov sbuf,@r0
wait3:
 ;cjne @r0,#'b',tttt
 ;ljmp startpoint
;tttt:
 ;cjne r0,#66h,wait3
 cjne r0,#46h,wait3
 mov r3,#0ffh
 anl ie,#00h
 clr tr1
popp:
 cjne r3,#00h,popp1
 dec r3
 sjmp popp
popp1:
mov a,#3ch
acall command
mov a,#0eh
```

```
acall command
mov a,#06h
acall command
mov a,#01h
acall command
mov a,#0fh   ;;;;to on the cursor blinking.
acall command
mov a,#80h
acall command
mov a,#'T'
acall display
mov a,#'R'
acall display
mov a,#'A'
acall display
mov a,#'N'
acall display
mov a,#'S'
acall display
mov a,#'M'
acall display
mov a,#'I'
acall display
mov a,#'S'
acall display
mov a,#'S'
acall display
mov a,#'O'
acall display
mov a,#'N'
acall display
mov a,#' '
acall display
mov a,#0c0h
acall command
mov a,#'C'
acall display
mov a,#'O'
acall display
mov a,#'P'
acall display
mov a,#'L'
acall display
mov a,#'E'
acall display
```

```
mov a,#'T'
acall display
mov a,#'E'
acall display
mov a,#'.'
acall display
mov a,#'.'
acall display
mov a,#'.'
acall display
mov a,#'.'
acall display
mov a,#'.'
acall display
mov a,#'.'
acall display
mov a,#'.'
acall display
mov a,#'.'
acall display
mov a,#dlylsb
mov b,#dlymsb
acall softime
mov a,#dlylsb
mov b,#dlymsb
acall softime
mov a,#dlylsb
mov b,#dlymsb
acall softime
;;acall softime
ljmp startpointx
end
```

Appendix B
Algorithm of Different Modules of TMS Software

Part A User login module

```
Algorithm:
            Select user type
            Get input user ID and password from the user
            Set Flag to True
            Check user type, user id and password from database
            If user ID = <data base user ID> and <password> then
                 Go to main form
              Else
                 Set Flag to False and warning message.
```

Part B Master Module

```
Algorithm:
            Display option wise six modules
            If option I then
               Display Load Map modules
            Else if option II then
               Display Create Tag modules
            Else if option III then
               Display Create Member modules
            Else if option IV then
               Tag allotment modules
            Else if option V then
               Create Sensor modules
            Else if option VI then
               Assign Shift time modules
            Else
            Set close the module
```

Part C Set site map and Router module

```
Algorithm:
            Add mine plan in the list
            Load mine plan
```

```
If select mine plan from the list then
    Display mine plan
    Set Flag to True
Else
   Set Flag to False and warning message
If Add router then
   Generate router index max no of the field
   Router location = map name, place area
   Top and Left
   Set X, Y coordinates from the user
   Get input S/W, H/W, user allocated code,
       actual location from the user
   Set choke point yes/no
   If yes then
      Set router from left panel and add to the
      right panel
   If no then
      Store the configuration
   Else If
      Reset to cancel unsaved configurations
Else
      Closed to unload the configuration
Else if move then
   Move the router from old position to new position
Else if configure then
   Get input S/W, H/W, user allocated code,
   actual location from the user
   Select choke point yes/no
   If yes then
      Select router from left panel and add to the
      right panel
   If no then
      Save finally store configuration
   Else If
      Reset to cancel your unsaved configurations
   Else
      Closed to unload the configuration
      Display Tree View
      Get source of Tree View node from existing
          routers
      Get source of tree view child nodes from
          Input Data
```

Part D Create Tag module

Algorithm:

```
Find the max tag Id from database
  If max tag Id=0 then
      Create tag with Tag Id =1
  Else
      Create tag with Tag Id = max tag Id+1
      Display the tag in list
  Else
      Set close the module
```

Part E **Create Member module**

Algorithm:

```
If create new member then
    Find the max Member ID from database
    If max member ID = 0 then
        Create member with member ID = 1
        Set member name, member details
    Else
        Create tag with Tag ID = max tag ID + 1
        Display the tag in list
Else if delete/update exiting member then
      Set the member no and get member details
      Set data for updating or delete data
Else
      Set close the module
```

Part F **Tag Allotment module**

Algorithm:

```
Get Miner ID and Tag ID
If Miner ID<>null and Tag ID<>null then
    Set Miner ID/Tag ID
    Display the tag status is Assign to miner
Else
    Set close the module
```

Part G **Shift Time module**

Algorithm:

```
Set time for shift
If shift=First then
     Set first shift start Time and first end time
Else If shift=Second then
     Set second shift start Time and second shift
     end time
Else if shift=third then
     Set third shift start Time and third shift
     end time
Else
     Set close the module
```

Part H **Tag Status module**

Algorithm:

```
If tag is assign then
    Display the tag status is Assign
Else
   Display the tag status is Return
Set close the module
```

Part I **Current Position module**

Algorithm:

```
If tag is assign then
     Display the tag status is Assign
Else
     Display the tag status is Return
Set close the module
```

Part J **Tracking module**

Algorithm:

```
Display option wise three modules
If option I then
       Display tabular data module
Else if option II then
       Graphical tracking view module
Else if option III then
       Graphical miner path view module
Else   Set close the module
```

Part K **Choke point details module**

Algorithm:

```
Select Mine Plan.
Get data from the router
Set Flag to True
If tag under choke point router then
   Check tag is authorized
   If no then
   Message and Set Flag to False
```

Part L **Tag location information module**

Algorithm:

```
Select active Tag
Get data from the router
Set Flag to True
Check tag location
If yes then
     Add to list
```

```
Else
      Message and Set Flag to False
```

Part M **Report module**

Algorithm:

```
Display option wise seven modules
If option I then
   Display tag report
Else if option II then
   Display miner report
Else if option III then
   Display router report
Else if option IV then
   Display current miner position report
Else if option V then
   Display block area miner report
Else if option VI then
   Display member details report
Else if option VII then
   Display message details report
Else
   Set close the module
```

Part N **Message from end device module**

Algorithm:

```
Display option wise seven modules
If option I then
   Display tag report
Else if option II then
   Display miner report
Else if option III then
   Display router report
Else if option IV then
   Display current miner position report
Else if option V then
   Display block area miner report
Else if option VI then
   Display member details report
Else if option VII then
   Display message details report
Else
Set close the module
```

Index

A

Active, 11, 37, 38, 40, 45, 47, 50, 53, 66, 67, 95, 96, 128, 141, 160, 161, 163, 165, 166, 169, 170, 175, 176, 243, 250, 263, 264, 285, 286, 288, 292, 324, 365
Advanced encryption standard, 164
Alert message, 280, 284, 292, 294, 297
Amplifier, 5, 6, 51, 74, 75, 77, 78, 79, 95, 96, 99, 100, 115, 116, 123, 125, 129, 130, 131–137, 176, 177, 216, 221, 228
Amplitude modulation, 30, 33, 155
Analog to digital, 176, 177, 178, 179, 192, 193, 194, 209, 210, 211, 219, 220, 380, 426, 428, 429, 439, 440, 441
Antenna, 7, 10, 14, 15, 16, 17, 19, 21, 22, 23, 24, 26, 29, 30, 35, 37, 38, 41, 44, 46, 47, 50, 52, 53, 67, 68, 70, 74, 75, 77, 78, 86, 91, 94, 105, 113, 114, 116, 121, 122, 123, 126, 127, 131, 132, 133, 134, 135–136, 172, 176, 177, 184, 188, 190, 191, 192, 193, 194, 201, 203–205, 206, 211, 219, 237, 245, 264
Antistatic, 314, 315, 368
Application layer, 314, 315, 368
Application software, 147, 155, 166, 246, 280, 283–284, 298
Area office, 146
Atmosphere explosive, 304, 305
Attenuation, 10, 22, 24, 25, 65, 68, 70, 106, 113, 114, 122, 123, 124, 126, 134
Audio frequency, 95

B

Back off exponent, 161
Band pass filter, 115, 116, 176
Bandwidth, 14, 28, 30, 32, 33, 34, 35, 41, 43, 100, 128, 155, 160, 161, 165
Barrier, 27, 65, 67, 69, 145, 311, 312, 313, 322, 354, 355, 356–359, 361, 367
Base station, 2, 10, 26, 28, 30, 31, 32, 34, 122, 123, 129
Battery, 3, 5, 8, 11, 17, 18, 22, 25, 38, 46, 48, 49, 74, 80, 82, 83, 84, 85, 87, 89, 91, 100, 101, 102, 153, 154, 161, 165, 166, 167, 170, 171, 176, 178, 179, 180, 181, 182, 184, 185, 186, 189, 191, 193, 194, 195, 197, 198, 199, 200, 203, 206, 207, 208, 209, 210, 211, 212, 214, 217, 218, 219, 220, 223, 224, 236, 238, 245, 264, 268, 270, 300, 311, 324, 325, 327, 348, 349, 354, 369, 370
Beacon order, 161, 277, 380, 391, 396, 404, 417, 426, 428, 429, 439, 440, 441
Bell signaling, 3, 4, 5, 91
Bharat Coking Coal Limited, 88, 102, 252
Bifilar, 122
Braided, 122, 128
Branch unit, 123, 124, 125, 129, 131
Bureau of Indian Standard (BIS), 303, 305, 307

C

Cables communication, 7, 10, 13, 23, 41, 121, 135, 137
Cage, 4, 8, 91, 96, 98, 100, 102, 103, 118
Capacitance, 83, 126, 128, 188, 203, 310, 312, 323, 324, 327, 332, 338, 339, 340, 341, 342, 343, 344, 345, 346, 347, 351, 357, 358, 361, 363, 365, 368, 369, 370, 373
Capacitor, 8, 24, 75, 83, 85, 95, 98, 128, 181, 183, 187, 194, 202, 207, 210, 212, 215, 217, 221, 223, 224, 228, 235, 244, 302, 312, 324, 327, 351, 353, 354, 363, 365, 367, 368, 369, 370–371
Cap-lamp, 14, 17, 18, 22, 74, 75, 80, 85, 87, 88, 89, 145, 178, 179, 180, 181, 182, 184, 187, 189, 192, 193, 194, 195, 198, 199, 200, 203, 207, 209, 211, 212, 214, 217, 218, 219, 220, 223, 224, 236

Carrier current system, 7–9
Category, 85, 86, 144, 147, 148, 306, 312, 320, 347, 348, 350
Cathode ray tube, 24
CC2430 component, 172, 178, 184, 194, 196, 202, 211, 212, 216, 219, 221, 224, 225, 228, 244, 269
CDMA, 28, 166, 167
Cell, 82, 169, 311, 348–349, 365
Central conductor, 132, 133, 134
Central processing unit, 261, 284
Clearance, 311, 318–319, 321, 322, 351, 357, 361, 363, 365, 366, 369, 372
Cluster tree, 271
Coal, 1, 2, 9, 11, 19, 22, 28, 48, 65, 66, 67, 68, 69, 70, 73, 74, 87, 88, 89, 105–113, 118, 121, 125, 129, 137, 141–146, 147, 149, 150, 245, 252, 300, 303, 305, 315
Coaxial, 2, 10, 13, 27, 53, 122, 126, 128
Combined system, 9
Complimentary code keying, 27
Computerized, 142, 143, 145, 146, 247, 250, 283, 284, 285, 298
Conducting bar, 132, 133, 134
Conductivity, 22, 65, 67
Conductor, 37, 93, 121, 122, 126, 128, 132, 133, 134, 171, 316, 317, 358, 361, 366
Connector, 132, 133, 134, 184, 187, 202, 210, 216, 223, 224, 236, 269, 320, 322, 347
Contention access period, 160, 161
Contention free period, 160, 161
Contention window, 161
Coordinator, 156, 157, 159, 160, 161, 162, 175, 178–191, 247, 248, 250, 251, 252, 253, 254, 255, 256, 257, 259, 260, 266, 269, 270, 271, 273, 274, 277, 279, 280, 283, 284, 286, 379–401, 414, 415, 416, 418, 425, 426, 427, 428, 429, 431, 439, 440, 441
Cordless telecommunications, 12–13
Coupling, 8, 10, 36, 51, 75, 80, 93–94, 95, 112, 114, 126
Coverage, 10, 15, 17, 24, 27, 29, 41, 48, 52, 53, 123, 125, 127, 128, 129, 131, 135, 137, 155, 156, 163, 164
Creepage distance, 311, 313, 318, 319, 321, 357, 361, 366, 369
CSMA, 13, 154, 155, 156, 160, 161, 265, 276
Current coupler, 91, 95, 96, 97, 98, 99, 100, 101, 102, 104
  limiting resistor (CLR), 189, 197, 203, 207, 217, 324, 325, 327, 348, 353, 356, 361, 367, 372

D

Data acquisition software (DAS), 147
Data-link-layer (DLL), 155
Data logger, 170, 279, 280, 281, 439–441
Debugging, 266, 268–269, 379
Decibel (dBs), 24, 113
Decision support, 141–151
Delogne system, 128
Delta electromagnetic, 18
Detecting system, 65, 70, 73, 74, 87, 88, 89
Device, 15, 19, 23, 24, 25, 28, 37, 38, 65, 73, 93, 157, 158, 160, 161, 162, 164, 165, 169, 170, 171, 172, 175, 176, 177, 178, 179, 180, 181, 182, 185, 186, 189, 190, 191, 192, 193, 194, 195, 197, 198, 199, 200, 201, 203, 204, 205, 206, 207, 208, 209, 210, 211, 212, 213, 214, 215, 216, 217, 219, 220, 221, 222, 223, 224, 225, 226, 227, 228, 229, 230, 231, 232, 233, 234, 235, 236, 237, 238, 239, 240, 242, 243, 244, 245, 246, 247, 250, 251, 252, 253, 254, 255, 256, 257, 258, 259, 260, 263, 264, 266, 268, 269, 270, 271, 273, 274, 275, 277, 279, 280, 281, 285, 288, 291, 292, 299, 300, 302, 311, 312, 323, 324, 347, 348, 350, 351, 352, 357, 358, 361, 363, 365, 373, 380, 381, 382, 383, 388, 391, 392, 393, 396, 397, 398, 401, 402, 403, 407, 410, 414, 415, 416, 417, 419, 421, 423, 425, 426, 428, 429, 430, 431, 433, 435, 437, 439, 440, 441, 442, 445, 447, 449, 451, 453, 455, 457, 459, 461, 463, 469
Dielectric, 22, 65, 66, 67, 105, 106, 107, 108, 110, 184, 192, 193, 353, 368, 372
Digital enhanced cordless telecommunications, 13–14
Digital signal processing, 14
Diode, 14, 80, 81, 83, 85, 86, 96, 100, 178, 179, 180, 181, 183, 184, 187, 192, 193, 194, 196, 202, 207, 209, 210, 211, 212, 215, 217, 219, 220, 221, 223, 224, 228, 244, 311, 322, 323, 324, 348, 350, 353, 354, 355, 356, 357, 361, 362, 363, 365, 367, 368, 369, 371, 372
Directorate general of mines safety (DGMS), 88
Direct sequence spread spectrum, 155, 162, 177
Disaster, 2, 4, 9, 18, 20, 40, 45–50, 73, 144–145, 175, 237, 247, 251, 261, 283, 298

Discrete Fourier transform (DFT), 33
Dynamic routing, 271, 278, 279, 281

E
EEPROM (electronically erasable programming read only memory), 37
Electric field, 93, 94, 106, 112, 357, 369, 372
Electromagnetic, 9, 14, 18, 20, 21–22, 24, 36, 65, 66, 67, 70, 78, 91, 93–94, 95, 104, 105, 106, 114, 115, 128, 135, 166, 263, 359–360
Electromotive force, 9
Embedded, 37, 38, 54, 172, 176, 178, 179, 184, 192, 193, 194, 202, 209, 210, 211, 212, 216, 219, 221, 224, 229, 263, 268, 269, 279, 281, 352, 379, 380
Emergency broadcast network (EBN), 19
Enclosures, 47, 73, 85, 100, 124, 181, 189, 197, 201, 213, 216, 224, 229, 237, 243, 304, 309, 313–315, 318, 322, 349, 357, 364, 367
End device, 157, 175, 176, 178, 192–206, 243, 246, 247, 248, 249, 250, 251, 252, 253, 254, 255, 256, 257, 258, 259, 260, 263, 266, 268, 269, 277, 279, 280, 281, 283, 285, 288, 291, 293, 402, 407, 415–426, 439–441
Entity parameters, 312, 313
Equipment, 1, 9, 11, 12, 13, 14, 17, 18, 19, 21, 24, 26, 40, 41, 44, 45, 48, 51, 53, 121, 123, 129, 142–143, 147, 168, 170, 175, 177, 237, 243, 245, 247, 249, 250, 260, 263, 279, 283, 298, 299, 300, 302, 304–305, 306, 307, 308, 309, 312, 313, 315, 317, 322, 357, 359–360, 362, 365–366
ERTL, 88, 308
Ethernet, 13–14, 27, 28, 51, 154, 273
Ex-equipment, 300, 302, 304–305, 307, 308, 312, 313, 317, 365–366
Explosion proof, 2, 50

F
FCC (Federal Communication Commission), 35, 129, 177
FFT (fast Fourier transform), 33, 34
Fiber optics, 9
Flameproof, 73, 149, 364, 366, 367
Frame check sequence (FCS), 159, 275
Frequency division multiplexing, 27, 32–34
Frequency-modulated (FM), 8
Full-function device (FFD), 269
Fuse, 80, 81, 82, 83, 85, 111, 181, 184, 187, 191, 194, 197, 202, 204, 207, 210, 212, 216, 217, 218, 220, 223, 224, 231, 244, 300, 311, 323, 347, 348, 350, 352, 355, 357, 358, 359, 361, 363, 367, 368, 369

G
Gain, 21, 30, 70, 95, 96, 99, 115, 116, 125, 127, 135, 136, 176, 191, 204
Galvanic, 358, 359
Gas, 12, 40, 82, 83, 84, 144, 191, 202, 205, 209, 212, 219, 247, 252, 280, 284, 299, 302, 303, 304, 305, 306, 308, 309, 311, 315, 332, 348, 349, 357, 358, 360, 367, 370
Global system for mobile communications, 28
GLON-LOP, 18
GLON-W, 18
Ground-penetrating radar (GPR), 144
Guaranteed time slots, 154, 160
Guide rope, 91, 95, 96, 97, 98, 99, 100, 101, 102, 103, 104

H
Head end, 16, 17, 123, 125, 129
Head office (HO), 146
Hoist rope radio, 7, 8, 9

I
IEC, 191, 205, 303, 304, 307, 308, 312, 349, 363
IEEE, 13, 27, 29, 32, 34, 35, 39, 42, 46, 47, 153, 154, 155, 156, 157, 158, 159, 160, 161, 162, 163, 164, 165, 166, 169, 170, 171, 172, 176, 264, 265, 268, 379
Impedance, 75, 79, 94, 95, 96, 114, 115, 116, 117, 124, 125, 126, 127, 128, 129, 191, 204, 369
Inductance, 91–93, 94, 188, 201, 202, 310, 311, 312, 322, 323, 324, 327, 351, 353, 357, 358, 361, 365, 368, 369, 370, 371, 373
Induction, 78, 91, 102, 103, 104, 128, 146
Infallible, 81, 309, 310, 313, 324, 327, 351–353, 354, 355, 356, 361, 374
Information, 1, 11, 12, 14, 18, 21, 23, 33, 37, 38, 40, 41, 43, 44, 49, 83, 88, 141–151, 154, 155, 158, 166, 169, 175–261, 263, 264, 274, 275, 280, 286, 287, 313, 365, 380, 382, 391, 394, 395, 402, 412, 415, 416, 422, 426, 428, 429, 435, 439, 440, 441, 468
Ingress protection, 2, 304, 313
Insertion loss, 105, 125

Installation, 29, 98, 100–101, 103, 104, 113, 129, 135, 247–250, 261, 285–297, 307, 349, 359, 360, 361, 365, 366, 374
Integrated circuit, 37, 162, 166, 172, 263, 350
Intrinsic safety, 65, 73, 190, 203, 213, 225, 232, 299–374
Inundation, 144
ISM, 35, 39, 153, 162, 172, 176, 177, 264, 268
IT, 141, 142, 143, 145, 146, 147, 149, 150, 155
ITU, 159

L
Leaky feeder, 10, 124–126
Line amplifiers, 123, 129, 137
Line-of-sight, 10, 30–32, 46, 48, 51, 54, 66, 105–119, 126, 247, 287
Line terminator, 123, 124, 129
Link quality indication, 154, 162
LLC, 153, 157, 158, 271, 273, 382
LMDS, 30
Local area network, 44, 154
Local oscillator, 115, 116, 176, 177
Location, 11, 12, 13, 14, 18, 19, 20, 22, 23, 25, 34, 40, 41, 42, 43, 45, 48, 49, 50, 51, 53, 54, 73, 85, 87, 89, 101, 125, 129, 135, 142, 144, 148, 149, 150, 154, 167, 169, 175, 190, 191, 204, 217, 218, 231, 232, 236, 245, 246, 247, 250, 257, 259, 260, 261, 276, 279, 283, 284, 285, 286, 287, 290, 292, 307, 322, 466, 468
Long-wire antennas, 127
Lower explosive limit, 301, 303
Low frequency, 3, 7, 14, 15, 16, 19, 20, 22, 37, 40, 49, 50, 52, 66, 67, 74, 91, 104, 115, 121, 144
Low noise amplifier, 115
Low rate wireless personal area network, 159

M
MAC common part sub-layer, 158
MAC footer, 158, 275
MAC header, 158, 159, 275
MAC layer management entity, 158
MAC protocol data unit, 158
MAC service data unit, 158
Mahanadi Coalfields Limited, 88, 129
MCIT, 88
Medium access control, 157, 275, 277, 379
Methanometer, 143, 145
Metropolitan area network, 154
Microcontroller, 114, 119, 171, 172, 177, 178, 190, 194, 204, 211, 213, 216, 218, 219, 220, 221, 228, 229, 231, 234, 236, 237, 264, 265, 280, 281, 443, 447, 457
Middleware, 146, 148–149, 167
Mine fire, 144
Miners, 1, 10, 12, 14, 18, 19, 20, 40, 49, 53, 54, 65, 73, 113, 118, 126, 142, 144, 145, 170, 175, 243, 245, 246–247, 249, 250, 260, 261, 279, 283, 284, 285, 286, 287, 290, 292, 294, 295, 296, 298
Miner's unsafe practice, 40, 284
Mine-wide communication, 15, 65, 66, 121–137
Minimum igniting current, 310, 325, 327
Minimum ignition energy, 300, 302
Mining, 1, 2, 8, 9, 10, 17, 18, 19, 20, 23, 24, 28, 39, 40, 41, 42, 50, 54, 101, 121, 141–151, 175, 243, 247, 260, 261, 264, 283, 303, 305, 308, 312, 363
Mining industry, 8, 19, 20, 39, 40, 42, 141–151, 175
MMDS, 30
Monitoring, 11–12, 13, 14, 36, 39, 40, 42, 46, 51, 54, 141, 142, 143, 144, 147, 154, 155, 163, 167, 170, 175, 178, 192, 206–214, 215, 216, 218, 219, 220, 221, 222, 223, 225, 226, 227, 229, 231, 232, 237, 243, 246, 247, 260, 263, 270, 280, 283–298
Monofilar, 122
MSHA, 1, 15, 18, 19, 40, 41, 46, 308, 352

N
National Electrical Code (NEC), 302, 303, 304, 307
Network layers, 155–157, 162, 165, 271–273, 280
Network topology, 155, 164, 251
Number of back off (NB), 161

O
OFDM, 27, 30, 32, 33, 34, 35
OFDMA, 30, 33
Off-line mode, 285
OMS, 243
Online, 40, 141, 142, 143, 144, 145, 147, 149, 283, 284, 292, 298
Opencast mines, 28, 40, 246, 247, 249, 250, 251, 284, 298
Open circuit, 8, 300, 310, 318, 324, 347, 348, 349, 350, 351, 353, 354, 357
OSI, 157, 158, 271, 273, 280

P

Passive amplifier, 131–137
Passive semiconductor, 37, 312
Passive tags, 37, 38, 264
Path loss, 113
Peer-to-peer, 36, 154, 156, 157, 166, 270, 271
Personal emergency device (PED), 14, 15–17
Phase shift keying, 30, 33, 171
PHY protocol data unit (PPDU), 163
PHY service data unit (PSDU), 163
Physical, 17, 21, 40, 41, 50, 93, 126, 135, 153, 155, 156, 157, 158, 162–164, 165, 243, 263, 271, 273, 280, 283, 299, 300, 318, 323
Physical layer, 156, 157, 158, 162–164, 165, 271, 273
Physical signaling sublayer (PSS), 273, 274
Pit bottom, 91, 96, 100, 102, 103, 104, 118
Pit top, 91, 96, 100, 102, 103, 104, 118, 129, 130, 176, 245, 261
Plugs, 319–320, 322, 347, 355, 363, 366
Point to hole (PTH), 74, 88
Point-to-multipoint (PMP), 29, 30, 31
Polarization, 70, 113, 114, 191, 203, 204
Postdisaster, 145
Power, 2, 3, 6, 7, 8, 10, 12, 16, 18, 21, 24, 25, 27, 29, 30, 34, 36, 37, 38, 42, 43, 47, 48, 49, 54, 65, 73, 74, 75, 77, 78, 80, 81, 82, 83, 84, 89, 95, 96, 98, 99, 100, 101, 105, 108, 110, 112, 113, 114, 116, 117, 118, 119, 123, 124, 125, 126, 127, 128, 129, 130, 131, 132, 135, 149, 153, 154, 155, 157, 162, 164, 165, 166, 167, 169, 170, 171, 172, 173, 174, 175, 176, 177, 178, 179, 180, 181, 184, 185, 186, 187, 189, 190, 191, 192, 193, 194, 195, 197, 200–201, 203, 204, 205, 206, 207, 208, 209, 210, 211, 212, 213, 214, 217, 218, 219, 220, 221, 222, 223, 225, 228, 231, 232, 234, 236, 237, 239, 245, 246, 264, 265, 267, 268, 276, 299, 310, 316, 317, 318, 322, 323, 324, 325, 327, 347, 348, 349, 350, 351, 352, 353, 354, 355, 357, 360, 361, 362, 363, 366, 367, 368, 369, 370, 371, 372, 373, 380, 392, 402, 415, 426, 428, 429, 439, 440, 441
Power coupler (PC), 123, 124, 125, 129, 131
Presentation layer, 271, 272
Printed circuit board (PCB), 74, 133, 134, 135, 299, 316, 321, 363, 370
Programming, 37, 114, 119, 179, 180, 184, 192, 193, 194, 195, 196, 197, 202, 209, 210, 211, 212, 216, 219, 220, 221, 224, 236, 263–281, 282, 379–463
Propagation, 8, 19–22, 24, 26, 52, 65, 66, 67, 68, 69, 70, 73, 74, 105, 106, 107, 109, 110, 111, 113–114, 115, 122, 128, 162
Pulse, 20, 43, 50, 81, 273, 360, 443

Q

Quadrature amplitude modulation (QAM), 30, 33, 155
Quality of service, 34, 272

R

Radiated power, 114, 119, 190, 193, 205, 206, 237, 245
Radio, 7, 8–9, 10, 14, 15, 17, 19, 20, 21, 22, 24, 25, 26, 27, 28, 29, 30, 34, 36, 37, 41, 43, 46, 47, 49, 51, 52, 53, 65, 66, 67, 70, 73, 74, 89, 94, 105–113, 121, 122, 123, 126, 161, 171, 172, 175, 176, 177, 178, 179, 184, 192, 193, 194, 202, 210, 211, 212, 216, 219, 221, 224, 229, 263, 264, 265, 270, 273, 276, 380, 402, 415, 426, 428, 429, 439, 440, 441
Radio frequency, 7, 11, 27, 36–41, 53, 65, 73, 123, 166, 176, 263, 360
Range, 1, 3, 8, 9, 10, 12, 13, 15, 17, 22, 24, 27, 28, 30, 31, 32, 35, 36, 37, 39, 40, 46, 54, 65, 70, 73, 105, 113–114, 115, 116, 117, 118, 119, 121, 123, 126, 127, 135, 137, 153, 155, 156, 162, 164, 165, 167, 170, 171, 172, 177, 180, 187, 190, 200, 205, 237, 245, 246, 247, 260, 263, 265, 270, 271, 279, 347, 350, 351, 370, 380, 402, 415, 426, 428, 429, 439, 440, 441
Rating, 80, 83, 84, 85, 172, 187, 189, 190, 191, 200, 203, 204, 213, 217, 218, 225, 231, 232, 236, 304, 311, 313, 314, 316, 322, 347, 348, 350, 352, 353, 355, 357, 361, 362, 363, 365, 367, 368, 369, 370, 371, 372
Reader, 11, 12, 36, 37, 38, 39, 40, 45, 51, 53, 147, 166, 167, 245, 246, 263, 264, 284, 288
Receiver unit, 73, 74, 77–79, 80, 81, 82, 84, 87, 89
Reduced functionality device (RFD), 165, 171, 269, 271, 280
Relay, 12, 14, 46–47, 48, 49, 86, 94, 245, 322, 355, 361, 369, 371
Remote access server (RAS), 148

Resistance, 14, 34, 80, 81, 126, 145, 180, 194, 212, 220, 221, 311, 315, 316, 321, 323, 324, 347, 348, 350, 351, 358, 363, 366, 370, 371
RFID, 11, 25, 36, 37, 38, 39, 40, 41, 44, 45, 50, 51, 53, 166, 167, 168, 169, 170, 175, 176, 179, 180, 181, 182, 183, 184, 185, 186, 187, 190, 191, 193, 194, 196, 198, 199, 200, 202, 205, 206, 211, 212, 213, 214, 215, 216, 218, 220, 221, 222, 223, 224, 225, 228, 231, 234, 236, 243, 245, 246, 247, 250, 251–260, 263–281, 285, 287, 292, 379–463
Roof fall, 14, 15, 22, 23, 24, 54, 65, 73, 87, 89, 144, 246
Router, 157, 165, 175, 178, 192–206, 243, 245, 246, 247, 248, 249, 250, 251, 252, 253, 254, 255, 256, 257, 258, 260, 263, 266, 268, 269, 277, 279, 280, 281, 283, 284, 285, 286, 287, 288, 290, 291, 292, 293, 296, 380, 392, 402–414, 415, 426, 429, 431, 433, 435, 437, 438, 441, 465, 466, 468, 469
Route tracking, 284

S

Safety, 1, 2, 11, 17, 35, 41, 42, 45, 50, 53, 65, 73, 80, 82–84, 85, 88, 89, 93, 114, 119, 141, 143, 144, 145, 146, 147, 175–261, 283, 284, 299–374
Schweizerischer Elektrotechnischer Verein, 308
Security, 14, 17, 29, 35, 36, 41, 43, 147, 153, 154, 155, 156, 163, 164–165, 167, 168, 169, 170, 263, 265, 275, 285
Seismic, 20, 21, 54
Semi-passive tags, 38
Semi-wireless communication system, 10–14
Sensitivity, 10, 38, 81, 100, 113, 117, 126, 164, 171, 177, 265
Sensor, 12, 13, 14, 36, 38, 39, 41, 42, 43, 44, 48, 53, 54, 95, 103, 141, 142, 143, 144, 145, 147, 149, 150, 153, 154, 155, 165, 166, 169, 170, 172, 176, 180, 210, 211, 212, 213, 214, 215, 218, 219, 221, 224, 225, 226, 227, 228, 229, 230, 232, 247, 248, 250, 251, 265, 266, 270, 279, 280, 281, 284, 286, 291, 429–439, 465
Service access points, 158
Service-specific convergence sub-layer (SSCS), 157, 158
Session layer, 271, 272
Shaft, 4, 22, 66, 91–104, 114, 118, 119, 247, 248
Sheath, 126, 127, 366
Short circuit, 8, 17, 80, 82, 83, 84, 95, 189, 203, 300, 302, 318, 320, 324, 325, 327, 332, 333, 334, 335, 336, 337, 348, 349, 350, 351, 352, 353, 354, 357, 361, 362, 367, 372, 373
Short-range communication, 263
Shovel-dumper, 246, 251, 284, 298
Shunt, 311, 324, 348, 349, 350, 353, 354, 355, 357
Signal-to-noise ratio, 15
Signal strength, 3, 24, 43, 50, 66, 68, 69, 70, 74, 113, 129, 137
Skin depth effect, 123
Slotted shield, 128
Socket, 13, 319–320, 355, 363
Software, 16, 17, 28, 44, 45, 47, 70, 114, 119, 146, 147–148, 153, 155, 166, 172, 176, 246, 247, 263, 268, 269, 272, 279, 283–298, 360, 442, 465–469
South African Bureau of Standard (SABS), 308
Spectrum analyzer, 68, 82, 102, 135, 136
SQL, 147, 283, 285, 286
Standard Association (SA), 308
Static routing, 271, 277–278, 279
Strata, 15, 16, 17, 65, 66, 67, 70, 73, 74, 89, 144, 147, 150
Super frame order, 277
Sveriges Provnings-Och Forskingsinstitut, 309
Synchronization, 27, 34, 43, 162, 163, 271, 273, 275
System, 16, 17, 28, 44, 45, 47, 70, 114, 119, 146, 147–148, 153, 155, 166, 172, 176, 246, 247, 263, 268, 269, 272, 279, 280, 283–298, 360, 442, 465–469
System on chip, 153, 176, 264

T

Tags, 11, 23, 37, 38, 39, 40, 44, 45, 53, 166, 175, 245, 246, 247, 253, 256, 259, 260, 264, 284, 286, 287, 291, 292, 296, 465, 466, 467, 468, 469
Telephone, 1, 3, 5, 6, 7, 9, 12, 13, 25, 26, 49, 145, 148, 263
Temperature, 12, 38, 117, 125, 126, 143, 144, 147, 150, 156, 170, 171, 172, 177, 252, 255, 256, 265, 280, 299, 300, 302, 303, 304–305, 311, 313, 315, 316, 317, 318, 322, 323, 347, 349, 350, 357, 362, 365, 366, 370
Text message communication, 283, 284
Through-the-air (TTA), 2, 23
Through-the-wire (TTW), 2

TI-MAC, 271, 277–279, 379, 380
Tracking, 1, 14, 15, 18, 37, 40, 41, 42, 45, 50, 51, 52, 53, 54, 142, 146, 162, 167, 168, 170, 175, 178, 192, 237–245, 246, 247, 258, 259, 260, 270, 279, 280, 283–298, 319, 468
Tracking and monitoring software, 283–298
Transceivers, 7, 8, 26, 73, 91, 96, 99, 100, 101, 102, 104, 113, 114, 117, 118, 119, 123, 126, 130, 146, 150, 171, 172, 247
Transformer, 17, 94, 95, 96, 97, 98, 99, 100, 101, 128, 347, 350, 351–353, 354, 358, 361, 363, 369, 372
Transmitter unit, 73, 74–77, 78, 80, 81, 82, 85, 87, 88
Transmission Control Protocol/Internet Protocol, 13
Transport layer, 271
Transverse electromagnetic, 106
Trapped miner, 12, 14, 20, 21, 22, 23, 49, 54, 65, 66, 73–89, 145, 146, 261, 283
Trolley carrier phone, 7, 8
Twin feeder, 127
Two-way, 1, 6, 10, 13, 15, 19, 26, 28, 46, 49, 51, 121, 122, 126

U
Ultra high frequency, 3, 105, 146
Ultra-low frequency, 15, 19, 20
Ultra wide band, 25, 36, 41–43, 154
Underground mine, 1, 2, 3, 4, 5, 6, 9, 12, 14, 17, 19, 21, 23, 24, 40, 42, 45, 46, 47, 48, 50, 65, 66, 67, 70, 73, 74, 87, 88, 89, 91, 102, 103, 104, 105, 114, 117, 119, 123, 124, 129, 131, 135, 137, 146, 153, 175, 243, 246, 247, 248, 250, 251, 252, 253, 255, 257, 258, 259, 260, 263, 279, 280, 283, 284, 285, 286, 292, 298, 299, 300, 305, 313
Uninterrupted power supply, 149
Upper explosive limit, 301, 303
USART, 265
USBM, 20

V
Vehicle collisions, 170, 280, 284
Ventilation, 16, 19, 143, 145
Very high frequency, 3, 10, 121, 146
Very low frequency, 3, 14, 19–22, 121
Visual Basic, 148, 283
Voice internet protocol, 13, 28
Voltage, 4, 5, 6, 8, 18, 20, 24, 46, 53, 54, 75, 78, 79, 80, 82, 83, 84, 85, 86, 95, 96, 97, 99, 100, 101, 115, 116, 125, 130, 171, 172, 177, 178, 179, 180, 181, 182, 184, 185, 186, 187, 190, 191, 192, 193, 194, 195, 196, 197, 198, 199, 200, 202, 203, 204, 205, 207, 208, 209, 210, 211, 212, 213, 216, 217, 218, 219, 220, 221, 223, 224, 228, 229, 230, 265, 266, 268, 299, 300, 302, 310, 312, 321, 322, 323, 324, 325, 327, 332, 333, 334, 335, 336, 337, 338, 339, 340, 341, 342, 343, 344, 346, 347, 348, 349, 350, 351, 352, 353, 354, 355, 356, 357, 358, 359, 361, 363, 366, 367, 368, 369, 370, 371, 372, 373

W
Waveguide, 105, 106, 108, 110, 111, 112, 113, 122
Wavelength, 2, 3, 37, 95, 105, 107, 111, 113, 127, 132, 133, 134, 245
Web-based information, 141–151
Weight-o-meter, 142, 150
Wide-band, 41, 43, 128
Wired communication services, 2–7, 9
Wireless application protocol, 273–275
Wireless fidelity, 25, 27–29, 154
Wireless local area networks, 154
Wireless network program, 279
Wireless personal area network, 27, 154, 159
Wireless sensor networks, 41, 153, 172, 270
World interoperability for microwave access, 25, 29–36, 154

Z
ZigBee, 36, 39, 40, 49, 153–173, 175, 176, 177, 228, 243, 245, 247, 248, 250, 251, 252, 255, 256, 258, 264, 267, 269, 270, 285, 379, 382, 395, 403, 416, 430
Zone, 2, 18, 26, 40, 41, 44, 51, 53, 65, 179, 184, 263, 292, 302, 303, 304, 305, 306, 307, 365, 372

FSC
www.fsc.org
MIX
Papier | Fördert gute Waldnutzung
FSC® C083411

Kolibri 360 GmbH
Ferdinand-Jühlke-Straße 7
99095 Erfurt, Deutschland
produktsicherheit@kolibri360.de